Bioactive Compounds in Fruits

Biomedical Applications and Functional Foods

Editors

Shandesh Bhattarai

Faculty of Science
Nepal Academy of Science and Technology (NAST)
Khumaltar, Lalitpur, Nepal

Edivan Carvalho Vieira

Department of Chemistry
Federal University of Piaui (UFPI)
Teresina - PI, Brazil

Mahendra Rai

Department of Biotechnology
SGB Amravati University
Amravati, Maharashtra, India

CRC Press
Taylor & Francis Group
Boca Raton London New York

CRC Press is an imprint of the
Taylor & Francis Group, an **informa** business
A SCIENCE PUBLISHERS BOOK

Cover credit: Photo: Babassu tree and fruits (*Attalea speciosa* Mart ex. Spreng) by Dr. Mahendra Rai, place-Timon, Maranhão forest, Brazil.

First edition published 2027
by CRC Press
2385 NW Executive Center Drive, Suite 320, Boca Raton FL 33431

and by CRC Press
4 Park Square, Milton Park, Abingdon, Oxon, OX14 4RN

CRC Press is an imprint of Taylor & Francis Group, LLC

Library of Congress Cataloging-in-Publication Data (applied for)

ISBN: 978-1-041-08544-7 (hbk)
ISBN: 978-1-041-13127-4 (pbk)
ISBN: 978-1-003-66816-9 (ebk)

DOI: 10.1201/9781003668169

Typeset in Times New Roman
by Prime Publishing Services

Preface

Fruits are essential not only for their nutritional value but also for their rich array of bioactive compounds, which hold immense therapeutic potential. Natural phytochemicals, such as polyphenols, flavonoids, carotenoids, triterpenoids, sterols, anthocyanins, essential oils, and bioactive polysaccharides, have garnered growing scientific and public interest due to their diverse health-promoting properties. In an era where consumers increasingly prioritize fitness, clean-label foods and fruit-based functional foods have emerged as a vital avenue for promoting public health. Recognizing this shift, global health authorities and policymakers are actively encouraging dietary strategies that emphasize the role of functional foods, particularly those derived from fruits.

This book, *Bioactive Compounds in Fruits: Biomedical Applications and Functional Foods*, brings together global expertise to provide a holistic and up-to-date overview of the biomedical applications of fruit-derived compounds.

The volume is organized into three sections that collectively offer a comprehensive understanding of fruit bioactives, spanning from foundational chemistry to practical applications and future directions.

Section I introduces the fundamentals of bioactive compounds in fruits, establishing the basis by exploring the various classes of phytochemicals, their natural sources, and their biological relevance. It also highlights the role of both native and introduced fruit species in urban landscapes, setting the stage for understanding their nutritional and functional significance.

Section II focuses on bioactivities, therapeutic potential, and functional food applications, examining how these compounds contribute to human health through antioxidant, anti-inflammatory, and disease-preventing mechanisms. It delves into key aspects such as bioavailability, metabolism, and the therapeutic promise of regionally significant fruits and bioactives—including peptides, polysaccharides, and carotenoids—demonstrating their integration into functional food systems.

Section III addresses regulations and future perspectives. It discusses regulatory frameworks necessary for optimizing and commercializing these compounds. This section outlines in vitro cultivation strategies to enhance phytochemical production, reviews global safety standards, and envisions the evolving trajectory of fruit-based functional foods.

This interdisciplinary volume features contributions from leading researchers across the globe, including from Brazil, Pilar, Paraguay, India, the United States, Sri Lanka, and Nepal, offering a truly international perspective. It is intended to serve as a valuable reference for researchers, graduate students, food technologists, health professionals, nutritionists, and policymakers who are committed to harnessing the potential of fruits to combat disease and promote human well-being.

We hope this book inspires continued research and innovation at the intersection of food science, biotechnology, and nutrition, and illuminates new pathways for developing safe, effective, and sustainable functional foods from nature's most vibrant resources.

We extend our sincere gratitude to all contributors for their valuable insights, tireless efforts, and dedication in bringing this volume to fruition. We also express our heartfelt thanks to the Commissioning Editor at CRC Press (Taylor & Francis) for his unwavering support and collaboration throughout this project. Mahendra Rai wishes to thank Federal University of Piaui at Teresina, Brazil for Senior Visiting Professorship.

Shandesh Bhattarai, Nepal
Edivan Carvalho Vieira, Brazil
Mahendra Rai, India

Contents

Section III: Regulations and Future Functional Foods

Section I

Fundamentals of Bioactive Compounds in Fruits

Chapter 1

Introduction to Bioactive Compounds in Fruits

Varsha S. Dhoran,[1] *Vishal P. Deshmukh*[2,*] and *Mahendra Rai*[3,4]

1.1 Introduction

Fruits are one of the richest sources of vital nutrients. They mainly account for macronutrients such as carbohydrates, fats, and proteins, as well as micronutrients like various bioactive compounds, minerals, and vitamins (Patra et al. 2022). Fruits are recognized as basic foods for healthy human beings as they add numerous flavors and are linked with a superior quality of life (Soquetta et al. 2018). In recent times, consciousness regarding the advantages of balanced dietary content has boosted the current pattern in the masses for healthier diet practices (Rodríguez et al. 2021). Fruits are universally accepted food with huge potential for health benefits. Because of their low cost and extensive availability, fruits are the most accessible and affordable food component for many people. Use of fruits in daily diet can assist in reducing the risk of certain types of cancer, heart-related problems, stroke, and many other severe diseases (Goldberg 2008). In fact, in 2010, the dietary guidelines devised for Americans strongly advocate keeping fruits and vegetables to half of their plate during meals (Slavin and Lloyd 2012). Although in 2007, the estimated global per capita consumption of fresh fruits was approximately 69.9 kilograms/year

[1] Department of Botany, Sant Gadge Baba Amravati University, Amravati-444602, Amravati, Maharashtra, India.

[2] Department of Botany, Jagadamba Mahavidyalaya, Achalpur City-444806, Amravati, Maharashtra, India.

[3] Department of Biotechnology, Sant Gadge Baba Amravati University, Amravati-444602, Amravati, Maharashtra, India.

[4] Department of Chemistry, Federal University of Piaui, Piaui, Brazil.

* Corresponding author: vishalpdeshmukh@yahoo.com

DOI: 10.1201/9781003668169-2

(FAOSTAT 2011), in 2017, 3.9 million deaths were attributed globally to insufficiency of fruit consumption (WHO 2023). The historical perspective of dietary guidance holds a prime place for fruits pertaining to their nutrition-rich contents (Naseer 2018). Numerous bioactive compounds such as carotenoids, flavonoids, lignans, phenolic acids, stilbenes, and tannins have been abundantly reported from various fruits. Additionally, fruits are recognized as vital sources of vitamins like A, C, E, and K. Besides this, fruits are also considered a prime source of electrolytic minerals like calcium, magnesium, and potassium as well as dietary fibers (Liu 2004). More than 5000 different phytochemicals were reported from fruits and other sources, but a significant number of them have not yet been characterized (Liu 2013). Fruit by-products like seeds and peel/skin also turn out to be exceptional reservoirs of bioactive compounds, mainly phenolic compounds (such as carotenoids, flavonoids, phenolic acids), biologically active proteins (like purified peptides, amino acids), fatty acids, dietary fiber, and other important entities. Among the cited by-products, seeds derived from ripened fruits are rich in essential oils, different phytochemicals, and phytosterol content. Likewise, the peels are recognized for their high content of minerals, pectin, and valuable fibers (Meghwal et al. 2018). These fruit-based by-products serve as a precursor for extracting various bioactive compounds (Figure 1) by opting for various technologies assisting in the preparation of various fruit-based products (Patra et al. 2022).

Naturally occurring bioactive compounds observed from fruits have most frequently served as important therapeutic agents regardless of their origin. These compounds reported a significant impact on human health via direct and indirect approaches and revealed immense medicinal benefits (Karasawa and Mohan 2018). Biosynthesis of bioactive compounds is triggered by light; hence, they are confined to skin part of fruits (Zorrati et al. 2014). The concentration of bioactive compounds ranges in different types of fruit; besides this, genetic factors and external factors like light exposure, stage of ripeness, and postharvest handling practices affect them. Stage of fruit maturity while harvesting the fruit can also be a vital factor altering bioactive compounds levels (Vallejo et al. 2003, Deepa et al. 2007). As fruits contain bioactive compounds along with vitamins and minerals in liquid form, it ensures their bioavailable to humans through diet (Yildiz et al. 2010).

1.2 Overview of Bioactive Compounds

Bioactive compounds are the lively organic compounds found in the cellular hierarchy of all life forms. Bioactive compounds were alternatively termed “secondary metabolites” as they are not directly related to the natural growth and development of the concerned organism (Kumar and Naraian 2019). These bioactive compounds reflect medicative as well as noxious consequences toward humans and pets. They are also regarded as plant-based secondary metabolites (Bernhoft 2010). According to the National Cancer Institute (NCI) Dictionary of Cancer Terms (2025), mentioned on the U.S. official website of the National Cancer Institute, bioactive compounds are described as “A type of chemical found in small amounts in plants and certain foods (such as fruits, vegetables, nuts, oils, and whole grains)”. Fruit is the prime source of bioactive compounds; however, they are also derived from vegetables and whole

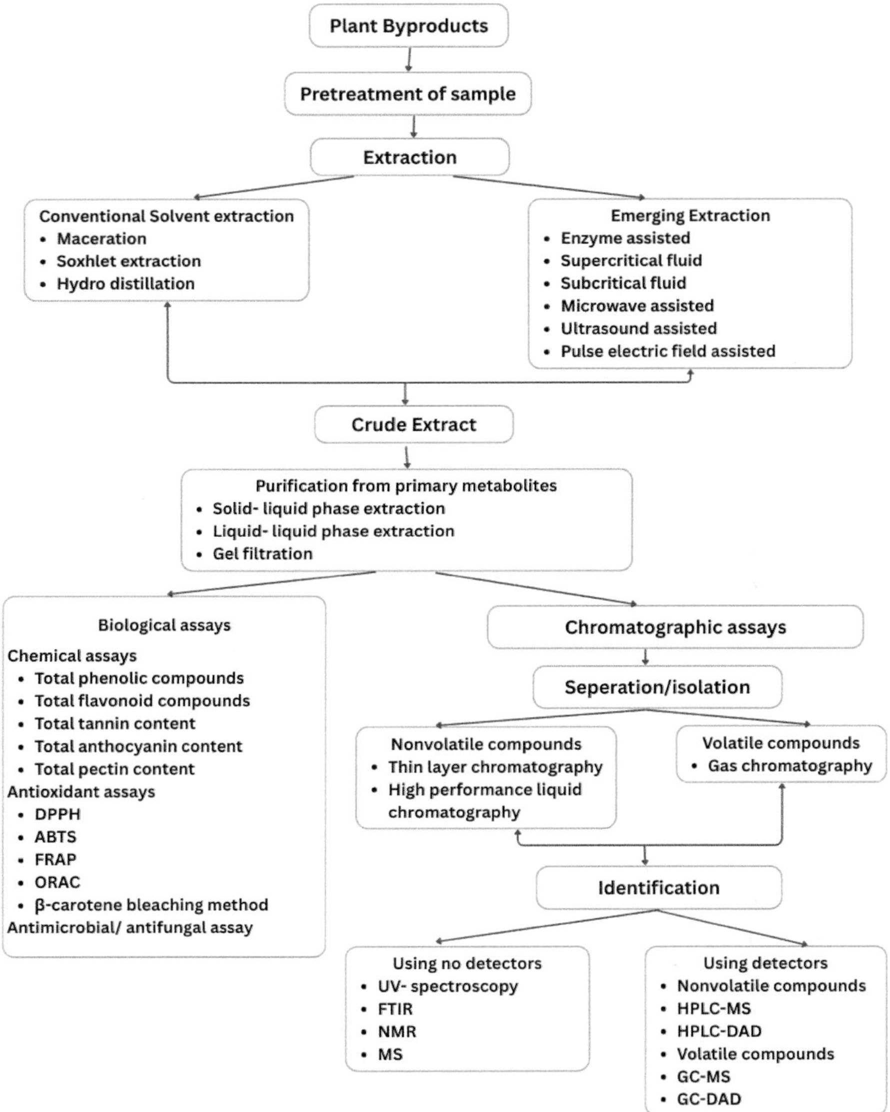

Figure 1. Flow diagram depicting the extraction process of bioactive compounds from fruit-based by-products. Reprinted from Patra et al. (2022) under the terms and conditions of the Creative Commons Attribution (CC By) license (https://creativecommons. org/licenses/by/4.0/).

grains (Gani et al. 2012, Shashirekha et al. 2015). These bioactive compounds were defined with special qualities attributed and can be applied as a remedy for a large range of diseases. Bioactive compounds exhibit their actions through behavioral, immunological, and physiological effects. Currently, a large number of bioactive compounds are reported from fruits and they are mainly categorized as phenolics, flavonoids, phytoestrogens, stilbenes, carotenoids, and vitamins. These bioactive

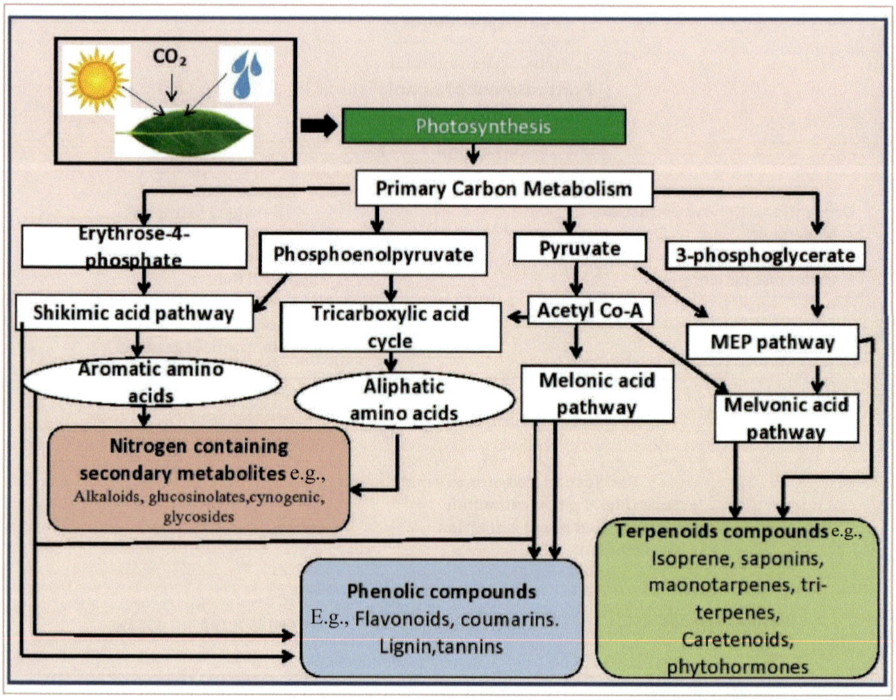

Figure 2. Schematic presentation of biosynthetic pathways for the synthesis of bioactive compounds. Reprinted from Jamloki et al. (2021) under the terms and conditions of the Creative Commons Attribution (CC By) license (https://creativecommons.org/licenses/by/4.0/).

compounds are recognized worldwide for their pharmacological potential, revealing anticarcinogenic, antimicrobial, anti-inflammatory, and antioxidant properties (Hamzalıoğlu and Gökmen 2016). Primary metabolites in plants are the prime source for the synthesis of bioactive compounds. Primary metabolites incorporate organic substances responsible for the development and growth of plants. Amino acids, carbohydrates, fatty acids, lipids, nucleotides, and steroids are some of the prominent primary metabolites reported from the plant kingdom, having their influence on plant growth. Proteins and lipids are largely concerned with the control of plant maturation (Batra and Sharma 2013). These primary metabolites provide the substrate for the synthesis of bioactive compounds by following various pathways (Figure 2).

1.3 Classification and Sources of Bioactive Compounds from Fruits

Bioactive compounds isolated from fruit resources have been categorized based on their chemical nature. Some of the main classes of bioactive compounds and their sources in fruits are categorized in Table 1.

Table 1. Diverse bioactive compounds and their fruit sources.

Major group Bioactive compounds	**Name of bioactive compounds**	**Source fruit**	**Reference**
Alkaloids	Caffeine and trigonelline	Coffee beans (*Coffea Arabica* L.), Family-*Rubiaceae*	Ashihara 2006
	p-synephrine	Sour orange (*Citrus aurantium* L.), Sweet orange (*C. sinensis* L. Osbeck), Mandarin orange (*C. deliciosa*), Lemon (*C. limon* (L.) Burm. f.) and Rangpur lime (*C. limonia* L.), Family-*Rutaceae*	Arbo et al. 2008
	α-tomatine and α-solanine	Tomato (*Solanum lycopersicum* L.), Family- *Solanaceae*	Chen and Lin 2019
	Piperine	Black pepper (*Piper nigrum* L.), Family-*Piperaceae*	Chen and Lin 2019
	Arecoline	Black Cutch (*A. catechu* (L.f.) Willd), Family- *Fabaceae*	Chen and Lin 2019
	Four forms of tetrahydro-β-carboline	Apple (*Malus domestica* Borkh.), grape (*Vitis vinifera* L.), kiwi (*Actinidia deliciosa* (A. Chev.) C.F. Liang & A.R. Ferguson.), orange (*Citrus × sinensis* (L.) Osbeck.), peach (*Prunus persica* (L.) Batsch), pear (*Pyrus communis* L.), pineapple (*Ananas comosus* (L.) Merr.) and tomato (*Solanum lycopersicum* L.)	Herraiz and Galisteo 2003
	Clausenanisine A-F and, other 12 identified analogues	Kayumanis (*Clausena anisum-olens* (Blanco) Merr.) Family-*Rutaceae*	Liu et al. 2021
	Synephrine, tyramine and octopamine	Citrus hybrids	He et al. 2011

Table 1 contd. ...

...Table 1 contd.

Major group Bioactive compounds	**Name of bioactive compounds**	**Source fruit**	**Reference**
Flavonoids			
Anthocyanidins	Cyanidin-3,5-O-dihexoside, Cyanidin-3-O-glucoside, Cyanidin-3-O-rutinoside, Peonidin-3-O-rutinoside	Sweet cherry (*Prunus avium* L.), Family- *Rosaceae*	Acero et al. 2019
	Cyanidin-3-glucoside	Black Cherry (*Prunus serotinai* Ehrh.) Family- *Rosaceae*	Brozdowski et al. 2021
Flavonols	Isorhamnetin-O-hexoside, Quercetin-3-O-rutinoside	Sweet cherry (*Prunus avium* L.) Family- *Rosaceae*	Acero et al. 2019
	Quercetin-3-rutinoside, Quercetin-3-glucoside, Quercetin-3-xyloside, Quercetin-3-Arabinopyranoside, Quercetin-3-Arabinofuranoside, Kaempferol-3-rutinoside, Kaempferol pentoside, Kaempferol hexoside, Kaempferol rhamnoside Hexoside, Isorhamnetin-3-rutinoside, Isorhamnetin hexoside, Isorhamnetin pentoside 1, Isorhamnetin pentoside 2	Black Cherry (*Prunus serotinai* Ehrh.) Family- *Rosaceae*	Brozdowski et al. 2021
Flavanones	Eriodictyol hexoside 1, Eriodictyol hexoside 2	Black Cherry (*Prunus serotinai* Ehrh.) Family- *Rosaceae*	Brozdowski et al. 2021
	Naringin, Hesperidin, and Neohesperidin	Citrus hybrids	He et al. 2011
Flavones	Vicenin-2, Lucenin-2, Stellarin-2, Isovitexin, Scoparin, Lucenin-2 4’-methyl ether, Rhoifolin, Neodiosmin, Rhoifolin 4’-O-glucoside, Orientin 4’-methyl ether, Chrysoeriol 7-O-neoesperidoside, Chrysoeriol 7-O-neohesperidoside -4’-O-glucoside	Juice of Bergamot (*Citrus bergamia* Risso) Family- *Rutaceae*	Gattuso et al. 2007a, b
Flavanols	Procyanidin dimer (1-4), catechin, epicatechin, Procyanidin trimer (1-4) and Procyanidin tetramer	Fruit of black cherry (*Prunus serotinai* Ehrh.) Family- *Rosaceae*	Brozdowski et al. 2021
	Procyanidin B1, Procyanidin B2, (-)-Epicatechin, Procyanidin A2	Fruits of peach (*Prunus persica* L.) cultivars, Family- *Rosaceae*	Nowicka et al. 2023

Table 1 contd. ...

...Table 1 contd.

Major group Bioactive compounds	**Name of bioactive compounds**	**Source fruit**	**Reference**
Isoflavones	Osajin, pomiferin, warangalone, and auriculasin	Osage orange (*Maclura pomifera* (Raf.) C.K.Schneid.), Family-*Moraceae*	Bailly 2024
	biochanin A	Golden Shower Tree (*Cassia fistula* L.), Family-*Fabaceae*	Sartorelli et al. 2009
Proanthocyanidins	Derivatives of quercetin and kaempferol	Andean blackberry (*Rubus glaucus* Benth), capulí cherry (*Prunus salicifolia* HBK.), plum (*Prunus domestica* L.), and strawberry (*Fragaria × ananassa* Duchesne ex Rozier), Family-*Rosaceae*	Vasco et al. 2009
Phenolic acids	3-O-Caffeoylquinic acid, 4-O-Caffeoylquinic acid, 3-p-Coumarylquinic acid	Sweet cherry (*Prunus avium* L.), Family-*Rosaceae*	Acero et al. 2019
	Vanillic acid hexoside, *p*-coumaric acid hexoside, 5-feruloylquinic acid 1, 5-feruloylquinic acid 2, 5-caffeoylquinic acid 1, 5-caffeoylquinic acid 2, 5-*p*-coumaroylquinic acid 1, 5-*p*-coumaroylquinic acid 2, 4-*p*-coumaroylquinic acid, Caffeic acid hexoside, Dicaffeoylquinic acid	Black cherry (*Prunus serotina*), Family-*Rosaceae*	Brozdowski et al. 2021
Phytoestrogens			
Coumestans	Coumestrol	Dried dates (*Phoenix dactylifera* L.), Dried apricots (*Prunus armeniaca* L.), Dried prunes (*Prunus domestica* L.), Oranges (*Citrus sinensis* (L.) Osbeck.), Grapefruit (*Citrus × paradisi* Macfad.)	Thompson et al. 2006

Table 1 contd. ...

...Table 1 contd.

Major group Bioactive compounds	**Name of bioactive compounds**	**Source fruit**	**Reference**
Isoflavonoid	Genistein, Orobol, 7,4'-Dimethoxy-5-hydroxyisoflavone, Genistin, Oroboside, 3'-O-Methylorobol-7-glucoside, Sphaerobioside, Wighteone, Gancaonin A, 4',5,7-Trihydroxyisoflavonone, 5,7,3',4'-Tetrahydroxy-6,8-diprenylisoflavone, Alpinumisoflavone, 4'-O-Methylalpinumisoflavone, 5,3',4'-Trihydroxy-6",6"-dimethylpyrano-[2",3";7,6] isoflavone, Scandenone, Derrone, Derrone-4'-O-methylether, Isochandalone, Ulexin B, and Ulexone B	Mandarin melon berries (*Cudrania tricuspidata* (Carr.) Bur. ex Lavallee), Family: Moraceae	Jo et al. 2017
Lignans	(+)-isolariciresinol; (+)-5-methoxy-isolariciresinol; *erythro*-1-(4-hydroxy-3-methoxyphenyl)-2-[4-(3-hydroxypropyl)-2-methoxyphenoxy]-1,3-propanediol; *threo*-1-(4-hydroxy-3-methoxyphenyl)-2-[4-(3-hydroxypropyl)-2-methoxyphenoxy]-1,3-propanediol; (-)-(7*R*,8*S*)-dihydrodehydroconiferyl Alcohol; (+)-(7*R*,8*S*)-5-methoxy-dihydrodehydroconiferyl Alcohol; (+)-lariciresinol; (+)-pinoresinol; (+)-syringaresinol	Açai (*Euterpe oleracea* Mart.), Family-*Arecaceae*	Chin et al. 2008
Stilbenes	Z- and E-astringin, Z- and E-piceid, piceatannol, pallidol, pallidol-3-*O*-glucoside, parthenocissin A, trans-resveratrol, hopeaphenol, ampelopsin H, caraphenol B, vaticanol C isomer, isohopeaphenol, E- and *Z*-ε-viniferin, *E* and *Z* miyabenol C, E and Z δ-viniferin, trans-resveratrol-4'-methyl ether	Grapes (*Vitis vinifera* L.) varieties *Raboso Piave* and *Primitivo*, Family-*Vitaceae*	Flamini et al. 2013

Table 1 contd. ...

...Table 1 contd.

Major group Bioactive compounds	**Name of bioactive compounds**	**Source fruit**	**Reference**
Carotenoids	Lutein, zeaxanthin, anhydrolutein, anhydrozeaxanthin, zeinoxanthin, β-cryptoxanthin, α-carotene, β-carotene, zeinoxanthin myristate, β-cryptoxanthin myristate, zeinoxanthin palmitate, β-cryptoxanthin palmitate, lutein dimyristate, lutein Myristoylpalmitate, and lutein dipalmitate	Arazá Fruit (*Eugenia stipitata* McVaugh), Family-*Myrtaceae*	Garzón et al. 2012
Vitamins	Ascorbic acid	Black Currants (*Ribes nigrum* L.), Family-*Grossulariaceae*; Mango (*Mangifera indica* L.), Family-*Anacardiaceae*	Vagiri et al. 2013, Olale et al. 2019
	α, β, γ, and δ tocopherols	Seeds of Canadian Prairie fruits	Anwar et al. 2008
	α–Tocopherol, γ–Tocopherol, δ-Tocopherol	Red Açaí (*Euterpe oleracea* M.); Inajá (*Maximiliana maripa* D.), Buriti (*Mauritia flexuosa* L.) and Uxi (*Endopleura uchi* C.), Family-*Arecaceae*	da Costa et al. 2010
	γ–Tocopherol, δ-Tocopherol and α–Tocopherol	Wood apple (*Limonia acidissima* L.), Family-*Rutaceae*	Lamani et al. 2021
	α, β, γ, and δ tocopherols and α, β, γ, and δ tocotrienols	Strawberry (*Fragaria* × *ananassa* Duchesne ex Rozier.), raspberry (*Rubus idaeus* L.), blackcurrant (*Ribes nigrum* L.), and apple (*Malus domestica* Borkh.) seeds	Pieszka et al. 2015
Triterpenoids	α-amyrin/lupeol, β–amyrin, α–amyrenone, oleanolic aldehydes, ursolic aldehydes, erythrodiol, uvaol (Neutral triterpenoids); oleanolic acid, ursolic acid, betulinic acid, maslinic acid, corosolic acid, pomolic acid (Triterpenoid acids)	Rugosa Rose (*Rosa rugosa* Thunb.), Family-*Rosaceae*	Dashbaldan et al. 2021

Table 1 contd. ...

...Table 1 contd.

Major group Bioactive compounds	**Name of bioactive compounds**	**Source fruit**	**Reference**
Sterols	Campesterol, Stigmasterol, β-Sitosterol, Δ^5-Avenasterol, Gramisterol, Cycloartenol, Δ^7-Stigmasterol, Δ^7-Avenasterol, 24-Methylenecycloartanol, and Citrostadienol	Seeds of Canadian Prairie fruits	Anwar et al. 2008
	Brassicasterol, Campesterol, Stigmasterol, β-Sitosterol + sitostanol, Δ^5-Avenasterol + Δ^7-stigmasterol, Δ^7-Avenasterol	Mucajá (*Couma rigida* M.), Family-*Apocynaceae*; Red Açaí (*Euterpe oleracea* M.), Family-*Arecaceae*; Inajá (*Maximiliana maripa* D.), Family-*Arecaceae*; Jenipapo (*Genipa americana* L.), Family-*Rubiaceae*; Buriti (*Mauritia flexuosa* L.), Family-*Arecaceae*; and Uxi (*Endopleura uchi* C.), Family- *Humiriaceae*	da Costa et al. 2010
	Campesterol, cholesterol, cholesta-3,5-dien-7-one, isofucosterol, 24-methylenecycloartanol, obtusifoliol, sitosterol, sitostenone, stigmasterol, stigmasta-3.5-dien-7-one	Rugosa Rose (*Rosa rugosa* Thunb.), Family- *Rosaceae*	Dashbaldan et al. 2021

1.3.1 Alkaloids

The common chemical feature of alkaloids includes the presence of at least one nitrogen atom (Dostál 2000). They primarily encompass different heterocyclic nitrogen-rich compounds serving as defense tactics against herbivore attack and pathogenic infections (Ali et al. 2019). Due to their structural diversity, it is quite imprecise to categorize alkaloids; however, they can be grouped into three major divisions, viz. true alkaloids, protoalkaloids, and pseudoalkaloids, based on their primary precursor and final structure (Eguchi et al. 2019). True alkaloids comprise numerous amino acid derivatives having heterocyclic rings with nitrogen. In contrast to this, heterocyclic rings without any nitrogen moiety are called protoalkaloids with similar amino acid derivation. Pseudoalkaloids are derived from those amino acids that have nitrogen in the heterocyclic ring (Singh 2017). Herraiz and Galisteo (2003) showed that fruits like apple, grape, kiwi, orange, peach, pear, pineapple, and tomato are the prime source of four different forms of tetrahydro-β-carboline alkaloid. Coffee beans are also the richest source of alkaloids like caffeine and trigonelline (Ashihara

2006). Unripe fruits of *Citrus aurantium*, *C. sinensis*, *C. deliciosa*, *C. limon*, and *C. limonia* reported 0.012% to 0.099% of the alkaloid p-synephrine (Arbo et al. 2008). Tomato fruits contain poisonous alkaloids such as α-tomatine and α-solanine. The concentration of α-tomatine in green immature tomatoes measures nearly 500 mg/kg of fresh weight. During the ripening process, it gradually degrades, and when it turns red, it lowers to 5 mg/kg of fresh weight. On the contrary, the fruits of *Piper nigrum* revealed alkaloids like piperin, which imparts spiciness and is used along with salt for seasoning purposes. Arecoline is an important alkaloid extracted from the fruits of *A. catechu* (Chen and Lin 2019). Recently, fruits of *Clausena anisum-olens* have been found to harbor eighteen carbazole alkaloids showing promising prospects in treating diabetes mellitus (Liu et al. 2021).

1.3.2 Phenolics

Phenolic compounds comprise a unique class of bioactive compounds produced through both shikimic acid and pentose phosphate pathways involved in phenylpropanoid metabolization (Lin et al. 2016). Their core design is made up of benzene rings with one or several hydroxyl substituents. Structurally, they have simple phenolic compounds, whereas their complexity extends to highly polymerized compounds (Reis Giada 2013). More than 8000 diverse phenolic structures were reported from plant sources, including fruits (Tungmunnithum et al. 2018). Fruits and beverages are often considered prime dietary sources for phenolic compounds. Among fruits, apples, grapes, pears, cherries, and berries are considered prime sources of phenolics, accounting for 0.2–0.3% per 100 gms of fresh weight. The diet-based consumption of phenolic compounds was estimated at 1 gram per day (Manach et al. 2004). Based on phenolic contents in food, Vasco et al. (2008) proposed a classification system identifying it as low (< 100 mg GAE/100 g), medium (100–500 mg GAE/100 g), and high (> 500 mg GAE/100 g). This system of categorization confines blackberry and strawberry under the high phenolic content category. On the contrary, red raspberries and blueberries came under the medium category and were recognized as fine sources of phenols. Phenolics are mainly classified into two types, viz. flavonoids and phenolic acids (Tsao 2010).

1.3.3 Flavonoids

Flavonoids have the structural skeleton of C6-C3-C6 with two aromatic rings joined by a three-carbon link (Mandal 2022). Flavonoids obtained from fruits can be subdivided into many groups such as anthocyanidins (comprising compounds like cyanidin, delphinidin, and pelargonidin), flavonols (including kaempferol and quercetin), flavanones (comprising hesperetin and naringenin), flavones (containing nobiletin and tangeretin), flavanols (represented by catechin, epicatechin, and procyanidins), isoflavones (including biochanin A and formononetin), and proanthocyanidins (Singla et al. 2019).

Anthocyanins are a highly valuable class of flavonoids soluble in water and are the main cause for imparting an array of colors like blue, purple, and vibrant red to fruits, helping fruit plants attract birds (Schaefer et al. 2008, He and Giusti 2010).

Various glycosylated forms are found in fruits; among them, cyanidin, delphinidin, malvidin, petunidin, pelargonidin, and peonidin are widely observed (Khoo et al. 2017). Cyanidin-3-glucoside, cyanidin-3-rutinoside, delphinidin-3-glucoside, and others are predominantly reported from various types of berries (Wu et al. 2006). Cyanidin-3,5-O-dihexoside, cyanidin-3-O-glucoside, cyanidin-3-O-rutinoside, and peonidin-3-O-rutinoside were also isolated from *Prunus avium* cherry fruits (Acero et al. 2019).

The most common forms of flavonols found in fruits like apples, berries and citrus include quercetin and kaempferol. Besides this, isorhamnetin-O-hexoside and quercetin-3-O-rutinoside were also extracted from the fruits of *Prunus avium* cherry (Acero et al. 2019).

Among different flavanones, hesperetin is found in a family of citrus fruits, characterized by low bioavailability, sparing solubility in water, and a short period of biological activity. These properties cease its use for therapeutic application; rather, consumption of raw citrus fruit eliminates its deficiency. Likewise, hesperetin and naringenin are also identified as naturally occurring flavanones, with predominance in fruits like grapes and citrus (Erlund 2004). Although the evaluation of eight *Citrus* hybrids reported a strong presence of naringin, hesperidin, and neohesperidin, hesperidin was still the prime flavanone in *Citrus*, with a range of 108.9 ± 0.8–7995 ± 253 µg/g (He et al. 2011).

Flavones are a prominent subclass of flavonoids represented by nobiletin and tangeretin, which are predominantly reported from the peels of citrus fruit. Cultivars of bergamot (*Citrus bergamia* Risso) were found rich in twelve flavones with predominance of vicenin-2 (Gattuso et al. 2007a, b).

Flavanols are often referred to as flavan-3-ols or catechins due to the consistent position of a hydroxyl group (OH) at the third carbon position of the C-ring (Kumar and Pandey 2013). Under the umbrella of flavanols, numerous compounds are included, such as catechin, epicatechin, epicatechin gallate, gallocatechin, epigallocatechin, gallocatechin gallate, and epigallocatechin gallate. Amongst these compounds, perception of taste in fruits is due to epicatechin gallate, whereas catechins serve as precursors of tannins (Mattivi et al. 2009). Fully ripe fruit of *Prunus serotina* was observed for major flavanols like procyanidin dimer (1-4), catechin, epicatechin, procyanidin trimer (1-4), and procyanidin tetramer (Brozdowski et al. 2021).

Isoflavones, represented by molecules like daidzein, genistein, and glycitein, are comprised under phytoestrogens, revealing significant estrogenic action. These compounds are mainly reported from plants belonging to the Leguminosae family, with predominance in *Glycine max* (L.) Merr. (soybeans) and *Trifolium* spp. (Křížová et al. 2019). Still, trace amounts of daidzein and genistein are reported in apples, berries, pomegranates, and plums. Recently, isoflavones like osajin, pomiferin, warangalone, and auriculasin have been identified in the fruits of Osage orange (Bailly 2024).

Proanthocyanidins are also called condensed tannins, belonging to the flavonoids family. Several units of flavan-3-ols, mainly containing catechin and epicatechin, join each other through interflavan bonds to comprise oligomeric or polymeric chains. Mainly, these units show that these monomers reveal C4–C8 or C4–C6 types of linkages. Cranberries, Chinese wolfberries, grape seeds, red goji berries, and pomegranates are prime sources of proanthocyanidins (Gu et al. 2004, Qi et al. 2023). Proanthocyanidins are also predominantly found in other fruits like Andean blackberry, capulí cherry, plum, and strawberry, with predominance of quercetin derivatives and trace quantities of kaempferol derivatives (Vasco et al. 2009).

1.3.4 Phenolic Acids

Phenolic acids are the other main class of phenolics, mainly categorized into hydroxybenzoic acids and hydroxycinnamic acids. Gallic acid, protocatechuic acid, and p-hydroxybenzoic acid are some of the important structural derivatives obtained from benzoic acid; caffeic, coumaric, ferulic, and sinapic acids belong to a group of cinnamic acid derivatives (Valanciene et al. 2020). Except for certain red-colored fruits, trace amounts of hydroxybenzoic acid are reported in palatable plants. On the contrary, hydroxycinnamic acids are widely found in fruits (Bento-Silva et al. 2020). Caffeic acid is the most predominant occurring form, reflecting the highest hydroxycinnamic acid content of fruits. It is mostly confined to peripheral parts of fruits like apple, blueberry, cherry, and plum (Huang et al. 2022). Acero et al. (2019) isolated cyanidin-3,5-O-dihexoside, cyanidin-3-O-glucoside, cyanidin-3-O-rutinoside, and peonidin-3-O-rutinoside as derivatives of phenolic acid from fruits of sweet cherry (*Prunus avium* L.).

1.3.5 Phytoestrogens

Phytoestrogens are plant-based bioactive compounds that reveal estrogen-like activity. Phytoestrogens comprise phenolic compounds covered under four major classes based on their chemical architecture, viz., coumestans, isoflavonoids, lignans, and stilbenes (Nikolić et al. 2017).

1.3.5.1 Coumestans

Coumestans are depicted with very high similarities in terms of structure and functions with isoflavones and are most frequently categorized alongside isoflavones due to their parallel estrogenic properties. Amongst widely available coumestans, most fail to illustrate any estrogenic activity; however, coumestrol and 4-methoxycoumesterol are exceptions in terms of their strong estrogenic potential. Although coumestans are predominantly reported from vegetables, fruits like oranges, dried dates, dried apricots, dried prunes, and grapefruit have also proved to be minor sources of them (Thompson et al. 2006, Kuhnle et al. 2009).

1.3.5.2 Isoflavonoids

Isoflavonoids show structural similarities with flavonoids; hence, many times they are recognized as a subgroup of the flavonoid family. Plant-based isoflavonoids depict structural resemblance with human endogenous estrogens, making it quite easy for them to bind with estrogen receptors. Some of the notable isoflavones are genistein, daidzein, and glycitein, which are predominantly reported from soybean-related products (Whitten et al. 2020). Besides soybeans, twenty different isoflavonoids were determined in mandarin melon berries (*Cudrania tricuspidata*), immature as well as ripe fruit (Jo et al. 2017).

1.3.5.3 Lignans

Lignans are important plant-based bioactive compounds and are regarded as the most promising dietary source of phytoestrogens. Some of the prominent lignans isolated from foodstuffs are identified as secoisolariciresinol, matairesinol, lariciresinol, and pinoresinol (Nadeem et al. 2020). Mostly, their occurrence is confined to seeds, grains, vegetables, and certain fruits (Chin et al. 2008). Fruit of açaí (*Euterpe oleracea*) showed nine lignans belonging to aryl tetrahydronaphthalene, dihydrobenzofuran, furofuran, 8-O-4′-neolignan, and tetrahydrofuran structural types (Chin et al. 2008).

1.3.5.4 Stilbenes

Stilbenes are an important class of polyphenolic compounds belonging to phytoestrogens. Compared to other phytoestrogenic compounds, stilbenes lack wide distribution across different foods. These compounds are reported in trace amounts in grapes, grape juice, red wine, etc. (El Khawand et al. 2018). Two red wine varieties of grapes (Raboso Piave and Primitivo) reported eighteen stilbene derivatives (Flamini et al. 2013). Within fruits, trace amounts of stilbenes are found in skin and flesh; however, they are specifically produced by plants under biotic and abiotic stress conditions like fungal infection, exposure to UV radiation, or physical damage (Valletta et al. 2021). Among the available stilbenes, resveratrol is the most rigorously evaluated compound with many biological activities. It is mainly reported from grapevines, berries, and pine-like trees (Tian et al. 2020). Resveratrol exists in both cis- or trans-isomeric forms; however, trans-resveratrol forms are recognized as more biologically active compared to others. In addition to resveratrol, other important stilbenes were recognized as pterostilbene, revealing structural alignment with resveratrol and found as a naturally dimethylated analog of resveratrol.

1.3.6 Carotenoids

Carotenoids are a diverse set of lipophilic, water-insoluble pigment molecules that are extensively reported on the Earth's surface, mainly concerned with the addition of vibrant yellow, orange, and red shades to different fruits and vegetables (Maghsoudi et al. 2022). Plants, a few microorganisms, and fungi are the prime producers of these compounds; however, mammals and other animals are unable to synthesize them on their own and fulfill the need through dietary food sources (Alcaíno et al. 2016). The molecular structure of carotenoids reveals eight isoprene units in its backbone, thus forming a conjugated polyene chain accountable for their array of coloration and antioxidant potential. Some of the common examples of carotenoids include β-carotene, β-cryptoxanthin, canthaxanthin, capsanthin, capsorubin, flavoxanthin, lutein, lycopene, and zeaxanthin. Arazá (*Eugenia stipitata* McVaugh), an important fruit found in the Amazon forest, reported fifteen carotenoid compounds. The peel part of Arazá fruit revealed a very high portion of total carotenoid (2484 ± 421 μg/100 g FW) compared to pulp (806 ± 348 μg/100 g FW), recognizing its role in providing colored shades to fruits (Garzón et al. 2012). In addition to their roles in photosynthesis, carotenoids are considered extremely important in human nutrition, pertaining to their antioxidant potential and as vitamin A precursors (Riaz et al. 2021). Visual appeal of edible fruits is mainly due to carotenoids, which impart attractive colors to them. Besides providing fruit coloration, they also act as precursor molecules for numerous volatile compounds, adding aroma and flavor to fruits (Young and Britton 2004, Meléndez-Martínez et al. 2015).

1.3.7 Vitamins

For a healthy life, regular consumption of fruits is highly essential as it complies with the need of the human body for natural antioxidants like vitamin C (ascorbic acid) and vitamin E (tocopherols). Nearly all plants and animals can generate water-soluble vitamins from D-glucose or D-galactose; however, absence of L-gulonolactone oxidase enzyme prevents humans from synthesizing vitamin C (Linster and Van Schaftingen 2007). Tropical fruits, along with certain vegetables, are the richest sources of vitamin C on Earth's surface (Popova 2019). Some of the prime sources of vitamin C include citrus fruits (like oranges and grapefruit), strawberries, tomatoes, and peppers (Santhi et al. 2020). Globally, fruits of *Terminalia ferdinandiana* Exell (vernacular name- Kakadu plum), *Myrciaria dubia* (Kunth) McVaugh (vernacular name- Camu-camu) and *Malpighia emarginata* DC. (vernacular name- Acerola) were recognized as abundant sources of vitamin C with a range of 820–22,490 mg/100 g (Doseděl et al. 2021). In the Indian context, analysis of ten wild fruits from Manipur showed *Emblica officinalis* as the richest source (379.7 ± 20 mg/100 g) of vitamin C (Khomdram and Devi 2010).

Tocopherol, an important form of vitamin E, is a highly rated bioactive compound reported from numerous fruits and vegetables. The chemical structure of vitamin E reveals eight distinct isoforms, comprising four tocopherols named α, β, γ, and δ, respectively, and four tocotrienols, differentiated from each other based on their degree of methylation and saturation of the side chain (Szymańska et al. 2020). Lipid profiles of chokecherry, thorny buffaloberry, woods' rose, and hawthorn, collectively called prairie fruits, showed all four forms of tocopherols ranging 6.3 ± 1.5 to 2,673 ± 80.8 mg/kg of lipids (Anwar et al. 2008). Oil derived from apple seeds, blackcurrant, strawberry, and raspberry reported α, β, γ, and δ tocopherols and α, β, γ, and δ tocotrienols (Pieszka et al. 2015). Brazilian fruits like Red Açaí (*Euterpe oleracea* M.), Inajá (*Maximiliana maripa* D.), Buriti (*Mauritia flexuosa* L.), and Uxi (*Endopleura uchi* C.) were also enriched with α-, γ-, and δ-tocopherol, confirming them as excellent reservoirs of vitamin E. Seed oil of wood apple (*Limonia acidissima*) is also a main source of γ-tocopherol (Lamani et al. 2021).

1.3.8 Triterpenoids

Triterpenoids are a special group of bioactive compounds biologically obtained from a 30-carbon precursor molecule known as squalene. Compounds belonging to this group are profusely observed in living entities like plants, fungal species, and a few marine life forms. Triterpenoids are recognized through their critical C_{30} structure, composed of six isoprene units (C_5H_8) (Cárdenas et al. 2019). Fruits like apples, olives, and grapes are natural sources of triterpenoids like ursolic acid, oleanolic acid, and betulinic acid, respectively (Han and Bakovic 2015). Accessory fruit such as Rugosa rose (*Rosa rugosa* Thunb.) showed enhanced triterpenoid contents as the fruit ripened (Dashbaldan et al. 2021).

1.3.9 Sterols

Sterols belong to the steroid family, mainly composed of lipophilic molecules. They reveal a cyclopentanoperhydrophenanthrene ring structure, having four fused rings, and the C3 position is occupied by a hydroxyl group, keeping it stand-alone from other types of lipids (Talapatra et al. 2015). In nature, sterol synthesis is mainly performed by plants, animals, and microorganisms like fungi, but is completely lacking in prokaryotes. Plant-based sterols are called phytosterols and are obtained from various fruits (Jun-Hua et al. 2008). Seed oil of fruits like apple, grapes, and olive has higher percentages of phytosterols (Górnaś et al. 2014, 2019, Maestri et al. 2019). In addition to fruits of the Indian subcontinent, pulp of Brazilian fruits like Uxi (*Endopleura uchi* C.), Inajá (*Maximiliana maripa* D.), Buriti (*Mauritia flexuosa* L.), etc. also turn out to be a predominant source of phytosterols (da Costa et al. 2010). Rugosa rose (*Rosa rugosa* Thunb.), an accessory fruit, revealed numerous phytosterol compounds; however, sitosterol stands out with the highest content (208.08 ± 30.02 μg/g d.w.) (Dashbaldan et al. 2021).

1.4 Role in Human Health

Food not only supplies the vital nutrients necessary for sustaining life but also contains various bioactive compounds (Yalcin and Çapar 2017). In recent times, there has been increased attention towards healthy nutrition with a major emphasis on fruit-rich diets serving to resolve many human-health-related issues. Amongst the diverse bioactive compounds reported from fruit-based diets, alkaloids are the prime group of molecules directly affecting the central nervous system and have revealed numerous pharmacological activities. Likewise, alkaloids show several pharmacological activities related to human well-being, such as anti-cancer, anti-inflammatory, anti-hypertensive, anti-diabetic, etc. (Rajput et al. 2022). Another class of phenolic compounds gained scientific attention mainly due to their powerful antioxidant potential, huge occurrence in dietary foods, and their possible ability to control various chronic diseases (Pérez-Jiménez et al. 2010). In recent times, flavonoids extracted from fruits have revealed various pharmacological activities, including antioxidant potential and inhibitory action against a diverse array of microorganisms, and heart-related disorders (Rees et al. 2018, Perez-Vizcaino and Fraga 2018, Bakoyiannis et al. 2019). Anthocyanins are mainly responsible for imparting color to fruits and vegetables; however, they are also revealed to have anti-inflammatory, antioxidant, and anti-carcinogenic properties (Kong et al. 2003). Hesperetin and naringenin belonging to the flavanone class exhibited a role in better brain insulin signaling and support in cognitive functions (Erlund 2004). Flavones, represented mostly by compounds like nobiletin and tangeretin, depict antioxidant, neuroprotective, anti-inflammatory, and anti-cancer potential (Li et al. 2006, Nogata et al. 2006). Fruit-based isoflavones revealed antioxidant and phytoestrogenic potential (Liggins et al. 2000, Thompson et al. 2006). Phenolic acids, mainly found in some red-colored fruits, are recognized for their antioxidant, cardioprotective, and immunomodulatory properties (Oluwole et al. 2022). Currently, phytoestrogens are recognized for their beneficial role against many severe diseases like different types of cancers and heart-related diseases. In recent years, they have garnered significant scientific interest due to their potential role in reducing the hazards of various adverse conditions like different types of cancer, cardiovascular ailments, osteoporosis, and characteristic features related to women's menopause (Lagari and Levis 2014, Rietjens et al. 2017).

Stilbenes, including resveratrol, are categorized under phytoestrogens mainly pertaining to their potential to interact with estrogen-related receptors and alter hormone-linked pathways (Salehi et al. 2018). Additionally, pterostilbene occurs in fruits like blueberries, revealing notable health-promoting effects (Nagarajan et al. 2022). Due to their quenching nature, carotenoids prove extremely useful for human health, exhibiting antioxidant activity and assisting in relieving oxidative stress (Fiedor and Burda 2014). Vitamin C obtained from fruits in the form of ascorbic acid is extremely beneficial to human health, as it helps in developing human connective tissues, developing sturdy walls of blood vessels, enhancing immunity, and improving wound-healing potential (Khomdram and Devi 2010). Tocopherol, obtained mainly from fruit oil, is involved in supporting cell membranes, regulating proper blood circulation, and providing action against numerous diseases (Lamani et al. 2021). Many tropical fruits are rich sources of triterpenoids, and their consumption has a significant impact on human health as they exhibit strong anti-inflammatory and cardioprotective activity (Han and Bakovic 2015). Sterols obtained from plant origin assist in reducing cholesterol levels in the blood. This feat is due to their ability to retard absorption of extrinsic as well as intrinsic cholesterol in the human intestine (Trautwein et al. 2018).

1.5 Conclusion

Plants synthesize bioactive compounds as part of their defense system in response to attack by pathogens, animals, and environment-related stresses. Humans have wisely used these compounds to treat many chronic disorders. Bioactive compounds mainly include phenolics, carotenoids, flavonoids, phytoestrogens, vitamins, and sterols. As fruits are the prime source of many important bioactive compounds, many studies revealed that incorporation of a significant amount of fruits might help in avoiding severe diseases related to the heart, kidney, liver, and brain. This protective nature of fruit is mainly attributed to the numerous bioactive compounds it possesses.

References

Acero, N., Gradillas, A., Beltran, M., García, A. and Mingarro, D. M. (2019). Comparison of phenolic compounds profile and antioxidant properties of different sweet cherry (*Prunus avium* L.) varieties. Food Chemistry 279: 260–271. https://doi.org/10.1016/j.foodchem.2018.12.020.

Alcaíno, J., Baeza, M. and Cifuentes, V. (2016). Carotenoid distribution in nature. pp. 3–33. *In*: Stange, C. (ed.). Carotenoids in Nature: Biosynthesis, Regulation and Function. Springer International Publishing. https://doi.org/10.1007/978-3-319-39126-7_1.

Ali, A. H., Abdelrahman, M. and El-Sayed, M. A. (2019). Alkaloid role in plant defense response to growth and stress. Bioactive Molecules in Plant Defense: Signaling in Growth and Stress 145–158.

Anwar, F., Przybylski, R., Rudzinska, M., Gruczynska, E. and Bain, J. (2008). Fatty acid, tocopherol and sterol compositions of Canadian prairie fruit seed lipids. Journal of the American Oil Chemists' Society 85(10): 953–959. https://doi.org/10.1007/s11746-008-1292-1.

Arbo, M. D., Larentis, E. R., Linck, V. M., Aboy, A. L., Pimentel, A. L., Henriques, A. T. et al. (2008). Concentrations of p-synephrine in fruits and leaves of *Citrus* species (Rutaceae) and the acute toxicity testing of *Citrus aurantium* extract and p-synephrine. Food and Chemical Toxicology 46(8): 2770–2775. https://doi.org/10.1016/j.fct.2008.05.006.

Ashihara, H. (2006). Metabolism of alkaloids in coffee plants. Brazilian Journal of Plant Physiology 18: 1–8. https://doi.org/10.1590/S1677-04202006000100001.

Bailly, C. (2024). Pharmacological properties of extracts and prenylated isoflavonoids from the fruits of Osage orange (*Maclura pomifera* (Raf.) C. K. Schneid.). Fitoterapia 174: 106112. https://doi.org/10.1016/j.fitote.2023.106112.

Bakoyiannis, I., Daskalopoulou, A., Pergialiotis, V. and Perrea, D. (2019). Phytochemicals and cognitive health: Are flavonoids doing the trick? Biomedicine & Pharmacotherapy 109: 1488–1497. https://doi.org/10.1016/j.biopha.2018.10.155.

Batra, P. and Sharma, A. K. (2013). Anti-cancer potential of flavonoids: Recent trends and future perspectives. Biotech 3(6): 439–459. https://doi.org/10.1007/s13205-013-0167-7.

Bento-Silva, A., Koistinen, V. M., Mena, P., Bronze, M. R., Hanhineva, K., Sahlstrøm, S. et al. (2020). Factors affecting intake, metabolism and health benefits of phenolic acids: Do we understand individual variability? European Journal of Nutrition 59: 1275–1293. https://doi.org/10.1007/s00394-019-02088-5.

Bernhoft, A. (2010). A brief review on bioactive compounds in plants. pp. 11–23. *In*: Bernhoft, A. (ed.). Bioactive Compounds in Plants. The Norwegian Academy of Science and Letters.

Brozdowski, J., Waliszewska, B., Loffler, J., Hudina, M., Veberic, R. and Mikulic-Petkovsek, M. (2021). Composition of phenolic compounds, cyanogenic glycosides, organic acids and sugars in fruits of black cherry (*Prunus serotina* Ehrh.). Forests 12(6): 762. https://doi.org/10.3390/f12060762.

Cárdenas, P. D., Almeida, A. and Bak, S. (2019). Evolution of structural diversity of triterpenoids. Frontiers in Plant Science 10: 1523. https://doi.org/10.3389/fpls.2019.01523.

Chen, C. and Lin, L. (2019). Alkaloids in diet. pp. 1–35. In Handbook of Dietary Phytochemicals. Springer. https://doi.org/10.1007/978-981-13-7647-4_24-1.

Chin, Y. W., Chai, H. B., Keller, W. J. and Kinghorn, A. D. (2008). Lignans and other constituents of the fruits of *Euterpe oleracea* (Acai) with antioxidant and cytoprotective activities. Journal of Agricultural and Food Chemistry 56(17): 7759–7764. https://doi.org/10.1021/jf8014717.

da Costa, P. A., Ballus, C. A., Teixeira-Filho, J. and Godoy, H. T. (2010). Phytosterols and tocopherols content of pulps and nuts of Brazilian fruits. Food Research International 43(6): 1603–1606. https://doi.org/10.1016/j.foodres.2010.04.027.

Dashbaldan, S., Rogowska, A., Pączkowski, C. and Szakiel, A. (2021). Distribution of triterpenoids and steroids in developing rugosa rose (*Rosa rugosa* Thunb.) accessory fruit. Molecules 26(17): 5158. https://doi.org/10.3390/molecules26175158.

Deepa, N., Kaur, C., George, B., Singh, B. and Kapoor, H. (2007). Antioxidant constituents in some sweet pepper (*Capsicum annuum* L.) genotypes during maturity. LWT–Food Science and Technology 40(1): 121–129. https://doi.org/10.1016/j.lwt.2005.09.020.

Doseděl, M., Jirkovský, E., Macáková, K., Krčmová, L. K., Javorská, L., Pourová, J. et al. (2021). Vitamin C—Sources, physiological role, kinetics, deficiency, use, toxicity, and determination. Nutrients 13(2): 615. https://doi.org/10.3390/nu13020615.

Dostál, J. (2000). Two faces of alkaloids. Journal of Chemical Education 77(8): 993–998. https://doi.org/10.1021/ed077p993.

Eguchi, R., Ono, N., Hirai-Morita, A., Katsuragi, T., Nakamura, S., Huang, M. et al. (2019). Classification of alkaloids according to the starting substances of their biosynthetic pathways using graph convolutional neural networks. BMC Bioinformatics 20: 380. https://doi.org/10.1186/s12859-019-2964-7.

El Khawand, T., Courtois, A., Valls, J., Richard, T. and Krisa, S. (2018). A review of dietary stilbenes: Sources and bioavailability. Phytochemistry Reviews 17(5): 1007–1029. https://doi.org/10.1007/s11101-018-9552-3.

Erlund, I. (2004). Review of the flavonoids quercetin, hesperetin, and naringenin: Dietary sources, bioactivities, bioavailability, and epidemiology. Nutrition Research 24(10): 851–874. https://doi.org/10.1016/j.nutres.2004.07.005.

FAOSTAT. (2011). Food supply – Crops primary equivalent. Retrieved from http://faostat.fao.org/site/609/default.aspx#ancor.

Fiedor, J. and Burda, K. (2014). Potential role of carotenoids as antioxidants in human health and disease. Nutrients 6(2): 466–488. https://doi.org/10.3390/nu6020466.

Flamini, R., De Rosso, M., De Marchi, F., Dalla Vedova, A., Panighel, A., Gardiman, M. et al. (2013). An innovative approach to grape metabolomics: Stilbene profiling by suspect screening analysis. Metabolomics 9(6): 1243–1253. https://doi.org/10.1007/s11306-013-0550-y.

Gani, A., Wani, S. M., Masoodi, F. A. and Hameed, G. (2012). Whole-grain cereal bioactive compounds and their health benefits: A review. J. Food Process. Technol 3(3): 1000146.

Garzón, G. A., Narváez-Cuenca, C. E., Kopec, R. E., Barry, A. M., Riedl, K. M. and Schwartz, S. J. (2012). Determination of carotenoids, total phenolic content, and antioxidant activity of Arazá (*Eugenia stipitata* McVaugh), an Amazonian fruit. Journal of Agricultural and Food Chemistry 60(18): 4709–4717. https://doi.org/10.1021/jf300831h.

Gattuso, G., Barreca, D., Caristi, C., Gargiulli, C. and Leuzzi, U. (2007a). Distribution of flavonoids and furocoumarins in juices from cultivars of *Citrus bergamia* Risso. Journal of Agricultural and Food Chemistry 55(24): 9921–9927. https://doi.org/10.1021/jf071889n.

Gattuso, G., Barreca, D., Gargiulli, C., Leuzzi, U. and Caristi, C. (2007b). Flavonoid composition of citrus juices. Molecules 12(8): 1641–1673. https://doi.org/10.3390/12081641.

Goldberg, G. (ed.). (2008). Plants: Diet and Health. John Wiley & Sons.

Górnaś, P., Rudzińska, M. and Segliņa, D. (2014). Lipophilic composition of eleven apple seed oils: A promising source of unconventional oil from industry by-products. Industrial Crops and Products 60: 86–91. https://doi.org/10.1016/j.indcrop.2014.05.041.

Górnaś, P., Rudzińska, M., Grygier, A. and Lācis, G. (2019). Diversity of oil yield, fatty acids, tocopherols, tocotrienols, and sterols in the seeds of 19 interspecific grape crosses. Journal of the Science of Food and Agriculture 99(5): 2078–2087. https://doi.org/10.1002/jsfa.9390.

Gu, L., Kelm, M. A., Hammerstone, J. F., Beecher, G., Holden, J., Haytowitz, D. et al. (2004). Concentrations of proanthocyanidins in common foods and estimations of normal consumption. Journal of Nutrition 134(3): 613–617. https://doi.org/10.1093/jn/134.3.613.

Hamzalıoğlu, A. and Gökmen, V. (2016). Interaction between bioactive carbonyl compounds and asparagine and impact on acrylamide. pp. 355–376. *In*: Gökmen, V. (ed.). Acrylamide in Food. Academic Press.

Han, N. and Bakovic, M. (2015). Biologically active triterpenoids and their cardioprotective and anti-inflammatory effects. Journal of Bioanalysis & Biomedicine 12(5): 1948–59. https://doi.org/10.4172/1948-593X.S12.005.

He, D., Shan, Y., Wu, Y., Liu, G., Chen, B. and Yao, S. (2011). Simultaneous determination of flavanones, hydroxycinnamic acids, and alkaloids in citrus fruits by HPLC-DAD–ESI/MS. Food Chemistry 127(2): 880–885. https://doi.org/10.1016/j.foodchem.2011.01.037.

He, J. and Giusti, M. M. (2010). Anthocyanins: Natural colorants with health-promoting properties. Annual Review of Food Science and Technology 1: 163–187. https://doi.org/10.1146/annurev.food.080708.100754.

Herraiz, T. and Galisteo, J. (2003). Tetrahydro-β-carboline alkaloids occur in fruits and fruit juices: Activity as antioxidants and radical scavengers. Journal of Agricultural and Food Chemistry 51(24): 7156–7161. https://doi.org/10.1021/jf034837c.

Huang, R., Li, C., Guo, M., Liu, J., Qu, L., Fan, Y. et al. (2022). Caffeic acid enhances storage ability of apple fruit by regulating fatty acid metabolism. Postharvest Biology and Technology 192: 112012. https://doi.org/10.1016/j.postharvbio.2022.112012.

Jamloki, A., Bhattacharyya, M., Nautiyal, M. C. and Patni, B. (2021). Elucidating the relevance of high temperature and elevated CO_2 in plant secondary metabolites (PSMs) production. Heliyon 7(8).

Jo, Y. H., Kim, S. B., Liu, Q., Do, S. G., Hwang, B. Y. and Lee, M. K. (2017). Comparison of pancreatic lipase inhibitory isoflavonoids from unripe and ripe fruits of *Cudrania tricuspidata*. PLOS ONE 12(3): e0172069. https://doi.org/10.1371/journal.pone.0172069.

Jun-Hua, H., Yue-Xin, Y. and Mei-Yuan, F. (2008). Contents of phytosterols in vegetables and fruits commonly consumed in China. Biomedical and Environmental Sciences 21(6): 449–453.

Karasawa, M. M. and Mohan, C. (2018). Fruits as prospective reserves of bioactive compounds: A review. Natural Products and Bioprospecting 8: 335–346. https://doi.org/10.1007/s13659-018-0182-4.

Khomdram, S. and Devi, G. S. (2010). Determination of antioxidant activity and vitamin C of some wild fruits of Manipur. The Bioscan 5(3): 501–504.

Khoo, H. E., Azlan, A., Tang, S. T. and Lim, S. M. (2017). Anthocyanidins and anthocyanins: Colored pigments as food, pharmaceutical ingredients, and the potential health benefits. Food & Nutrition Research 61(1): 1361779. https://doi.org/10.1080/16546628.2017.1361779.

Kong, J. M., Chia, L. S., Goh, N. K., Chia, T. F. and Brouillard, R. (2003). Analysis and biological activities of anthocyanins. Phytochemistry 64(5): 923–933. https://doi.org/10.1016/S0031-9422(03)00438-2.

Křížová, L., Dadáková, K., Kašparovská, J. and Kašparovský, T. (2019). Isoflavones. Molecules 24(6): 1076.

Kuhnle, G. G., Dell'Aquila, C., Aspinall, S. M., Runswick, S. A., Joosen, A. M., Mulligan, A. A. et al. (2009). Phytoestrogen content of fruits and vegetables commonly consumed in the UK based on LC–MS and 13C-labelled standards. Food Chemistry 116(2): 542–554. https://doi.org/10.1016/j.foodchem.2009.02.061.

Kumar, A. and Naraian, R. (2019). Producers of bioactive compounds. pp. 205–221. *In*: Gupta, V. K. and Pandey, A. (eds.). New and Future Developments in Microbial Biotechnology and Bioengineering. Elsevier. https://doi.org/10.1016/B978-0-444-63501-3.00014-9.

Kumar, S. and Pandey, A. K. (2013). Chemistry and biological activities of flavonoids: An overview. The Scientific World Journal 2013: 162750. https://doi.org/10.1155/2013/162750.

Lagari, V. S. and Levis, S. (2014). Phytoestrogens for menopausal bone loss and climacteric symptoms. The Journal of Steroid Biochemistry and Molecular Biology 139: 294–301. https://doi.org/10.1016/j.jsbmb.2013.02.001.

Lamani, S., Anu-Appaiah, K. A., Murthy, H. N., Dewir, Y. H. and Rihan, H. Z. (2021). Fatty acid profile, tocopherol content of seed oil, and nutritional analysis of seed cake of wood apple (*Limonia acidissima* L.), an underutilized fruit-yielding tree species. Horticulturae 7(9): 275. https://doi.org/10.3390/horticulturae7090275.

Li, S., Lo, C. Y. and Ho, C. T. (2006). Hydroxylated polymethoxyflavones and methylated flavonoids in sweet orange (*Citrus sinensis*) peel. Journal of Agricultural and Food Chemistry 54(12): 4176–4185. https://doi.org/10.1021/jf060291x.

Liggins, J., Bluck, L. J., Runswick, S., Atkinson, C., Coward, W. A. and Bingham, S. A. (2000). Daidzein and genistein content of fruits and nuts. The Journal of Nutritional Biochemistry 11(6): 326–331. https://doi.org/10.1016/S0955-2863(00)00100-0.

Lin, D., Xiao, M., Zhao, J., Li, Z., Xing, B., Li, X. et al. (2016). An overview of plant phenolic compounds and their importance in human nutrition and management of type 2 diabetes. Molecules 21(10): 1374. https://doi.org/10.3390/molecules21101374.

Linster, C. L. and Van Schaftingen, E. (2007). Vitamin C: Biosynthesis, recycling and degradation in mammals. The FEBS Journal 274(1): 1–22. https://doi.org/10.1111/j.1742-4658.2006.05607.x.

Liu, R. H. (2004). Potential synergy of phytochemicals in cancer prevention: Mechanism of action. The Journal of Nutrition 134(12): 3479S–3485S. https://doi.org/10.1093/jn/134.12.3479S.

Liu, R. H. (2013). Dietary bioactive compounds and their health implications. Journal of Food Science 78(s1): A18–A25. https://doi.org/10.1111/1750-3841.12101.

Liu, Y. P., Li, Y. J., Zhao, Y. Y., Guo, J. M., Liu, Y. Y., Wang, X. P. et al. (2021). Carbazole alkaloids from the fruits of *Clausena anisum-olens* with potential PTP1B and α-glucosidase inhibitory activities. Bioorganic Chemistry 110: 104775. https://doi.org/10.1016/j.bioorg.2021.104775.

Maestri, D., Barrionuevo, D., Bodoira, R., Zafra, A., Jiménez-López, J. and Alché, J. D. D. (2019). Nutritional profile and nutraceutical components of olive (*Olea europaea* L.) seeds. Journal of Food Science and Technology 56: 4359–4370. https://doi.org/10.1007/s13197-019-03989-2.

Maghsoudi, S., Taghavi Shahraki, B., Rabiee, N., Fatahi, Y., Bagherzadeh, M., Dinarvand, R. et al. (2022). The colorful world of carotenoids: A profound insight on therapeutics and recent trends in nano delivery systems. Critical Reviews in Food Science and Nutrition 62(13): 3658–3697. https://doi.org/10.1080/10408398.2020.1860931.

Manach, C., Scalbert, A., Morand, C., Rémésy, C. and Jiménez, L. (2004). Polyphenols: Food sources and bioavailability. The American Journal of Clinical Nutrition 79(5): 727–747. https://doi.org/10.1093/ajcn/79.5.727.

Mandal, B. (2022). Flavonoids: The miracle molecules of nature. Acta Scientific Pharmacology 3(10): 60–62.

Mattivi, F., Vrhovsek, U., Masuero, D. and Trainotti, D. (2009). Differences in the amount and structure of extractable skin and seed tannins amongst red grape varieties. Australian Journal of Grape and Wine Research 15(1): 27–35. https://doi.org/10.1111/j.1755-0238.2008.00031.x.

Meghwal, M., Divyashree, K. and Kumar, N. (2018). Fruits, vegetables, and grains: their peels and fiber for better human health. State-Art Technol Food Sci Human Health Emerg Issues Spec Topics 3: 10.

Meléndez-Martínez, A. J., Mapelli-Brahm, P., Benítez-González, A. and Stinco, C. M. (2015). A comprehensive review on the color of carotenoids: From color description to color stability in foods. Trends in Food Science & Technology 44(2): 149–171. https://doi.org/10.1016/j.tifs.2015.04.006.

Nadeem, M., Taj Khan, I., Khan, F., Shah, M. A., Silva, A. S., Nabavi, S. F. et al. (2020). Lignans and flavonolignans. pp. 98–116. *In*: Silva, A. S., Nabavi, S. F., Saeedi, M. and Nabavi, S. M. (eds.). Recent Advances in Natural Products Analysis. Elsevier.

Nagarajan, S., Mohandas, S., Ganesan, K., Xu, B. and Ramkumar, K. M. (2022). New insights into dietary pterostilbene: Sources, metabolism, and health promotion effects. Molecules 27(19): 6316. https://doi.org/10.3390/molecules27196316.

Naseer, M. (2018). Fruits of medicinal importance. International Journal of Unani and Integrative Medicine 2(1): 6–10.

National Cancer Institute. (2025). NCI Dictionary of Cancer Terms. Bethesda, MD: National Cancer Institute. Retrieved April 13, 2025, from https://www.cancer.gov/publications/dictionaries/cancer-terms/def/bioactive-compound.

Nikolić, I. L., Savić-Gajić, I. M., Tačić, A. D. and Savić, I. M. (2017). Classification and biological activity of phytoestrogens: A review. Advances in Technology 6: 96–106. https://doi.org/10.12955/at.v6.111.

Nogata, Y., Sakamoto, K., Shiratsuchi, H., Ishii, T., Yano, M. and Ohta, H. (2006). Flavonoid composition of fruit tissues of citrus species. Bioscience, Biotechnology, and Biochemistry 70(1): 178–192. https://doi.org/10.1271/bbb.70.178.

Nowicka, P., Wojdyło, A., Tkacz, K. and Turkiewicz, I. P. (2023). Quantitative and qualitative determination of carotenoids and polyphenolic compounds in selected cultivars of *Prunus persica* L. and their ability to *in vitro* inhibit lipoxygenase, cholinesterase, α-amylase, α-glucosidase, and pancreatic lipase. Food Chemistry: X 17: 100619. https://doi.org/10.1016/j.fochx.2023.100619.

Olale, K., Walyambillah, W., Mohammed, S. A., Sila, A. and Shepherd, K. (2019). FTIR-DRIFTS-based prediction of β-carotene, α-tocopherol, and L-ascorbic acid in mango (*Mangifera indica* L.) fruit pulp. SN Applied Sciences 1(9): 1066. https://doi.org/10.1007/s42452-019-1117-3.

Oluwole, O., Fernando, W. B., Lumanlan, J., Ademuyiwa, O. and Jayasena, V. (2022). Role of phenolic acid, tannins, stilbenes, lignans, and flavonoids in human health—A review. International Journal of Food Science and Technology 57(10): 6326–6335. https://doi.org/10.1111/ijfs.15750.

Patra, A., Abdullah, S. and Pradhan, R. C. (2022). Review on the extraction of bioactive compounds and characterization of fruit industry by-products. Bioresources and Bioprocessing 9(1): 14. https://doi.org/10.1186/s40643-022-00487-w.

Pérez-Jiménez, J., Neveu, V., Vos, F. and Scalbert, A. (2010). Identification of the 100 richest dietary sources of polyphenols: An application of the Phenol-Explorer database. European Journal of Clinical Nutrition 64(3): S112–S120. https://doi.org/10.1038/ejcn.2010.221.

Perez-Vizcaino, F. and Fraga, C. G. (2018). Research trends in flavonoids and health. Archives of Biochemistry and Biophysics 646: 107–112. https://doi.org/10.1016/j.abb.2018.03.011.

Pieszka, M., Migdał, W., Gąsior, R., Rudzińska, M., Bederska-Łojewska, D., Pieszka, M. et al. (2015). Native oils from apple, blackcurrant, raspberry, and strawberry seeds as a source of polyenoic fatty acids, tocochromanols, and phytosterols: A health implication. Journal of Chemistry 2015: 659541. https://doi.org/10.1155/2015/659541.

Popova, A. (2019). Comparison of vitamin C content of commercially available fresh fruits. Asian Food Science Journal 13(1): 1–6. https://doi.org/10.9734/AFSJ/2019/v13i130075.

Qi, Q., Chu, M., Yu, X., Xie, Y., Li, Y., Du, Y. et al. (2023). Anthocyanins and proanthocyanidins: Chemical structures, food sources, bioactivities, and product development. Food Reviews International 39(7): 4581–4609. https://doi.org/10.1080/87559129.2022.2074208.

Rajput, A., Sharma, R. and Bharti, R. (2022). Pharmacological activities and toxicities of alkaloids on human health. Materials Today: Proceedings 48: 1407–1415. https://doi.org/10.1016/j.matpr.2021.06.167.

Rees, A., Dodd, G. F. and Spencer, J. P. (2018). The effects of flavonoids on cardiovascular health: A review of human intervention trials and implications for cerebrovascular function. Nutrients 10(12): 1852. https://doi.org/10.3390/nu10121852.

Reis Giada, M. L. (2013). Food phenolic compounds: Main classes, sources and their antioxidant power. pp. 87–112. *In*: Morales-Gonzalez, J. A. (ed.). Oxidative Stress and Chronic Degenerative Diseases—A Role for Antioxidants. IntechOpen.

Riaz, M., Zia-Ul-Haq, M. and Dou, D. (2021). Chemistry of carotenoids. pp. 43–76. *In*: Carotenoids: Structure and Function in the Human Body. Springer. https://doi.org/10.1007/978-3-030-66572-5_4.

Rietjens, I. M., Louisse, J. and Beekmann, K. (2017). The potential health effects of dietary phytoestrogens. British Journal of Pharmacology 174(11): 1263–1280. https://doi.org/10.1111/bph.13622.

Rodríguez, L. G., Gasga, V. M., Pescuma, M., Van Nieuwenhove, C., Mozzi, F. and Burgos, J. A. (2021). Fruits and fruit by-products as sources of bioactive compounds: Benefits and trends of lactic acid fermentation in the development of novel fruit-based functional beverages. Food Research International 140: 109854. https://doi.org/10.1016/j.foodres.2020.109854.

Salehi, B., Mishra, A. P., Nigam, M., Sener, B., Kilic, M., Sharifi-Rad, M. et al. (2018). Resveratrol: A double-edged sword in health benefits. Biomedicines 6(3): 91. https://doi.org/10.3390/biomedicines6030091.

Santhi, V. P., Poomaruthai, M., Sarasu, V. P., Gurusamy, K., Parthiban, S., Geetha, K. et al. (2020). Fruits: A potential source of vitamin C as essential human nutrition and immunity development: A review. The Pharma Innovation Journal 9(9): 132–144.

Sartorelli, P., Carvalho, C. S., Reimão, J. Q., Ferreira, M. J. P. and Tempone, A. G. (2009). Antiparasitic activity of biochanin A, an isolated isoflavone from fruits of *Cassia fistula* (Leguminosae). Parasitology Research 104(2): 311–314. https://doi.org/10.1007/s00436-008-1214-4.

Schaefer, H. M., McGraw, K. and Catoni, C. (2008). Birds use fruit colour as honest signal of dietary antioxidant rewards. Functional Ecology 22(2): 303–310. https://doi.org/10.1111/j.1365-2435.2007.01363.x.

Shashirekha, M. N., Mallikarjuna, S. E. and Rajarathnam, S. (2015). Status of bioactive compounds in foods, with focus on fruits and vegetables. Critical Reviews in Food Science and Nutrition 55(10): 1324–1339.

Singh, I. (2017). Antimicrobials in higher plants: Classification, mode of action and bioactivities. Chemical Biology Letters 4(2): 48–62.

Singla, R. K., Dubey, A. K., Garg, A., Sharma, R. K., Fiorino, M., Ameen, S. M. et al. (2019). Natural polyphenols: Chemical classification, definition of classes, subcategories, and structures. Journal of AOAC International 102(5): 1397–1400. https://doi.org/10.5740/jaoacint.18-0418.

Slavin, J. L. and Lloyd, B. (2012). Health benefits of fruits and vegetables. Advances in Nutrition 3(4): 506–516.

Soquetta, M. B., Terra, L. D. and Bastos, C. P. (2018). Green technologies for the extraction of bioactive compounds in fruits and vegetables. CyTA–Journal of Food 16(1): 400–412. https://doi.org/10.1080/19476337.2017.1411978.

Szymańska, R., Nowicka, B., Trela, A. and Kruk, J. (2020). Vitamin E: Structure and forms. pp. 67–90. *In*: Molecular Nutrition. Academic Press. https://doi.org/10.1016/B978-0-12-812109-2.00004-5.

Talapatra, S. K., Talapatra, B., Talapatra, S. K. and Talapatra, B. (2015). Steroids: Cholesterol and other phytosterols. pp. 553–583. In Chemistry of Plant Natural Products: Stereochemistry, Conformation, Synthesis, Biology, and Medicine. Springer. https://doi.org/10.1007/978-3-319-16528-3_13.

Thompson, L. U., Boucher, B. A., Liu, Z., Cotterchio, M. and Kreiger, N. (2006). Phytoestrogen content of foods consumed in Canada, including isoflavones, lignans, and coumestans. Nutrition and Cancer 54(2): 184–201. https://doi.org/10.1207/s15327914nc5402_5.

Tian, B. and Liu, J. (2020). Resveratrol: A review of plant sources, synthesis, stability, modification and food application. Journal of the Science of Food and Agriculture 100(4): 1392–1404. https://doi.org/10.1002/jsfa.10106.

Trautwein, E. A., Vermeer, M. A., Hiemstra, H. and Ras, R. T. (2018). LDL-cholesterol lowering of plant sterols and stanols—Which factors influence their efficacy? Nutrients 10(9): 1262. https://doi.org/10.3390/nu10091262.

Tsao, R. (2010). Chemistry and biochemistry of dietary polyphenols. Nutrients 2(12): 1231–1246. https://doi.org/10.3390/nu2121231.

Tungmunnithum, D., Thongboonyou, A., Pholboon, A. and Yangsabai, A. (2018). Flavonoids and other phenolic compounds from medicinal plants for pharmaceutical and medical aspects: An overview. Medicines 5(3): 93. https://doi.org/10.3390/medicines5030093.

Vagiri, M., Ekholm, A., Öberg, E., Johansson, E., Andersson, S. C. and Rumpunen, K. (2013). Phenols and ascorbic acid in black currants (*Ribes nigrum* L.): Variation due to genotype, location, and year. Journal of Agricultural and Food Chemistry 61(39): 9298–9306. https://doi.org/10.1021/jf402364j.

Valanciene, E., Jonuskiene, I., Syrpas, M., Augustiniene, E., Matulis, P., Simonavicius, A. et al. (2020). Advances and prospects of phenolic acids production, biorefinery and analysis. Biomolecules 10(6): 874. https://doi.org/10.3390/biom10060874.

Vallejo, F., García-Viguera, C. and Tomás-Barberán, F. A. (2003). Changes in broccoli (*Brassica oleracea* L. var. italica) health-promoting compounds with inflorescence development. Journal of Agricultural and Food Chemistry 51(13): 3776–3782. https://doi.org/10.1021/jf030073u.

Valletta, A., Iozia, L. M. and Leonelli, F. (2021). Impact of environmental factors on stilbene biosynthesis. Plants 10(1): 90. https://doi.org/10.3390/plants10010090.

Vasco, C., Riihinen, K., Ruales, J. and Kamal-Eldin, A. (2009). Phenolic compounds in Rosaceae fruits from Ecuador. Journal of Agricultural and Food Chemistry 57(4): 1204–1212. https://doi.org/10.1021/jf803443w.

Vasco, C., Ruales, J. and Kamal-Eldin, A. (2008). Total phenolic compounds and antioxidant capacities of major fruits from Ecuador. Food Chemistry 111(4): 816–823. https://doi.org/10.1016/j.foodchem.2008.04.054.

Whitten, P. L., Kudo, S. and Okubo, K. K. (2020). Isoflavonoids. pp. 117–137. In Handbook of Plant and Fungal Toxicants. CRC Press.

World Health Organization (WHO). (2023). Increasing fruit and vegetable consumption to reduce the risk of noncommunicable diseases. Retrieved April 13, 2025, from https://www.who.int.

Wu, X., Beecher, G. R., Holden, J. M., Haytowitz, D. B., Gebhardt, S. E. and Prior, R. L. (2006). Concentrations of anthocyanins in common foods in the United States and estimation of normal consumption. Journal of Agricultural and Food Chemistry 54(11): 4069–4075. https://doi.org/10.1021/jf060300l.

Yalçin, H. and Çapar, T. D. (2017). Bioactive compounds of fruits and vegetables. pp. 723–745. *In*: Beaulieu, R. C., Mattheis, J. P. and DellaPenna, D. P. (eds.). Minimally Processed Refrigerated Fruits and Vegetables. Springer. https://doi.org/10.1007/978-1-4939-7012-8_29.

Yildiz, F., Ozcan, B. and Guneser, O. (2010). Phytochemicals in fruits and vegetables. pp. 3–36. *In*: Yildiz, F. (ed.). Phytochemicals in Fruits and Vegetables: Health Promotion and Disease Prevention. CRC Press.

Young, A. J. and Britton, G. (2004). Carotenoids and the human eye. Nutrition and Health 18(1): 21–32. https://doi.org/10.1177/026010600401800103.

Zoratti, L., Karppinen, K., Luengo Escobar, A., Häggman, H. and Jaakola, L. (2014). Light-controlled flavonoid biosynthesis in fruits. Frontiers in Plant Science 5: 534.

Chapter 2

Fruit Species in the Urban Landscape

Native and Introduced Species as Functional Foods

Nelida Soria,[1,*] *Lida Ramoa de Guerrero,*[1] *María Elena López de Silva,*[1] *Miguel Ángel Delpino*[1] and *Pasionaria Ramos*[2]

2.1 Introduction

The origin of fruit species within urban environments dates to the earliest periods of human history, serving not only ornamental functions but also fulfilling various practical purposes. Recently, cities have encountered considerable challenges related to sustainability, climate change, and food security. Within this framework, fruit species have emerged as an innovative and multifunctional solution in urban landscape design. These plants, besides enhancing the aesthetic appeal of cities, provide nutritious food and contribute significantly to both human and environmental well-being. Additionally, the integration of native and introduced species increases the diversity of urban ecosystems and improves their resilience (Castillo Rodríguez and Ferro Cisneros 2015).

Due to population growth and increased construction using heat-retaining cement, cities are becoming increasingly less hospitable. This rise in urban temperatures

[1] Faculty of Applied Sciences, National University of Pilar.
[2] Faculty of Medical Sciences, National University of Asunción. Researcher SISNI-CONACYT.
* Corresponding author: nsoria2000@yahoo.com

DOI: 10.1201/9781003668169-3

necessitates rethinking urban planning by increasing green spaces through the cultivation of tree species. Urban trees can contribute to making cities more livable by providing several benefits to residents, such as reducing noise and air pollution, creating habitats for birds, and improving the microclimate with lower temperatures. Additionally, these plants could offer food for the community, emphasizing that fruit and vegetable consumption is essential for nutritional balance. Therefore, cultivating fruit-bearing trees along city streets presents a healthy alternative (Alfonso et al. 2022).

Integrating fruit species into urban landscapes offers a novel approach to tackling current urban issues such as climate change, food insecurity, and biodiversity loss. These species contribute aesthetically by enhancing the appearance of streets, parks, and squares, while also offering various environmental, social, and economic benefits that improve the livability and sustainability of urban areas.

Growing fruit crops can transform agri-food systems by altering spatial patterns of food demand and influencing consumer preferences, thus impacting how, where, and what food is produced, supplied, and consumed. These changes present both challenges and opportunities for agrifood systems to ensure access to affordable and healthy diets for all.

Fruit trees can be classified as indigenous, native, or introduced. Introduced fruit species have varying origins. For instance, citrus fruits are believed to originate from Asia. They were introduced to Europe, specifically Spain, around the 7th century by the Arabs. The citron was the first citrus fruit introduced, followed by the bitter orange (Ancillo and Medina 2007, Ibañez et al. 2015). From Spain, these were brought to Latin America during the Conquest. Some scholars suggest that this era marked a significant biological event due to the substantial exchange of plants and animals between the continents (Machuca 2013).

To analyze the origin of the species cultivated in cities, we use the city of Pilar, the capital of the Department of Ñeembucú in the Republic of Paraguay, as an example. This city is in southern Paraguay, on the banks of the Paraguay River and the Ñeembucú stream, approximately 230 km south of Asunción, the capital of the country. It has minimal elevation changes, with a maximum altitude of 100 m above sea level. The soil is moist due to numerous wetlands, making it susceptible to flooding (Atlas Censal del Paraguay 2016, Contreras et al. 2007). These topographic characteristics restrict the variety of species suitable for urban cultivation. While the cultivation of fruit species in cities often lacks planning criteria, both native and introduced species remain commonly found throughout the urban landscape.

2.2 Native and Introduced Species in the Urban Area of the City of Pilar

In this city, both native and introduced species can be found. These species generally do not adhere to a tree planting policy but rather reflect the preferences of the residents. Introduced species increase the variety of fruits available, though they require additional management to prevent pest invasions, providing nutritional and cultural benefits. Common examples of introduced species growing on streets

Table 1. Native and introduced species in the urban area of the city of Pilar, Paraguay.

Common name	Scientific name	Family	Origin
Acerola	*Malpighia emarginata* DC.	Malpighiaceae	I
Araticui	*Annona emarginata* (Schltdl.) H. Rainer	Annonaceae	N
Avocado, palta	*Persea americana* L.	Lauraceae	I
Bitter Orange	*Citrus x aurantium* L	Rutaceae	I
Guava	*Psidium guajava* L.	Myrtaceae	I
Inga	*Inga marginata* Willd.	Fabaceae	N
Lemon	*Citrus × limon* (L.) Osbeck	Rutaceae	I
Loquat	*Eriobotrya japonica* (Thunb.) Lindl	Rosaceae	I
Mango	*Mangifera indica* L	Anacardiaceae	I
Mulberry	*Morus nigra* L.	Moraceae	I
Mbokaya or Macaúba Palm	*Acrocomia aculeata* (Jacq.) Lodd. ex Mart.	Arecaceae	N
Pindo Palm	*Syagrus romanzoffiana* (Cham.) Glasman	Arecaceae	N
Yvapovo	*Melicoccus lepidopetalus* Radlk	Sapindaceae	N

Note: I = Introduced, N = Native

in Pilar include the mango tree *Mangifera indica* L., citrus trees such as the bitter orange *Citrus x aurantium*, and mulberry *Morus nigra* L. (García et al. 2006, Parrotta 1993, Ipek Suntar et al. 2018, Stohs 2017).

Native species need fewer resources due to their adaptation to local conditions. Integrating them into urban landscapes preserves ecological identity, promotes ecosystem balance, and maintains remnants of pre-city forest formations. For example, *Annona emarginata* (Schltdl.) H. Rainer, commonly known as araticui, is notable for its high functional compounds and climate adaptability (Table 1).

From a nutritional perspective, urban species' fruits are often rich in vitamins (such as vitamin C), minerals (calcium, magnesium, and potassium), and dietary fiber, which are essential for bodily functions. Many of these plants produce secondary metabolites with antioxidant, anti-inflammatory, and antimicrobial properties. Bioactive compounds such as polyphenols, flavonoids, and carotenoids have potential benefits in preventing chronic diseases like diabetes, cardiovascular disease, and certain types of cancer.

When studying the fruit species distributed in the urban area of Pilar, Paraguay, there is a predominance of introduced species, as shown in Table 1. In many cases, native species are remnants of the forest formations that existed prior to urban development, such as *Annona emarginata* or the pindo palm *Syagrus romanzoffiana* (Cham.) Glassman.

2.3 Characteristics of Edible Fruits

Generally, fruits can be consumed fresh or prepared in a culinary recipe to take advantage of their nutritional properties. They can be sweet, like the inga or araticui, or sour-sweet like the loquat. They can have different colors, aromas, flavors, and textures. The edible part can be the mesocarp, and sometimes the seeds are coated with an aril that is edible and sweet, like the inga, or can be fibrous like the mango or the palm of the mbokaya (Figure 1, Figure 2, Figure 3, Figure 4).

Figure 1. Araticui (*Annona emarginata*) in the streets of Pilar.

Figure 2. Inga (*Inga marginata*) in the streets of Pilar.

Figure 3. Loquat (*Eryobotria japonica*) in the streets of Pilar.

Figure 4. Mango (*Mangifera indica*) in the streets of Pilar.

2.4 Functional Foods: A Contribution from the Urban Landscape

Urban plants contribute significantly to environmental enhancement and air quality improvement, while also offering essential nutrients and bioactive compounds. Both native and introduced species can facilitate healthier, sustainable diets.

The fruits of these plants are abundant in vitamins (such as vitamin C), minerals (including calcium, magnesium, and potassium), and dietary fiber crucial for bodily functions. Their secondary metabolites exhibit antioxidant, anti-inflammatory, and antimicrobial properties. Bioactive compounds such as polyphenols, flavonoids, and carotenoids may play a role in preventing chronic diseases such as diabetes, cardiovascular disease, and certain cancers (Sarmiento Rubiano 2006).

Functional foods provide health benefits that extend beyond fundamental nutrition, including immune support and disease prevention. Urban fruit species serve as a significant source of these foods due to their bioactive components. Incorporating these foods into urban diets promotes overall health and strengthens community connections with nature (Silveira Rodríguez et al. 2003).

In this article, we will examine several examples of functional foods. Indeed, common urban species like mulberry (*Morus nigra*), guava (*Psidium guajava* L.), and citrus (*Citrus* spp.) demonstrate substantial potential. Native species such as inga (*Inga marginata* Willd.) and araticui (*Annona emarginata*) are receiving increased attention for their rich functional compounds and adaptability to various climates (Sung and Chang 2019, Leal 2023, Marquina et al. 2008).

2.5 Functional Foods

Fruits such as oranges, guavas, acerola, and inga contain antioxidants, which help neutralize free radicals in the body. Free radicals are unstable molecules that can cause cell damage and contribute to premature aging and the development of chronic diseases like cancer and heart disease. These fruits also contain fiber, anti-inflammatory compounds, prebiotics, and other essential nutrients that support physiological functions such as digestion, the immune system, and stress reduction. Additionally, their accessibility and low cost make these fruits an important source of nutrition, especially in urban communities where access to fresh, nutritious food is limited.

Fruits with anti-inflammatory properties, such as acerola and araticui found in the urban area of the city of Pilar, contain natural compounds like flavonoids and polyphenols. These compounds could reduce inflammation in the body, which may help prevent and treat chronic inflammatory diseases such as arthritis and cardiovascular disease. Consuming fruits with anti-inflammatory properties could be beneficial for joint health, blood circulation, and reducing the risk of inflammation-related diseases.

Fruits rich in fiber, such as mango and loquat, are excellent sources of dietary fiber. Fiber is important for the proper functioning of the digestive system and can help prevent constipation, improve intestinal motility, and reduce the risk of gastrointestinal diseases. Additionally, a high-fiber diet is associated with a reduced risk of developing type 2 diabetes, heart disease, and certain types of cancer. Fiber-rich fruits may also help maintain healthy cholesterol levels, regulate blood sugar, and promote a balanced gut microbiota, supporting optimal digestion and overall health.

Oranges and lemons are great sources of water and electrolytes like potassium. They help maintain hydration and provide quick energy due to their natural sugars such as fructose. These fruits are ideal for staying hydrated in hot cities like Pilar or during intense physical activities, supporting fluid balance and cellular function (Hirahara 2004).

Local fruits can be turned into functional products. Fruits like guava, mango, and loquat can be made into jams, juices, preserves, smoothies, and snacks. Adding probiotics, extra fiber, or antioxidants increases their nutritional value, making it easier to include functional foods in daily diets (Dhiman et al. 2021, Mezadri et al. 2006, Galán 2009).

Urban tree fruits provide affordable and healthy food, offering a free or low-cost option for those who struggle to access fresh produce. Including these species in urban design enhances biodiversity and fosters sustainable harvesting programs, raising awareness of their nutritional and cultural value. This strengthens local food systems and improves cities' ecological resilience.

The strategic utilization of urban species as sources of essential nutrients and bioactive compounds offers significant potential to enhance food security, improve community health, and reinforce the connection between urban environments and nature. Table 2 presents the nutritional contributions of various species within the urban area of Pilar.

Growing fruit species in urban landscapes is a strategy that offers environmental, social, and economic advantages. Integrating fruit trees in cities can enhance the environment, improve air quality, provide shade, foster biodiversity, and support food security.

Several benefits can be obtained from growing fruit species in urban areas, including:

- Improving air quality, as fruit trees can filter pollutants and produce oxygen, leading to cleaner air.
- Supporting food security by providing access to nutrient-rich foods for communities.
- Contributing to urban biodiversity by encouraging the presence of pollinators and enhancing the ecological resilience of cities.

It can also contribute to nutrition education by promoting knowledge about functional foods and their incorporation into healthy diets, as it allows communities to access fresh and healthy produce, promoting food self-sufficiency.

Table 2. Nutritional contribution of the species.

Popular name	Scientific name	Composition, use, nutritional contribution
Acerola	*Malpighia emarginata* DC. Family: Malpighiaceae	Also known as Barbados cherry, acerola is one of the richest natural sources of vitamin C—far exceeding that of citrus fruits. It contains anthocyanins, flavonoids, and carotenoids with strong antioxidants and anti-inflammatory effects, beneficial in preventing cardiovascular and neurodegenerative diseases. It is often used in juices, jams, and even nutraceuticals due to its high ascorbic acid content (Mezadri and Fernández 2006).
Araticuí	*Annona emarginata* (Schltdl.) H. Rainer Family: Annonaceae	This native fruit contains bioactive compounds called acetogenins, which have shown anticancer, cytotoxic, antiparasitic, and antidiabetic properties. It's traditionally used for medicinal purposes (Benítez and Bertoni 2016, Padmanabhan and Paliyath 2016).
Avocado	*Persea americana* L. Family: Lauraceae	Avocado (or palta) is highly valued in urban gardens for its dense nutritional profile. Rich in monounsaturated fats—especially oleic acid—it supports heart health by improving cholesterol balance. It also contains vitamin E, K, lutein, and zeaxanthin, beneficial for skin and eye health. Its phenolic compounds also exhibit anti-inflammatory effects (Teliz-Ortíz 2000, Pérez Rosales 2005).
Bitter Orange	*Citrus* × *aurantium* Family: Rutaceae	Bitter orange has demonstrated antimicrobial, antioxidant, anxiolytic, antidiabetic, and anti-obesity effects, particularly due to active compounds like p-synephrine, which is considered safe at commonly used doses (Stohs 2017, Ipek Suntar et al. 2018).
Guava	*Psidium guajava* L. Family: Myrtaceae	Guava is rich in vitamin C, lycopene, and antioxidants that promote cardiovascular health and strengthen the immune system. Its fresh fruit and derivatives serve as affordable sources of polyphenols and flavonoids, making guava a valuable ally in preventing diseases related to oxidative stress, such as metabolic syndrome. Fresh guava has significantly higher antioxidant capacity than processed forms like jam. It is recommended to consume the pulp directly to retain its polyphenol content (Marquina et al. 2008).
Inga	*Inga marginata* Family: Fabaceae	The sweet, white aril is consumed fresh and is rich in antioxidants such as flavonoids and catechins. Its seeds can be ground into flour rich in protein, fiber, and carbohydrates (Sánchez Mendoza et al. 2006, Leal 2023)
Lemon	*Citrus* × *limon* Family: Rutaceae	Lemon is renowned for its medicinal and antibacterial properties, thanks to its high vitamin C content. It helps prevent respiratory infections and is used both in cooking and natural cosmetic products (Borja López 2022, Ibáñez et al. 2015).

Table 2 contd. ...

...Table 2 contd.

Popular name	Scientific name	Composition, use, nutritional contribution
Loquat	*Eriobotrya japonica* Family: Rosaceae	Loquat is mainly consumed fresh or in fruit salads, pastries, or jams. It contains simple sugars (fructose, glucose), vitamin A, potassium, and soluble fiber (pectin), which supports digestion and helps regulate intestinal transit. The seeds contain cyanogenic compounds and should not be consumed (Bononad Gascón and Escrivá Fandos 1969, Dhiman et al. 2021).
Mango	*Mangifera indica* L. Family: Anacardiaceae	Mango is a tropical fruit widely cultivated in urban areas due to its adaptability and nutritional richness. It provides vitamin C, beta-carotene (provitamin A), fiber, and potassium. Its polyphenols, such as mangiferin, have shown antioxidant and anti-inflammatory properties, and are being studied for their potential in managing metabolic disorders like type 2 diabetes (Parrotta 1993, Galán 2009).
Mulberry	*Morus nigra* L. Family: Moraceae	Traditionally used in sericulture, mulberry fruits are rich in anthocyanins, polyphenols, and flavonoids—making them ideal as functional ingredients in nutraceuticals, jams, juices, and cosmetics (Kattil et al. 2024, Sung Ho Lim and Chang Ik Choi 2019, García et al. 2006).
Mbokaya or Macaúba Palm	*Acrocomia aculeata* Family: Arecaceae	All parts of the plant are used—for food, forage, and traditional medicine. Its seeds yield highly valued oil for food and soap production. Its pulp and kernel contain proteins, fiber, and essential fatty acids, suitable for products like cereal bars and cookies (Benítez and Bertoni 2016, Rodríguez Bonet et al. 2020).
Pindo Palm	*Syagrus romanzoffiana* (Cham.) Glasman Family: Arecaceae	This native fruit is rich in essential fatty acids (omega-3 and omega-6 precursors) and fiber. Its seed oil has also been studied for biodiesel production due to its stability and low acidity (Moreira et al. 2013, Mandon et al. 2023, Rompato et al. 2015).
Yva Povo	*Melicoccus lepidopetalus* Family: Sapindaceae	The fruit contains phenolic compounds with potential benefits for preventing diabetes, cardiovascular diseases, and gastrointestinal disorders. It is commonly consumed fresh (Pérez et al. 2009, Wilson et al. 2019).

2.6 Benefits of Growing Fruit Species in Urban Landscapes

Growing fruit in cities enriches landscapes and provides environmental, social, and economic benefits. For instance, cultivating fruits like mango, avocado, and acerola enhances human health and connects communities to local resources, helping transform cities into sustainable food systems.

2.6.1 Environmental Benefits

Tree species' traits influence their ecosystem services. For instance, larger leaves help reduce urban heat, colorful foliage adds ornamental value, and fragrant flowers can soothe residents. Cultivating fruit trees in urban areas improves food supply and offers relaxing environmental benefits (Khaniabadi et al. 2017, Ministry of Ecological Environment of the People's Republic of China 2023, Bellver 2016).

Fruit species significantly enhance the ecological balance of cities. They capture carbon dioxide, release oxygen, and provide shade to reduce urban temperatures. Their roots stabilize soil, prevent erosion, and improve rainwater infiltration, helping to mitigate flooding risks. Additionally, fruit trees attract pollinators, increasing biodiversity and fostering a resilient ecosystem (Rivas Torres 2006).

2.6.2 Social Benefits

Growing fruit species in urban spaces not only enhances the aesthetics of cities but also provides numerous social benefits that improve the quality of life for city inhabitants. As cities expand and encounter challenges such as climate change, limited green space, and pollution, street fruit trees have become a valuable means to promote sustainability and reinforce the social fabric.

Furthermore, green spaces with fruit species offer educational and recreational opportunities, encouraging learning about urban agriculture and sustainability. They can also serve as community gathering points, strengthening social ties and fostering greater cohesion within neighborhoods. Fruit trees along streets can function as a focal point that unites residents. The opportunity to harvest the fruits from these trees not only provides a source of food but also creates avenues for social interaction. Neighbors can organize themselves to care for and harvest the fruits, thereby promoting collaboration and a sense of community. Moreover, activities such as local markets or produce exchanges between neighbors encourage cooperation and solidarity within the neighborhood (Padmanabhan and Paliyath 2016).

Growing fruit trees on the streets educates citizens, especially children, about nature and sustainability. Communities can host workshops on urban agriculture, increasing environmental awareness and promoting responsible behaviors like recycling and waste reduction.

Street fruit trees also boost the local economy by allowing fallen fruits to be sold at markets, creating small businesses through products like jams and juices. This promotes local self-sufficiency and generates jobs in the harvesting and processing of fruit products.

Access to fresh fruit presents significant public health advantages for urban residents. Consumption of fresh fruit is associated with the prevention of diseases such as obesity, diabetes, and cardiovascular issues. Furthermore, fruit trees contribute to enhanced air quality, thereby mitigating the adverse effects of urban pollution on citizens' respiratory health. The presence of trees and vegetation within urban environments is also correlated with reduced stress levels and improved mental well-being. Engaging with nature, such as picking fruit or walking along tree-lined streets, can positively impact individuals' mental health.

The planting of fruit trees in urban streets enhances the aesthetic appeal of neighborhoods. Tree-lined streets become more attractive, fostering a more pleasant living and walking environment. This aesthetic enhancement can lead to the revitalization of distressed or neglected neighborhoods, drawing more people to the area, improving perceptions of safety, and promoting local tourism. Additionally, green spaces, including urban gardens, serve as vital "green lungs" that improve air quality and provide shade in areas experiencing high temperatures.

Street-grown fruit species hold cultural significance, tying traditions to local practices. Cultivating these fruits in public spaces can strengthen neighborhood identities and promote generational knowledge transfer. Additionally, fruit trees symbolize community history and heritage, preserving them for future generations.

Growing fruit species in urban streets can have environmental benefits and social impacts. It may enhance community cohesion, improve food security, promote public health, and revitalize urban spaces. Additionally, it can provide educational, economic, and cultural opportunities for urban residents. The integration of fruit trees into urban design may contribute to the development of healthier, more inclusive, and sustainable cities.

2.6.3 Economic Benefits

Fruit species have economic potential because the fruits can be marketed or used by the community, generating income and reducing maintenance costs of green areas by providing them with a productive purpose. Additionally, trees in urban areas can increase real estate value, attract investment, and improve residents' quality of life.

In summary, fruit species contribute to the urban landscape by promoting a healthier, more inclusive, and sustainable environment. Their integration into cities should be considered in urban planning strategies to maximize benefits and address management challenges (Rivas Torres 2006).

2.7 Challenges and Opportunities

The integration of fruit trees into the urban landscape encounters several challenges, such as limited space in densely populated areas and the management of fruit waste to prevent hygiene issues. Fruit trees grown in urban spaces such as streets and avenues can enhance the quality of life in cities. However, fallen fruit that is not properly managed can become a breeding ground for bacteria, fungi, and pests. Rotting fruit generates unpleasant odors, attracts insects and animals, and may damage urban infrastructure like pavements and sidewalks (Bellver 2016).

To prevent these hygiene problems, measures such as the regular collection of fallen fruit can be implemented. This task should be performed more frequently during the harvesting season of fruit trees, which varies by species but generally occurs during warmer months. Harvesting can be part of clean-up days organized by local authorities or done in cooperation with residents.

Instead of discarding fruit in landfills, an environmentally friendly alternative is composting. Fruit unsuitable for human consumption can be converted into compost, a valuable organic fertilizer. This may involve setting up organic waste collection points for neighbors to drop off their fallen fruit. Composting reduces waste volume and contributes to the improvement of community gardens and green spaces.

Community education and participation are vital for urban hygiene. Information campaigns can highlight the importance of collecting fallen fruit, using it as functional food, and managing waste. Involving neighbors in cleaning and composting fosters collective responsibility. Properly managing street fruit tree waste enhances urban hygiene, environmental sustainability, and residents' quality of life. Collaboration between city authorities and citizens prevents waste accumulation and promotes a cleaner environment.

Urban fruit trees offer opportunities for urban agriculture and environmental education projects. Cities can implement policies to encourage planting, promoting food security and sustainable practices within the community.

2.8 Conclusion

The integration of fruit species in urban areas represents a comprehensive solution to the challenges of modern cities. Their ability to beautify, nurture, and balance urban ecosystems makes them essential elements for sustainable development. The combination of native and introduced species allows for the best of both perspectives, strengthening biodiversity and promoting public health.

Among the environmental benefits of fruit species is their ability to generate oxygen and remove carbon dioxide, a greenhouse gas, during photosynthesis. They can help reduce soil erosion, decrease wind speed, moderate temperatures caused by the "heat island" effect in areas with concrete structures, improve rainwater runoff, and enrich biodiversity, especially when it comes to native species.

A common problem in many cities is the management of waste that accumulates on streets when fruits are not utilized. However, plans can be made to mitigate this issue, such as organizing activities to train interested parties in fruit collection, identifying public spaces where fruit trees are located, understanding their fruiting seasons, and educating people on their chemical composition and nutritional content. This would enable better utilization of these fruits as functional foods and teach people how to consume and cook them effectively.

Numerous products can be made from fruit. They can be eaten fresh, made into sweets and jams, and some species, like the mbocaya palm, can be used to produce soaps, cosmetics, and oils. Others can be used to make liqueurs, preserves, flavored infusions, natural juices, vinegars, etc. The peels of some fruits, especially citrus ones, can be used to make teas and even crafts. It is crucial for cities to recognize the value of these species and work on strategies to ensure their effective integration. In this way, urban fruit trees can fulfill their potential as functional foods and pillars of a greener and healthier urban future.

Additionally, the distribution of fruit trees can be a tourist attraction, especially during their flowering and fruiting seasons.

The strategic use of urban species as sources of essential nutrients and bioactive compounds has the potential to contribute to food security, improve community health, and strengthen the connection between urban spaces and nature. Cultivating fruit species in urban streets is not only an environmentally beneficial practice but also has a profound social impact. The integration of fruit trees into urban design is undoubtedly a strategy that contributes to the creation of healthier, more inclusive, and sustainable cities.

Therefore, the cultivation of fruit species in urban landscapes is a sustainable strategy that provides environmental, social, and economic benefits. Integrating fruit trees into cities not only beautifies the surroundings but also improves air quality, provides shade, fosters biodiversity, and promotes food security. There are multiple benefits to cultivating fruit species in cities, such as improving air quality since fruit trees can help filter pollutants and produce oxygen, contributing to cleaner air. Urban fruit species enhance food security by offering nutrient-rich and bioactive foods accessible to communities and contribute to urban biodiversity, as they encourage the presence of pollinators and strengthen the ecological resilience of cities.

References

Alfonso, J., Balmori, D., Benitez, S., Franco, B., Escobar, J., Rafael, C. et al. (2022). Árboles frutales para fortalecer la Seguridad Alimentaria del Asentamiento Santa Librada de la ciudad de Limpio. Revista Multidisciplinar.

Ancillo, G. and Medina, A. (2007). Monografías botánicas LOS CÍTRICOS Jardín Botánico de la Universitat de València. ISBN: 978-84-370-9553-0.

Atlas Censal del Paraguay. (2016). https://www.ine.gov.py/Publicaciones/Biblioteca/Atlas%20Censal%20del%20Paraguay/1%20Presentacion%20Atlas.pdf.

Bellver, J. (2016). CIUDADES SOSTENIBLES. DEL SUEÑO A LA ACCIÓN., FUHEM–Icaria. file:///C:/Users/Downloads/JBellver_WorldwatchInstitute_Ciudades-sostenibles.pdf.

Benítez, B. and Bertoni, S. (2016). Algunos aspectos ecológicos y de aprovechamiento de ciertas especies arbóreas y arbustivas en el área de influencia de los Arroyos Caañabé y Mbaey, Departamento de Paraguarí-Paraguay. Steviana 8(1): 9–30. https://doi.org/10.56152/StevianaFacenV8N1A2_2016.

Borja Lopez. (2022). Accedido 4-01-2024 https://www.mundodeportivo.com/uncomo/comida/articulo/origen-e-historia-del-limon-44075.html.

Bononad Gascón, S. and Escrivá Fandos, A. (1969). E L N IS P E R O. Hojas divulgadoras. No. 5-6. Accedido 8-01-2024, https://www.mapa.gob.es/ministerio/pags/biblioteca/hojas/hd_1969_05-06.pdf.

Castillo Rodríguez, Larisa and Ferro Cisneros, Sergio. (2015). The problem of designing trees on urban roads: green with black stitching. Arquitectura y Urbanismo 36(1): 5–24. Recuperado en 11 de marzo de 2025, de http://scielo.sld.cu/scielo.php?script=sci_arttext&pid=S1815-58982015000100002&lng=es&tlng=es.

Contreras Roqué, J. R., Contreras Chialchia A. O. and Delpino Aguayo, M. A. (2007). Estudios bioecológicos sobre los humedales del Ñeembucú. Desarrollo urbano y antropización creciente del medio natural en el Departamento del Ñeembucú al sur del río Tebicuary, República del Paraguay. 1ª ed.- Buenos Aires.

Dhiman, A., Suhag, R., Thakur, D., Gupta, V. and Prabhakar, P. K. (2021). Estado actual del níspero (Eriobotrya Japonica Lindl.): funciones bioactivas, enfoques de conservación y productos procesados. Food Reviews International 38(sup1): 286–316. https://doi.org/10.1080/87559129.2020.1866007.

Galán, V. (2009). El cultivo del mango. Mundi-Prensa. 2da edición. 340 pp. https://www.mango.org/es/mangocrushmonday-historia-del-mango/.

García, D., Noda, Y., Medina, M., Martín, G. and Soca, M. (2006). La morera: una alternativa viable para los sistemas de alimentación animal en el trópico. Avances en Investigación Agropecuaria 10(1): 55–72.

Hirahara, T. (2004). Key factors for the success of functional foods. Biofactors 22: 289–294. DOI: 10.1002/biof.5520220156.

Ibáñez, V., García-Usach, A., Carbonell-Caballero, J., Alonso, R., Terol, J., Dopazo, J. et al. (2015). EL ORIGEN DE LAS ESPECIES CULTIVADAS DE CITRICOS. Accedido 4-01-2024. https://redivia.gva.es/bitstream/handle/20.500.11939/4176/2015_Iba%C3%B1ez_El%20Origen.pdf?sequence=1&isAllowed=y.

Ipek Suntar, Haroon Khan, Seema Patel, Rita Celano and Luca Rastrelli. (2018). Citrus aurantium L. Overview: Its Functions as a Food Ingredient and Therapeutic Agent https://doi.org/10.1155/2018/7864269.

Kattil, A., Dash, K. K., Shams, R. and Sharma, S. (2024). Nutritional composition, phytochemical extraction, and pharmacological potential of mulberry: A comprehensive review. Future Foods 9: 100295. https://doi.org/10.1016/j.fufo.2024.100295.

Leal, K. (2023). https://www.tuasaude.com/es/inga/.

Mandon, E. and Campagna, M. E. (2023). Estudio de los frutos de Butia yatay y Syagrus romanzoffiana (Arecaceae) cultivadas con fines ornamentales como fuente de nutrientes y compuestos bioactivos. Bol. Soc. Argent. Bot. 58: 533–546. DOI: https://doi.org/10.31055/1851.2372.v58.n4.41049.

Marquina, V., Araujo, L., Ruíz, J., Rodríguez-Malaver, A. and Vit, P. (2008). Composición química y capacidad antioxidante en fruta, pulpa y mermelada de guayaba (Psidium guajava L.). Archivos Latinoamericanos de Nutrición 58(1): 98–102. Recuperado en 09 de enero de 2025, de http://ve.scielo.org/scielo.php?script=sci_arttext&pid=S0004-06222008000100014&lng=es&tlng=es.

Mezadri, T., Fernández-Pachón, M. S., Villaño, D., García-Parrilla, M. C. and Troncoso, A. M. (2006). El fruto de la acerola: composición y posibles usos alimenticios. Archivos Latinoamericanos de Nutrición 56(2): 101–109. Recuperado en 09 de enero de 2024, de http://ve.scielo.org/scielo.php?script=sci_arttext&pid=S0004-06222006000200001&lng=es&tlng=es.

Ministry of Ecological Environment of the People's Republic of China. (2023). Newsletter on China's Ecological and Environmental Conditions in 2022. Available online: https://www.mee.gov.cn/hjzl/sthjzk/zghjzkgb/.

Moreira, M. A. C., Payret Arrúa, M. E., Antunes, A. C., Fiuza, T. E. R., Costa, B. J., Weirich, P. H. et al. (2013). Characterization of *Syagrus romanzoffiana* oil aiming at biodiesel production, Industrial Crops and Products 48: 57–60. https://doi.org/10.1016/j.indcrop.2013.04.006.

Padmanabhan, P. and Paliyath, G. (2016). Annonaceous fruits. pp. 169–173. Editor(s): Benjamin Caballero, Paul M. Finglas and Fidel Toldrá (eds.). Encyclopedia of Food and Health, Academic Press, ISBN 9780123849533, https://doi.org/10.1016/B978-0-12-384947-2.00031-3.

Parrotta, John A. (1993). Mangifera indica L. Mango. SO-ITF-SM-63. New Orleans, LA: U.S. Department of Agriculture, Forest Service, Southern Forest Experiment Station. Accedido 5-01-2024 file:///C:/Users/Nelida%20Soria/Downloads/Mangiferaindica.pdf.

Pérez Rosales, R., Villanueva Rodríguez, S. and Cosío Ramírez R. (2005). Avocado oil and its nutritional properties. e-Gnosis (3). Available in: https://www.redalyc.org/articulo.oa?id=73000310.

Pérez, C., Hilda, C., Gómez, P., Manuel and Vila, Josefina. (2009). Physical characteristics of mamón (Melicoccus bijugatus Jacq.) fruits according to their location on the tree and storage. Bioagro 21(3): 189–194. Retrieved April 19, 2025, from http://ve.scielo.org/scielo.php?script=sci_arttext&pid=S1316-33612009000300007&lng=es&tlng=es.

Rivas Torres, D. (2006). Beneficio de los árboles Urbanos. file:///C:/Users/lucas/Downloads/Beneficios_de_los_Arboles_Urbanos.pdf, accessed March 11, 2025.

Rodríguez Bonet, S., Chaparro Aguilera, L., Benítez Villalba, J., Villalba, D., Arévalos Rotela, L. and Benítez Franco, R. (2020). Caracterización fisicoquímica de la pulpa y almendra de Acrocomia aculeata. Investigación Agraria 22(1): 46–52. https://doi.org/10.18004/investig.agrar.

Rompato, K., Franco, R., Somoza, S. N. and Santiago Rompato, L. S. (2015). Nutritional composition of Syagrus romanzoffiana (pindo) fruits native to Formosa- Argentina. Boletim do Centro de Pesquisa de Processamento de Alimentos 33(2): 105–112. DOI: 10.5380/cep.v33i2.46238.

Sánchez Mendoza, N. A., Jiménez Martínez, C., Cardador Martínez, A., Martín del Campo Barba, S. and Dávila Ortiz, G. (2016). Caracterización física, nutricional y no nutricional de las semillas de Inga paterno. Revista chilena de nutrición 43(4): 400–407. https://dx.doi.org/10.4067/S0717-75182016000400010.

Sarmiento Rubiana, L. A. (2006). Functional foods, a new feeding alternative. Revista ORINOQUIA 10(1): 16–23.

Silveira Rodríguez, M. Belén, Monereo Megías, Susana and Molina Baena, Begoña. (2003). Alimentos funcionales y nutrición óptima: ¿Cerca o lejos?. Revista Española de Salud Pública 77(3): 317–331. Recuperado en 17 de abril de 2025, de http://scielo.isciii.es/scielo.php?script=sci_arttext&pid=S1135-57272003000300003&lng=es&tlng=es.

Stohs, S. J. (2017). Safety, efficacy and mechanism studies on Citrus aurantium (bitter orange) extract and p-synephrine. Phytotherapy Research 31(10): 1463–1474. https://doi.org/10.1002/ptr.5879.

Sung Ho Lim and Chang Ik Choi. (2019). Pharmacological properties of Morus nigra L. (Black Mulberry) as a Promising Nutraceutical Resource Nutrients 11(2): 437; https://doi.org/10.3390/nu11020437.

Teliz-Ortíz, D., Mora-Aguilera, G. and Morales-García, L. (2000). Importancia histórica y socioeconómica del aguacate. Mundi Prensa. México.

Wilson, J., Goldson-Barnaby, A. and Bailey, D. (2019). Melicoccus Bijugatus (guinep): Phytochemical properties, associated health benefits and commercial applications international. Journal of Fruit Science 20(4): 659–666. https://doi.org/10.1080/15538362.2019.166951<u>17.</u>

Section II

Bioactivities, Therapeutic Potential, and Functional Food Applications

Chapter 3

Bioavailability and Bioactivities of Fruit-Derived Compounds

Mikael Kélvin de Albuquerque Mendes,[1,*]
Carla Mariana da Silva Medeiros,[2]
Luíza Santa Brígida de Barros Góes,[1] *Michely da Silva Sousa,*[2]
Renata Santos Vieira da Conceição,[1] *Edivan Carvalho Vieira,*[2]
Francisco Xavier Nobre[3] and *Emanuel Carrilho*[1]

3.1 Introduction

A balanced and nutritious diet is essential for good health and quality of life. The nutritional pyramid recommends the consumption of at least five servings of fruits and vegetables daily, as they provide essential nutrients and a wide variety of bioactive compounds with therapeutic and preventive potential, such as polyphenols, carotenoids, antioxidant vitamins, and dietary fibers (Arias 2022). The health benefits and bioactivity of these compounds are directly dependent on their bioavailability, which refers to the fraction that is effectively released from the food matrix, absorbed, metabolized, and utilized by the organism (Mendes 2024).

Several factors determine the bioavailable levels of these compounds, including their origin, the food matrix, processing conditions, digestion, and cellular metabolism. Another key factor is that most fruits are not consumed raw and undergo different types of processing, such as cooling, heating, drying, fermentation, among others

[1] Instituto de Química de São Carlos, Universidade de São Paulo, São Carlos, SP, 13566-590, Brazil.
[2] Departamento de Química, Universidade Federal do Piauí – UFPI, P.O. Box 64049-550, Teresina, PI, Brasil.
[3] Grupo de Recursos Energéticos e Nanomateriais (GREEN), Instituto Federal do Amazonas, Campus Manaus Centro (IFAM-CMC), Avenida Sete de Setembro, 1975, 9020-120, Manaus, AM, Brazil.
* Corresponding author: mikaelmendes@usp.br

DOI: 10.1201/9781003668169-5

(Arfaoui 2021). The plant species and the low water solubility of certain bioactive compounds are also relevant factors that interfere with oral absorption, thereby reducing their bioavailability (Arfaoui 2021, Zhang et al. 2022a). For instance, thermal processing techniques such as blanching, pasteurization, sterilization, and drying can preserve the quality of antioxidant compounds, and heat treatment may alter protein structures and enhance the release of bioactive peptides (Zhang et al. 2023a, Wu et al. 2024a).

The estimation of bioactive compounds directly from the food matrix does not predict their safety, efficacy, or *in vivo* activity, since many of these molecules undergo enzymatic and microbial metabolism during digestion and intestinal absorption. These processes often lead to the formation of metabolites that differ from their original structures, consequently modifying their function. For example, bioactive peptides may be degraded by digestive enzymes and intestinal peptidases, which alter their structure, biological activity, and therefore, their bioavailability (Amigo and Hernández-Ledesma 2020, Mendes et al. 2024).

To accurately characterize the bioavailability of dietary bioactives, a combination of *in vitro* and *in vivo* models is required (Barros et al. 2019, Álvarez-Olguín et al. 2022). *In vitro* digestion models simulate physiological conditions and provide a controlled environment to study the release and solubilization of compounds, while *in vivo* studies—including animal models and clinical trials—offer valuable insights into absorption kinetics, metabolism, and systemic effects. Additionally, advanced analytical techniques such as mass spectrometry and omics-based approaches have emerged as essential tools for tracking bioactive metabolites in biological fluids and tissues, helping to elucidate their mechanisms of action and potential health benefits (Demarque et al. 2020, Collins et al. 2021).

The complexity of the mechanisms involved in elucidating the bioavailability of fruit-derived bioactive compounds encompasses absorption mechanisms, dietary and physiological factors, and fruit processing technologies. Therefore, it is crucial to address the main variables required to translate the functional potential of bioactive compounds from fruits into measurable outcomes in health. This chapter discusses the major classes of bioactive compounds found in fruits, the factors that influence their release and absorption, the impact of food processing, the methodologies used to assess bioavailability, and their implications for human health.

3.2 Applications and Implications for Human Health

3.2.1 Antioxidant and Anti-inflammatory Effects

Some of the bioactive compounds derived from fruit have antioxidant and anti-inflammatory effects, playing a fundamental role in the prevention of chronic diseases. This effect is associated with phenolic compounds, flavonoids, vitamin C, and carotenoids, which act as natural antioxidants. These compounds help prevent oxidative damage caused by reactive oxygen species, neutralizing free radicals and protecting cells from structural damage. Their structure allows them to act as reducing agents and hydrogen donors, combating oxidative species and inhibiting the initiation and formation of new free radicals, as well as binding to metal ions to reduce their reactivity (Paz et al. 2015).

Studies show that several Brazilian fruits, such as acerola (*Malpighia emarginata*), cashew (*Bertholletia excelsa*), and Formosa papaya (*Carica papaya*), have a high antioxidant capacity and are potent sources of these bioactive compounds. In addition to their antioxidant action, many of these compounds have anti-inflammatory properties, modulating inflammatory pathways and contributing to the prevention of diseases such as atherosclerosis and cancer. The phenolic compounds present in fruit can also prevent chronic non-communicable diseases, such as cardiovascular diseases and diabetes, by reducing inflammatory processes in the body (Lim et al. 2007, Paz et al. 2015).

Citrus fruits such as oranges, lemons, and acerola are rich in vitamin C and flavonoids and have antioxidant and anti-inflammatory properties. Studies show that orange juice consumption can promote metabolic changes and protect tissues from damage resulting from the accumulation of reactive oxygen species (Martinez et al. 2021).

The study by Reis et al. (2015) identified the presence of compounds such as myricetin, fisetin, and quercetin in papaya that have antioxidant activities and can reduce the risk of developing pathologies such as atherosclerosis and cancer. Brazilian fruit exhibited antioxidant values related to phenolic compound content and flavonoid content in the study by Paz et al. (2015). The author points out that acerola and açaí fruit had the highest antioxidation values, while tamarind extracts could be further studied for their anti-inflammatory functions, in the prevention or adjuvant treatment of diseases caused by *Pseudomonas aeruginosa, Escherichia coli, Listeria monocytogenes, Salmonella* sp., and *Staphylococcus aureus.*

Morais et al. (2022) evaluated the antioxidant and anti-inflammatory activities of the ripe fruits of *Solanum lycocarpum* by GC-MS and LC-DAD-MS. The chemical compounds detected, derived from caffeoylquinic acid, spermidine, stigmasterol, and β-sitosterol, may be correlated with these activities. The ripe fruits of this species could be a dietary alternative rich in bioactive compounds and beneficial for human health.

Thus, the regular inclusion of fruit in the diet is an effective strategy to take advantage of the benefits of bioactive compounds with antioxidant and anti-inflammatory properties. These compounds play a crucial role in preventing chronic diseases and promoting general health.

3.2.2 Impact on the Prevention of Non-communicable Diseases (NCDs)

Bioactive compounds have several impacts on preventing chronic diseases. Regularly consuming fruits rich in bioactive compounds provides significant health benefits, reducing the risk of chronic diseases through antioxidant, anti-inflammatory, and metabolic regulatory mechanisms.

3.2.2.1 Cardiovascular Disease

Cardiovascular diseases, such as atherosclerosis, hypertension, and heart attack, are associated with oxidative stress and chronic inflammation. Bioactive compounds present in fruits help protect cardiovascular health through different mechanisms (Din et al. 2023).

In general, when it comes to cardiovascular diseases, some bioactive compounds are effective in the treatment of cardiovascular diseases. Carotenoids increase cardiac function and decrease cytotoxicity and oxidative stress. Flavonoids decrease nitric oxide, reactive oxygen species, platelet aggregation, lipid peroxidation, and proinflammatory cytokines. Polyphenols increase endothelial function and decrease platelet aggregation, nitric oxide, oxidative stress, and reactive oxygen species (Sindhu et al. 2021, Anguera-Tejedor et al. 2024).

Kaempferol, another bioactive compound, has been shown to have potential cardioprotective effects and anti-inflammatory activity. The *in vitro* and *in vivo* study by Kim et al. (2022) on human tissue cells or model animals supports this concept. Mortality from ischemic heart disease tended to be lower at higher intakes of quercetin and kaempferol (Knekt et al. 2002). Although kaempferol belongs to the flavonol family, like quercetin, it has a different intestinal metabolism. Unlike quercetin, it does not have a catechol group, and this structural difference may be related to changes in intestinal metabolism in mice (Biat et al. 2022, Kim et al. 2022).

Zuraini et al. (2021) reviewed commonly consumed nutritious fruits (apple, avocado, grapes, mango, orange, kiwi, pomegranate, papaya, pineapple, and watermelon), which have significant beneficial effects for the management and treatment of cardiovascular diseases. Among the 10 fruits, pomegranate and grapes were well explored. Grape extracts significantly reduced oxidative stress induced by cardiac and cerebral ischemia, since grapes contain compounds that help prevent oxidative stress, resulting in a significant impact on blood lipids, such as a decrease in low-density lipoprotein oxidation.

For pomegranate fruit, Zuraini et al. (2021) demonstrated that the fruit has abundant antioxidants and strong anti-hypertonic properties, whether in the peel, seed, or juice. Punicalagin, the main bioactive component of pomegranate, has been shown to have cardiovascular protection through its potent antioxidant activity. Pomegranate seed extract also improved motor and cognitive deficits caused by chronic cerebral hypoperfusion ischemia, which was probably due to its antioxidant and free radical scavenging properties.

The consumption of fruits, whose active constituents vary depending on the food source, could contribute to the development of nutritional supplements that provide effective and consistent protection against cardiovascular diseases.

3.2.2.2 Diabetes

Insulin resistance and inflammation are key factors in the development of type 2 diabetes. The bioactive compounds in fruit help with glycemic control, for example through the action of soluble fibers, which help slow down glucose absorption and improve insulin sensitivity (de Lima Teixeira 2020).

The main results derived from the literature, particularly obtained in *in vitro* and *in vivo* models, are that anthocyanins (açaí, blueberry, and blackberry) have the potential to reduce diabetes and hyperglycemia. Different mechanisms have been reported, such as the ability to regulate targets such as digestive enzymes (α-amylase and α-glucosidase). Anthocyanins also regulate blood glucose by normalizing insulin secretion and resistance (Les et al. 2020, de Lima Teixeira 2020).

Chlorogenic acid (found in berries, coffee, and apples) reduces glucose absorption in the intestine. Chronic and acute chlorogenic acid can improve glucose tolerance and insulin sensitivity. The results showed that chlorogenic acid decreased body weight and improved glucose tolerance and insulin resistance (Yan et al. 2022). Polyphenols, such as flavonoids, phenolic acids, and tannins, found in fruits such as grapes, apples, berries, and citrus fruits, can improve insulin sensitivity. They inhibit enzymes such as α-amylase and α-glucosidase, reducing the absorption of carbohydrates and preventing glycemic spikes after meals. In addition, polyphenols reduce oxidative stress and inflammation, which are factors involved in insulin resistance (Carvalho 2020, Costa et al. 2020).

Regular consumption of fruit rich in bioactive compounds can help regulate blood glucose, improve insulin sensitivity, and reduce the risk of complications associated with type 2 diabetes. However, it is important to moderate the intake of fruits with a high glycemic index, such as overripe bananas, watermelon, and grapes, especially for individuals who already have insulin resistance. Insulin resistance and inflammation are central factors in the development of type 2 diabetes. The bioactive compounds in fruit help with glycemic control, for example, through the help of soluble fibers, which help slow down glucose absorption and improve insulin sensitivity (de Lima Teixeira 2020).

Evidence from *in vitro* and *in vivo* studies indicates that anthocyanins—present in fruits such as açaí, blueberries, and blackberries—exert antidiabetic effects through multiple mechanisms. These include the inhibition of digestive enzymes like α-amylase and α-glucosidase, as well as the modulation of insulin secretion and sensitivity, contributing to improved glycemic control (Les et al. 2020, de Lima Teixeira 2020). Regular consumption of fruits rich in such bioactive compounds may therefore help regulate blood glucose levels, enhance insulin responsiveness, and reduce the risk of complications associated with type 2 diabetes. However, it is important to moderate the intake of fruits with a high glycemic index, particularly among individuals with insulin resistance, to prevent postprandial glycemic spikes.

3.2.2.3 Cancer

Oxidative stress and chronic inflammation play a fundamental role in carcinogenesis. Several bioactive compounds in fruits have anti-carcinogenic properties, such as polyphenols, which have anti-proliferative effects and can induce apoptosis (programmed cell death) in tumor cells.

The study by Akermi et al. (2025) and Tiwari and Mishra (2020) discussed the use of flavonoids as radioprotective and radiosensitizing agents, killing malignant tumor cells without affecting normal cells. Recent studies have shown that flavonoids can inhibit the proliferation of cancer cells by targeting key signaling pathways, such as the epidermal growth factor receptor (EGFR)/protein kinase axis. Notably, aberrant expression of the human epidermal growth factor receptor 2 (HER2) and mutations leading to constitutive activation of its tyrosine kinase domain are strongly associated with tumor aggressiveness and resistance to therapy, particularly in breast cancer. Although HER2-targeted therapies exist, their efficacy is often limited by adverse side effects and the development of acquired resistance. In this context, flavonoids have gained increasing attention for their antiproliferative and pro-apoptotic properties, positioning them as promising chemopreventive agents. Current research focuses on identifying flavonoids capable of directly inhibiting HER2 tyrosine kinase activity and validating their antitumor potential in HER2-positive cancers (Chandrika et al. 2016).

Another study in the literature reported that the antioxidant effect of tomato fruits is due to the presence of polyphenols (flavonoids and hydroxycinnamic acids), which are capable of eliminating peroxyl radicals. A high cytotoxic effect of α-tomatine in green tomato extracts on normal and cancer cell lines was positively evaluated, resulting in high antioxidant power (Del Giudice et al. 2015). Resveratrol exerts anticancer effects by modulating pathways that regulate cell proliferation, limiting the growth of tumor cells. In addition, the work by Ko et al. (2017) highlights pre-clinical and clinical studies that demonstrate the potential of resveratrol as a complementary therapeutic agent in the fight against cancer.

3.2.2.4 Obesity and Metabolic Syndrome

Obesity and metabolic syndrome are risk factors for several chronic diseases, including diabetes and cardiovascular disease. The bioactive compounds in fruits help control weight and metabolism. Supplementation of fruits with catechins prevented high-fat diet (HFD)-induced body weight gain by partially inhibiting subcutaneous and epididymal adipose tissue weight gain in rodents (Kim et al. 2022). In addition, a study conducted on HFD-fed mice demonstrated that catechins played a role in inhibiting intestinal lipid absorption and promoting lipid excretion, which may contribute to their anti-obesity effects.

Specific carotenoids and carotenoid derivatives restrict adipogenesis and adipocyte hypertrophy while increasing fat oxidation and energy dissipation in brown and white adipocytes, and counteract obesity in animal models, indicating a beneficial effect of *β*-cryptoxanthin supplementation (Zhang et al. 2020a).

Gallic acid has shown positive effects in mice with metabolic syndrome and obesity, improving triglyceride and blood glucose levels, as well as glucose tolerance and lipid metabolism. It acts on adipose tissue, reducing fat formation, improving insulin signaling, and combating inflammation and oxidative stress. In addition, it can help prevent cardiovascular disease by restoring antioxidants and reducing damage to the heart. In a small clinical study, gallic acid supplementation prevented oxidative damage to DNA and reduced inflammatory markers in patients with type 2 diabetes (Kim et al. 2022, Chao et al. 2014).

Supplementation with guava polysaccharides showed positive effects in reducing diet-induced obesity and associated metabolic disorders in a mouse model. The anti-obesity effect of these polysaccharides is directly related to their ability to remodel the intestinal microbiota and modulate its metabolites. In addition, guava polysaccharides are considered promising prebiotics with potential benefits against obesity (Pinheiro et al. 2024).

On the other hand, baru pulp showed strong prebiotic potential, promoting the growth and metabolism of probiotics such as *Bifidobacterium animalis* BB-12, *Lactobacillus acidophilus* LA-05, and *Lactobacillus casei* L-26. In addition, its consumption has had a positive impact on the microbiota of the human colon, increasing the abundance of *Lactobacillus*, *Enterococcus*, *Bifidobacterium*, and *Bacteroides*, *Prevotella* (Pinheiro et al. 2024).

3.3 Potential Therapeutic Uses and Supplementation

The guidelines for fruit consumption in disease prevention follow the recommendations of health bodies such as the World Health Organization (WHO). In Brazil, the food guide for the Brazilian population suggests that fruit should form a significant part of the daily diet, preferably fresh (BRASIL 2014). For the German Nutrition Society (Schäfer et al. 2024), the ideal consumption would be 250 g of fruit; in France (Santé-publique France 2025), at least 5 portions of fruit with serving sizes of 80 g to 100 g. The 2020–2025 Dietary Guidelines for Americans advise incorporating more fruits and vegetables into U.S. residents' diets as part of healthy dietary patterns. Adults should consume 1.5–2 cup-equivalents of fruits (Lee et al. 2019).

Based on these guidelines, supplements and functional foods are developed with bioactive compounds. Among the main classes of functional foods are probiotics, live microorganisms that help balance the intestinal microbiota. Prebiotics, on the other hand, are dietary fibers or non-digestible oligosaccharides that contribute to gastrointestinal regulation and metabolism. Antioxidants include various substances such as vitamins A (carotenoids), C (ascorbic acid) and E (tocopherol), as well as phenolic compounds such as flavonoids and phenolic acids, which play a fundamental role in protecting against free radicals. Fatty acids, such as omega-3, omega-6, and conjugated linoleic acid, have therapeutic properties, including anti-inflammatory and anticoagulant action, as well as helping to reduce cholesterol (Couri 2021).

An example of a product developed with bioactive compounds was the "Ready to Use Food" (RUF) energy–protein supplement using inajá oil and watermelon seed protein hydrolysate. The hydrolysis of watermelon seed cake was performed with bromelain, optimizing variables such as time, temperature, and enzyme concentration. Four versions of the supplement were formulated with different concentrations of hydrolysate. The formulations contained vitamins C and E, in addition to minerals such as iron, zinc, and manganese, meeting 30 to 70% of the daily recommendations (Oliveira 2024).

Another example is the development of an Energy Gel from Buriti (*Mauritia flexuosa*) and Guarana (*Paullinia cupana*) for Physical Activity Practitioners (Lima et al. 2020), which addresses the creation of an energy gel using Amazonian ingredients to meet the needs of physical activity practitioners. The study highlights the potential of buriti and guarana as sources of carbohydrates and bioactive compounds such as carotenoids and caffeine, suggesting that this combination may result in a natural and functional energy supplement.

However, fruit processing can significantly impact the preservation of bioactive compounds, depending on the method used. Some techniques preserve or even increase the bioavailability of these compounds, while others may degrade them. Some processes can degrade bioactive compounds, such as high temperatures due to pasteurization, boiling, or dehydration. Polyphenols and carotenoids can degrade due to prolonged exposure to heat and oxygen. Oxidation also degrades some compounds in fruits when cut. Some techniques have been developed to preserve these bioactive compounds, such as freeze-drying, cold dehydration, microencapsulation, and fermentation, among others. Therefore, choosing the right processing method is essential to ensure that processed fruits maintain their nutritional and functional benefits (Lima et al. 2024).

3.4 Main Bioactive Compounds in Fruits

Fruits are natural sources of bioactive compounds, such as biologically active proteins, phenolic compounds, phytosterols, carotenoids, and some vitamins, which can bring benefits to human health in the treatment or prevention of diseases (Santos et al. 2022). This biological potential offered to the human body by the consumption of different food matrices is related to the diversity of these metabolites present in food (Borgonovi et al. 2022). Bioactive compounds are widely distributed in nature, especially in fruits. Therefore, increasing the consumption of foods that are sources of these metabolites is encouraged for the population, as it can contribute to reducing the risk of weight gain, diabetes, and some coronary diseases, such as myocardial infarction (Borgonovi et al. 2022, Santos et al. 2022).

The in-depth study of these compounds is of great importance as it makes it possible to understand their mechanism of action and their applications in the food, pharmaceutical, and nutraceutical sectors. Thus, the identification and characterization of the main bioactive compounds in fruits, including phenolic compounds, carotenoids, vitamins with antioxidant potential, and dietary fibers (Figure 1), represents a promising field for the development of health promotion and disease prevention strategies.

3.4.1 Phenolic Compounds

Phenolic compounds are the most abundant class of biologically active compounds found in various food matrices, with approximately 8000 phenolic compounds reported in the literature (Mamun et al. 2024, Davinelli et al. 2025). These metabolites are characterized by the presence of one or more hydroxyl groups attached to aromatic rings and play a crucial role in various biological processes, including antioxidant potential, anti-diabetic, anti-inflammatory, acetylcholinesterase inhibition, and anticarcinogenic properties (Guarniz et al. 2019, Fernández et al. 2020, Wang et al. 2020, Pan et al. 2025, Yu et al. 2025a).

Due to their chemical and biological properties, phenolic compounds are of particular interest because of their unique structures and functions that contribute

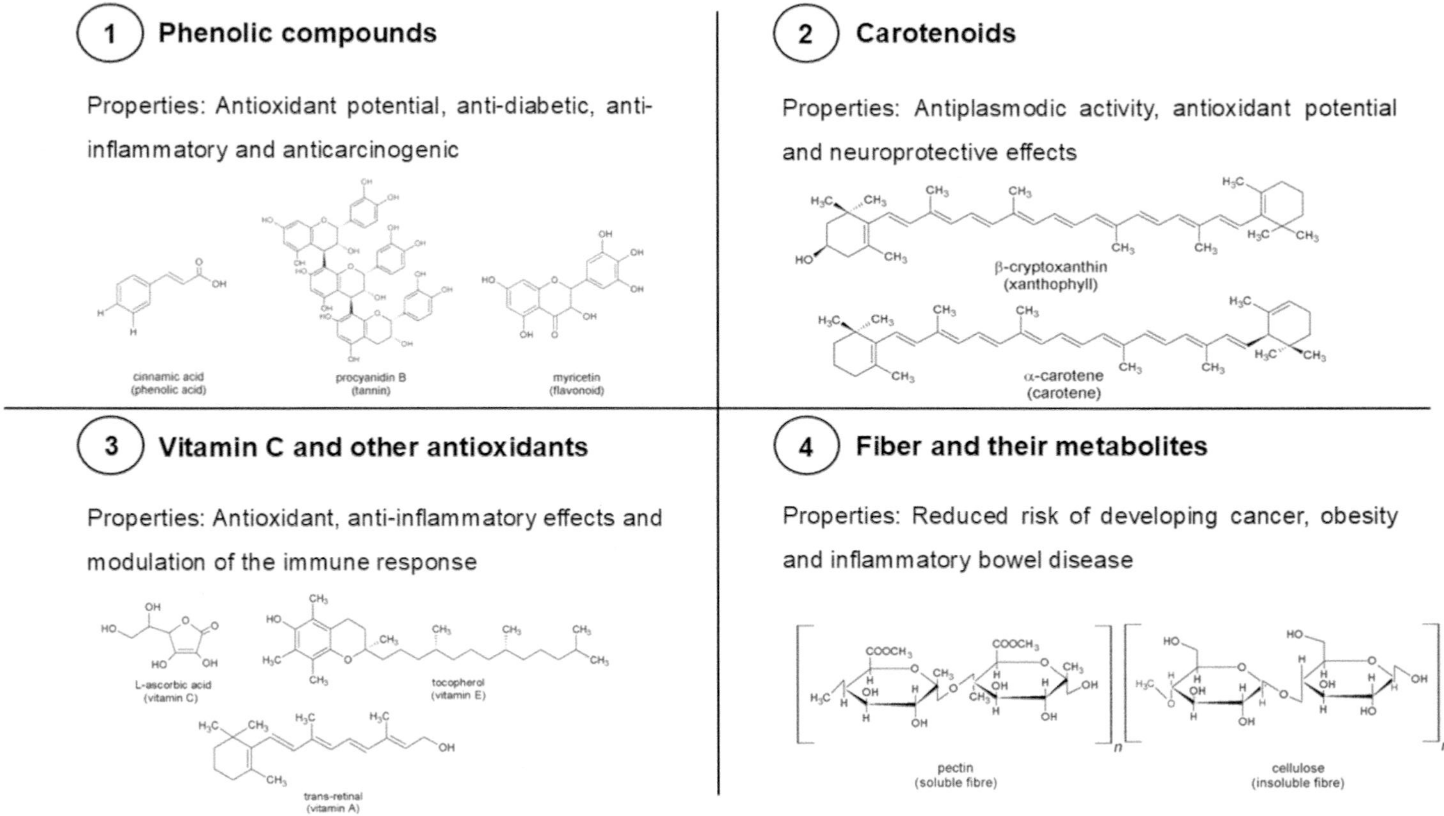

Figure 1. Main Bioactive Compounds in Fruits.

to their health-promoting effects. Structurally, these metabolites can be classified according to their chemical structure, biosynthetic pathway, and degree of polymerization (Davinelli et al. 2025). In general, phenolic compounds can be divided into three main groups: phenolic acids, tannins, and flavonoids.

Phenolic acids, such as cinnamic acid, have a relatively simple structure containing at least one hydroxyl group (–OH) directly linked to an aromatic ring and the carboxyl group (Mamari 2021). According to the literature, phenolic acids can be associated with fruit flavor, being responsible for acidity and sweetness (Xiao et al. 2024). In addition to their sensory function, these acids are also associated with the antioxidant potential of these foods, showing potential in human health (Oliveira et al. 2020).

Flavonoids are phenolic compounds with a more complex structure, with a C_6-C_3-C_6 skeleton, i.e. two aromatic rings linked by a three-carbon chain, often with two or more hydroxyl groups (Cetinkaya et al. 2025). The main subgroups include flavanones, flavonols, flavones, isoflavones, anthocyanins and flavan-3-ols (Davinelli et al. 2025). Flavonoids are increasingly associated with beneficial biological activities, and a growing body of scientific evidence links the consumption of flavonoid-rich fruits with anti-inflammatory, anti-diabetic and antioxidant effects, encouraging the inclusion of these foods in the diet (Ojo and Adanlawo 2024). In addition, these compounds are often associated with the aroma, color and flavor of the fruit, key factors that encourage consumption (Cetinkaya et al. 2025).

Finally, tannins are high molecular weight polyphenolic compounds ranging from 500 to 20,000 Da, soluble in water, and characterized by their ability to form complexes with proteins and metals (Bobasa et al. 2021). They are traditionally divided into three groups according to their structure: hydrolysable tannins, which consist of a polyol core, usually d-glucose esterified with gallic acid or hexahydroxydiphenic acid; condensed tannins (or proanthocyanidins), formed by polymers of flavan-3-ol units; and complex tannins, composed of hydrolysable tannins and flavan-3-ol, i.e., condensed tannins (Silva et al. 2020). Recently, some studies have associated the presence of hydrolysable tannins with high antioxidant properties and strong antimicrobial effects in fruits (Bobasa et al. 2021). Table 1 summarizes the class of phenolic compounds identified in fruit and the biological properties attributed to these metabolites.

In the study conducted by Guarniz et al. (2019), the authors evaluated the antimicrobial potential of *Momordica charantia* L. *In vitro* antimicrobial assays demonstrated that acetate extracts from the leaves had a good antibacterial effect against *Staphylococcus aureus* strains, moderate activity against *Staphylococcus epidermidis* strains, and weak antifungal activity against *Candida albicans*, with MICs of 0.25, 1.0, and 2.0 $mg \cdot mL^{-1}$, respectively. The authors also evaluated the chemical profile of the ethanol extract of *Momordica charantia* L. by ultra-performance liquid chromatography–quadrupole time-of-flight mass spectrometry (UPLC-QTOF-ESI-MS). The authors identified phenolic compounds such as quercetin-O-hexoside and kaempferol-O-hexoside, as well as compounds from other classes, including oleanane-triterpenoid saponin, epoxycurcubitatrienediol, and hydroxycucurbitatetraenal.

Table 1. Phenolic compounds identified in fruits and their biological properties.

Fruit	Bioactive compounds	Bioactivities	Reference
Olive (*Olea europaea* L.)	Phenolic acids, flavanol, flavonoids, and simple phenolics	Antioxidant activity	Özcan et al. 2019
Bitter melon (*Momordica charantia* L.)	Flavonoids	Antimicrobial activity	Guarniz et al. 2019
Terebinth (*Pistacia terebinthus*)	Phenolic acid, simple phenolic, flavonol and flavanol	Antioxidant activity	Özcan et al. 2020
Pepper (*Capsicum chinense*)	Phenolic acid, flavonol, flavone, flavanone	Antioxidant activity	Santos et al. 2021
Chinese jujube (*Ziziphus jujuba* Mill.)	Flavonols, flavanols, flavones, flavanones and chalcones	Antibacterial and anti-hepatoma activity	Shi et al. 2022
Goldenberry (*Physalis pubescens* L.)	Chalcone, flavonol and flavanol	Gastrointestinal action	Zhu et al. 2023
Tamarillo (*Solanum betaceum* Cav.)	Flavonol, chalcone, phenolic acid and phenylpropanoids	Gastrointestinal action	Zhu et al. 2023
Purple passion fruit (*Passiflora edulis* Sims)	Phenolic acids, flavonols and phenolic aldehydes	Antioxidant activity	Costa et al. 2023
Sumba (*Berberis aristata*)	Phenolic acids, anthocyanins and flavanol	Antioxidant, antimicrobial, antidiabetic and cytotoxin	Mueed et al. 2024
Wampee (*Clausena lansium* Lour.)	Phenolic acids, flavonoids, tannin, stilbenes, lignans and coumarins	Anti-inflammatory activity	Zeng et al. 2024
Kiwi (*Actinidia eriantha*)	Benzoic acids, hydroxycinnamic acids, flavanols and flavonols	Antioxidant activity	Jia et al. 2024
Sea Buckthorn-monk (*Hippophae rhamnoides*)	Flavonoids, and phenolic acids	Antidiabetic, antihypertensive, and antioxidant	Mubeen et al. 2025
Guabiroba (*Campomanesia xanthocarpa*)	Flavonoids, isoflavones, flavanones, flavanes, flavones, flavan-3-ols, and proanthocyanidins	Antioxidant activity	Machado et al. 2025

In a study conducted by Fernández et al. (2020), the authors evaluated the inhibitory acetylcholinesterase (AChE) and antimicrobial potential of Amazonian fruit oils. The oils evaluated were from the following fruits: abiu (*Pouteria caimito*), acerola (*Malpighia emarginata*), araçá (*Psidium cattleianum*), bacuparí (*Rheedia gardneriana*), biribá (*Rollinia mucosa*), camu-camu (*Myrciaria dubia*), fruta-do-conde (*Annona squamosa*), graviola (*Annona muricata*) and taperebá (*Spondias mombin* L.). They tested the oils' antimicrobial activity against *Candida albicans* ATCC 18804, *Staphylococcus aureus* ATCC 29212, *Bacillus cereus* ATCC 11778, *Escherichia coli* ATCC 25922, and *Salmonella typhimurium* ATCC 14028. The best inhibition results were obtained for the yeast *Candida albicans*, with inhibition percentages of 94.46%, 87.12%, 85.23%, and 85.22% by the taperebá bark, acerola bark, araçá seed, and taperebá pulp extracts, respectively. AChE inhibition results indicated that several extracts, such as biribá bark, taperebá seeds, and acerola pulp, demonstrate a strong inhibitory capacity of this enzyme. The authors associated the potential for AChE inhibition and antimicrobial activity with phenolic compounds. However, studies linking this potential to isolated compounds are limited in the literature.

3.4.2 Carotenoids

Carotenoids are lipophilic organic pigments that are widely distributed in nature (Zhang et al. 2024). Belonging to the terpene class, these compounds have been extensively studied. Currently, more than 1,000 carotenoids have been described in the literature, of which about 40 are present in the human diet, mainly through the consumption of fruits and vegetables (Bohn et al. 2023). These bioactive compounds are divided into two main groups according to their structure and function: xanthophylls, compounds containing oxygen functional groups, such as lutein, zeaxanthin and β-cryptoxanthin, and carotenes, such as lycopene, α-carotene and β-carotene, which are hydrocarbon compounds formed by polyisoprene chains and may have a cyclic or acyclic structure at both ends, without the presence of functional groups (Zhang et al. 2024a).

Xanthophylls are an important part of the naturally occurring carotenoids. These compounds are oxygenated derivatives of carotenoids and are characterized by a yellowish color (Aziz et al. 2020). Widely distributed in nature, xanthophylls have been extensively studied for their biological properties, including antiplasmodic activity, neuroprotective effects, and preventive action in the progression of age-related macular degeneration (AMD) (Aziz et al. 2020).

Among the different types of xanthophylls, lutein and zeaxanthin stand out as plant pigments that are associated with selective accumulation in the macula of the retina, thus playing an important role in the prevention of AMD (Perry et al. 2009, Hu et al. 2024). This bioactive activity occurs through the absorption of blue light and the antioxidant potential of these metabolites, promoting eye protection. The main food sources of these xanthophylls include squash, maize and some tropical fruits, such as monkey fruit (*Garcinia intermedia*) (Irías-Mata et al. 2018).

Carotenes, molecules composed of hydrocarbons, are natural orange pigments that occur in nature mostly in the *trans*-isomeric form (Schieber and Carle 2005).

Among them, β-carotene, a lipophilic polyphenol, is one of the most important carotenoids in the diet and an important precursor for vitamin A biosynthesis, and is used as a vitamin supplement (El-baz et al. 2023). In addition, several health benefits have been attributed to β-carotene, such as antioxidant activity, strengthening of the immune system, improvement of gastric functions and potential in the prevention of chronic diseases, including macular degeneration and cancer (Altunay et al. 2022). Therefore, given its therapeutic potential, it is recommended to consume foods rich in β-carotene. Among the main fruits that stand out as sources of this metabolite are mangoes, peppers and tomatoes, with contents of 14.79, 21.67 and 17.10 $\mu g{\cdot}g^{-1}$, respectively (Li et al. 2013, Mercado-Mercado et al. 2018, Rodríguez-Rodríguez et al. 2020). Another very important carotenoid is lycopene, a compound with high antioxidant activity and a remarkable ability to neutralize free radicals (Kang et al. 2023). Its effect is thought to be about twice as strong as that of β-carotene, and it contributes significantly to reducing the risk of atherosclerosis and some types of cancer, such as stomach and lung cancer (Yu et al. 2023, Zhou et al. 2023). In addition to these effects related to its antioxidant potential, several studies point to the therapeutic effect of lycopene in neurological and metabolic disorders, highlighting its role in inhibiting inflammatory processes, cell apoptosis and restoring mitochondrial function (Zhou et al. 2023). Regular consumption of foods rich in this bioactive compound is therefore highly beneficial for human health. Lycopene, a natural reddish pigment, is bioavailable in various fruits, especially tomatoes, guava and watermelon (Kang et al. 2023).

3.4.3 Vitamin C and Other Antioxidants

Vitamins are part of a large group of low-molecular-weight bioactive compounds involved in numerous metabolic reactions and cellular processes in the human body (Mateeva et al. 2023). They perform various biological functions that are beneficial to the body. Among the biological potential of vitamins, the antioxidant property stands out, an activity associated with protecting the body against oxidative stress, which protects against diseases such as diabetes, neurodegenerative disorders, cardiovascular disease and cancer (Suwanwong and Boonpangrak 2021). Thus, exogenous antioxidants play an essential role in disease prevention, particularly vitamin C (L-ascorbic acid), vitamin E (tocopherols and tocotrienols), and vitamin A (provitamins).

L-ascorbic acid, or vitamin C, is a water-soluble micronutrient that is essential to the human body and is fundamental to physical and neurological health (Hassuna et al. 2023). This metabolite acts as an important bioactive compound with antioxidant properties, playing an important role in neutralizing reactive oxygen species and various free radicals such as superoxide, singlet oxygen, hydrogen peroxide, and hydroxyl radicals (Zhang et al. 2023b). As humans are unable to synthesize this vitamin, its supplementation is necessary as it is associated with a reduction in oxidative stress, anti-inflammatory activity, and modulation of the immune response (Sim et al. 2025).

L-ascorbic acid deficiency in the human body is associated with scurvy, a disease that compromises skin sensitivity, causes bleeding gums, and impairs the wound healing process (Pullar et al. 2017). In addition to scurvy, chronic respiratory diseases and systemic arterial hypertension are also linked to a deficiency of this micronutrient (Kwak and Kim 2023, Zhang et al. 2024b). Therefore, adequate intake of vitamin C is essential in the human diet. Citrus fruits such as oranges and lemons have high levels of this compound, as do tomatoes and potatoes (See et al. 2024).

Vitamin E, or tocopherol, is composed of a mixture of tocopherols (TOCs) and tocotrienols (TCTs) synthesized by plants from homogentisic phenolic acid (Mutalip et al. 2018). This fat-soluble micronutrient can be present in fruits and vegetables in different analogous forms, such as α-, β-, δ-tocopherols and α-, β-, δ-tocotrienols (Xiong et al. 2023). However, only α-tocopherol meets the needs of the human body as it has the greatest biological activity due to preferential retention and distribution in the body (Muthulakshmi et al. 2023). It integrates the structure of the cell membrane, exhibits antioxidant properties and plays an essential role in various biological functions, including protection against radiation damage, regulation of cholesterol levels, cardiovascular health, and prevention of neurological disorders (Latib et al. 2024, Najafabadi et al. 2025).

Vitamin E deficiency in the body is directly related to neurodegenerative diseases such as ataxia, age-related macular degeneration, and non-alcoholic fatty liver disease (Muthulakshmi et al. 2023). Therefore, getting enough vitamin E from the diet by consuming fruit and vegetables is essential, as the human body is unable to produce it on its own. Good sources of vitamin E from fruit include tomatoes, currants and avocados (Orsavová et al. 2019, Vincent et al. 2020, Mesa and Munné-Bosch 2025).

Trans-retinal, popularly known as vitamin A, is a fat-soluble vitamin composed of a series of retinol derivatives and is directly related to carotenoids—especially β-carotene and β-cryptoxanthin—which have provitamin A activity and are the main sources of this molecule (Karaźniewicz-Łada and Głõwka 2016, Meng et al. 2025). This micronutrient performs essential functions in the human body, particularly in the visual and immune systems, and is associated with the absorption of iron (Fe) from food (Davey et al. 2006, Chacón-Ordóñez et al. 2019). Given the body's inability to synthesize this nutrient, it is therefore necessary to supplement vitamin A through diet. Thus, mapote (*Pouteria sapota* Jacq.) and cantaloupe melon fruits stand out (Laur and Tian 2011, Chacón-Ordóñez et al. 2019). Vitamin A deficiency is a major public health problem, particularly in developing countries, leading to eye disease and increased mortality (Zhu et al. 2020). In addition, this deficiency is associated with reduced immunity, making the body more susceptible to a range of diseases, including diarrhea, measles and respiratory infections (Lee 2017).

3.4.4 Fiber and Their Metabolites

Dietary fibers are functional metabolites with beneficial health effects and are considered essential components of foods (Desloovere et al. 2024). Currently, the Codex Alimentarius Commission (CAC) defines dietary fiber as edible carbohydrate polymers composed of three or more monomers that are neither hydrolyzed nor absorbed in the small intestine and upper digestive system (Stephen et al. 2017).

Increased consumption of this nutrient is associated with a reduced risk of developing cancer, diabetes, obesity and inflammatory bowel disease. In addition, dietary fiber plays multiple roles in promoting health, such as regulating lipid metabolism and reducing the risk of chronic diseases (Li et al. 2023, Cen et al. 2025).

Dietary fibers can be classified in several ways. One approach considers their origin, dividing them into dietary fibers derived from plants, animals or microorganisms (Li et al. 2023). Another form of classification is based on molecular weight, dividing them into polysaccharide- or oligosaccharide-type fibers. However, the most common form of classification is based on water solubility, with fibers classified as soluble or insoluble (Yu et al. 2025b).

Soluble dietary fiber (SDF) is characterized by a structure composed of water-soluble non-cellulosic polysaccharides such as inulin, oligofructans, pectin and mucilages (Cen et al. 2025). Recent studies have associated soluble dietary fibers with greater potential to modulate the gut microbiota and prevent conditions such as hyperlipidemia and hyperglycemia, due to their high water-retention capacity, enhanced adsorption properties, enzyme inhibitory effects and probiotic activity (Zhuang et al. 2024, Cen et al. 2025).

Therefore, the consumption of fruits rich in soluble fiber is extremely important for maintaining quality of life and preventing metabolic and gastrointestinal diseases. The main dietary sources of this nutrient are foods of plant origin, such as whole grains, vegetables, fruits and their by-products, such as seeds, leaves, pomace and peels (Abboud et al. 2019). In the work by Abboud et al. (2019), the authors isolated soluble dietary fiber from sour passion fruit and evaluated its potential in the treatment of ethanol-induced ulcers in rats. The results showed that consumption of SDF from passion fruit peel exerted significant gastroprotective effects *in vivo* on experimentally induced gastric ulcers.

Insoluble dietary fibers (IDFs), which include cellulose, hemicellulose and lignin, are a type of carbohydrate whose main action in the body is associated with the gastrointestinal tract, as their biological mechanism is mainly characterized by slow or incomplete fermentation, which promotes intestinal peristalsis, increases stool volume and reduces intestinal transit time (Wang et al. 2022a, Cen et al. 2025). In addition, IDFs may also help reduce glucose absorption by inhibiting its diffusion into the gut through the barrier effect of the particles and their ability to adsorb glucose directly (Zhu et al. 2022).

Thus, several studies indicate that including fruits rich in insoluble fiber in the diet can reduce the incidence of diabetes, hyperlipidemia and other metabolic diseases, in addition to promoting gastrointestinal health (Liu et al. 2025). The main fruits providing IDF are pears, date fruit and soursop (Peng et al. 2022, Leivas et al. 2023, Almoumen et al. 2024).

3.5 Influence of Fruit Cell Structure

The cellular structure of citrus fruits has a significant influence on the release and bioavailability of bioactive compounds, such as hesperidin, a flavonoid concentrated mainly in the albedo (inner white part) of these fruits. Being a matrix rich in fiber, the release of hesperidin can be limited, resulting in its delayed absorption, predominantly in the colon, after hydrolysis by the intestinal microbiota (Manach et al. 2005).

Pharmacokinetic studies show that hesperidin has an absorption time (tmax) of 4.4 to 7 hours (Nielsen et al. 2006, Brett et al. 2009), reflecting the need for microbial metabolism before its passive absorption. However, structural modifications, such as enzymatic conversion to hesperetin-7-glucoside by cleavage of the rhamnose fraction, result in up to five times greater bioavailability (De Goede et al. 2010, Gonzàlez-Barrio et al. 2004), as well as a significantly reduced tmax (0.6 hours), indicating earlier absorption in the small intestine (Nielsen et al. 2006).

Although cell structure influences the release of hesperidin, studies show that factors such as the physical state of the food (fresh fruit or juice) and coingestion with other foods do not significantly alter its absorption (Mullen et al. 2008, Brett et al. 2009). On the other hand, the dose ingested has a direct impact on the amount absorbed (Manach et al. 2003, Nielsen et al. 2006). Thus, although the cellular matrix acts as a barrier, strategies such as chemical modifications can optimize the bioavailability of this functional compound.

3.6 Processing and Transformation of Bioactive Compounds

The growing consumer awareness of the health benefits provided by food-derived bioactive compounds has driven research into the behavior of these molecules throughout the digestion process. However, the journey of these compounds from ingestion to systemic circulation is complex and heavily influenced by the food matrix and processing methods. The food matrix, composed of structural elements such as cell walls, proteins, fibers and lipids, plays a decisive role in either promoting or inhibiting the release of bioactive compounds. Processing techniques, including thermal treatments, fermentation and mechanical disruption, play a crucial role in altering the integrity of the matrix, potentially enhancing or impairing the release of bioactive compounds (Carbonell-Capella et al. 2014).

The digestibility of proteins is often impaired by cross-linking reactions induced by heat or oxidizing agents during processing. These reactions compact protein structures and reduce enzyme accessibility, limiting the release of amino acids and peptides (Matthews and Laster 1965). Thermal processing can degrade heat-sensitive polyphenols, but it can also break down cell walls, enhancing the extraction of carotenoids (Tahergorabi and Hosseini 2017).

Similarly, high-pressure (HP) processing has been shown to modify protein conformations, improving enzymatic digestibility and the release of peptides (Chicón et al. 2008). This highlights the dual nature of processing effects: while it may promote the release of certain bioactive compounds, it can also cause irreversible losses or transformations that compromise their functional potential. Emerging technologies, such as nanoencapsulation and colloidal delivery systems, aim to overcome these limitations by protecting bioactives from degradation and targeting their release within the gastrointestinal tract (Fernández-García et al. 2009).

3.7 Absorption

3.7.1 Effect of Digestive Enzymes on Compound Release

Human digestion is a dynamic process involving the coordinated action of digestive enzymes in the release of bioactive compounds from the food matrix. Among the main enzyme classes involved are amylases, proteases and lipases, which act on the degradation of carbohydrates, proteins and lipids, respectively. The effectiveness of these enzymes in releasing bioactive compounds depends on several factors, including the chemical structure of the compounds, their interaction with the food matrix and the physiological conditions of the gastrointestinal tract, such as pH and the presence of cofactors (Karima et al. 2012).

In the gastric environment, the acidic pH promotes activation of pepsin, which cleaves proteins and can release bioactive peptides associated with local and systemic physiological effects. Additionally, the presence of gastric lipases facilitates the partial breakdown of lipids, impacting the release of lipophilic compounds. However, the stability of bioactive compounds varies considerably: while some, such as aglycone flavonoids, better resist acidic conditions, their glycosylated counterparts tend to be hydrolyzed or degraded before reaching the intestine (Aurélie et al. 2012).

The digestibility and bioavailability of bioactive compounds are also influenced by physicochemical factors such as temperature and thermal processing. Studies show that heat treatment can modify protein structure, making them more accessible to enzymatic action and facilitating the release of bioactive peptides (Lourdes and Blanca 2020). Similarly, bile-mediated lipid emulsification is essential for the efficient digestion of lipophilic compounds, enabling their subsequent intestinal absorption.

Understanding the effects of digestive enzymes on the release and bioavailability of bioactive compounds requires integrative methodological approaches. *In vitro* studies use dynamic digestion systems that replicate physiological parameters (pH, enzymes, time) to analyze nutrient degradation and the release of active molecules, such as fruit polyphenols. Advanced platforms, like gut-on-a-chip models, incorporate mechanical flow and cellular interactions to mimic the complexity of the intestine, allowing evaluation of the microenvironment's influence on absorption (Rodrigues and Failla 2021).

Meanwhile, Caco-2 cell cultures are employed to investigate transcellular transport mechanisms and modulation by membrane proteins, identifying factors that limit or enhance bioavailability (Feiran et al. 2017). *In vivo* assays, including animal models and clinical trials, validate the physiological relevance of these findings by correlating enzymatic activity with pharmacokinetic parameters such as plasma concentration and hepatic metabolism (Kong and Singh 2008, Harjirder et al. 2015). This multi-scale methodological approach enables elucidation of how enzymatic digestion influences compound accessibility, guiding applications in personalized nutrition and functional food development.

3.7.2 Transport Across the Intestinal Mucosa

The intestinal absorption of compounds occurs predominantly in the jejunum, where digestion products such as peptides, fatty acids, mono- and oligosaccharides, vitamins and minerals cross the intestinal mucosa and enter systemic circulation. Initially, it was believed that only free amino acids were absorbed, while peptides and proteins would be degraded before crossing the epithelial barrier. However, recent evidence shows that various bioactive peptides can be absorbed in their intact form and exert biological effects in the body (Foltz et al. 2007, Righard et al. 2014, Alice et al. 2015).

The absorption of bioactive compounds occurs through different mechanisms: passive diffusion, active transport and endocytosis. Passive diffusion can be paracellular, through tight junctions (TJs) between epithelial cells, or transcellular, across cell membranes, being favored by lipophilic compounds of low molecular weight. Active transport, in turn, depends on transporter proteins that allow compounds to pass against the concentration gradient with energy expenditure. Endocytosis involves the internalization of compounds via vesicles and is essential for the absorption of macromolecules and long-chain peptides (Daniela et al. 2010, Komin et al. 2017).

Transporter proteins play a fundamental role in the absorption of bioactive compounds. ABC transporters (ATP-binding cassette) are involved in the efflux of metabolites and drugs, modulating intestinal bioavailability. SGLT transporters (sodium-glucose linked transporters) facilitate monosaccharide absorption, while GLUT transporters (glucose transporters) regulate glucose and other carbohydrate transport (Paulus et al. 2018). In the case of bioactive peptides, the PepT1 transporter (peptide transporter 1) is primarily responsible for the absorption of di- and tripeptides, favoring neutral and hydrophobic compounds (Gilbert et al. 2008, Xu et al. 2014, 2018).

Several factors influence the intestinal bioavailability of bioactive compounds. Interaction with fiber can reduce absorption by forming insoluble complexes, while the presence of lipids can increase the solubility of lipophilic compounds and favor their transcellular absorption. The gut microbiota also plays an essential role by modulating the biotransformation and absorption of bioactive metabolites, potentially enhancing or inhibiting their effects (Nazila and Thomas 2005, Diana et al. 2010).

The bioavailability of bioactive compounds from plant sources depends on overcoming multiple physiological and biochemical barriers to reach systemic circulation in metabolically active forms. Compounds such as polyphenols and carotenoids face challenges from release from the food matrix to intestinal absorption, including degradation by gastric pH, interference from dietary fibers and metabolism by the colonic microbiota.

Absorption efficiency is modulated by molecular characteristics (e.g., hydrophilicity) and interactions with specific transporters, such as GLUT1 and MRP2, which regulate influx or efflux through enterocytes. Additionally, the food matrix acts as a dynamic modulator: lipids can facilitate the solubilization of lipophilic molecules, while proteins or polysaccharides can sequester compounds, reducing their accessibility (Rodrigues et al. 2022). First-pass hepatic metabolism also limits their plasma concentration, requiring encapsulation strategies or synergistic combinations with other nutrients to optimize therapeutic efficacy.

The absorption of bioactive compounds through the intestinal mucosa involves a complex interaction between transport mechanisms and physicochemical properties of the molecules, where passive diffusion predominates for low molecular weight lipophilic substances, while polar compounds, such as certain polyphenols, depend on membrane transporters (e.g., GLUT, ABC) or enzymatic modifications to become soluble and permeable (Koepsell 2020).

Carotenoids, for example, require emulsification by bile salts and micelle formation to be internalized via proteins such as NPC1L1, whose expression directly influences their intestinal uptake (Ko et al. 2020). The food matrix plays an ambivalent role: fibers can encapsulate compounds, inhibiting their release, but their colonic fermentation by symbiotic bacteria produces short-chain fatty acids, which improve epithelial integrity and indirectly facilitate the absorption of bioactive metabolites (Danneskiold-Samsøe et al. 2019).

Furthermore, intestinal and hepatic biotransformation modifies chemical structures, such as the glucuronidation of polyphenols, altering their biological activity and excretion. Dysfunctions in bile secretion or genetic variants in transporters (e.g., ABCB1) can limit the efficiency of these processes, explaining interindividual differences in bioavailability (Rodrigues et al. 2022). Strategies such as combining with lipids or encapsulating in nanoparticles emerge as alternatives to overcome physiological barriers, maximizing the therapeutic potential of these compounds.

3.7.3 Hepatic Metabolism and Enterohepatic Circulation

Hepatic metabolism and enterohepatic circulation play crucial roles in determining the effects of dietary bioactive compounds on human health. The liver is the primary organ responsible for metabolizing these compounds, where its enzymes modify them in various ways, impacting their bioavailability and bioefficacy. Enterohepatic circulation is an additional process in which bioactive compounds, after being secreted into the bile, are reabsorbed in the intestine and returned to the liver. This process can increase the retention time of these compounds in the body, enhancing their beneficial effects. However, it is also important to note that this mechanism can result in the formation of toxic metabolites, which may adversely affect health.

Hepatic metabolism involves several chemical transformations of bioactive compounds. Phase I reactions, catalyzed by cytochrome P450 (CYP450) enzymes, introduce or expose functional groups such as hydroxyl, carboxyl, or amino groups. These modifications can alter the compounds' structures, affecting their biological properties. In Phase II reactions, compounds are conjugated with other molecules like glucuronic acid, sulfate, or glutathione, making them more water-soluble, which facilitates their excretion from the body. Additionally, hepatic metabolism also involves cleavage of chemical bonds, which changes the biological activity of the compounds as well as their stability and excretion capacity (Han et al. 2022).

Resveratrol, found in grapes and red wine, is an example of a compound that benefits from enterohepatic circulation (Baur and Sinclair 2006). It is metabolized in the liver through Phase II reactions such as glucuronidation and sulfation, and its bioavailability is increased by enterohepatic circulation, allowing it to remain in the body longer. In contrast, curcumin, found in turmeric, exhibits limited enterohepatic circulation, which contributes to its low bioavailability. It is metabolized by both Phase I and Phase II reactions, but the lack of an efficient enterohepatic cycle reduces its efficacy. Other compounds, such as flavonoids found in fruits, vegetables, and tea, also undergo hepatic metabolism, and depending on the specific type of flavonoid, enterohepatic circulation may influence their bioavailability and bioefficacy (Padilha et al. 2017).

Hepatic metabolism and enterohepatic circulation are influenced by a range of factors. The chemical structure of the bioactive compound determines which hepatic enzymes can metabolize it and its ability to be absorbed via enterohepatic circulation. The dose of the compound also affects these processes, as it can influence enzyme saturation and the extent of enterohepatic circulation. Other factors, such as the presence of other dietary compounds, can modulate these processes. Liver and intestinal health are fundamental since conditions that impair these organs can hinder the metabolism and absorption of bioactive compounds. Additionally, the gut microbiota plays an important role by metabolizing these compounds and influencing their enterohepatic circulation (Neilson et al. 2017).

These processes have significant implications for human health, as they influence the efficacy of drugs and dietary supplements. The ability to absorb and act upon these compounds can be altered, directly impacting the prevention and management of chronic diseases such as heart disease, cancer, and diabetes (Couve et al. 2008, Hooper et al. 2008, Davi et al. 2010). A better understanding of hepatic metabolism and enterohepatic circulation will enable the development of new strategies to improve the bioavailability and bioefficacy of bioactive compounds, expanding their health benefits. Despite advances, much remains to be investigated, and future research is essential to deepen the understanding of these processes and enhance the use of these compounds in health promotion and disease prevention.

3.8 Interaction with the Gut Microbiota and Their Bioavailability

Microbiota describes the living microorganisms found in a defined environment, like the gut environment, which is considered the most important microbiota environment in human health. The gut microbiota (GM) is composed of 6 phyla, including Firmicutes, Bacteroidetes, Actinobacteria, Proteobacteria, Fusobacteria, and Verrucomicrobia, among which Firmicutes and Bacteroidetes represent 80–90% of the gut microbiota and play an important role in biological processes (Hou et al. 2022).

This environment may affect these processes via several mechanisms and is involved in nutrient extraction, metabolism, and immunity. Moreover, the biosynthesis of bioactive compounds such as vitamins, amino acids, and lipids is also highly dependent on the GM. When the GM isn't balanced, disorder occurs, and its structure and function will change; this causes the development of some diseases, such as inflammation, blood glucose imbalance, and metabolic disorders (Jin et al. 2023).

Active ingredients in fruits, such as polysaccharides, dietary fiber, phenolics, flavonoids, and carotenoids, could influence the GM (Alasalvar et al. 2023). Bioactive compounds in fruits interact with lactic acid bacteria, promoting the diversity of beneficial microbial groups, altering the Firmicutes/Bacteroides ratio, and suppressing the growth of harmful gut bacteria (Borgonovi et al. 2022).

Recent studies have shown that polysaccharides exhibit selective promotion depending on the bacteria present in the gut microbiota environment. Beneficial bacteria, for example, can influence glucose metabolism. A decrease in *Lactobacillus* and *Bifidobacterium* is closely associated with impaired glucose tolerance (Song et al. 2021). Some polysaccharides can enrich these bacteria and, therefore, may be used to improve the diabetic phenotype (Zhang et al. 2022b).

Polyphenols and other bioactive compounds found in fruits play a crucial role in maintaining gut health through their interaction with the intestinal microbiota. These compounds help strengthen the intestinal barrier. Additionally, gut bacteria can metabolize them, contributing to the prevention of various diseases. In particular, polyphenols derived from berries have been shown to modulate the gut microbiota, inhibiting inflammation, suppressing cancer development, and controlling associated metabolic disorders (Bouyahya et al. 2022, Alqudah and Claesen 2024). By modulating the composition of the gut microbiota, these compounds can help regulate immune responses by suppressing the secretion of pro-inflammatory cytokines, reducing harmful bacteria, and enriching populations of short-chain fatty acid–producing bacteria that promote a balanced immune system (Zhang et al. 2024c, Kiriyama et al. 2024).

The intestinal microbiota plays a key role in the transformation of bioactive compounds into more active or bioavailable metabolites. The absorption of these compounds depends on their chemical form and interaction with the microbiota. This process occurs mainly through hydrolysis and reduction reactions carried out by intestinal bacteria under anaerobic conditions, resulting in nonpolar products of low molecular weight. In addition, the intestinal microbiota is capable of metabolizing other compounds, such as non-digestible polysaccharides, converting them into short-chain fatty acids that serve as an energy source for intestinal cells (Bastos et al. 2009).

Fruit-derived flavonoids are metabolized by the intestinal microbiota into simpler compounds, such as phenolic acids, which can be more easily absorbed by the body and exert beneficial biological effects. This biotransformation process is mediated by enzymes produced by intestinal bacteria, which modify the structure of the original flavonoids, increasing their bioavailability and enhancing their antioxidant and anti-inflammatory properties (Kim et al. 2022, Ribeiro et al. 2024).

Some natural products contain glycosides that, when present in the intestinal microbiome, are catalyzed by enzymes present in the GM to produce the hemiacetal and corresponding free glycoside ligands. These enzymes are essential for modifying the structure and bioavailability of glycosides. Furthermore, this process benefits the microbiota, since the released free glycosides can serve as an energy source for the growth of specific microorganisms in the intestine (He et al. 2024).

Resveratrol, a polyphenol found in grapes and blueberries, can bind to various nutrients such as proteins, and its solubility plays a key role in determining whether it is absorbed or eliminated by the body. Its bioavailability can be enhanced by increasing its solubility and reducing intestinal efflux and metabolic degradation (Vasconcelos et al. 2021). Studies have shown that resveratrol's solubility improves when it binds to fibrinogen or is dissolved in peanut oil (Gligorijević et al. 2020, Li et al. 2022). These strategies not only increase its absorption but also enhance its antioxidant properties.

Recent research on berries, such as raspberries and blackberries, has highlighted their positive impact on gut and metabolic health. A study by Núñez-Gómez et al. (2021) showed that whole raspberry and its phenolic extract promoted the production of higher levels of short-chain fatty acids (SCFAs), such as acetate and propionate, and phenolic catabolites like urolithins, compared to fiber fractions alone. These effects varied depending on the host's body weight, suggesting that microbiota composition influences the metabolic response. Notably, individuals with normal weight produced more urolithins from non-extractable phenolics in fiber fractions, while those with overweight showed greater urolithin production from extractable polyphenols. This reinforces the idea that the prebiotic potential of raspberries is primarily driven by their phenolic content rather than fiber alone.

The composition of the intestinal microbiota varies between individuals, influencing the efficiency of conversion of these bioactive compounds and, consequently, their bioavailability and health effects. Factors such as diet, antibiotic use, and lifestyle can alter the composition of the microbiota, affecting the metabolism of bioactive compounds.

3.9 Methods for Evaluating Bioavailability of Dietary Bioactives

The bioavailability of dietary bioactive compounds is influenced by a range of factors, including their physicochemical properties, biological interactions, the food matrix, and host-related variables (Figure 2). Unlike pharmaceutical drugs, dietary bioactives such as polyphenols, carotenoids, and peptides often exhibit low bioavailability. This is primarily due to poor solubility, instability in the gastrointestinal tract, and metabolic transformations that occur before they reach systemic circulation (Neilson et al. 2017, Shahidi et al. 2019). As a result, understanding the mechanisms that govern their absorption and metabolism is essential for optimizing their health benefits.

Key physicochemical properties, such as solubility, lipophilicity, and molecular structure, determine a bioactive's ability to dissolve and pass through intestinal membranes. Biological factors, including gastric acidity and the composition of gut microbiota, also play significant roles in the metabolism and effectiveness of these compounds. The food matrix itself is critical in modulating bioavailability by influencing the release, stability, and absorption of bioactives. Lastly, host-specific factors, such as age, genetic background, and dietary habits, add to the variability in bioactive absorption and utilization across individuals (Amigo and Hernández-Ledesma 2020, Vieira and Conte-Junior 2024).

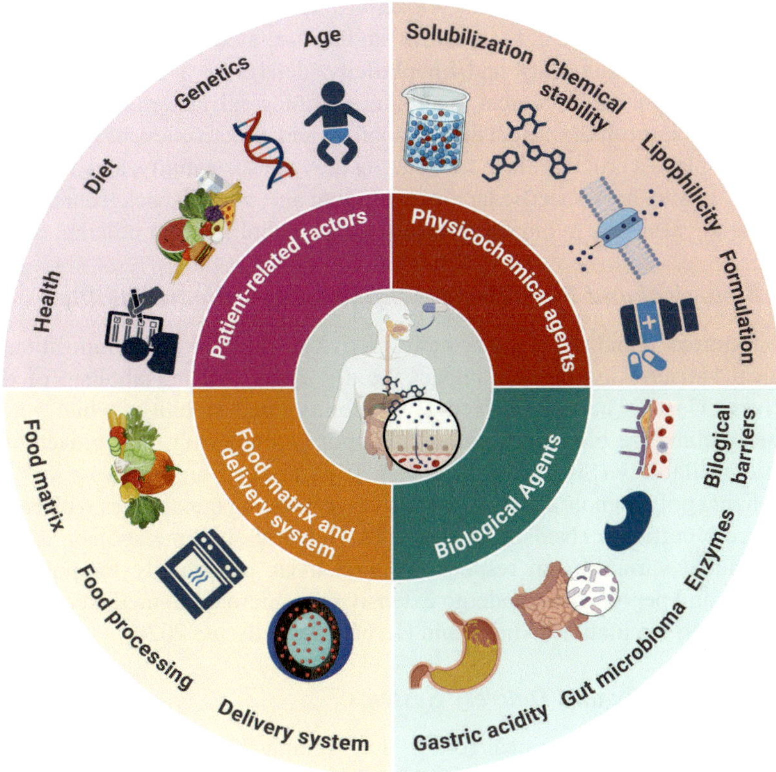

Figure 2. Key Factors Influencing Dietary Bioactive Compounds Bioavailability (Adapted from Wu et al. 2024b), created with Biorender.com.

3.9.1 Physicochemical Properties and Their Influence on Bioavailability

The physicochemical properties of dietary bioactives, including solubility, lipophilicity, and molecular structure, play a fundamental role in determining their bioavailability. Solubility is particularly important, as bioactive compounds must dissolve in the gastrointestinal fluids to be absorbed into the bloodstream (Neilson et al. 2017). Lipophilic compounds, such as carotenoids and curcumin, often require the presence of lipids or emulsifiers to form micelles, which enhance their solubility and facilitate absorption through intestinal membranes (Shahidi et al. 2019). The molecular structure also influences a bioactive's ability to permeate the gut lining, as larger, more complex molecules may face difficulties in crossing biological membranes. Additionally, bioactives with unstable chemical structures may degrade during digestion, reducing their bioavailability (Vieira and Conte-Junior 2024). Overall, the interaction of these physicochemical characteristics with the gastrointestinal environment determines the extent to which bioactives can be absorbed and utilized by the body (Amigo and Hernández-Ledesma 2020).

The bioavailability of dietary bioactives is influenced by the interplay of physicochemical, biological, food matrix, and host-related factors. Physicochemical properties, such as solubility and lipophilicity, determine a bioactive's ability to dissolve and permeate biological membranes. Biological factors, including gastric acidity and gut microbiota, affect the metabolism and effectiveness of bioactives. The food matrix plays a crucial role in modulating the release, stability, and absorption of bioactives. Lastly, host-specific factors, including age, genetic background, and diet, contribute to variations in bioactive absorption, metabolism, and utilization.

3.9.2 Biological and Physiological Influences on Bioavailability

Gastrointestinal conditions, enzymatic activity, microbiota composition, and biological barriers significantly impact the absorption and metabolism of dietary bioactives. Gastric acidity influences solubility and stability, while intestinal enzymes metabolize bioactives before absorption, modifying their bioactivity and systemic availability (Shahidi et al. 2019). The gut microbiota plays a key role in metabolizing polyphenols and other bioactives, producing metabolites with enhanced or reduced bioefficacy (Neilson et al. 2017). This microbial metabolism introduces interindividual variability in response to bioactives, particularly for polyphenols, flavonoids, and peptides that undergo extensive gut microbiota–mediated processing before entering circulation (Amigo and Hernández-Ledesma 2020).

3.9.3 Food Matrix and Delivery Systems

The food matrix plays a critical role in determining the bioavailability of dietary bioactives by affecting their release, stability, and absorption. The presence of macronutrients, such as lipids, proteins, and carbohydrates, can enhance or inhibit bioactive absorption by modifying solubility and interaction with transporters. Lipophilic bioactives such as curcumin and carotenoids demonstrate enhanced absorption when co-ingested with dietary fats due to micelle formation, while polyphenols may exhibit reduced bioavailability due to protein binding or fiber (Vieira and Conte-Junior 2024). The solubility and stability of bioactives are also influenced by their chemical structure; hydrophilic bioactives (e.g., flavonoids and anthocyanins) generally dissolve better in aqueous environments, while hydrophobic bioactives (e.g., carotenoids, curcumins, and phytosterols) require lipid carriers to improve bioaccessibility (Dima et al. 2024, Jia et al. 2024).

Food processing and formulation strategies have been developed to enhance bioactive bioavailability. Encapsulation techniques, including liposomal and polymeric nanoparticles, protect bioactives from enzymatic degradation, enhance solubility, and facilitate targeted release (Neilson et al. 2017). Other novel delivery approaches, such as emulsions, nanoemulsions, and complexation with cyclodextrins, have shown promise in improving stability and solubility in the gastrointestinal tract (Amigo and Hernández-Ledesma 2020). Additionally, food structure, particle size, and food matrix composition significantly impact bioactive release. Smaller particle sizes enhance bioaccessibility by increasing surface area and digestion rates, while certain carbohydrates, such as pectin, may hinder or enhance absorption depending on their structure and processing conditions (Shahidi et al. 2019, Liu et al. 2024).

Food processing methods also play a critical role in determining the bioaccessibility of dietary bioactives. Thermal processing, such as blanching, steaming, baking, frying, and pasteurization, can disrupt the food matrix, promoting the release and absorption of bioactives while also potentially degrading sensitive compounds (Jia et al. 2024). For instance, blanching can lead to the loss of phenolic compounds and carotenoids due to oxidation and isomerization, but under optimized conditions, it may enhance bioactive retention and improve bioaccessibility. Similarly, baking and frying have been shown to increase the bioaccessibility of flavonols in onions by up to 65.5% (Shim et al. 2011). Conversely, excessive heat exposure can lead to degradation of anthocyanins and other heat-sensitive bioactives, as observed in strawberry pasteurization (Nhan and Quyen 2023).

To overcome the limitations of thermal processing, non-thermal techniques such as high-pressure processing (HPP), pulsed electric field (PEF), ultrasound, and freeze-drying have been explored. These methods minimize bioactive degradation while enhancing stability and bioaccessibility. HPP has been shown to increase the bioaccessibility of carotenoids in juices and maintain polyphenol content in fruit purées (Lou et al. 2022, Jia et al. 2024). PEF treatment of carrot purée significantly improved carotenoid bioaccessibility, demonstrating its potential as an alternative processing method (López-Gámez et al. 2021). Additionally, ultrasound treatment has been found to increase the phenolic and flavonoid content of vegetables while preserving their structural integrity (Traore et al. 2020). Freeze-drying, although beneficial for retaining bioactives, may sometimes lower bioaccessibility compared to conventional drying methods due to alterations in food structure (Ozay-Arancioglu et al. 2022).

Pretreatments and combined treatments are also gaining attention for their ability to improve bioactive stability and bioaccessibility. Pretreatment methods, such as enzymatic treatment, ohmic heating, gas-phase plasma treatment, and microwave treatment, can enhance the retention of bioactives before further processing (Jia et al. 2024). For example, ethanol pretreatment of apple slices prior to drying was shown to improve the retention of phenolic compounds (Wang et al. 2022b). Ultrasound pretreatment has also been demonstrated to disrupt the food matrix, reduce viscosity, and decrease particle size, leading to improved bioaccessibility and cellular uptake of carotenoids and phenolic compounds in broccoli (Zhang et al. 2021a). High-pressure processing (HPP) pretreatment of peeled oranges increased the flavonoid, phytoene, and phytofluene content in juices while preserving vitamin C (Ancos et al. 2020). Additionally, ohmic heating before frying resulted in the highest retention of total polyphenols, flavonoids, and chlorogenic acid in potatoes compared to other pretreatments (Sari et al. 2024).

Combined treatments integrating different technologies have shown promising results in improving the antioxidant capacity and bioaccessibility of bioactives. Techniques such as high-pressure processing combined with heating, irradiation with heating, and electroporation with heating have been effective in inactivating microorganisms while preserving bioactives (Soni and Brightwell 2022, Jia et al.

2024). The combination of pulsed electric field and ultrasound significantly increased the total phenolic, flavonoid, and anthocyanin content while enhancing antioxidant activity (Manzoor et al. 2019). However, some combined treatments, such as ultrasound with ozone, may reduce polyphenol content despite their antibacterial benefits (Traore et al. 2020). Further research is needed to optimize processing conditions and identify the most suitable techniques for enhancing the bioavailability of specific dietary bioactives (Jia et al. 2024).

Protein interactions also influence bioactive bioavailability. While protein–polyphenol interactions can reduce free polyphenol content, they may also enhance stability and protect bioactives from enzymatic degradation. However, some protein-based food matrices, such as dairy products, have been shown to decrease the bioavailability of anthocyanins due to the formation of insoluble complexes (Oksuz et al. 2019, Muntaha et al. 2025). Additionally, minerals in the food matrix, such as iron, can chelate polyphenols, reducing their bioaccessibility and impacting their stability and sensory properties (Hider et al. 2001).

3.9.4 Host-Specific Factors and Interindividual Variability

Host-specific factors, including age, genetics, health status, and dietary habits, play a crucial role in the bioavailability of dietary bioactives. Age-related physiological differences, such as variations in enzymatic activity, gut microbiota composition, and intestinal permeability, significantly impact absorption and metabolism. For example, older individuals may experience reduced bioavailability of certain bioactives due to decreased gastric acid secretion and altered gut microbiota profiles (Vieira and Conte-Junior 2024). Genetic polymorphisms affecting metabolizing enzymes and transport proteins, such as P-glycoprotein (P-gp) and organic anion-transporting polypeptides (OATPs), can further contribute to interindividual differences in bioactive absorption and systemic distribution (Shahidi et al. 2019).

Dietary habits and co-consumption of foods also influence bioavailability. Certain dietary components can enhance bioavailability through synergistic interactions, such as the well-documented effect of black pepper–derived piperine on curcumin absorption (Neilson et al. 2017). Conversely, factors such as high fiber intake or the presence of antinutritional compounds may reduce bioavailability by interfering with absorption pathways (Amigo and Hernández-Ledesma 2020).

3.9.5 Methods for Evaluating Bioavailability

Accurately evaluating the bioavailability of dietary bioactives is essential for optimizing their health benefits and guiding formulation strategies, as it directly influences predictions about their absorption, metabolism, and overall effectiveness in the body. Recent advancements have led to more sophisticated methods that go beyond traditional models, offering more reliable insights into bioavailability. These include *in vitro* assays, *in vivo* studies, and advanced analytical techniques like mass spectrometry (Figure 3). *In vitro* models, such as simulated digestion systems and cell-based absorption assays, provide controlled environments to study how bioactives interact with biological factors, enabling bioavailability predictions without the

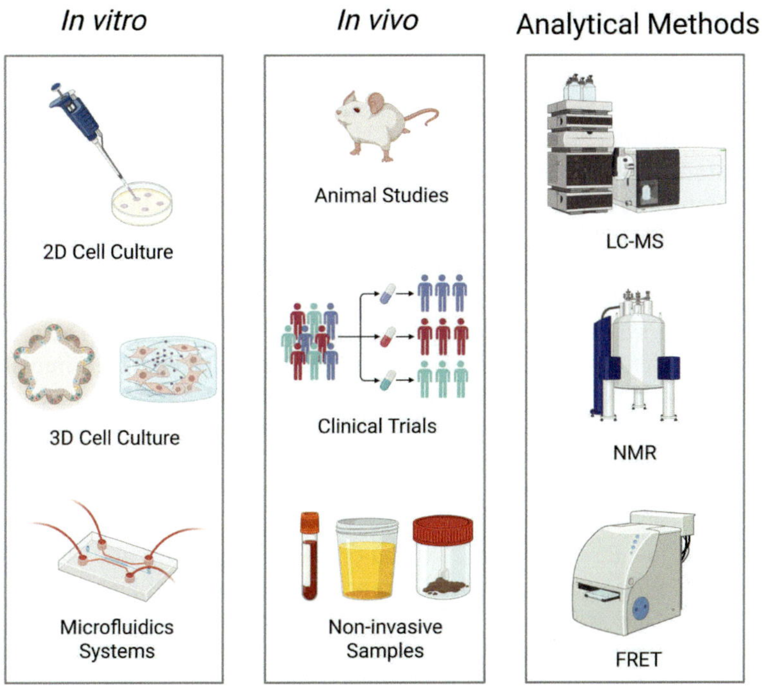

Figure 3. Relevant *in vitro*, *in vivo*, and analytical methods for bioavailability assessment. Source: Own authorship, created with Biorender.com.

ethical and logistical challenges of clinical trials. *In vivo* studies, using animal models and human trials, offer valuable real-world data on how bioactives perform within living organisms. Meanwhile, advanced techniques like mass spectrometry allow for precise quantification of bioactives and their metabolites, enabling detailed pharmacokinetic profiling and the identification of biomarkers associated with bioavailability. Together, these innovations create a comprehensive approach to bioavailability testing, driving progress in the development and optimization of dietary bioactive formulations. Each method has its strengths and limitations, and the choice of technique depends largely on the research context and specific objectives.

3.9.5.1 In Vitro Assays and Simulated Digestion Models

Traditional 2D cell culture models have long served as a cornerstone for assessing bioavailability, providing a cost-effective and simplified approach to evaluating bioactive absorption, permeability, and metabolism. In these models, cells are cultured as monolayers, enabling the study of molecule interactions with biological factors under controlled conditions. Notably, Caco-2 cell lines have been extensively used to simulate intestinal absorption and study bioactive permeability (Panse and Gerk 2022, Tsuchitani et al. 2022, Xu et al. 2016, Li et al. 2016). While these models offer valuable preliminary insights, they are inherently limited by their inability to replicate the complex, three-dimensional structure and dynamic physiological processes of human tissues, including cellular interactions, tissue architecture, and bioactive metabolism. These limitations have spurred the development of more sophisticated *in vitro* models that more accurately reflect the *in vivo* environment (Costa and Ahluwalia 2019, Jensen and Teng 2020).

Recent advancements in *in vitro* bioavailability testing have led to the development of more complex systems that improve the prediction of human bioactive absorption and metabolism. Traditional static models, while valuable for initial screening, are constrained by their inability to replicate the dynamic nature of gastrointestinal (GI) digestion, absorption, and metabolism. In response to these limitations, organ-on-a-chip systems have emerged as a promising alternative. These systems integrate cell cultures with microfluidic technology, enabling the simulation of human tissue architecture and physiological processes such as digestion, absorption, and metabolism. Organ-on-a-chip models are rapidly advancing, allowing for more accurate predictions of bioavailability by replicating *in vivo* conditions, including pH variations, enzyme secretion, and peristalsis (Bein et al. 2018, Marrero et al. 2021, Carvalho et al. 2023).

However, while promising, organ-on-a-chip models are still in the developmental phase and may not fully replicate all aspects of human physiology. Challenges include scaling up for high-throughput screening, long-term culture viability, and ensuring the stability of the model over time. Despite these limitations, organ-on-a-chip systems provide a more accurate reflection of human bioactive absorption and metabolism by recreating *in vivo*-like conditions. They can be tailored to simulate specific organ systems and interactions, offering a more physiologically relevant environment (Skardal 2024, Srivastava et al. 2024).

The evolution of cell-based absorption models, particularly Caco-2 cell lines, has also contributed to the refinement of bioavailability testing. By incorporating advanced 3D culture techniques and co-culturing with human gut microbiota, these models now offer a more accurate representation of human absorption processes. The use of 3D cell cultures better mimics the complexity and architecture of human tissues, providing a more physiologically relevant environment for studying bioactive absorption and metabolism. Co-culture systems, which integrate additional cell types such as endothelial cells or fibroblasts, further enhance the predictive power of these models, improving their ability to simulate the human gastrointestinal system and predict bioavailability (Sugimoto and Sato 2017, Zhang et al. 2020b, Pimenta et al. 2022, Shin and Kim 2022).

The integration of the gut microbiome into *in vitro* bioavailability studies has gained significant attention as it enhances the accuracy of bioactive metabolism simulations (Beterams et al. 2021, Xu et al. 2024). Innovations in microbiome-based co-culture models enable a more realistic representation of how gut bacteria influence the metabolism of dietary bioactives, providing a better understanding of interindividual variability in response to these substances (Sardelli et al. 2021). These models reflect the diverse microbial populations in the human gut that affect the efficacy and toxicity of bioactive molecules. Recent advances, such as the development of the Microbiome Derived Metabolism (MDM)-Screen, have enabled more precise predictions of how the microbiome metabolizes these compounds. This approach considers microbial communities as a whole, offering a more accurate reflection of variability in responses across individuals. Furthermore, factors like diet and genetics contribute to the variation in microbial activity, which in turn influences metabolism and patient-specific responses to bioactives (Zhang et al. 2020a, Średnicka et al. 2023, Jimonet et al. 2024).

In silico gastrointestinal (GI) models, such as GastroPlus, have significantly advanced in predicting bioavailability by simulating critical physiological factors. These models replicate human digestion with high precision by accounting for variables such as pH fluctuations, transit time, and enzymatic degradation, which are essential for understanding the absorption and metabolism of dietary compounds. For instance, GastroPlus integrates parameters like pH-dependent solubility, dissolution rates, and intestinal permeability to accurately forecast the absorption of orally administered substances (Hussain et al. 2023, Wang et al. 2023). Furthermore, dynamic simulations incorporating aspects like peristalsis and enzyme secretion enhance the representation of the human GI tract's motility and secretory functions, providing a more comprehensive understanding of how bioactive compounds behave. By integrating these factors, *in silico* models offer valuable insights into bioactive behavior without extensive *in vivo* experimentation, thereby enhancing the efficiency of pharmaceutical development (Sjögren et al. 2016, Singh 2024).

In summary, while traditional 2D cell culture models, including Caco-2 assays, remain essential for evaluating permeability, the advancement of organ-on-a-chip systems, dynamic GI models, microbiome-based co-cultures, and 3D cell culture techniques has significantly enhanced the accuracy of *in vitro* bioavailability assessments. These innovative models provide a more comprehensive and physiologically relevant platform for predicting the absorption and metabolism of bioactive compounds, enabling more reliable development and therapeutic optimization.

3.9.5.2 In Vivo Studies and Clinical Trials

In vivo studies are crucial for validating the predictions made by *in vitro* models, offering vital insights into the real-world behavior of bioactive compounds. Animal models continue to play a key role in bioavailability testing, providing essential data on the absorption, distribution, metabolism, and excretion (ADME) of these compounds. However, animal models may not always be predictive of human bioavailability due to interspecies differences in metabolism, microbiota composition, and physiological responses. Additionally, the ethical concerns associated with animal testing are a growing challenge. Human clinical trials, while providing essential data, are costly, time-consuming, and often require large cohorts to account for variability in response to bioactives (Van Norman 2019, Barroca et al. 2022).

Non-invasive monitoring techniques such as bioluminescent imaging have been developed to enhance the precision and non-invasiveness of *in vivo* studies. For example, non-invasive monitoring techniques such as bioluminescent imaging enable real-time observation of bioactive absorption and distribution in live animals. These innovations enhance the accuracy of pharmacokinetic (PK) assessments while reducing reliance on invasive procedures, improving both the ethical standards of testing and the efficiency of development (Pellegatti 2012, Zou et al. 2020).

Alongside these advancements, the integration of sophisticated pharmacokinetic (PK) and pharmacodynamic (PD) modeling tools has played a crucial role in translating animal data to human predictions (Wang and Zhou 2016). These models have become increasingly sophisticated by incorporating a wide array of biological factors, enabling more accurate predictions of human bioavailability (Kato et al. 2019, Walter et al. 2024). One particularly promising development in this area is the integration of precision medicine into clinical trials, which takes into account individual patient characteristics such as genetic polymorphisms, microbiome composition, and environmental factors. This approach enables the optimization of bioactive efficacy and bioavailability across diverse patient populations, contributing to more personalized therapeutic strategies (Heckman-Stoddard and Smith 2014, Fountzilas et al. 2022, Duan et al. 2024).

Additionally, there have been significant advancements in clinical trial designs, particularly those focused on optimizing pharmacokinetic and therapeutic outcomes. Innovations like the PKBOIN-12 model, which incorporates pharmacokinetic outcomes in early-phase clinical trials, are helping to identify the optimal biological dose by balancing efficacy and safety (Sun and Tu 2024). These advancements aim to optimize the risk-benefit trade-off, improving the accuracy of bioavailability assessments during early-phase trials.

Finally, non-invasive biomarkers for monitoring bioavailability have gained significant attention in clinical trials. Techniques such as saliva and breath analysis have emerged as viable alternatives to traditional blood sampling, offering a less invasive and more patient-friendly way to monitor bioactive concentrations and metabolism. These methods enable continuous, real-time tracking of pharmacokinetics, reducing patient burden and improving the efficiency of bioavailability assessments during clinical trials. By enabling non-intrusive and real-time data collection, these innovations are transforming bioavailability evaluation, ensuring a more patient-centered approach in clinical research (Qiu et al. 2023, Chen 2024, Moore et al. 2024).

3.9.5.3 Advanced Analytical Techniques for Bioavailability Assessment

Advanced analytical techniques are essential for accurately assessing the bioavailability of dietary bioactives, allowing for precise quantification of bioactives, metabolites, and biomarkers in biological systems. Among these, mass spectrometry (MS) remains a foundational method, with recent advancements in high-resolution mass spectrometry (HRMS) and liquid chromatography–mass spectrometry (LC-MS) significantly improving sensitivity, specificity, and real-time monitoring capabilities. These innovations enable ultra-trace detection of bioactives and metabolites, enhancing the accuracy of pharmacokinetic assessments, particularly in complex biological matrices (He et al. 2015, Klont and Hopfgartner 2021, Fernández-Ochoa et al. 2022).

In addition to MS, nuclear magnetic resonance (NMR) spectroscopy has gained prominence in bioavailability studies due to its ability to provide structural and quantitative data on bioactives and their metabolites without requiring extensive sample preparation. Recent applications of high-field NMR have enabled real-time tracking of bioactive metabolism and interactions with biomolecules, offering complementary insights to MS-based techniques (Redel et al. 2015, Li et al. 2021).

Chromatographic techniques, including high-performance liquid chromatography (HPLC) and ultra-high-performance liquid chromatography (UHPLC), remain essential in bioavailability studies, particularly when combined with tandem mass spectrometry (LC-MS/MS). These methods provide high-throughput, robust, and reproducible analysis of bioactive concentrations in plasma, enabling precise bioavailability profiling of novel formulations, such as those using nanocarriers or controlled-release systems (Silva et al. 2014, He et al. 2015, Songvut et al. 2023). However, MS and NMR spectroscopy, while highly sensitive and precise, require significant expertise and advanced instrumentation. Additionally, MS is limited by matrix effects in complex biological samples, which can interfere with the detection of low-concentration analytes. Chromatographic techniques like HPLC can be time-consuming and may require extensive sample preparation, limiting their use for high-throughput analyses.

Fluorescence-based techniques, such as fluorescence resonance energy transfer (FRET) and surface plasmon resonance (SPR), are increasingly applied in bioavailability studies to assess bioactive binding interactions, permeability, and transport kinetics in real time. These methods are particularly useful for studying membrane transporters and receptor-mediated bioactive uptake, complementing MS and chromatographic approaches in evaluating absorption mechanisms (Mariani and Minunni 2014, Zhang et al. 2021b, Guo et al. 2023, Butt 2025). Nevertheless, they do have limitations. For FRET, the need for specific fluorophore pairs can restrict its use in complex systems, and issues such as photobleaching and low signal-to-noise ratios can reduce its reliability. SPR, while useful for studying binding interactions in real time, requires careful sensor surface preparation and can be sensitive to variations in environmental conditions, limiting its versatility (Piston and Kremers 2007, Špringer et al. 2025).

At last, integrating multi-omics approaches, including metabolomics and lipidomics, has further advanced bioavailability studies by identifying biomarkers linked to bioactive metabolism and distribution. When combined with MS and NMR, these techniques provide a systems-level understanding of bioactive absorption, metabolism, and excretion, allowing for more accurate predictions of bioavailability and interindividual variability (Fernández-Ochoa et al. 2022).

3.10 Outlook

Research on the bioavailability of bioactive compounds from fruits is rapidly advancing, driven by growing interest in their potential benefits for human health. The future outlook for this field is broad and includes everything from deepening our understanding of the mechanisms of absorption and metabolism of these compounds to developing innovative strategies to optimize their bioavailability and bioefficacy.

One of the main focuses of future research will be a detailed understanding of the mechanisms by which bioactive compounds are released from the food matrix, absorbed by the body, and metabolized. Elucidating these processes at the molecular level will allow identification of key factors that influence bioavailability, such as the chemical structure of the compounds, the presence of other dietary components, and the gut microbiota. This understanding is essential for creating effective optimization strategies.

Based on knowledge of these mechanisms, new approaches to optimize bioavailability will be developed. Among potential strategies, the use of nanotechnology to encapsulate bioactive compounds stands out, protecting them from degradation in the gastrointestinal tract. Additionally, new food processing methods may be developed to enhance the release of compounds from the food matrix, and synergistic food combinations will be identified to improve the absorption of these compounds, maximizing their health benefits.

Future research will also focus on personalized nutrition, considering that individual responses to the bioavailability of bioactive compounds may vary based on genetic, epigenetic, and environmental factors. Identifying biomarkers that can predict this individualized response will be fundamental to customizing nutritional recommendations, optimizing health benefits, and promoting a more precise approach to the use of bioactive compounds in the diet.

Although *in vitro* and animal model studies provide important information on bioavailability, confirming the health benefits of bioactive compounds depends on well-designed clinical studies. Conducting randomized controlled clinical trials will be a priority to evaluate the bioefficacy of these compounds in the prevention and treatment of chronic diseases, allowing for more robust validation of observed benefits.

Furthermore, research will seek new sources of bioactive compounds beyond traditional fruits. Exotic fruits, medicinal plants, and byproducts from the food industry represent promising sources that may contain bioactive compounds with great potential to promote human health. Identifying and characterizing these new compounds will open new possibilities for functional nutrition.

An integrated approach combining food science and nutrition will be essential for advancing research on the bioavailability of bioactive compounds. Collaboration among experts in food chemistry, molecular biology, physiology, and medicine will be crucial to translating research findings into tangible health benefits for the population. This multidisciplinary approach will not only optimize the utilization of bioactive compounds but also integrate this knowledge into effective strategies for health promotion.

3.11 Conclusion

The beneficial effects of fruit-derived bioactive compounds on health depend on various factors that influence their bioavailability. Compounds such as polyphenols, carotenoids, vitamins, and fibers exhibit antioxidant, anti-inflammatory, and metabolic regulatory activities, but their effectiveness is related to several processes, including release from the food matrix, digestion and intestinal absorption, hepatic metabolism, and interaction with the gut microbiota, which is unique to each individual. The consumption of isolated compounds or multicomponent fruit extracts does not guarantee their bioavailability and bioactivity due to the complexity of matrix release and synergistic effects among their constituents. To assess bioavailability, *in vitro* methods are used, such as gastrointestinal digestion simulations and cell culture assays—particularly with the Caco-2 cell line, which is derived from human colorectal adenocarcinoma. For this purpose, *in vivo* assays, including animal models and clinical studies, may also be employed, alongside advanced analytical approaches such as mass spectrometry and omics-based techniques. The integration of these methods allows for the tracking of compounds, metabolites, and metabolic pathways associated with exposure, enabling the elucidation of mechanisms of action and efficacy in human physiological contexts. Moreover, strategies such as appropriate processing methods and nanoencapsulation of bioactive compounds have emerged as promising solutions to enhance their stability, intestinal absorption, and biological activity, contributing directly to the prevention and management of various chronic diseases.

References

Abboud, K. Y., da Luz, B. B., Dallazen, J. L., Werner, M. F. de P., Cazarin, C. B. B., Maróstica, J. M. R. et al. (2019). Gastroprotective effect of soluble dietary fibres from yellow passion fruit (*Passiflora edulis f. flavicarpa*) peel against ethanol-induced ulcer in rats. Journal of Functional Foods 54(February): 552–558. https://doi.org/10.1016/j.jff.2019.02.003.

Akermi, S., Chaari, M., Elhadef, K., Sharma, A., Dey, A., Choudhary, A. et al. (2025). Bioactive compounds and cancer prevention: a nutritional approach. pp. 257–270. In Unleashing the Power of Functional Foods and Novel Bioactives. Academic Press. http://dx.doi.org/10.1016/B978-0-443-28862-3.00013-3.

Alasalvar, C., Chang, S. K., Kris-Etherton, P. M., Sullivan, V. K., Petersen, K. S., Guasch-Ferré, M. et al. (2023). Dried fruits: bioactives, effects on gut microbiota, and possible health benefits—an update. Nutrients 15(7): 1611. https://doi.org/10.3390/nu15071611.

Alice, B. N. and Richard, J. F. G. (2015). The scientific evidence for the role of milk protein-derived bioactive peptides in humans: A review. Journal of Functional Foods 17: 640–656. https://doi.org/10.1016/j.jff.2015.06.021.

Almoumen, A., Mohamed, H., Ayyash, M., Yuliarti, O., Kamleh, R., Al-marzouqi, H. et al. (2024). Harnessing date fruit pomace: Extraction of high fibre dietary ingredient and its impact on high fibre wheat flour dough. NFS Journal 35: 100178. https://doi.org/10.1016/j.nfs.2024.100178.

Alqudah, S. and Claesen, J. (2024). Mechanisms of gut bacterial metabolism of dietary polyphenols into bioactive compounds. Gut Microbes 16(1): 2426614. https://doi.org/10.1080/19490976.2024.2426614.

Altunay, N., Tuzen, M., Farooque, M., Reza, M. and Mogaddam, A. (2022). Food composition and analysis optimization of a rapid and sensitive ultrasound-assisted liquid-liquid microextraction using switchable hydrophilicity solvent for extraction of β-carotene in fruit juices and vegetables. Journal of Food Composition and Analysis 114(May): 104791. https://doi.org/10.1016/j.jfca.2022.104791.

Álvarez-Olguín, M. A., Beltrán-Barrientos, L. M., Hernandez-Mendoza, A., González-Córdova, A. F. and Vallejo-Cordoba, B. (2022). Current trends and perspectives on bioaccessibility and bioavailability of food bioactive peptides: *in vitro* and *ex vivo* studies. Journal of the Science of Food and Agriçulture 102(15): 6824–6834.

Amigo, L. and Hernández-Ledesma, B. (2020). Current evidence on the bioavailability of food bioactive peptides. Molecules 25(19): 4479. https://doi.org/10.3390/molecules25194479.

Anguera-Tejedor, M., Garrido, G., Garrido-Suárez, B. B., Ardiles-Rivera, A., Bistué-Rovira, À., Jiménez-Altayó, F. et al. (2024). Exploring the therapeutic potential of bioactive compounds from selected plant extracts of mediterranean diet constituents for cardiovascular diseases: a review of mechanisms of action, clinical evidence, and adverse effects. Food Bioscience 105487. https://doi.org/10.1016/j.fbio.2024.105487.

Arfaoui, L. (2021). Dietary plant polyphenols: Effects of food processing on their content and bioavailability. Molecules 26(10): 2959. https://doi.org/10.3390/moléculas26102959.

Arias, A., Feijoo, G. and Moreira, M. T. (2022). Exploring the potential of antioxidants from fruits and vegetables and strategies for their recovery. Innov. Food Sci. Emerg. Technol. 77(10.1016). https://doi.org/10.1016/j.ifset.2022.102974.

Aurélie, G., Lucie, É., Valérie, L., Silvestre D., Stéphanie B. and Monique A. (2012). Relevance and challenges in modeling human gastric and small intestinal digestion. Trends Biotechnol 30: 591–600. http://dx.doi.org/10.1016/j.tibtech.2012.08.001.

Aziz, E., Batool, R., Akhtar, W., Rehman, S., Shahzad, T., Malik, A. et al. (2020). Xanthophyll: Health benefits and therapeutic insights. Life Sciences 240(September 2019). https://doi.org/10.1016/j.lfs.2019.117104.

Barbosa, M. B. L., Assis, F. S. de, Santos, M. M. de O., Santos, P. C. dos, Moreira, F. I., Gomes, R. A. B. et al. (2024). Efeito do processamento e da adição de prebióticos nos teores de compostos bioativos em sucos de seriguela (*Spondias purpurea* L.). RBONE 18(116). https://www.rbone.com.br/index.php/rbone/article/view/2519.

Barroca, N. C. B., Della, G. S., Suchecki, D., García-Cairasco, N. and Umeoka, E. H. de L. (2022). Challenges in the use of animal models and perspectives for a translational view of stress and psychopathologies. Neuroscience & Biobehavioral Reviews 140: 104771. https://doi.org/10.1016/j.neubiorev.2022.104771.

Barros, H. D. D. F. Q. and Junior, M. R. M. (2019). Phenolic compound bioavailability using *in vitro* and *in vivo* models. pp. 113–126. In Bioactive Compounds. Woodhead Publishing. https://doi.org/10.1016/B978-0-12-814774-0.00006-2.

Bastos, D. H., Rogero, M. M. and Arêas, J. A. G. (2009). Effects of dietary bioactive compounds on obesity induced inflammation. Arquivos Brasileiros de Endocrinologia & Metabologia 53: 646–656. https://doi.org/10.1590/S0004-27302009000500017.

Baur, J. and Sinclair, D. (2006). Therapeutic potential of resveratrol: the *in vivo* evidence. Nature Reviews Drug Discovery 5: 493–506. https://doi.org/10.1038/nrd2060.

Bein, A., Shin, W., Jalili-Firoozinezhad, S., Park, M. H., Sontheimer-Phelps, A., Tovaglieri, A. et al. (2018). Microfluidic organ-on-a-chip models of human intestine. Cellular and Molecular Gastroenterology and Hepatology 5(4): 659–668. https://doi.org/10.1016/j.jcmgh.2017.12.010.

Beterams, A., De Paepe, K., Maes, L., Wise, I. J., De Keersmaecker, H., Rajkovic, A. et al. (2021). Versatile human *in vitro* triple coculture model coincubated with adhered gut microbes reproducibly mimics pro-inflammatory host-microbe interactions in the colon. The FASEB Journal 35(12). https://doi.org/10.1096/fj.202101135R.

Bian, Y., Lei, J., Zhong, J., Wang, B., Wan, Y., Li, J. et al. (2022). Kaempferol reduces obesity, prevents intestinal inflammation, and modulates gut microbiota in high-fat diet mice. Journal of Nutritional Biochemistry 99. https://doi.org/10.1016/j.jnutbio.2021.108840.

Bobasa, E. M., Phan, A. D. T., Netzel, M. E., Cozzolino, D. and Sultanbawa, Y. (2021). Hydrolysable tannins in Terminalia ferdinandiana Exell fruit powder and comparison of their functional properties from different solvent extracts. Food Chemistry 358(April): 129833. https://doi.org/10.1016/j.foodchem.2021.129833.

Bohn, T., Balbuena, E., Ulus, H., Iddir, M., Wang, G., Crook, N. et al. (2023). Carotenoids in health as studied by omics-related endpoints. Advances in Nutrition 14(6): 1538–1578. https://doi.org/10.1016/j.advnut.2023.09.002.

Borgonovi, T. F., Virgolin, L. B., Janzantti, N. S., Casarotti, S. N. and Penna, A. L. B. (2022). Compostos bioativos de frutas: Efeito sobre bactérias lácticas e sobre a microbiota intestinal. Food Research International 161: 111809. Elsevier Ltd. https://doi.org/10.1016/j.foodres.2022.111809.

Bouyahya, A., Omari, N. E., El Hachlafi, N., Jemly, M. E., Hakkour, M., Balahbib, A. et al. (2022). Chemical compounds of berry-derived polyphenols and their effects on gut microbiota, inflammation, and cancer. Molecules 27(10): 3286. https://doi.org/10.3390/molecules27103286.

BRASIL. (2014). Ministério da Saúde. Secretaria de Atenção Primária à Saúde. Guia alimentar para a população brasileira. 2. ed. Brasília: Ministério da Saúde.

Brett, G. M., Hollands, W., Needs, P. W., Teucher, B., Dainty, J. R., Davis, B. D. et al. (2009). Absorption, metabolism and excretion of flavanones from single portions of orange fruit and juice and effects of anthropometric variables and contraceptive pill use on flavanone excretion. Br J. Nutr. 101: 664–675. http://doi.org/10.1017/S000711450803081X.

Butt, M. A. (2025). Surface plasmon resonance-based biodetection systems: principles, progress and applications—a comprehensive review. Biosensors 15(1): 35. https://doi.org/10.3390/bios15010035.

Carbonell-Capella, J. M., Buniowska, M., Barba, F. J., Esteve, M. J. and Frìgola, A. (2014). Analytical methods for determining bioavailability and bioaccessibility of bioactive compounds from fruits and vegetables: a review. In Food Science and Food Safety 13: 155–171. http://doi.org/10.1111/1541-4337.12049.

De Carvalho, G. A. (2021). Compostos Fenólicos De Frutas Tropicais E Suas Bioatividades Na Obesidade E Diabetes Tipo II (Doctoral Dissertation, Universidade De São Paulo).

Carvalho, M. R., Yan, L. P., Li, B., Zhang, C. H., He, Y. L., Reis, R. L. et al. (2023). Gastrointestinal organs and organoids-on-a-chip: advances and translation into the clinics. Biofabrication 15(4): 042004. https://doi.org/10.1088/1758-5090/acf8fb.

Cen, Q., Fan, J., Hui, F., Hu, W., Yu, S., Liu et al. (2025). Influence and underlying mechanism of soluble dietary fiber derived from Ganoderma Lucidum-fermented sweet potato residue on the physicochemical-digestive characteristics of wheat starch. International Journal of Biological Macromolecules 142801. https://doi.org/10.1016/j.ijbiomac.2025.142801.

Chicón, R., Belloque, J., Alonso, E. and López-Fandiño, R. (2008). Immunoreactivity and digestibility of highpressure-treated whey proteins. International Dairy Journal 18(4): 367–376. http://doi.org/10.1016/j.idairyj.2007.11.010.

Cetinkaya, A., Yayla, S., Hurkul, M. M. and Ozkan, S. A. (2025). Comprehensive review on chromatographic analysis of flavonoids in fruits. Journal of Chromatography Open 7(December 2024): 100209. https://doi.org/10.1016/j.jcoa.2025.100209.

Chao, J., Huo, T. I., Cheng, H. Y., Tsai, J. C., Liao, J. W., Lee, M. S. et al. (2014). Gallic acid ameliorated impaired glucose and lipid homeostasis in high fat diet-induced NAFLD mice. PloS One 9(6): e96969. https://doi.org/10.1371/journal.pone.0096969.

Chacón-Ordóñez, T., Esquivel, P., Quesada, S., Jiménez, R. R., Cordero, A., Carle, R. et al. (2019). Mamey sapote fruit and carotenoid formulations derived thereof are dietary sources of vitamin A – A comparative randomized cross-over study. Food Research International 122(April): 340–347. https://doi.org/10.1016/j.foodres.2019.04.009.

Chandrika, B. B., Steephan, M., Kumar, T. S., Sabu, A. and Haridas, M. (2016). Hesperetin and naringenin sensitize HER2 positive cancer cells to death by serving as HER2 tyrosine kinase inhibitors. Life Sciences 160: 47–56. https://doi.org/10.1016/j.lfs.2016.07.007.

Chen, D. (2024). Biomarkers navigate drug development: Pharmacology, effectiveness and safety. Medicine in Drug Discovery 21: 100174. https://doi.org/10.1016/j.medidd.2023.100174.

Collins, S. L., Koo, I., Peters, J. M., Smith, P. B. and Patterson, A. D. (2021). Current challenges and recent developments in mass spectrometry–based metabolomics. Annual Review of Analytical Chemistry 14(1): 467–487. https://doi.org/10.1146/annurev-anchem-091620-015205.

Conlon, M. A. and Bird, A. R. (2015). The impact of diet and lifestyle on gut microbiota and human health. pp. 17–44. In Nutrients (Vol. 7, Issue 1). https://doi.org/10.3390/nu7010017.

Costa, C. A. R. da, Machado, G. G. L., Rodrigues, L. J., de Barros, H. E. A., Natarelli, C. V. L. and Boas, E. V. de B. V. (2023). Phenolic compounds profile and antioxidant activity of purple passion fruit's pulp, peel and seed at different maturation stages. Scientia Horticulturae 321(March). https://doi.org/10.1016/j.scienta.2023.112244.

Costa, J. and Ahluwalia, A. (2019). Advances and current challenges in intestinal *in vitro* model engineering: a digest. Frontiers in Bioengineering and Biotechnology 7. https://doi.org/10.3389/fbioe.2019.00144.

Costa, J. T. G., Lira, S. M., Holanda, M. O., Marques, C. G., Lima, C. L. S., Coelho, L. C. et al. (2020). Evidências do uso de frutas no tratamento complementar da Ansiedade e do Diabetes. Research, Society and Development 9(9): e464997432–e464997432. https://doi.org/10.33448/rsd-v9i9.7432.

Couri, S. Alimentos Funcionais. (2021). Disponível em: Alimentos funcionais - Portal Embrapa. Available at: https://www.embrapa.br/agencia-de-informacao-tecnologica/tematicas/tecnologia-de-alimentos/qualidade/nutricional/alimentos-funcionais. Accessed on March 10, 2025.

Couve, A., Gawande, S. and Kotwal, S. (2008). Cancer phytotherapeutics: role for flavonoids at the cellular level. Phytotherapy Research 22: 567–577. https://doi.org/10.1002/ptr.2283.

Daniela, R., Daniel, M., Gianfranco, G., Daniel, T., Didier, D., Joëlle, L. et al. (2010). The (193–209) 17-residues peptide of bovine β-casein is transported through Caco-2 monolayer. Molecular Nutrition & Food Research 54: 1428–1435. http://doi.org/10.1002/mnfr.200900443.

Danneskiold-Samsøe, N. B., Barros, H. D. F. Q., Santos, R., Bicas, J. L., Cazarin, C. B. B., Madsen, L. et al. (2019). Interplay between food and gut microbiota in health and disease. Food Research International 115: 23–31. https://doi.org/10.1016/j.foodres.2018.07.043.

Davey, M. W., Keulemans, J. and Swennen, R. (2006). Methods for the efficient quantification of fruit provitamin A contents. Journal of Chromatography A 1136(2): 176–184. https://doi.org/10.1016/j.chroma.2006.09.077.

Davi, G., Santilli, F. and Patrono, C. (2010). Nutraceuticals in diabetes and metabolic syndrome. Cardiovascular Therapeutics 28: 216–226. https://dx.doi.org/10.1111/j.1755-5922.2010.00179.x.

Davinelli, S., Medoro, A., Hu, F. B. and Scapagnini, G. (2025). Dietary polyphenols as geroprotective compounds: From Blue Zones to hallmarks of ageing. Ageing Research Reviews 108: 102733. https://doi.org/10.1016/j.arr.2025.102733.

De Ancos, B., Rodrigo, M. J., Sánchez-Moreno, C., Pilar Cano, M. and Zacarías, L. (2020). Effect of high-pressure processing applied as pretreatment on carotenoids, flavonoids and vitamin C in juice of the sweet oranges "Navel" and the red-fleshed "Cara Cara." Food Research International 132: 109105. https://doi.org/10.1016/j.foodres.2020.109105.

De Goede, J., Geleijnse, J. M., Boer, J. M. A., Kromhout, D. and Verschuren, W. M. M. (2010). Marine (n-3) fatty acids, fish consumption, and the 10-year risk of fatal and nonfatal coronary heart disease in a large population of Dutch adults with low fish intake. The Journal of Nutrition 140: 1023–1028. http://doi.org/10.3945/jn.109.119271.

de Lima Teixeira, A. P. (2020). Uso da antocianina na abordagem terapêutica da diabetes mellitus tipo 2: uma revisão bibliográfica. Trabalho De Conclusão De Curso. Universidade Federal De Minas Gerais.2020.

Del Giudice, R., Raiola, A., Tenore, G. C., Frusciante, L., Barone, A., Monti, D. M. et al. (2015). Antioxidant bioactive compounds in tomato fruits at different ripening stages and their effects on normal and cancer cells. Journal of Functional Foods 18: 83–94. https://doi.org/10.1016/j.jff.2015.06.060.

Demarque, D. P., Dusi, R. G., de Sousa, F. D., Grossi, S. M., Silvério, M. R., Lopes, N. P. et al. (2020). Mass spectrometry-based metabolomics approach in the isolation of bioactive natural products. Scientific Reports 10(1): 1051. https://doi.org/10.1038/s41598-020-58046-y.

Desloovere, A., Polderman, N., Renken-Terhaerdt, J., Shaw, V., Anderson, C., Greenbaum, L. A. et al. (2024). The management of dietary fiber intake in children with chronic kidney disease – clinical practice recommendations from the pediatric renal nutrition taskforce. Journal of Renal Nutrition 35(1): 207–220. https://doi.org/10.1053/j.jrn.2024.05.008.

Diana, H. O., Simon, B. L., Carsten, U. N., Bente, S., Lars, O., Flemming, S. J. et al. (2010). A quantitative structure-activity relationship for translocation of tripeptides via the human proton-coupled peptide transporter, hPEPT1 (SLC15A1). The AAPS Journal 12: 385–396. http://doi.org/10.1208/s12248-010-9195-z.

Dima, C., Assadpour, E., Nechifor, A., Dima, S., Li, Y. and Jafari, S. M. (2024). Oral bioavailability of bioactive compounds; modulating factors, *in vitro* analysis methods, and enhancing strategies. Critical Reviews in Food Science and Nutrition 64(24): 8501–8539. https://doi.org/10.1080/10408398.2023.2199861.

Din, S. R. U., Saeed, S., Khan, S. U., Kiani, F. A., Alsuhaibani, A. M. and Zhong, M. (2023). Bioactive compounds (BACs): A novel approach to treat and prevent cardiovascular diseases. Current Problems in Cardiology 48(7): 101664. https://doi.org/10.1016/j.cpcardiol.2023.101664.

Duan, X.-P., Qin, B.-D., Jiao, X.-D., Liu, K., Wang, Z. and Zang, Y.-S. (2024). New clinical trial design in precision medicine: discovery, development and direction. Signal Transduction and Targeted Therapy 9(1): 57. https://doi.org/10.1038/s41392-024-01760-0.

El-baz, F. K., Ali, S. I., Elgohary, R. and Salama, A. (2023). Natural β-carotene prevents acute lung injury induced by cyclophosphamide in mice. PLOS One 1–17. https://doi.org/10.1371/journal.pone.0283779.

Fernàndez-Garcìa, E., Carvajal-Lérida, I. and Pérez-Gàlvez, A. (2009). *In vitro* bioaccessibility assessment as a prediction tool of nutritional efficiency. Nutrition Research 29: 751–760. http://doi.org/10.1016/j.nutres.2009.09.016.

Fernández, I. M., Chaves, E. A., Maldonado, S. A. S., Takahashi, J. A., Alemán, R. S., Filho, A. A. de M. et al. (2020). Antimicrobial activity and acetilcolinesterase inhibition of oils and Amazon fruit extracts. Journal of Medicinal Plants Research 14(3): 88–97. https://doi.org/10.5897/JMPR2019.6790.

Fernández-Ochoa, Á., Cádiz-Gurrea, M. de la L., Fernández-Moreno, P., Rojas-García, A., Arráez-Román, D. and Segura-Carretero, A. (2022). Recent analytical approaches for the study of bioavailability and metabolism of bioactive phenolic compounds. Molecules 27(3): 777. https://doi.org/10.3390/molecules27030777.

Foltz, M., Meynen, E. E., Bianco, V., Platerink, C. V., Koning, T. M. and Kloek, J. (2007). Angiotensin converting enzyme inhibitory peptides from a lactotripeptide-enriched milk beverage are absorbed intact into the circulation. The Journal of Nutrition 137: 953–958. https://doi.org/10.1093/jn/137.4.953.

Fountzilas, E., Tsimberidou, A. M., Vo, H. H. and Kurzrock, R. (2022). Clinical trial design in the era of precision medicine. Genome Medicine 14(1): 101. https://doi.org/10.1186/s13073-022-01102-1.

Feiran, X., Lifeng, W., Xingrong, J. Jing, Z., Shi, Y., Jiayi, Shi. et al. (2017). Transepithelial transport of YWDHNNPQIR and its metabolic fate with cytoprotection against oxidative stress in human intestinal Caco-2 cells. Journal of Agricultural and Food Chemistry 65: 2056–2065. http://doi.org/10.1021/acs.jafc.6b04731.

Guarniz, W. A. S., Canuto, K. M., Ribeiro, P. R. V., Dodou, H. V., Magalhaes, K. N., Sá, K. M. et al. (2019). *Momordica Charantia* L. Variety from Northeastern Brazil: Analysis of antimicrobial activity and phytochemical components. Pharmacognosy Journal 11(6). https://doi.org/10.5530/pj.2019.11.203.

Gilbert, E. R., Wong, E. A. and Webb, K. E. (2008). Board-invited review: Peptide absorption and utilization: Implications for animal nutrition and health. Journal of Animal Science 86: 2135–2155. http://doi.org/10.2527/jas.2007-0826.

Gligorijević, N., Radomirović, M., Rajković, A., Nedić, O. and Ćirković Veličković, T. (2020). Fibrinogen increases resveratrol solubility and prevents it from oxidation. Foods 9(6): 780. https://doi.org/10.3390/foods9060780.

Guia alimentar da população brasileira. (2021). Disponível em: https://www.gov.br/saude/pt-br/assuntos/saude-brasil/publicacoes-para-promocao-a-saude/guia_alimentar_populacao_brasileira_2ed.pdf/view Acesso em 10 março de 2025.

Guo, C., Yuan, H., Yu, Y., Gao, Z., Zhang, Y., Yin, T. et al. (2023). FRET-based analysis on the structural stability of polymeric micelles: Another key attribute beyond PEG coverage and particle size affecting the blood clearance. Journal of Controlled Release 360: 734–746. https://doi.org/10.1016/j.jconrel.2023.07.026.

Han, D. G., Seo, S. W., Choi, E., Kim, M. S., Yoo, J. W., Jung, Y. et al. (2022). Impact of route-dependent phase-II gut metabolism and enterohepatic circulation on the bioavailability and systemic disposition of resveratrol in rats and humans: A comprehensive whole body physiologically-based pharmacokinetic modeling. Biomedicine & Pharmacotherapy 151: 113–141. https://doi.org/10.1016/j.biopha.2022.113141.

Harjirder, S., Aiqian, Y. and Maria, J. F. (2015). Aspects of food structures in the digestive tract. Current Opinion in Food Science 3: 85–93. https://doi.org/10.1016/j.cofs.2015.06.007.

Hassuna, N. A., Rabie, E. M., Mahd, W. K. M., Refaie, M. M. M., Yousef, R. K. M. and Abdelraheem, W. M. (2023). Antibacterial effect of vitamin C against uropathogenic *E. coli in vitro* and *in vivo*. BMC Microbiology 23(1): 1–12. https://doi.org/10.1186/s12866-023-02856-3.

He, J., Liu, X., Zhang, J., Wang, R., Cao, X. and Liu, G. (2024). Gut microbiome-derived hydrolases - an underrated target of natural product metabolism. In Frontiers in Cellular and Infection Microbiology 14. Frontiers Media SA. https://doi.org/10.3389/fcimb.2024.1392249.

Heckman-Stoddard, B. M. and Smith, J. J. (2014). Precision medicine clinical trials: defining new treatment strategies. Seminars in Oncology Nursing 30(2): 109–116. https://doi.org/10.1016/j.soncn.2014.03.004.

Hider, R. C., Liu, Z. D. and Khodr, H. H. (2001). Metal Chelation of Polyphenols (pp. 190–203). https://doi.org/10.1016/S0076-6879(01)35243-6.

Hodgkinson, A. J., Wallace, O. A. M., Boggs, I., Broadhurst, M. and Prosser, C. G. (2018). Gastric digestion of cow and goat milk: Impact of infant and young child *in vitro* digestion conditions. Food Chemistry 245: 275–81. http://doi.org/10.1016/j.foodchem.2017.10.028.

Hou, K., Wu, Z. X., Chen, X. Y., Wang, J. Q., Zhang, D., Xiao, C. et al. (2022). Microbiota in health and diseases. Signal Transduction and Targeted Therapy 7(1): 13. https://doi.org/10.1038/s41392-022-00974-4.

Hooper, I., Kroon, P. A., Rimm, E. B., Cohn, J. S., Harvey, I., Cornu, K. A. L. et al. (2008). Flavonoids, flavonoid-rich foods, and cardiovascular risk: a meta-analysis of randomized controlled trials. American Journal of Clinical Nutrition 88: 38–50. https://doi.org/10.1093/ajcn/88.1.38.

Hu, W., Shankar, P., Yao, Y., Su, X. and Kim, J. E. (2024). Effect of xanthophyll-rich food and supplement intake on visual outcomes in healthy adults and those with eye disease: a systematic review, meta-analysis, and meta-regression of randomized controlled trials. Nutrition Reviews 82(1): 34–46. https://doi.org/10.1093/nutrit/nuad037.

Hussain, A., Altamimi, M. A., Ramzan, M., Mirza, M. A. and Khuroo, T. (2023). GastroPlus- and HSPiP-oriented predictive parameters as the basis of valproic acid-loaded mucoadhesive cationic nanoemulsion gel for improved nose-to-brain delivery to control convulsion in humans. Gels 9(8): 603. https://doi.org/10.3390/gels9080603.

Irías-Mata, A., Jiménez, V. M., Steingass, C. B., Schweiggert, R. M., Carle, R. and Esquivel, P. (2018). Carotenoids and xanthophyll esters of yellow and red nance fruits (*Byrsonima crassifolia* (L.) Kunth) from Costa Rica. Food Research International 111(March): 708–714. https://doi.org/10.1016/j.foodres.2018.05.063.

Jensen, C. and Teng, Y. (2020). Is It time to start transitioning from 2D to 3D cell culture. Frontiers in Molecular Biosciences 7. https://doi.org/10.3389/fmolb.2020.00033.

Jia, D., Liao, G., Ye, B., Zhong, M., Huang, C. and Xu, X. (2024). Changes in fruit quality, phenolic compounds, and antioxidant activity of kiwifruit (*Actinidia eriantha*) during on-vine ripening. Lwt 206(July): 116564. https://doi.org/10.1016/j.lwt.2024.116564.

Jia, H., Ren, F. and Liu, H. (2024). Evaluation of bioaccessibility and bioavailability of dietary bioactives and their application in food systems. Food Bioscience 62: 105428. https://doi.org/10.1016/j.fbio.2024.105428.

Jimonet, P., Druart, C., Blanquet-Diot, S., Boucinha, L., Kourula, S., Le Vacon, F. et al. (2024). Gut microbiome integration in drug discovery and development of small molecules. Drug Metabolism and Disposition 52(4): 274–287. https://doi.org/10.1124/dmd.123.001605.

Jin, Y., Chen, L., Yu, Y., Hussain, M. and Zhong, H. (2023). Bioactive components in fruit interact with gut microbes. Biology 12(10): 1333. https://doi.org/10.3390/biology12101333.

Kang, J., Li, Y., Ma, Z., Wang, Y., Zhu, W. and Jiang, G. (2023). Protective effects of lycopene against zearalenone-induced reproductive toxicity in early pregnancy through anti-inflammatory, antioxidant and anti-apoptotic effects. Food and Chemical Toxicology 179(July): 113936. https://doi.org/10.1016/j.fct.2023.113936.

Karaźniewicz-Łada, M. and Głõwka, A. (2016). A review of chromatographic methods for the determination of water- and fat-soluble vitamins in biological fluids. Journal of Separation Science 39(1): 132–148. https://doi.org/10.1002/jssc.201501038.

Karima, B., François, M., Isabelle, C., Cécille, B., Julien, J., Isabelle, L. et al. (2012). *In vivo* digestion of infant formula in piglets: Protein digestion kinetics and release of bioactive peptides. British Journal of Nutrition 108: 2105–2114. https://doi.org/10.1017/S000711451200027X.

Kato, T., Watanabe, T., Nakamura, K. and Ando, S. (2019). Integration of *in silico* pharmacokinetic modeling approaches into *in vitro* dissolution profiles to predict bioavailability of a poorly soluble compound. Journal of Pharmaceutical Sciences 108(11): 3723–3728. https://doi.org/10.1016/j.xphs.2019.06.026.

Kim, E., Cui, J., Zhang, G. and Lee, Y. (2022). Physiological effects of green-colored food-derived bioactive compounds on cardiovascular and metabolic diseases. Applied Sciences 12(4): 1879. https://doi.org/10.3390/app12041879.

Kiriyama, Y., Tokumaru, H., Sadamoto, H., Kobayashi, S. and Nochi, H. (2024). Effects of phenolic acids produced from food-derived flavonoids and amino acids by the gut microbiota on health and disease. Molecules 29(21): 5102. https://doi.org/10.3390/molecules29215102.

Klont, F. and Hopfgartner, G. (2021). Mass spectrometry based approaches and strategies in bioanalysis for qualitative and quantitative analysis of pharmaceutically relevant molecules. Drug Discovery Today: Technologies 40: 64–68. https://doi.org/10.1016/j.ddtec.2021.10.004.

Knekt, P., Kumpulainen, J., Järvinen, R., Rissanen, H., Heliövaara, M., Reunanen, A. et al. (2002). Flavonoid intake and risk of chronic diseases. The American Journal of Clinical Nutrition 76(3): 560–568. https://doi.org/10.1093/ajcn/76.3.560.

Ko, C. W., Qu, J., Black, D. D. and Tso, P. (2020). Regulation of intestinal lipid metabolism: current concepts and relevance to disease. Nature Reviews Gastroenterology & Hepatology 17: 169–183. https://doi.org/10.1038/s41575-019-0250-7.

Ko, J. H., Sethi, G., Um, J. Y., Shanmugam, M. K., Arfuso, F., Kumar, A. P. et al. (2017). The role of resveratrol in cancer therapy. In International Journal of Molecular Sciences 18(12). MDPI AG. https://doi.org/10.3390/ijms18122589.

Koepsell, H. (2020). Glucose transporters in the small intestine in health and disease. Pflugers Arch - European Journal of Physiology 472: 1207–1248. https://doi.org/10.1007/s00424-020-02439-5.

Komin, A., Russell, L. M., Hristova, K..A. and Searson, P. C. (2017). Peptide-based strategies for enhanced cell uptake, transcellular transport, and circulation: Mechanisms and challenges. Advanced Drug Delivery Reviews 110–111: 52–64. http://doi.org/10.1016/j.addr.2016.06.002.

Kong, F. and Singh, R. P. (2008). Disintegration of solid foods in human stomach. Journal of Food Science 73: 67–80. https://doi.org/10.1111/j.1750-3841.2008.00766.x.

Kwak, J. H. and Kim, H. J. (2023). The Association between air pollutants exposure with pre- and hypertension by vitamin C intakes in korean adults: a cross-sectional study from the 2013–2016 korea national health and nutrition examination. Journal of Nutrition, Health and Aging 27(1): 21–29. https://doi.org/10.1007/s12603-022-1872-y.

Laur, L. M. and Tian, L. (2011). Provitamin A and vitamin C contents in selected California-grown cantaloupe and honeydew melons and imported melons. Journal of Food Composition and Analysis 24(2): 194–201. https://doi.org/10.1016/j.jfca.2010.07.009.

Leal, D., Ribciro, S. and Fonseca, K. Z. (2022). Interactions between dietary flavonoids and the gut microbiome: a comprehensive review. In Revista Inova Saúde, Criciúma, 14.

Lee, S. H., Latetia, M. V., Park, S., Harris, D. M. and Blanck, H. M. (2022). Morbidity and Mortality Weekly Report Adults Meeting Fruit and Vegetable Intake Recommendations-United States, 2019. https://www.myplate.gov/eat-healthy/fruits.

Lee, H. (2017). Transgenic pro-vitamin a biofortified crops for improving vitamin A deficiency and their challenges. The Open Agriculture Journal 11(1): 11–23. https://doi.org/10.2174/187433150171101 0011.

Leivas, C. L., Cantu-jungles, T. M., Luz, B. B. da, Werner, M. F. de P., Iacomini, M. and Cordeiro, L. M. C. (2023). Investigation of the chemical structure and analgesic and anti-inflammatory properties of polysaccharides that constitute the dietary fibers of soursop (*Annona muricata*) fruit. Food Research International 166(August 2022): 112588. https://doi.org/10.1016/j.foodres.2023.112588.

Les, F., Cásedas, G., Gómez, C., Moliner, C., Valero, M. S. and López, V. (2021). The role of anthocyanins as antidiabetic agents: from molecular mechanisms to *in vivo* and human studies. Journal of Physiology and Biochemistry 77(1): 109–131. https://doi.org/10.1007/s13105-020-00739-z.

Li, H., Deng, Z., Liu, R., Loewen, S. and Tsao, R. (2013). Carotenoid compositions of coloured tomato cultivars and contribution to antioxidant activities and protection against H_2O_2-induced cell death in H9c2. Food Chemistry 136(2): 878–888. https://doi.org/10.1016/j.foodchem.2012.08.020.

Li, H., Li, J., Liu, L., Zhang, Y., Luo, Y., Zhang, X. et al. (2016). Elucidation of the intestinal absorption mechanism of celastrol using the Caco-2 cell transwell model. Planta Medica 82(13): 1202–1207. https://doi.org/10.1055/s-0035-1568597.

Li, L., Yan, S., Liu, S., Wang, P., Li, W., Yi, Y. et al. (2023). In-depth insight into correlations between gut microbiota and dietary fiber elucidates a dietary causal relationship with host health. Food Research International 172(April): 113133. https://doi.org/10.1016/j.foodres.2023.113133.

Li, M., Xu, W. and Su, Y. (2021). Solid-state NMR spectroscopy in pharmaceutical sciences. TrAC Trends in Analytical Chemistry 135: 116152. https://doi.org/10.1016/j.trac.2020.116152.

Li, T., Guo, Q., Qu, Y., Li, Y., Liu, H., Liu, L. et al. (2022). Solubility and physicochemical properties of resveratrol in peanut oil. Food Chemistry 368: 130687. https://doi.org/10.1016/j.foodchem.2021.130687.

Lim, Y. Y., Lim, T. T. and Tee, J. J. (2007). Antioxidant properties of several tropical fruits: A comparative study. Food Chemistry 103(3): 1003–1008. https://doi.org/10.1016/j.foodchem.2006.08.038.

Lima, J. da S., dos Santos, I. L. and Fanaro, G. B. (2020). Desenvolvimento de Gel Energético a partir do Buriti (Mauritia flexuosa) e do Guaraná (Paullinia cupana) para Praticantes de Atividade Física: Development of carbohydrate gel from buriti (Mauritia flexuosa) and guaraná (Paullinia cupana). Revista Ensino, Saúde e Biotecnologia da Amazônia 03–09.

Lima, M. B., de Assis, F. S., de Oliveira Santos, M. M., dos Santos, P. C., Moreira, F. I., Gomes, R. A. B. et al. (2024). Efeito do processamento e da adição de prebióticos nos teores de compostos bioativos em sucos de seriguela (Spondias purpurea L.). RBONE-Revista Brasileira de Obesidade, Nutrição e Emagrecimento 18(116): 974–990.

Liu, T., Zhen, X., Zheng, N., Ma, Y., Zhang, Y., Lei, H. et al. (2025). Functional properties and application potential of black soybean meal insoluble dietary fiber remodeled by ultrasound combined with ultrahigh pressure. LWT 221(8326): 117594. https://doi.org/10.1016/j.lwt.2025.117594.

Liu, Y., Deng, J., Zhao, T., Yang, X., Zhang, J. and Yang, H. (2024). Bioavailability and mechanisms of dietary polyphenols affected by non-thermal processing technology in fruits and vegetables. Current Research in Food Science 8: 100715. https://doi.org/10.1016/j.crfs.2024.100715.

López-Gámez, G., Elez-Martínez, P., Martín-Belloso, O. and Soliva-Fortuny, R. (2021). Pulsed electric field treatment strategies to increase bioaccessibility of phenolic and carotenoid compounds in oil-added carrot purees. Food Chemistry 364: 130377. https://doi.org/10.1016/j.foodchem.2021.130377.

Lou, X., Xiong, J., Tian, H., Yu, H., Chen, C., Huang, J. et al. (2022). Effect of high-pressure processing on the bioaccessibility of phenolic compounds from cloudy hawthorn berry (*Crataegus pinnatifida*) juice. Journal of Food Composition and Analysis 110: 104540. https://doi.org/10.1016/j.jfca.2022.104540.

Lourdes, A. and Blanca, H. (2020). Current evidence on the bioavailability of food bioactive peptides. Molecules 25: 4479. https://doi.org/10.3390/molecules25194479.

Machado, P. G., Londero, D. S., Alves, E. E., Farias, C. A. A., Pudenzi, M. A., Barcia, M. T. et al. (2025). Combining data from two mass spectrometric analyzers coupled to liquid chromatography (LC-ESI-QToF-MS/MS and LC-ESI-QqQ-MS/MS) to characterize the free and bound phenolic compounds from peel, pulp, and seeds of guabiroba fruits. Microchemical Journal 209(January): 112757. https://doi.org/10.1016/j.microc.2025.112757.

Mamari, H. H. Al. (2021). Phenolic compounds: classification, chemistry, and updated techniques of analysis and synthesis. IntechOpen 11: 13. https://www.intechopen.com/books/advanced-biometric-technologies/liveness-detection-in-biometrics.

Mamun, M. A. Al, Rakib, A., Mandal, M., Kumar, S., Singla, B. and Singh, U. P. (2024). Polyphenols: Role in modulating immune function and obesity. Biomolecules 14(2). https://doi.org/10.3390/biom14020221.

Manach, C., Williamson, G., Morand, C., Scalbert, A. and Remesy, C. (2005). Bioavailability and bioefficacy of polyphenols in humans. I. Review of 97 bioavailability studies. Am. J. Clin. Nutr. 81(1 Suppl.): 230S–242. https://doi.org/10.1093/ajcn/81.1.230S.

Manzoor, M. F., Zeng, X.-A., Rahaman, A., Siddeeg, A., Aadil, R. M., Ahmed, Z. et al. (2019). Combined impact of pulsed electric field and ultrasound on bioactive compounds and FT-IR analysis of almond extract. Journal of Food Science and Technology 56(5): 2355–2364. https://doi.org/10.1007/s13197-019-03627-7.

Mariani, S. and Minunni, M. (2014). Surface plasmon resonance applications in clinical analysis. Analytical and Bioanalytical Chemistry 406(9–10): 2303–2323. https://doi.org/10.1007/s00216-014-7647-5.

Marrero, D., Pujol-Vila, F., Vera, D., Gabriel, G., Illa, X., Elizalde-Torrent, A. et al. (2021). Gut-on-a-chip: Mimicking and monitoring the human intestine. Biosensors and Bioelectronics 181: 113156. https://doi.org/10.1016/j.bios.2021.113156.

Martinez, P. F., Carvalho, M. R. D., Mendonça, M. L. M., Okoshi, M. P. and Oliveira-Junior, S. A. D. (2021). Antioxidant and anti-inflammatory effects of orange juice. Arquivos Brasileiros De Cardiologia 116: 1137–1138. https://doi.org/10.36660/abc.20210418.

Mateeva, A., Kondeva-Burdina, M., Peikova, L., Guncheva, S., Zlatkov, A. and Georgieva, M. (2023). Simultaneous analysis of water-soluble and fat-soluble vitamins through RP-HPLC/DAD in food supplements and brewer's yeast. Heliyon 9(1): e12706. https://doi.org/10.1016/j.heliyon.2022.e12706.

Matthews, D. M. and Laster, L. (1965). Absorption of protein digestion products: A review. Gut 6(5): 411–426. http://doi.org/10.1136/gut.6.5.411.

Mendes, M. K. A., dos Santos Oliveira, C. B. dos S., da Silva Medeiros, C. M. da S., dos Santos, L. R., Lopes Júnior, C. A. L. and Vieira, E. C. (2024). Challenges and strategies for bioavailability of curcumin. pp. 21–37. In Curcumin and Neurodegenerative Diseases. Traditional to Translational Medicines. Singapore: Springer Nature Singapore. https://doi.org/10.1007/978-981-99-7731-4_2.

Meng, W., Huang, X., Xing, L., Zhang, J. and Dong, P. (2025). Hepatoprotective activity of vitamin A and provitamin A purified from scallop (*Azumapecten farreri*) viscera waste using macroporous resin. Food Bioscience 65(December 2024). https://doi.org/10.1016/j.fbio.2025.106077.

Mercado-Mercado, G., Montalvo-González, E., González-Aguilar, G. A., Alvarez-Parrilla, E. and Sáyago-Ayerdi, S. G. (2018). Ultrasound-assisted extraction of carotenoids from mango (*Mangifera indica* L.'Ataulfo') by-products on *in vitro* bioaccessibility. Food Bioscience 21(September 2017): 125–131. https://doi.org/10.1016/j.fbio.2017.12.012.

Mesa, T. and Munné-Bosch, S. (2025). Vitamin E, total antioxidant capacity and potassium in tomatoes: A triangle of quality traits on the rise. Food Chemistry 475(August 2024). https://doi.org/10.1016/j.foodchem.2025.143375.

Moore, K., Grégoire, S., Eilstein, J., Delgado-Charro, M. B. and Guy, R. H. (2024). Reverse iontophoresis: noninvasive assessment of topical drug bioavailability. Molecular Pharmaceutics 21(1): 234–244. https://doi.org/10.1021/acs.molpharmaceut.3c00791.

Morais, M. G., Saldanha, A. A., Azevedo, L. S., Mendes, I. C., Rodrigues, J. P. C., Amado, P. A. et al. (2022). Antioxidant and anti-inflammatory effects of fractions from ripe fruits of Solanum lycocarpum St. Hil.(Solanaceae) and putative identification of bioactive compounds by GC–MS and LC-DAD-MS. Food Research International 156: 111145. https://doi.org/10.1016/j.foodres.2022.111145.

Mubeen, B., Aregbe, A. Y. and Ma, Y. (2025). Lactic acid fermentation of a novel Sea buckthorn-monk fruit beverage: Phenolic profile, volatile compounds, and antidiabetic and antihypertensive potential. Food Research International 207(October 2024): 116033. https://doi.org/10.1016/j.foodres.2025.116033.

Mueed, A., Aljahdali, S. M., Albalawi, M., Altarjami, L. R., Aljhdli, M. O., Al-Hoshani, N. et al. (2024). Phenolic compounds and biological activities of berberis fruit: Enhancing role on physiochemical and antioxidant properties of yogurt. Lwt 211(April). https://doi.org/10.1016/j.lwt.2024.116834.

Mullen, W., Archeveque, M. A., Edwards, C. A., Matsumoto, H. and Crozier, A. (2008). Biodisponibilidade e metabolismo de flavonóides do suco de laranja em humanos: impacto de um iogurte integral. J. Agric Food Chem. 56: 11157–11164. https://doi.org/10.1021/jf801974v.

Muntaha, S. T., Rakha, A., Rasheed, H., Fatima, I., Butt, M. S., Abdi, G. et al. (2025). Polyphenol-protein particles: A nutraceutical breakthrough in nutrition and food science. Journal of Agriculture and Food Research 19: 101641. https://doi.org/10.1016/j.jafr.2025.101641.

Muthulakshmi, M. V., Srinivasan, A. and Srivastava, S. (2023). Antioxidant green factories: toward sustainable production of vitamin E in plant *in vitro* cultures. ACS Omega 8(4): 3586–3605. https://doi.org/10.1021/acsomega.2c05819.

Nhan, M. T. and Quyen, D. K. (2023). Effects of heating process on kinetic degradation of anthocyanin and vitamin C on hardness and sensory value of strawberry soft candy. Acta Scientiarum Polonorum Technologia Alimentaria 22(2): 227–236. https://doi.org/10.17306/J.AFS.1133.

Nielsen, I. L., Chee, W. S., Poulsen, L., Offord-Cavin, E., Rasmussen, S. E., Frederiksen, H. et al. (2006). Bioavailability is improved by enzymatic modification of the citrus flavonoid hesperidin in humans: a randomized, double-blind, crossover trial. The Journal of Nutrition 136: 404–408. https://doi.org/10.1093/jn/136.2.404.

Neilson, A. P., Goodrich, K. M. and Ferruzzi, M. G. (2017). Bioavailability and metabolism of bioactive compounds from foods. pp. 301–319. In Nutrition in the Prevention and Treatment of Disease. Elsevier. https://doi.org/10.1016/B978-0-12-802928-2.00015-1.

Núñez-Gómez, V., Periago, M. J., Navarro-González, I., Campos-Cava, M. P., Baenas, N. and González-Barrio, R. (2021). Influence of raspberry and its dietary fractions on the *in vitro* activity of the colonic microbiota from normal and overweight subjects. Plant Foods for Human Nutrition 76: 494–500. https://doi.org/10.1007/s11130-021-00923-6.

Ojo, A. B. and Adanlawo, I. G. (2024). Antioxidant, antidiabetic, and anti-inflammatory activities of flavonoid-rich fractions of Solanum anguivi Lam. fruit: *In vitro* and *ex vivo* studies. Heliyon 10(11): e31895. https://doi.org/10.1016/j.heliyon.2024.e31895.

Oksuz, T., Tacer-Caba, Z., Nilufer-Erdil, D. and Boyacioglu, D. (2019). Changes in bioavailability of sour cherry (*Prunus cerasus* L.) phenolics and anthocyanins when consumed with dairy food matrices. Journal of Food Science and Technology 56(9): 4177–4188. https://doi.org/10.1007/s13197-019-03888-2.

Oliveira, A. M. de, Malunga, L. N., Perussello, C. A., Beta, T. and Ribani, R. H. (2020). Phenolic acids from fruits of Physalis angulata L. in two stages of maturation. South African Journal of Botany 131(2020): 448–453. https://doi.org/10.1016/j.sajb.2020.02.029.

Oliveira, L. G. D. (2024). Desenvolvimento de suplemento alimentar calórico proteico pronto para uso a base de baru (*Dipteryx Alata*), hidrolisado proteéico de semente de melancia (*Crimson Sweet*) e inajá (*Maximiliana Maripa*). 2024. 86f. Dissertação (Mestrado em Ciência e Tecnologia de Alimentos) – Universidade Federal do Tocantins, Programa de Pós-Graduação em Ciência e Tecnologia de Alimentos, Palmas.

Orsavová, J., Hlaváčová, I., Mlček, J., Snopek, L. and Mišurcová, L. (2019). Contribution of phenolic compounds, ascorbic acid and vitamin E to antioxidant activity of currant (*Ribes* L.) and gooseberry (*Ribes uva-crispa* L.) fruits. Food Chemistry 284(September 2018): 323–333. https://doi.org/10.1016/j.foodchem.2019.01.072.

Ozay-Arancioglu, I., Bekiroglu, H., Karadag, A., Saroglu, O., Tekin-Çakmak, Z. H. and Karasu, S. (2022). Effect of different drying methods on the bioactive, microstructural, and *in-vitro* bioaccessibility of bioactive compounds of the pomegranate arils. Food Science and Technology 42. https://doi.org/10.1590/fst.06221.

Özcan, M. M., Al Juhaimi, F., Uslu, N., Ahmed, I. A. M., Babiker, E. E., Osman, M. A. et al. (2020). Effect of sonication process of terebinth (*Pistacia terebinthus* L.) fruits on antioxidant activity, phenolic compounds, fatty acids and tocopherol contents. Journal of Food Science and Technology 57(6): 2017–2025. https://doi.org/10.1007/s13197-019-04235-1.

Özcan, M. M., Fındık, S., AlJuhaimi, F., Ghafoor, K., Babiker, E. F. E. and Adiamo, O. Q. (2019). The effect of harvest time and varieties on total phenolics, antioxidant activity and phenolic compounds of olive fruit and leaves. Journal of Food Science and Technology 56(5): 2373–2385. https://doi.org/10.1007/s13197-019-03650-8.

Padilha, C. V. S., Miskinis, G. A., Souza, M. E. A. O., Pereira, G. E., Oliveira, D., Luiz, M. T. B. et al. (2017). Rapid determination of flavonoids and phenolic acids in grape juices and wines by RP-HPLC/DAD: Method validation and characterization of commercial products of the new Brazilian varieties of grape. Food Chemistry 228: 106–115. http://dx.doi.org/10.1016/j.foodchem.2017.01.137.

Pan, L., Zhou, T., Maimaiti, K., Xu, H. and Ma, G. (2025). Anti-rheumatoid arthritis effects of Caragana acanthophylla Kom. on collagen-induced arthritis and the anti-inflammatory activity of polyphenols as main active components. Journal of Ethnopharmacology 346(March): 119637. https://doi.org/10.1016/j.jep.2025.119637.

Panse, N. and Gerk, P. M. (2022). The Caco-2 Model: Modifications and enhancements to improve efficiency and predictive performance. International Journal of Pharmaceutics 624: 122004. https://doi.org/10.1016/j.ijpharm.2022.122004.

Paulus. G. M. J., João, G., Antonius, M. V. K., Rosalinde, M. and Prescilla V. J. (2018). Evaluating human intestinal cell lines for studying dietary protein absorption. Nutrients 10: 322. http://doi.org/10.3390/nu10030322.

Paz, M., Gúllon, P., Barroso, M. F., Carvalho, A. P., Domingues, V. F., Gomes, A. M. et al. (2015). Brazilian fruit pulps as functional foods and additives: Evaluation of bioactive compounds. Food Chemistry 172: 462–468. https://doi.org/10.1016/j.foodchem.2014.09.102.

Pellegatti, M. (2012). Preclinical *in vivo* ADME studies in drug development: a critical review. Expert Opinion on Drug Metabolism & Toxicology 8(2): 161–172. https://doi.org/10.1517/17425255.2012.652084.

Peng, F., Ren, X., Du, B., Niu, K., Yu, Z., Yang, Y. et al. (2022). Insoluble dietary fiber of pear fruit pomace (*Pyrus ussuriensis* Maxim) consumption ameliorates alterations of the obesity-related features and gut microbiota caused by high-fat diet. Journal of Functional Foods 99(360). https://doi.org/10.1016/j.jff.2022.105354.

Perry, A., Rasmussen, H. and Johnson, E. J. (2009). Xanthophyll (lutein, zeaxanthin) content in fruits, vegetables and corn and egg products. Journal of Food Composition and Analysis 22(1): 9–15. https://doi.org/10.1016/j.jfca.2008.07.006.

Pimenta, J., Ribeiro, R., Almeida, R., Costa, P. F., da Silva, M. A. and Pereira, B. (2022). Organ-on-chip approaches for intestinal 3D *in vitro* modeling. Cellular and Molecular Gastroenterology and Hepatology 13(2): 351–367. https://doi.org/10.1016/j.jcmgh.2021.08.015.

Pinheiro, D. F., Maciel, G. M., Lima, N. P., Lima, N. F., Ribeiro, I. S. and Haminiuk, C. W. I. (2024). Impact of fruit consumption on gut microbiota: benefits, contaminants, and implications for human health. Trends in Food Science & Technology 104785. https://doi.org/10.1016/j.tifs.2024.104785.

Piston, D. W. and Kremers, G.-J. (2007). Fluorescent protein FRET: the good, the bad and the ugly. Trends in Biochemical Sciences 32(9): 407–414. https://doi.org/10.1016/j.tibs.2007.08.003.

Pullar, J. M., Carr, A. C. and Vissers, M. C. M. (2017). The roles of vitamin C in skin health. Nutrients 9(8). https://doi.org/10.3390/nu9080866.

Qiu, S., Cai, Y., Yao, H., Lin, C., Xie, Y., Tang, S. et al. (2023). Small molecule metabolites: discovery of biomarkers and therapeutic targets. Signal Transduction and Targeted Therapy 8(1): 132. https://doi.org/10.1038/s41392-023-01399-3.

Redel, Y. D., Escudey, M., Alvear, M., Conrad, J. and Borie, F. (2015). Effects of land use change on P bioavailability determined by chemical fractionation and 31P-NMR spectroscopy in a Nothofagus forest and adjacent grassland. Journal of Soil Science and Plant Nutrition, ahead, 0–0. https://doi.org/10.4067/S0718-95162015005000074.

Reis, R. C., Viana, E. de S., de Jesus, J. L., Lima, L. F., das Neves, T. T. and da Conceição, E. A. (2015). Bioactive compounds and antioxidant activity of improved varieties of papaya. Ciencia Rural 45(11): 2076–2081. https://doi.org/10.1590/0103-8478cr20140776.

Righard, L., Carlsson-Jonsson, A. and Nyberg, F. (2014). Enhanced levels of immunoreactive β-casomorphin-8 in milk of breastfeeding women with mastitis. Peptides 51: 54–58. http://doi.org/10.1016/j.peptides.2013.10.027.

Rodrigues, D. B. and Failla, M. L. (2021). Intestinal cell models for investigating the uptake, metabolism and absorption of dietary nutrients and bioactive compounds. Current Opinion in Food Science 41: 169–179. https://doi.org/10.1016/j.cofs.2021.04.002.

Rodrigues, D. B., Marques, M. C., Hacke, A., Filho, P. S. l., Cazarin, C. B. B. and Mariutti, L. R. B. (2022). Trust your gut: Bioavailability and bioaccessibility of dietary compounds. Currente Research in Food Science 5: 228–233. https://doi.org/10.1016/j.crfs.2022.01.002.

Rodríguez-Rodríguez, E., Sánchez-Prieto, M. and Olmedilla-Alonso, B. (2020). Assessment of carotenoid concentrations in red peppers (*Capsicum annuum*) under domestic refrigeration for three weeks as determined by HPLC-DAD. Food Chemistry 6(May): 100092. https://doi.org/10.1016/j.fochx.2020.100092.

Santé-publique France. (2025). Disponível em: Augmenter les fruits et légumes Acesso em 10 de março de 2025. Disponível em https://www.dge.de/gesunde-ernaehrung/gut-essen-und-trinken/dge-empfehlungen/%20 Acesso em 10 março de 2025.

Santos, D. C. dos, Oliveira, F. J. G. de, Sousa, T. L. de, Ribeiro, C. B. and Egea, M. B. (2022). Ameliorating effects of metabolic syndrome with the consumption of rich-bioactive compounds fruits from Brazilian Cerrado: a narrative review. Critical Reviews in Food Science and Nutrition 62(27): 7632–7649. https://doi.org/10.1080/10408398.2021.1916430.

Santos, L. S., Fernandes, C. C., Santos, L. S., de DEUS, I. P. B., de SOUSA, T. L. and Miranda, M. L. D. (2021). Ethanolic extract from *capsicum chinense* jacq. Ripe fruits: Phenolic compounds, antioxidant activity and development of biodegradable films. Food Science and Technology (Brazil) 41(2): 497–504. https://doi.org/10.1590/fst.08220.

Sardelli, L., Perottoni, S., Tunesi, M., Boeri, L., Fusco, F., Petrini, P. et al. (2021). Technological tools and strategies for culturing human gut microbiota in engineered *in vitro* models. Biotechnology and Bioengineering 118(8): 2886–2905. https://doi.org/10.1002/bit.27816.

Sari, F., Incedayi, B., Turkmen Erol, N., Akpinar, P. and Copur, O. U. (2024). Impact of ohmic heating and ultrasound pretreatments on oil absorption and other quality parameters of fried potato. Potato Research. https://doi.org/10.1007/s11540-024-09740-1.

Schieber, A. and Carle, R. (2005). Occurrence of carotenoid cis-isomers in food: Technological, analytical, and nutritional implications. Trends in Food Science & Technology 16: 416–422. https://doi.org/10.1016/j.tifs.2005.03.018.

See, X. Z., Yeo, W. S. and Saptoro, A. (2024). A comprehensive review and recent advances of vitamin C: Overview, functions, sources, applications, market survey and processes. Chemical Engineering Research and Design 206(February): 108–129. https://doi.org/10.1016/j.cherd.2024.04.048.

Shahidi, F., Ramakrishnan, V. V. and Oh, W. Y. (2019). Bioavailability and metabolism of food bioactives and their health effects: a review. Journal of Food Bioactives 6–41. https://doi.org/10.31665/JFB.2019.8204.

Shi, Q., Li, X., Zhu, D., Jiang, J. and Li, X. (2022). Comprehensive analysis of antibacterial and anti-hepatoma activity of metabolites from jujube fruit. Food Bioscience 47(July 2021). https://doi.org/10.1016/j.fbio.2021.101511.

Shim, S.-M., Yi, H.-L. and Kim, Y.-S. (2011). Bioaccessibility of flavonoids and total phenolic content in onions and its relationship with antioxidant activity. International Journal of Food Sciences and Nutrition 62(8): 835–838. https://doi.org/10.3109/09637486.2011.552881.

Shin, W. and Kim, H. J. (2022). 3D *in vitro* morphogenesis of human intestinal epithelium in a gut-on-a-chip or a hybrid chip with a cell culture insert. Nature Protocols 17(3): 910–939. https://doi.org/10.1038/s41596-021-00674-3.

Silva, L. N., Rigo, G. V., Silva, D. B., Carollo, C. A., Trentin, D. S., Silva, M. V. et al. (2020). Hydrolyzable tannins from Poincianella (*Caesalpinia*) microphylla fruits: Metabolite profiling and anti-Trichomonas vaginalis activity. Food Research International 134(April). https://doi.org/10.1016/j.foodres.2020.109236.

Silva, S. C. R., Rezende, G. R. de and Boralli, V. B. (2014). Quick and simple LC-MS/MS method for the determination of simvastatin in human plasma: application to pharmacokinetics and bioequivalence studies. Brazilian Journal of Pharmaceutical Sciences 50(3): 543–550. https://doi.org/10.1590/S1984-82502014000300013.

Sim, M., Hong, S., Jung, M. H., Choi, E. Y., Hwang, G. S., Shin, D. M. et al. (2025). Gut microbiota links vitamin C supplementation to enhanced mental vitality in healthy young adults with suboptimal vitamin C status: A randomized, double-blind, placebo-controlled trial. Brain, Behavior, and Immunity 128(March): 179–191. https://doi.org/10.1016/j.bbi.2025.03.032.

Sindhu, R. K., Goyal, A., Algın Yapar, E. and Cavalu, S. (2021). Bioactive compounds and nanodelivery perspectives for treatment of cardiovascular diseases. Applied Sciences 11(22): 11031. https://doi.org/10.3390/app112211031.

Singh, R. P. (2024). Progress and challenges in designing dynamic *in vitro* gastric models to study food digestion. Frontiers in Nutrition 11. https://doi.org/10.3389/fnut.2024.1399534.

Sjögren, E., Thörn, H. and Tannergren, C. (2016). *In silico* modeling of gastrointestinal drug absorption: predictive performance of three physiologically based absorption models. Molecular Pharmaceutics 13(6): 1763–1778. https://doi.org/10.1021/acs.molpharmaceut.5b00861.

Skardal, A. (2024). Grand challenges in organoid and organ-on-a-chip technologies. Frontiers in Bioengineering and Biotechnology 12. https://doi.org/10.3389/fbioe.2024.1366280.

Song, Q., Wang, Y., Huang, L., Shen, M., Yu, Y., Yu, Q. et al. (2021). Review of the relationships among polysaccharides, gut microbiota, and human health. Food Research International 140: 109858. https://doi.org/10.1016/j.foodres.2020.109858.

Songvut, P., Pholphana, N., Suriyo, T., Rangkadilok, N., Panomvana, D., Puranajoti, P. et al. (2023). A validated LC–MS/MS method for clinical pharmacokinetics and presumptive phase II metabolic pathways following oral administration of Andrographis paniculata extract. Scientific Reports 13(1): 2534. https://doi.org/10.1038/s41598-023-28612-1.

Soni, A. and Brightwell, G. (2022). Effect of hurdle approaches using conventional and moderate thermal processing technologies for microbial inactivation in fruit and vegetable products. Foods 11(12): 1811. https://doi.org/10.3390/foods11121811.

Špringer, T., Bocková, M., Slabý, J., Sohrabi, F., Čapková, M. and Homola, J. (2025). Surface plasmon resonance biosensors and their medical applications. Biosensors and Bioelectronics 278: 117308. https://doi.org/10.1016/j.bios.2025.117308.

Średnicka, P., Roszko, M. Ł., Popowski, D., Kowalczyk, M., Wójcicki, M., Emanowicz, P. et al. (2023). Effect of *in vitro* cultivation on human gut microbiota composition using 16S rDNA amplicon sequencing and metabolomics approach. Scientific Reports 13(1): 3026. https://doi.org/10.1038/s41598-023-29637-2.

Srivastava, S. K., Foo, G. W., Aggarwal, N. and Chang, M. W. (2024). Organ-on-chip technology: Opportunities and challenges. Biotechnology Notes 5: 8–12. https://doi.org/10.1016/j.biotno.2024.01.001.

Stephen, A. M., Champ, M. M. J., Cloran, S. J., Fleith, M., Van Lieshout, L., Mejborn, H. et al. (2017). Dietary fibre in Europe: Current state of knowledge on definitions, sources, recommendations, intakes and relationships to health. In Nutrition Research Reviews 30(2). https://doi.org/10.1017/S095442241700004X.

Sugimoto, S. and Sato, T. (2017). Establishment of 3D Intestinal Organoid Cultures from Intestinal Stem Cells. Humana Press (pp. 97–105). https://doi.org/10.1007/978-1-4939-7021-6_7.

Sun, H. and Tu, J. (2024). <scp>PKBOIN</scp> -12: A Bayesian Optimal Interval Phase I/<scp>II</scp> Design incorporating pharmacokinetics outcomes to find the optimal biological dose. Pharmaceutical Statistics. https://doi.org/10.1002/pst.2444.

Suwanwong, Y. and Boonpangrak, S. (2021). Phytochemical contents, antioxidant activity, and anticancer activity of three common guava cultivars in Thailand. European Journal of Integrative Medicine 42(January): 101290. https://doi.org/10.1016/j.eujim.2021.101290.

Tahergorabi, R. and Hosseini, S. V. (2017). Chapter 2 - Proteins, peptides, and amino acids. nutraceutical and functional food components. Effects of Innovative Processing Techniques pp. 15–38. http://doi.org/10.1016/B978-0-12-805257-0.00002-8.

Tiwari, P. and Mishra, K. P. (2020). Flavonoids sensitize tumor cells to radiation: molecular mechanisms and relevance to cancer radiotherapy. International Journal of Radiation Biology 96(3): 360–369. https://doi.org/10.1080/09553002.2020.1694193.

Traore, M. B., Sun, A., Gan, Z., Long, W. Y., Senou, H., Zhu, Y. et al. (2020). Assessing the impact of the combined application of ultrasound and ozone on microbial quality and bioactive compounds with antioxidant attributes of cabbage (*Brassica oleracea* L. Var. Capitata). Journal of Food Processing and Preservation 44(10). https://doi.org/10.1111/jfpp.14779.

Tsuchitani, T., Akiyoshi, T., Imaoka, A. and Ohtani, H. (2022). Mechanistic bottom-up estimation of passive drug absorption from the gastrointestinal tract: Comparison among primary cultured human intestinal cells, Caco-2 cells, artificial membrane, and animal scale-up. Int. Journal of Clinical Pharmacology and Therapeutics 60(05): 217–224. https://doi.org/10.5414/CP204172.

Van Norman, G. A. (2019). Limitations of animal studies for predicting toxicity in clinical trials. JACC: Basic to Translational Science 4(7): 845–854. https://doi.org/10.1016/j.jacbts.2019.10.008.

Vasconcelos, T., Prezotti, F., Araújo, F., Lopes, C., Loureiro, A., Marques, S. et al. (2021). Third-generation solid dispersion combining Soluplus and poloxamer 407 enhances the oral bioavailability of resveratrol. International Journal of Pharmaceutics 595: 120245. https://doi.org/10.1016/j.ijpharm.2021.120245.

Vieira, I. R. S. and Conte-Junior, C. A. (2024). Dietary bioactive compounds and human health: the role of bioavailability. Nutrients 17(1): 48. https://doi.org/10.3390/nu17010048.

Vincent, C., Mesa, T. and Munne - Bosch, S. (2020). Identification of a new variety of avocados (*Persea americana* mill. cv. bacon) with high vitamin E and impact of cold storage on tocochromanols composition. Antioxidants 9(5): 1–13. https://doi.org/10.3390/antiox9050403.

Walter, M., Aljayyoussi, G., Gerner, B., Rapp, H., Tautermann, C. S., Balazki, P. et al. (2024). *In silico* PK predictions in Drug Discovery: Benchmarking of strategies to integrate machine learning with empiric and mechanistic PK modelling. Clinical and Translational Science. https://doi.org/10.1101/2024.07.30.605777.

Wang, J., Chen, Y., Wang, H., Wang, S., Lin, Z., Zhao, L. et al. (2022b). Ethanol and blanching pretreatments change the moisture transfer and physicochemical properties of apple slices via microstructure and cell-wall polysaccharides nanostructure modification. Food Chemistry 381: 132274. https://doi.org/10.1016/j.foodchem.2022.132274.

Wang, L., Chen, J., Chen, W., Ruan, Z., Lou, H., Yang, D. et al. (2023). *In silico* prediction of bioequivalence of atorvastatin tablets based on GastroPlusTM software. BMC Pharmacology and Toxicology 24(1): 69. https://doi.org/10.1186/s40360-023-00689-4.

Wang, S., Fang, Y., Xu, Y., Zhu, B., Piao, J., Zhu, L. et al. (2022a). The effects of different extraction methods on physicochemical, functional and physiological properties of soluble and insoluble dietary fiber from *Rubus chingii* Hu. fruits. Journal of Functional Foods 93(February): 105081. https://doi.org/10.1016/j.jff.2022.105081.

Wang, W. and Zhou, H. (2016). Pharmacological considerations for predicting PK/PD at the site of action for therapeutic proteins. Drug Discovery Today: Technologies 21–22: 35–39. https://doi.org/10.1016/j.ddtec.2016.09.006.

Wang, Z., Peng, S., Peng, M., She, Z., Yang, Q. and Huang, T. (2020). Adsorption and desorption characteristics of polyphenols from *Eucommia ulmoides* Oliv. leaves with macroporous resin and its inhibitory effect on α-amylase and α-glucosidase. Annals of Translational Medicine 8(16): 1004–1004. https://doi.org/10.21037/atm-20-5468.

Wu, K., Kwon, S. H., Zhou, X., Fuller, C., Wang, X., Vadgama, J. et al. (2024b). Overcoming challenges in small-molecule drug bioavailability: a review of key factors and approaches. International Journal of Molecular Sciences 25(23): 13121. https://doi.org/10.3390/ijms252313121.

Wu, Y., Liu, Y., Jia, Y., Feng, C. H., Zhang, H., Ren, F. et al. (2024a). Effects of thermal processing on natural antioxidants in fruits and vegetables. Food Research International 114797.7.

Xiao, Q., Ye, S., Wang, H., Xing, S., Zhu, W., Zhang et al. (2024). Soluble sugar, organic acid and phenolic composition and flavor evaluation of plum fruits. Food Chemistry 24(August): 101790. https://doi.org/10.1016/j.fochx.2024.101790.

Xiong, Z., Liu, L., Jian, Z., Ma, Y., Li, H., Jin, X. et al. (2023). Vitamin E and multiple health outcomes: an umbrella review of meta-analyses. Nutrients 15(15). https://doi.org/10.3390/nu15153301.

Xu, F.-F., Chen, Y.-S., Lin, X.-Q., Zhong, A.-H., Zhao, M., Li, Y.-Q. et al. (2024). Bioaccessibility and bioavailability assessment of cadmium in rice: *In vitro* simulators with/without gut microbiota and validation through *in vivo* mouse and human data. Science of The Total Environment 953: 175980. https://doi.org/10.1016/j.scitotenv.2024.175980.

Xu, Q., Liu, Z., Liu, H., Zhao, F., Huang, X., Wu, Y. et al. (2018). Functional characterization of oligopeptide transporter 1 of dairy cows. Journal of Animal Science and Biotechnology 9: 7. http://doi.org/10.1186/s40104-017-0219-8.

Xu, X. K., Chen, Z. Y., Li, Z. Q., Zhang, Y. M., Liao, L. P., Zhou, Q. et al. (2016). Absorption mechanism of neobavaisoflavone in Caco-2 cell monolayer mode. China Journal of Chinese Materia Medica 41(15): 2922–2926. https://doi.org/10.4268/cjcmm20161528.

Xu, Q., Wu, Y., Liu, H., Xie, Y., Huang, X. and Liu, J. (2014). Establishment and characterization of an omasal epithelial cell model derived from dairy calves for the study of small peptide absorption. PLOS ONE 9: 88993. http://doi.org/10.1371/journal.pone.0088993.

Yan, Y., Li, Q., Shen, L., Guo, K. and Zhou, X. (2022). Chlorogenic acid improves glucose tolerance, lipid metabolism, inflammation and microbiota composition in diabetic db/db mice. Frontiers in Endocrinology 13. https://doi.org/10.3389/fendo.2022.1042044.

Yu, J., Chen, X., Chen, B., Mao, Y. and Shao, P. (2023). Lycopene in hydrophobic deep eutectic solvent with natural catalysts: A promising strategy to simultaneously promote lycopene Z-isomerization and extraction. Food Chemistry 426(February). https://doi.org/10.1016/j.foodchem.2023.136627.

Yu, W., Zhang, Y., Lu, Y., Ouyang, Z., Peng, J., Tu, Y. et al. (2025a). Recent research on the bioactivity of polyphenols derived from edible fungi and their potential in chronic disease prevention. Journal of Functional Foods 124(October 2024). https://doi.org/10.1016/j.jff.2024.106627.

Yu, Z., Hou, C., Shi, C., Huang, Y., Liu, H. and Wang, F. (2025b). Different solubility dietary fibers regulate thermogenesis in rats through modulating adipose tissue activity and colonic Clostridiales. Food Bioscience 68(December 2024): 106468. https://doi.org/10.1016/j.fbio.2025.106468.

Zahid, H. F., Ranadheera, C. S., Fang, Z. and Ajlouni, S. (2021). Utilization of mango, apple and banana fruit peels as prebiotics and functional ingredients. Agriculture (Switzerland) 11(7). https://doi.org/10.3390/agriculture11070584.

Zeng, F., Chen, M., Yang, S., Li, R., Lu, X., Zhang, L. et al. (2024). Distribution profiles of phenolic compounds in a cultivar of wampee (*Clausena lansium* (Lour.) Skeels) fruits and *in vitro* anti-inflammatory activity. Journal of Ethnopharmacology 319(P1): 117168. https://doi.org/10.1016/j.jep.2023.117168.

Zhang, D., Liu, J., Cheng, H., Wang, H., Tan, Y., Feng, W. et al. (2022b). Interações entre polissacarídeos e microbiota intestinal: uma revisão metabolômica e microbiana. Food Research International 160: 111653. https://doi.org/10.1016/j.foodres.2022.111653.

Zhang, F., He, F., Li, L., Guo, L., Zhang, B., Yu, S. et al. (2020b). Bioavailability based on the gut microbiota: a new perspective. Microbiology and Molecular Biology Reviews 84(2): 10–1128. https://doi.org/10.1128/MMBR.00072-19.

Zhang, G., Wang, Y., Zhang, Z., He, Z., Liu, Y. and Fu, Q. (2021b). FRET imaging revealed that nanocrystals enhanced drug oral absorption by dissolution rather than endocytosis: A case study of coumarin 6. Journal of Controlled Release 332: 225–232. https://doi.org/10.1016/j.jconrel.2021.02.025.

Zhang, J., Liu, P., Huang, S., Chen, Q., Wang, X. and Liu, H. (2024a). Association between rheumatoid arthritis and serum vitamin C levels in Adults: Based on the National health and Nutrition Examination survey database. Preventive Medicine Reports 44(July 2023): 102793. https://doi.org/10.1016/j.pmedr.2024.102793.

Zhang, M., Liu, Y. and Chen, Y.-G. (2020a). Generation of 3D human gastrointestinal organoids: principle and applications. Cell Regeneration 9(1): 6. https://doi.org/10.1186/s13619-020-00040-w.

Zhang, S., Li, Z. M., Feng, Y. C., Wang, C. Y. and Zhang, D. J. (2023a). Processing enhances coix seed prolamins structure and releases functional peptides after digestion: *In silico* and *in vitro* studies. Foods 12(13): 2500.

Zhang, Q., Xu, Y., Bukvicki, D., Peng, Y., Li, F., Zhang, Q. et al. (2024c). Phenolic compounds in dietary target the regulation of gut microbiota: Role in health and disease. Food Bioscience 62: 105107. https://doi.org/10.1016/j.fbio.2024.105107.

Zhang, W., Lin, M., He, H., Wang, Y., Wang, J. and Liu, H. (2023b). Toward achieving rapid estimation of vitamin C in citrus peels by NIR spectra coupled with a linear algorithm. Molecules 28(4). https://doi.org/10.3390/molecules28041681.

Zhang, Y., Lai, Y. and Zheng, M. (2024b). Ultrasound-assisted intensification of Pickering interfacial biocatalysis preparation of vitamin A aliphatic esters. Ultrasonics Sonochemistry 107(April): 106929. https://doi.org/10.1016/j.ultsonch.2024.106929.

Zhang, Z., Li, X., Sang, S., McClements, D. J., Chen, L., Long, J. et al. (2022a). Polyphenols as plant-based nutraceuticals: health effects, encapsulation, nano-delivery, and application. Foods 11(15): 2189.

Zhang, Z., Nie, M., Xiao, Y., Zhu, L., Gao, R., Zhou, C. et al. (2021a). Positive effects of ultrasound pretreatment on the bioaccessibility and cellular uptake of bioactive compounds from broccoli: Effect on cell wall, cellular matrix and digesta. LWT 149: 112052. https://doi.org/10.1016/j.lwt.2021.112052.

Zhou, Y., Fu, R., Yang, M., Liu, W. and Tong, Z. (2023). Lycopene suppresses gastric cancer cell growth without affecting normal gastric epithelial cells. The Journal of Nutritional Biochemistry 116: 109313. https://doi.org/10.1016/j.jnutbio.2023.109313.

Zhu, C., Zhang, M., Wang, S., Gao, X., Lin, T., Yu, J. et al. (2023). Phenolic compound profile and gastrointestinal action of Solanaceae fruits: Species-specific differences. Food Research International 170(December 2022): 112968. https://doi.org/10.1016/j.foodres.2023.112968.

Zhu, K., Zheng, X., Ye, J., Jiang, Q., Chen, H., Mei, X. et al. (2020). Building the synthetic biology toolbox with enzyme variants to expand opportunities for biofortification of provitamin a and other health-promoting carotenoids. Journal of Agricultural and Food Chemistry 68(43): 12048–12057. https://doi.org/10.1021/acs.jafc.0c04740.

Zhu, Y., Ji, X., Yuen, M., Yuen, T., Yuen, H., Wang, M. et al. (2022). Effects of ball milling combined with cellulase treatment on physicochemical properties and in vitro hypoglycemic ability of sea buckthorn seed meal insoluble dietary fiber. Frontiers in Pharmacology 8(January): 1–12. https://doi.org/10.3389/fnut.2021.820672.

Zhuang, J., Zhu, J., Cheung, P. C. K. and Li, C. (2024). The physical and chemical interactions between starch and dietary fiber: Their impact on the physicochemical and nutritional properties of starch. Trends in Food Science and Technology 149(February): 104566. https://doi.org/10.1016/j.tifs.2024.104566.

Zou, H., Banerjee, P., Leung, S. S. Y. and Yan, X. (2020). Application of pharmacokinetic-pharmacodynamic modeling in drug delivery: development and challenges. Frontiers in Pharmacology 11. https://doi.org/10.3389/fphar.2020.00997.

Zuraini, N. Z. A., Sekar, M., Wu, Y. S., Gan, S. H., Bonam, S. R., Mat Rani, N. N. I. et al. (2021). Promising nutritional fruits against cardiovascular diseases: An overview of experimental evidence and understanding their mechanisms of action. Vascular Health and Risk Management 739–769. https://doi.org/10.2147/VHRM.S328096.

Chapter 4

Nutritional Composition and Bioactive Potential of Brazilian Nuts

Carla Mariana da Silva Medeiros,[1] *Mahendra Rai*[2] and *Edivan Carvalho Vieira*[1,*]

4.1 Introduction

Nuts are dietary foods with high nutritional value and positive health effects (Zhai et al. 2025). Consumed either raw or roasted, nuts are incorporated into the diet as a nutritional supplement or snack and are considered a good source of unsaturated fatty acids, such as oleic acid, with their consumption being recommended due to their health benefits (Polmann et al. 2021). In addition to their desirable lipid profile, nuts exhibit an interesting mineral composition, a high protein content, and phenolic compounds associated with antioxidant activity (Polmann et al. 2021, Campidelli et al. 2020).

Brazil has a rich biodiversity, boasting one of the most significant diversities of native plants in the world, along with an extensive genetic variety (Polmann et al. 2021). The Brazilian vegetation is divided into six ecoregions: the Amazon, Cerrado, Atlantic Forest, Caatinga, Pampas, and Pantanal (Lucas et al. 2023). Among these, the Amazon and Cerrado—biomes of great ecological importance—harbor a wide variety of nuts and seeds, playing a key role in the large-scale production of certain nuts, such as cashew and Brazil nuts, and contributing significantly to the Brazilian economy (Polmann et al. 2021).

[1] Department of Chemistry, Federal University of Piaui, Piaui, Brazil.
[2] Department of Biotechnology, Sant Gadge Baba Amravati University, Amravati, Maharashtra, India.
* Corresponding author: edivanvieira@ufpi.edu.br

DOI: 10.1201/9781003668169-6

In this context, studies have been conducted to explore Brazilian nuts and characterize their nutritional profile, correlating their consumption with health benefits. These efforts aim to expand dietary sources of essential nutrients necessary for maintaining quality of life. Among these nuts, baru (*Dipteryx alata* Vog.), cashew (*Anacardium occidentale* L.), Brazil nut (*Bertholletia excelsa*), chicha (*Sterculia striata*), gurgueia (*Dipteryx lacunifera* Ducke), pracaxi (*Pentaclethra macroloba*), and sapucaia (*Lecythis pisonis* Cambess) have been studied due to their high protein content, essential macro- and micronutrients, and bioactive compounds with antioxidant potential (Alves-Santos et al. 2021, Goetz et al. 2023, Polmann et al. 2021, Teixeira et al. 2018).

Therefore, this chapter aims to compile data on the nutritional composition and bioactive potential of baru, cashew, Brazil nut, chicha, gurgueia, pracaxi, and sapucaia nuts, with a focus on phenolic compounds. Here, we present macronutrient composition, macro- and micronutrient mineral content, phenolic profile, and antioxidant potential through hydrogen atom transfer (HAT) and single-electron transfer (SET) reactions. The objective of this chapter is to provide readers with a comprehensive overview of the nutritional and bioactive potential of Brazilian nuts, highlighting their value within Brazil's biodiversity and emphasizing the need for further studies to explore their full potential.

4.2 Brazilian Nuts

4.2.1 Baru Nut

The baruzeiro (*Dipteryx alata* Vog.) is a native plant from the Brazilian Cerrado, belonging to the Fabaceae family (Oliveira-Alves et al. 2020). Its tree can reach heights between 8 and 15 meters, flowering from November to May and producing fruit from July to October. In a productive harvest, a mature tree can yield up to 150 kg of fruit over two years (Lemos et al. 2012, Polmann et al. 2021). The fruit, known as baru nut, has a brown color and an oval shape, with dimensions ranging from 4 to 5 centimeters in length (Alves-Santos et al. 2021).

Classified as a drupe, the fruit has a fleshy layer surrounding an edible seed, and its flavor is reminiscent of peanuts. It is widely consumed in Brazil (Polmann et al. 2021). Due to its distinctive taste and nutritional qualities, such as high lipid content, unsaturated fatty acids, and proteins, baruzeiro nuts are used in the production of cookies, cereals, granola, and baked goods (Alves-Santos et al. 2021, Lima et al. 2021).

Economically, the baru nut represents an important source of income for local populations as well as for the global market, with 50% of the Brazilian production being exported (Carvalho et al. 2024). Moreover, the harvest of baru nut is carried out by local communities through sustainable management practices, contributing to the environmental preservation of the Brazilian Cerrado.

4.2.2 Brazil Nut

The Brazil nut (*Bertholletia excelsa*) is an emerging species native to the Amazon region and belongs to the Lecythidaceae family (Fontoura et al. 2023). The tree can reach up to 60 meters in height, and as a result, local extractivists wait for the fruit to fall to collect it. The fruits can weigh up to 1 kg and contain 12 to 20 nuts, which are enclosed in a thick, woody shell with a brown-grayish color (Chunhieng et al. 2004). The Brazil nut itself measures between 3 and 5 centimeters in length and has a slightly yellowish color.

Nutritionally, the Brazil nut is known for its high selenium content, potential anti-inflammatory and antioxidant activity, and its consumption is associated with improvements in lipid profiles (Hou et al. 2021). In addition to its nutritional benefits, the pleasant flavor of the Brazil nut makes it highly attractive for incorporation into diets, either in its raw form or integrated into biocomposites, flour, and oil (Paiva et al. 2023b).

In terms of profitability, most of the Brazil nuts are produced in conservation units, primarily by traditional peoples and communities, including extractivists, Indigenous people, and quilombolas. This nut is a significant economic resource and contributes to the subsistence of traditional populations (Fonseca et al. 2021).

4.2.3 Cashew Nut

Cashew tree (*Anacardium occidentale* L.) is a plant from the Anacardiaceae family, native to Latin America, that has continuously expanded since the colonial era. It is characterized by the significant agroindustrial culture of its fruit, especially in Brazil, Africa, India, and Vietnam (Zafeer and Bhat 2023). The cashew nut, the fruit of the cashew tree, is a drupe consisting of a dicotyledonous seed surrounded by a testa and protected by an external shell that constitutes approximately 70% of the fruit (Cruz-Reina et al. 2024, Zafeer and Bhat 2023). Its physical appearance is commonly associated with a kidney or heart shape, with colors ranging from grayish brown to green.

The pleasant flavor and the associated health benefits have been key factors contributing to the increased consumption of cashew nuts. Benefits such as antioxidant potential and the supplementation of minerals and vitamins expand its application beyond human consumption of roasted nuts and its inclusion in dietary snacks, such as its incorporation into ruminant diets for nutritional supplementation (Goetz et al. 2023, Akintunde et al. 2022, Semporé et al. 2021).

Among all nuts, cashew is the most produced and consumed. The cultivation of this fruit represents significant economic, social, and environmental interest at the global level. Global production of raw cashew nuts is estimated at 132.7 thousand tons in 2024, marking a 13% increase compared to 2023 (Fagundes 2024). In Brazil, the Northeast region stands out in cashew nut production, with the states of Ceará and Piauí accounting for 55.2% and 21.4% of total production and export sales, respectively (Fagundes 2024). The cultivation of this fruit holds considerable economic, social, and environmental significance globally.

4.2.4 Chicha Nut

The chicha tree (*Sterculia striata*) is a Brazilian endemic species found in the North, Northeast, Central-West, and Southeast regions of Brazil (Campos et al. 2022). Belonging to the Malvaceae family, the chichá tree can reach up to 30 meters in height, with an erect stem, thick branches, and fruit that matures between September and December (Policarpi et al. 2018, Silva and Fernandes 2011). The fruit, known as the chichá nut, is enclosed in a woody capsule that contains 8 to 20 seeds, which remain closed until the nut matures. These seeds are surrounded by a thin, dark outer membrane, easily removable, and a harder inner shell that contains the seed (Polmann et al. 2021, Quesada et al. 2021).

The seeds of chicha are oval, approximately 2 centimeters in length, and dark in color. They can be consumed raw, cooked, or roasted (Santos et al. 2020). Nutritionally, chicha nuts are lower in fat compared to other commonly consumed nuts, such as cashews and Brazil nuts. Additionally, they are a good source of protein, fiber, and minerals (de Britto Policarpi et al. 2018). The chicha nut is being explored beyond raw consumption in culinary applications. Carvalho et al. (2011) investigated the formulation and sensory acceptability of cereal bars made with chicha nuts, demonstrating its potential use in food recipes.

Economically, data related to the production, sale, and export of chicha nuts are limited, as the fruit is still little known and underexplored. It is classified as a non-conventional nut and, despite its interesting nutritional profile, is mainly consumed by local populations in the regions where it is cultivated, without large-scale production.

4.2.5 Gurgueia Nut

The gurgueia (*Dipteryx lacunifera* Ducke) is a species from the Leguminosae family distributed across the Central-Northern region of Brazil, particularly in the Southwest of Piauí and Maranhão (Polmann et al. 2024). Depending on the genotype, the gurgueia tree can reach heights ranging from 2 to 13 meters. Its fruit, similar to the baru, has a resistant, light brown unilocular capsule that encloses an edible dark brown almond, known as the gurgueia nut (Polmann et al. 2021, Carvalho et al. 2008).

The nut has a rounded shape, approximately 2 to 3 centimeters in length, with an average weight of 2 grams. It is rich in lipids, essential oils, proteins, minerals, and carbohydrates, making it an attractive option for human consumption (Ribeiro et al. 2012). Typically, the gurgueia nut is consumed raw after roasting, but it can also be used in confectionery for the production of sweets, cookies, and cereal bars, as well as in the pharmaceutical and cosmetic industries (Campos et al. 2024, Cavalcante et al. 2015).

Like the chicha nut, the gurgueia nut is a non-conventional nut, extensively utilized in the regions where it is cultivated but remaining largely unknown in other areas.

4.2.6 Pracaxi Nut

The pracaxi tree (*Pentaclethra macroloba*), Fabaceae, is an oilseed tree native to the Amazon region with wide geographical distribution, extending across several countries, such as Brazil, Peru, Colombia, Venezuela, Trinidad and Tobago, Guyana, and Suriname in South America, as well as Nicaragua, Honduras, Jamaica, Cuba, Costa Rica, and Panama in Central America (Teixeira et al. 2020). The tree ranges in height from 8 to 14 meters and produces pod-shaped fruits that can reach up to 25 centimeters in length, containing 4 to 8 seeds within (Silva et al. 2023, Polmann et al. 2021).

The seeds, known as pracaxi nuts, are flattened and brown. The consumption of these seeds is mainly associated with the oil, which constitutes approximately 45 to 48% of their composition. This oil is rich in fatty acids, especially unsaturated ones, and exhibits potential bioactivity, including wound-healing, anti-inflammatory, and larvicidal activities (Mohammadnezhad et al. 2023, Teixeira et al. 2020).

The oil from pracaxi nuts has multiple uses in the food, cosmetic, and pharmaceutical industries. The nuts are consumed raw, while the oil is used as a substitute for cooking oils and as an ingredient in soaps, shampoos, and creams. However, the commercialization of these products remains on a small scale.

4.2.7 Sapucaia Nut

The sapucaia tree (*Lecythis pisonis* Cambess) is a plant native to the Brazilian Amazon biome and belongs to the Lecythidaceae family (Rosa et al. 2020). The tree can reach an average height of 12 to 30 meters, and its fruiting period depends on the region, typically occurring between June and December (Polmann et al. 2021). The fruits are large with a hard shell, enclosing light brown and white seeds, and contain between 30 to 50 nuts. The sapucaia nuts weigh between 1 to 2.5 kilograms, with an average yield of 75 kilograms of nuts per tree (Demoliner et al. 2018).

The sapucaia nut has a pleasant flavor and is widely consumed raw by local populations. In addition to its taste, the nut has interesting properties for human health, as its almonds are rich in proteins, essential fatty acids, and minerals (Rosa et al. 2020, Demoliner et al. 2018). Thus, beyond raw consumption, the nut also holds potential for use in its defatted flour and in milk made from the nut cake (Demoliner et al. 2020, Teixeira et al. 2020).

Despite its potential, the sapucaia nut is relatively unknown and considered a non-conventional nut. As a result, its production is limited to the local populations in the North and Northeast of Brazil (Polmann et al. 2021). Figure 1 shows the distribution of the nuts studied in this chapter in the Brazilian biomes.

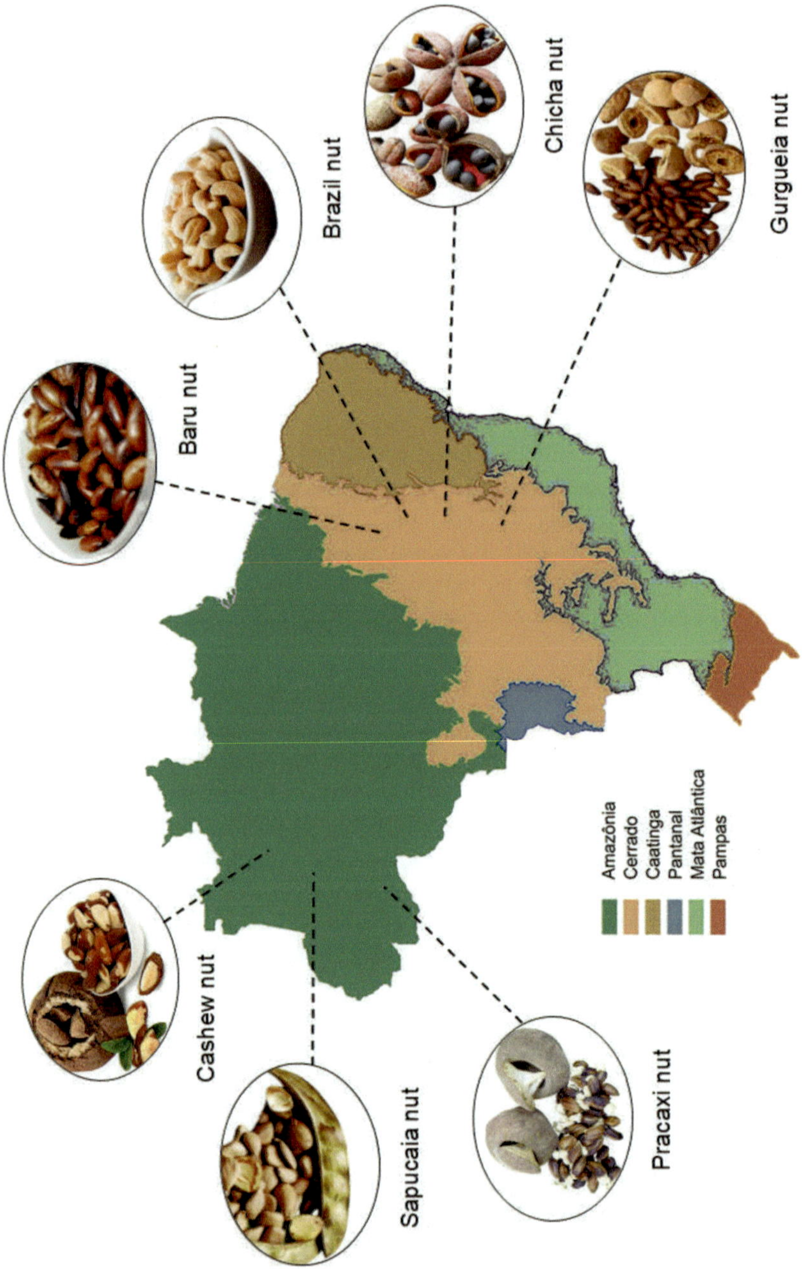

Figure 1. Brazilian biomes and the distribution of baru, Brazil, cashew, chicha, gurgueia, pracaxi, and sapucaia nut.

4.3 Nutritional Composition

4.3.1 The Importance of Macronutrients and Micronutrients in Nuts

Nuts are highly nutritious foods, providing a rich combination of macronutrients and micronutrients that are essential for human health. Macronutrients—carbohydrates, proteins, and lipids—are compounds that must be consumed in large quantities to provide energy for the body (Goveia 2007). In contrast, micronutrients are nutrients required in smaller amounts but are crucial for the proper functioning of the body, such as vitamins and minerals (Goveia 2007). The composition of nuts in terms of macronutrients and micronutrients varies depending on factors such as the region of origin, year of production, soil composition, fertilizers, harvest conditions, and others (Polmann et al. 2021). Studies have shown that including nuts in the human daily diet is beneficial for health. Therefore, studies to determine the levels of macro- and micronutrients in nuts are of significant interest in evaluating how well the consumption of these foods meets or exceeds daily nutritional needs. Table 1 summarizes the nutritional profile of the baru, Brazil, cashew, chicha, gurgueia, pracaxi, and sapucaia nuts.

4.3.1.1 Macronutrients in Nuts

Dietary macronutrients—proteins, lipids, and carbohydrates—are essential substances that significantly impact human health and life expectancy by providing energy and fulfilling structural functions in the body (Wang et al. 2014). Studies suggest that macronutrient composition in foods also plays a central role in appetite, satiety, and cognitive performance, particularly in mood regulation (Wang et al. 2014).

Among macronutrients, carbohydrates—also known as sugars or saccharides—are the most abundant class of metabolites worldwide and are essential for living organisms. They perform fundamental structural roles in cells, supply metabolic energy, and contribute to cell surface recognition processes (Chaves et al. 2016). The primary dietary source of carbohydrates is food consumption, obtained from both animal- and plant-based origins. Nuts contain a moderate amount of carbohydrates, typically ranging from 10% to 15% of their composition (Polmann et al. 2021), with some varieties exhibiting higher levels. Chicha nuts, an unconventional fruit, stand out due to their higher carbohydrate content compared to widely commercialized and consumed nuts, such as cashew nuts (29.70 g.100 g^{-1}) and pistachios (24.5 g.100 g^{-1}), making them a nutritionally interesting alternative (Cardoso et al. 2017).

In addition to carbohydrates, lipids are another highly relevant macronutrient present in nuts. These biologically derived compounds exhibit limited solubility in water but dissolve in organic solvents (Chaves et al. 2016). Lipids perform essential physiological roles, acting as energy reserves, structural components of cell membranes, and precursors for hormone synthesis. Furthermore, they possess antioxidant properties, contribute to metabolic regulation, and are involved in processes such as thermal insulation and emulsification in the digestive tract (Chaves et al. 2016).

Table 1. Nutritional composition macronutrients and micronutrients of Brazilian nuts.

Component ($g.100\ g^{-1}$)	Baru[1]	Brazil[2]	Cashew[3]	Chicha[4]	Gurgueia[5]	Pracaxi[6]	Sapucaia[7]
Ashe	3.24 ± 0.03	3.51 ± 1.14	4.87 ± 0.35	3.50 ± 0.04	2.29 ± 0.35	1.90 ± 0.04	2.89 ± 0.01
Moisture	3.20 ± 0.40	2.87 ± 0.17	7.77 ± 0.04	8.20 ± 0.08	8.25 ± 0.89	4.02 ± 0.06	4.19 ± 0.01
Carbohydrate	20.01 ± 0,02	30.00 ± 2.00	20.90	40.50	34.42 ± 1.12	25.17 ± 1.14	13.76 ± 0.70
Energy value (kcal/100 g)	603.28	705.03	598.00	463.03	560.00	644.00	665.22
Total fiber	6.01 ± 0.09	-	4.29 ± 0.21	5.80 ± 1.50	41.86 ± 1.37	-	-
Lipids	45.80 ± 0.05	66.11 ± 0.32	48.30 ± 1.6	24.05 ± 0.02	13.28 ± 0.24	53.42 ± 1.75	58.71 ± 0.30
Protein	27.06 ± 0.08	17.31 ± 0.13	29.78 ± 0.05	18.50 ± 0.14		15.50 ± 0.01	20.45 ± 0.41
Mineral							
Al	-	-	-	6.22 ± 3.58	1523.90	-	4.56 ± 0.66
Ca	1109.14	9.50 ± 0.12	3.28 ± 0.01	149.05 ± 0.79	-	8.90	1168.00 ± 0.40
Cd	-	-	-	0.37 ± 0.03	-	-	0.30 ± 0.01
Cr	-	-	-	0.039 ± 0.05	12.40	-	0.50 ± 0.01
Cu	-	0.89 ± 0.07	61.28 ± 0.01	7.48 ± 0.36	49.20	-	16.90 ± 0.63
Fe	35.70	2.15 ± 0.07	1.85 ± 0.01	218.20 ± 0.05	9394.20	1.70	-
K	9803.50	581.21 ± 14.12	39.52 ± 0.11	8718.00 ± 0.39	1718.60	-	-
Mg	1648.10	412.34 ± 1.14	36.87 ± 0.04	1327.00 ± 1.40	1523.90	5.60	1572.00 ± 4.38
Mn	-	-	-	32.10 ± 1.86	-	-	41.40 ± 1.87
Na	-	-	19.02 ± 0.01	3.62 ± 0.42	-	-	3.90 ± 0.03
Ni	-	-	-	0.35 ± 0.30	3611.70	-	0.50 ± 0.04
P	8328.00	-	129.52 ± 0.25	-	-	18.10	-
Pb	-	-	-	0.52 ± 0.50	-	-	0.04 ± 0.01
Se	-	-	-	-	-	-	46.90 ± 3.87
Sn	-	-	-	16.60 ± 0.92	-	-	11.10 ± 0.03
Zn	42.90	3.61 ± 0.13	2.24	24.16 ± 0.86	38.20	-	20.90 ± 2.26

[1] Polmann et al. 2021, Oliveira-Alves et al. 2020, [2] Maldonado et al. 2020, Brufau et al. 2006, [3] Nwosu et al. 2023, Rico et al. 2016, [4] Policarpi et al. 2018, [5] Queiroga-Neto et al. 2009, [6] Teixeira et al. 2020, [7] Demoliner et al. 2018, Teixeira et al. 2018

[1,5,7] Mineral expressed in micrograms per gram and [2,3,4,6] Mineral expressed in milligrams per 100 grams

Fatty acids, a key class of lipids, are essential nutrients that directly influence human health and disease risk (Marques et al. 2024). Studies indicate that the consumption of functional lipids, particularly polyunsaturated fatty acids, is associated with the prevention and management of non-communicable chronic diseases while also promoting overall well-being (Marques et al. 2024, Sant´Ana 2004). Nuts represent an excellent source of these beneficial lipids, including monounsaturated (MUFAs) and polyunsaturated fatty acids (PUFAs). Among oilseeds, gurgueia nuts exhibit the highest oleic acid content (68.2%), followed by pracaxi (54.3%) and baru nuts (52%), reinforcing the importance of these species in a healthy diet (Polmann et al. 2021).

Finally, proteins also play a crucial role in the composition of nuts. These macronutrients, formed by amino acid chains, regulate various physiological and metabolic functions in the body (Silva et al. 2012). Proteins contribute to cellular structure, catalyze biochemical reactions, facilitate molecular transport, and regulate metabolism (Cao et al. 2018). In addition to their health benefits, proteins exhibit valuable functional properties for the food industry, including solubility, water-holding capacity (WHC), oil-holding capacity (OHC), emulsification, gel formation, and foam stability (Gentile 2020, Silva et al. 2012).

Therefore, the consumption of nuts can significantly contribute to a balanced intake of macronutrients, providing carbohydrates, lipids, and proteins with essential nutritional and functional benefits (Cao et al. 2018).

4.3.1.2 Mineral Macronutrients and Micronutrients in Nuts

Micronutrients are essential compounds for the proper functioning of the human body. Although required in small amounts, they play vital roles in maintaining health, including metabolic regulation, immune system support, and cellular integrity.

Minerals, despite being needed in minimal quantities, are essential for the growth and proper functioning of human tissues, contributing to the optimal operation of the entire system (Shokunbi et al. 2023). Among macronutrients and micronutrients, nuts can contain aluminum (Al), calcium (Ca), cadmium (Cd), chromium (Cr), copper (Cu), iron (Fe), potassium (K), manganese (Mn), sodium (Na), nickel (Ni), phosphorus (P), lead (Pb), selenium (Se), tin (Sn), and zinc (Zn).

Aluminum (Al) is a mineral found in most foods consumed daily, including fresh fruits and vegetables, drinking water, and industrially processed foods such as seasonings and teas (Paiva et al. 2023a). The Brazilian Health Regulatory Agency (ANVISA), through Resolution RDC No. 285 of May 21, 2019, prohibits the use of food additives containing aluminum. This restriction is due to aluminum's association with adverse health effects, including accumulation in the lungs, liver, kidneys, thyroid, and brain. Its consumption has been linked to anemia, bone malformation, digestive disorders, Alzheimer's disease, multiple sclerosis, and neurotoxic effects (Mohamed et al. 2024, Paiva et al. 2023b).

Calcium (Ca) is an essential macronutrient for life, required for bone mass acquisition and maintenance (French et al. 2013). Studies indicate that dietary calcium intake helps meet nutritional recommendations and may protect against prevalent chronic diseases such as osteoporosis, with minimal reported adverse effects (Wadolowska et al. 2018, French et al. 2013). Primary dietary sources of calcium include dairy products, fish, vegetables, beans, and certain nuts, such as baru and sapucaia nuts (Polmann et al. 2021).

Like aluminum, cadmium (Cd), lead (Pb), and tin (Sn) are toxic to the human body. Consequently, the Agency for Toxic Substances and Disease Registry recommends a cadmium and lead limit of 0.05 and 0.2 mg/kg of body weight per day, respectively, in vegetables (FAO 2015). The presence of these metals in food is associated with factors such as geographic location, soil composition, pesticide use, and other agricultural conditions (Demoliner et al. 2018).

Chromium (Cr) is an essential trace element responsible for regulating carbohydrate and lipid metabolism. It is present in various foods, particularly meats, grains, nuts, fruits, and brewer's yeast (Morvaridzadeh et al. 2022, Asbaghi et al. 2021). Chromium intake is associated with health benefits, including improved glucose metabolism, insulin sensitivity, lipid profile, weight loss, body composition, inflammation reduction, and antioxidant effects (Asbaghi et al. 2021, Feng et al. 2015). The recommended daily intake is 30–35 mg for men and 20–25 mg for women.

Copper (Cu) is an essential micronutrient that plays a critical role in biological functions such as antioxidant enzyme formation, collagen synthesis, and iron metabolism (Faia et al. 2016). It is found in a variety of foods, including seafood, organ meats, nuts, seeds, and whole grains (Baierle et al. 2010, Macêdo et al. 2010). Adequate copper intake is associated with health benefits such as improved immune function, protection against oxidative stress, cardiovascular health maintenance, and nervous system development (Macêdo et al. 2010, Medicine 2002). The recommended daily intake ranges from 0.9 to 2.2 mg for adults, depending on age and physiological condition (Baierle et al. 2010).

Iron (Fe) is an essential micronutrient crucial for hemoglobin and myoglobin production, proteins responsible for oxygen transport in blood and muscles, respectively (World Health Organization 2021, Balci et al. 2012). It also plays a fundamental role in enzymatic processes and cellular energy production. Iron is present in various food sources, particularly red meat, poultry, fish, legumes, dark leafy greens, enriched cereals, and dried fruits (Chung et al. 2019). Iron deficiency can lead to iron-deficiency anemia, impairing oxygen transport and causing fatigue, weakness, and immune system dysfunction. The Institute of Medicine recommends a daily intake of 8–18 mg for adults, depending on sex and age.

Potassium (K) is an essential mineral involved in fluid balance regulation, muscle function, nervous system activity, and blood pressure maintenance (Farapti et al. 2022). Major dietary sources of potassium include bananas, potatoes, leafy vegetables, and beans, where it is commonly found as potassium citrate (D'Elia 2024). Potassium deficiency can lead to cardiovascular and muscular disorders, such as muscle cramps and decreased heart rate (D'Elia 2024, Farapti et al. 2022). The recommended daily intake is 3,500 mg.

Manganese (Mn) is a trace element required for bone mineralization, protein metabolism, glycosaminoglycan formation, and antioxidant enzyme production, such as superoxide dismutase, which neutralizes reactive oxygen species (Willians et al. 2022). Dietary sources include nuts, brown rice, legumes, and tea. The recommended daily intake is 1.8–2.3 mg for adults, and inadequate intake may impair nervous system function and metabolism (Ghosh et al. 2020).

Sodium (Na) is a crucial mineral for acid-base balance, blood pressure regulation, and nerve impulse transmission (Gagnon and Delpire 2021). It is primarily obtained from table salt and processed foods. Although necessary in small amounts, excessive sodium intake can lead to hypertension and cardiovascular diseases, making dietary control essential. The recommended daily intake is 1,500 mg (Xu et al. 2024).

Nickel (Ni) is a trace element essential for bacterial, plant, and animal growth. It plays a crucial role in cellular energy production and fatty acid transport (Babaahmadifooladi et al. 2020). It is found in most dietary sources, with high concentrations in green beans, broccoli, peas, canned foods, nuts, cocoa, and chocolate (Zambeli and Ciurli 2013). However, some individuals exhibit sensitivity to nickel, leading to systemic reactions known as systemic contact dermatitis (SCD) (Babaahmadifooladi et al. 2020, Zambeli and Ciurli 2013).

Phosphorus (P) is an essential nutrient for bone health and tooth formation. It is also a key component in energy metabolism, acid-base balance, and intracellular signaling (Vorland et al. 2017). Phosphorus is abundant in food, making deficiencies rare. Primary dietary sources include dairy products, meats, fish, and nuts (Serna and Bergwitz 2020, Vorland et al. 2017, Calvo et al. 2014). The recommended intake for adults aged 19–70 is 1,500–1,700 mg/day for men and 1,000–1,200 mg/day for women.

Selenium (Se) is a metabolically essential micronutrient with antioxidant properties, playing a crucial role in glutathione peroxidase activity, which protects cells from oxidative damage (Brown and Arthur 2007). Selenium deficiency is associated with neuromuscular disorders such as nutritional myodegeneration and cardiac dysfunction (Bazzano et al. 2019). The recommended daily intake is 55 μg, with major dietary sources including Brazil nuts, seafood, and meats (Genchi et al. 2023).

Zinc (Zn) is vital for immune function, protein synthesis, and wound healing (Forouzesh et al. 2024). It is primarily obtained from red meat, seafood, beans, and certain seeds. The recommended daily intake for adults is 8–11 mg (Roohani et al. 2013). Despite diverse food sources, zinc deficiency is common and may lead to immune dysfunction and growth delays (Jurowski et al. 2014).

Given the importance of these nutrients for human health, understanding the macro- and micronutrient mineral composition of Brazilian nuts allows for the incorporation of additional dietary sources of these elements. This knowledge facilitates the assessment of adequate consumption levels to address deficiencies and even estimate potential toxicity risks. Table 1 provides information on the macro- and micronutrient minerals present in the Brazilian nuts discussed in this chapter.

4.4 Phenolic Compounds

Phenolic compounds belong to a vast class of secondary metabolites produced by plants in response to biotic and abiotic factors, constituting one of the largest and most abundant groups of bioactive compounds (Polmann et al. 2021). These compounds

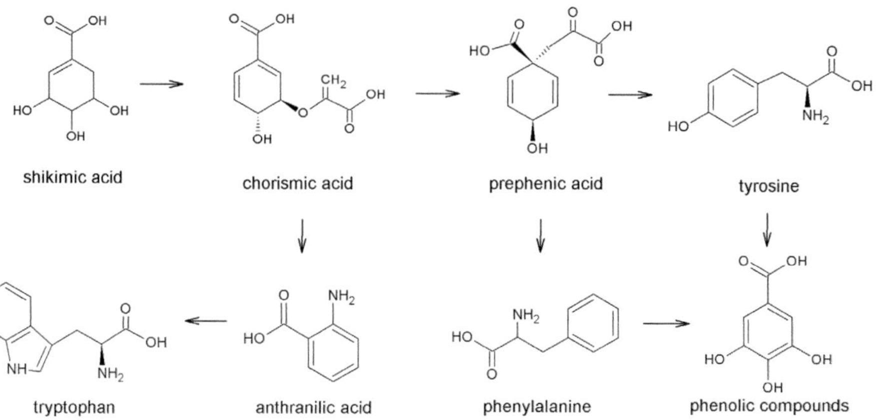

Figure 2. Shikimic acid route, which leads to the formation of phenolic compounds that act in plant defense.

are classified based on the biosynthetic pathway from which they originate, with acetyl-CoA or shikimic acid as their precursor. The shikimic acid pathway converts carbohydrate precursors derived from glycolysis and the pentose phosphate pathway into phenylalanine, tyrosine, and tryptophan, with phenylalanine being the most prevalent precursor in plants (Figure 2) (Albergaria et al. 2020). Subsequent reactions lead to the formation of simple phenolic compounds, including phenylpropanoids, lactones, and benzoic acid derivatives. Other phenolic compounds produced by plant species include polyphenols, such as phenolic acids, coumarins, flavonoids, anthocyanidins, and procyanidins (Chaves et al. 2016).

In terms of structure, simple phenolic compounds are characterized by a single phenol unit and the presence of one or more hydroxyl (-OH) groups. Their basic structure is relatively simple, which confers specific properties in terms of solubility and bioactivity (Nascimento 2016). Due to this structure and the hydroxyl groups, these compounds exhibit good solubility in polar solvents, with high dissolution in water, which is particularly important for intestinal absorption and the biological activity of simple phenols in the human body (Nascimento 2016).

Polyphenols, phenolic compounds containing multiple phenolic units, have a more complex chemical structure, which enhances their bioactive properties while limiting their solubility in water (Silva et al. 2022, Nascimento 2016). These compounds are soluble in organic solvents such as methanol and ethanol and can form complexes with other substances, which alters their solubility. For example, the solubility of flavonoids, tannins, and lignans can be influenced by factors such as the presence of hydroxyl groups and the formation of complexes with metal cations. These factors can affect both the absorption and distribution of these compounds in the body, as solubility, digestion, molecular size, degree of polymerization, conjugation with other phenolics, food matrix, and processing can significantly impact the metabolism and bioavailability of phenolic compounds (Aalim et al. 2025).

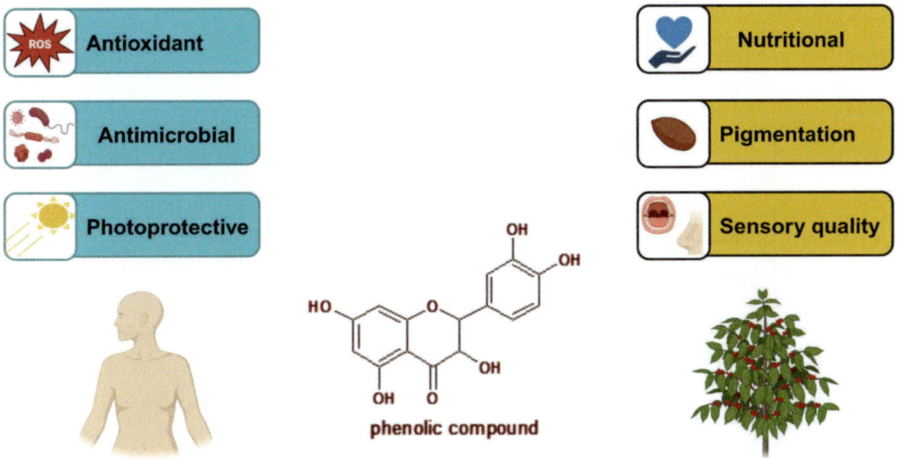

Figure 3. Phenolic compounds' activities and properties.

The importance of understanding the bioavailability of phenolic compounds is directly related to their chemical structure, as these compounds are generally associated with antimicrobial, photoprotective, and antioxidant properties, playing a crucial role in protecting plant tissues against ultraviolet radiation and pathogenic infections (Figure 3) (Bodoira and Maestri 2020, Greco et al. 2024). Studies have reported that incorporating sources of phenolic compounds into products or increasing dietary intake provides the human body with protective effects similar to those acquired by plants or fruits against stressors (Lemos et al. 2012).

In general, phenolic compounds stand out for their antioxidant capacity, as they act as free radical scavengers, preventing the oxidation of lipids, proteins, and DNA, processes associated with aging and the development of various diseases. This antioxidant activity is attributed to the presence of hydroxyl groups, which can donate a hydrogen atom or an electron, neutralizing free radicals (Teixeira et al. 2020, Alu'datt et al. 2017). In addition to their biological activities, phenolic compounds contribute to plant pigmentation, sensory and nutritional quality, influencing bitterness, astringency, flavor, odor, and oxidative stability (Figure 3) (Polmann et al. 2021, Demoliner et al. 2018).

Due to their bioactive properties in the human body, the consumption of sources rich in phenolic compounds has been extensively studied and recommended. Among these sources, nuts exhibit a rich phytochemical profile composed of flavonoids, organic acids, alkaloids, glycosides, monounsaturated fatty acids, tannins, phytates, and oils (Bhattacharya 2023). Owing to their diverse composition, their consumption is recommended, as they provide metabolites that contribute to protection against oxidative species. Table 2 summarizes the available literature data on the phenolic compound composition of Brazilian nuts.

Table 2. Phenolic compounds in Brazilian nuts.

Nut	**Total Phenolic (mg GAE/100 g sample)**	**Types Phenolic compounds**	**Reference**
Baru	250.4 ± 8.7	Anthocyanins, (+)-catechin, (-)-epicatechin, *p*-coumarin acid, ellagic acid, caffeic acid, gallic acid, *p*-hydroxybenzoic acid and ferulic acid	Lemos et al. 2012
Brazil	406.83 ± 29.28	Gallic acid, gallocatechin, protocatechuic acid, catechin, vanillic acid, taxifolin, ellagic acid, quercetin, protocatechuic acid, myricetin-3-o-rhamnoside, ellagic acid, taxifolin and vanillic acid	John and Shahidi 2010
Cashew	202.98 ± 8.4	(+)-Catechin, (-)-epicatechin, epicatechin gallate, condensed tannins, flavones, flavanoids, phenolic acids, (-)-epigallocatechin, procyanidin B1, (epi)-gallocatechin-(epi)-catechin gallate, procyanidin dimers, procyanidin trimer and naringenin	Sruthi et al. 2023
Chicha	16.85 ± 1.31	(+)-Catechin, salicylic acid, vanilic acid, ferulic acid, ellagic acid and rosmarinic acid	Policarpi et al. 2018
Gurgueia	4624.7 ± 0.04	-	Polmann 2023
Pracaxi	-	-	-
Sapucaia	20.5 ± 0.18	(+)-Catechin, (-)-epicatechin, myricetin, vanillic acid, ferulic acid and allagic acid	Demoliner et al. 2018

4.4.1 Phenolic Compounds and Antioxidant Activity

Phenolic compounds encompass a variety of classes synthesized by plants in response to external stressors such as weather conditions, predators, pollution, and extreme temperatures (Ferreira et al. 2024). In addition to protecting plants, these metabolites also contribute to human health. For this reason, the potential health effects of phenolic compounds have been investigated in both *in vitro* and *in vivo* studies. Research has reported that the consumption of foods rich in these compounds is associated with a reduced risk of metabolic disorders, such as type 2 diabetes mellitus, as well as anti-aging, antioxidant, anti-inflammatory, and antiproliferative effects (Aalim et al. 2025, Chamali et al. 2023, Elgadir et al. 2023, Ruskovska et al. 2020).

Among the studies evaluating the bioactive potential of phenolic compounds, the research conducted by Ruskovska et al. (2020) demonstrated that a diet rich in polyphenols (approximately 3 grams per day) has a positive impact on the human body by significantly increasing microbial diversity, which is linked to improved metabolic health. The biological properties of phenolic compounds are extensive, and therefore, investigations into their bioactivity and influence on health have been widely explored, particularly regarding their antioxidant potential, given that plants are natural sources of phenolic compounds with excellent antioxidant properties (Aalim et al. 2025, Ruskovska et al. 2020, Ferrera et al. 2016).

The antioxidant potential of plants is related to their reducing power and chemical structure, as the intermediates formed by the action of phenolic compounds are stabilized by the resonance effect of the aromatic ring in their structure (Sousa et al. 2007). Antioxidant-rich sources are beneficial for the body since reactive oxygen species (ROS), which are integral to human physiology, play important biological roles, such as in phagocytosis, where these species are produced to eliminate harmful agents (Santos 2016, Domej et al. 2014). However, when produced in excess, these highly reactive and unstable molecules cause oxidative stress, and many diseases or their progression are linked to an imbalance between pro-oxidant and antioxidant substances in the human body (Domej et al. 2014, Vasconcelos et al. 2007).

As a response to oxidative stress, the body produces antioxidants, which prevent ROS from impairing cellular function by inhibiting and reducing damage caused by free radicals. These substances can be generated endogenously or obtained through a diet rich in phenolic compounds (Vizzotto 2017). According to the literature, there is a strong correlation between the phenolic compound content and antioxidant activity in various food sources, including Brazilian nuts (Souza 2018, Costa et al. 2010).

Different methods for quantifying antioxidant content in plants and fruits are reported in the literature. *In vitro* methods are classified based on the reaction type: hydrogen atom transfer (HAT) and single electron transfer (SET) (Chaves et al. 2020, Apak et al. 2016). Among HAT-based methods, the most commonly used are the oxygen radical absorbance capacity (ORAC) assay, total peroxyl radical-trapping antioxidant parameter (TRAP), and lipid peroxidation suppression capacity (LPSC) (Apak et al. 2016). For SET-based methods, the most frequently applied include the 2,2-diphenyl-1-picrylhydrazyl (DPPH) assay, ferric reducing antioxidant power (FRAP), Trolox equivalent antioxidant capacity (TEAC or ABTS), copper reducing power (CUPRAC), and total reducing power (RP) assays (Chaves et al. 2020). Table 3 provides data on the HAT and SET methods used to quantify the antioxidant

Table 3. Quantification method and antioxidant potential of Brazilian nuts.

Nut	Methodology	Antioxidant potential	Reference
Baru	DPPH (%)	69.02 ± 2.86	Campidelli et al. 2020
	β-carotene/linoleic acid (%)	91.72 ± 3.35	
	ORAC ($\mu M.g^{-1}$)	4.06 ± 0.76	
Brazil	DPPH (%)	70.04 ± 0.07	Maldonado et al. 2020
Cashew	DPPH (%)	65.35 ± 2.24	Sruthi et al. 2023
	ORAC ($\mu M.g^{-1}$)	14089 ± 1651	
	TEAC (CE μmol/g)	19.69 ± 0.25	
Chicha	ABTS (μmol $TEAC.g^{-1}$)	13.25 ± 0.50	Policarpi et al. 2018
	FRAP (μmol $TEAC.g^{-1}$)	84.65 ± 0.56	
Gurgueia	-	-	-
Pracaxi	-	-	-
Sapucaia	ABTS (μmol $TEAC.g^{-1}$)	19.38 ± 1.04	Demoliner et al. 2018
	FRAP (μmol $TEAC.g^{-1}$)	48.13 ± 0.27	
	DPPH (μmol $TEAC.g^{-1}$)	0.00	

DPPH: 2,2-diphenyl-1-picrylhydrazyl; ORAC: Oxygen Radical Absorbance Capacity; TEAC: Trolox equivalent antioxidant capacity; ABTS: 2,2'-azino-bis(3-ethylbenzothiazoline-6-sulfonic acid); FRAP: Ferric Reducing Antioxidant Power; CE: Catechin equivalent.

potential in Brazilian nuts explored in this chapter, as well as their corresponding antioxidant content.

4.5 Outlook and Challenges

The nutritional composition of Brazilian nuts is an important aspect to be addressed in food analysis, providing essential information on nutritional content and enabling the development of functional food products. These nuts are not only a valuable source of essential nutrients but also a promising source of bioactive compounds with potential to promote health and prevent chronic diseases.

The growing appreciation of natural foods has been of interest to people seeking a healthier lifestyle. Therefore, this chapter has summarized information on the nutritional composition of some Brazilian nuts, including walnuts baru (*Dipteryx alata* Vog.), cashew (*Anacardium occidentale* L.), Brazil nut (*Bertholletia excelsa*), chicha (*Sterculia striata*), gurgueia (*Dipteryx lacunifera* Ducke), pracaxi (*Pentaclethra macroloba*), and sapucaia (*Lecythis pisonis* Cambess), with emphasis on the presence of macronutrients, moisture content, carbohydrates and lipids, proteins and dietary fiber, and macro- and micronutrient minerals. In addition to nutritional value, the phenolic composition and antioxidant potential of oilseeds have also been discussed.

Brazilian nuts are considered good sources of protein, especially cashew and baru nuts. Brazilian oilseeds are rich in minerals such as Mg, Zn, and Ca, which are important for maintaining good health and play an essential role in various physiological functions, including enzyme cofactors, enzymatic and neurological activity, and structural components. They also have important biological activities such as antioxidant, antifungal, and antimicrobial activities, especially antioxidant activity, which is directly related to their phenolic composition.

The gurgueia nut has been shown to have a high phenolic content, but there are no data in the literature to quantify the antioxidant potential of this nut. Nuts with antioxidant properties have various health benefits, as they are capable of slowing down the cellular oxidation reaction and acting to control chronic non-communicable diseases such as hypertension, diabetes, Alzheimer's, and even some types of cancer.

Methods for determining *in vitro* antioxidant or radical scavenging potential reported in the literature include the 2,2-diphenyl-1-picrylhydrazyl (DPPH) reaction mechanism, oxygen radical absorbance capacity (ORAC), Trolox equivalent antioxidant capacity (TEAC), 2,2'-azino-bis(3-ethylbenzothiazoline-6-sulfonic acid) (ABTS), ferric reducing antioxidant power (FRAP), and β-carotene/linoleic acid.

However, information on the nutritional composition, biological properties, and the influence of bioavailability on the activity of compounds in the human body is still quite limited. This understanding makes it possible to use well-established methodologies to improve oral bioavailability and, consequently, bioactivity through strategies for trapping compounds in nanostructured nanoparticles or combined delivery with other sources of metabolites. Innovative solutions must be proposed and, therefore, studies on the bioavailability and real benefits for human health of Brazil nuts are still needed.

Knowledge of the health potential attributed to the consumption of Brazil nuts could encourage their cultivation and commercialization on a large scale, favoring local populations native to the nuts, boosting extractive reserves, and improving the Brazilian economy. Incorporating these raw materials and their co-products, which are rich in nutrients and bioactive compounds, can enrich the formulation of food products.

4.6 Conclusions and Perspectives

Brazilian nuts provide essential nutrients for human health, contributing to the maintenance and quality of life through dietary inclusion. Baru nuts exhibit a high protein content, while cashew nuts are rich in phenolic compounds and possess significant antioxidant potential. The composition of macro- and micronutrients is particularly noteworthy, especially concerning calcium, magnesium, and zinc, which play crucial roles in maintaining health and preventing diseases. However, data on gurgueia, pracaxi, and sapucaia nuts remain scarce, particularly regarding their bioactive compound composition. Further studies exploring the bioactive potential and bioavailability of these compounds in the human body are necessary. Given the efforts of ongoing research, it is conceivable that the consumption of these nuts in their natural form or their incorporation as nutrient- and bioactive-rich raw materials into food formulations could enhance product development, encourage consumption, promote large-scale cultivation and commercialization, and ultimately drive positive economic impacts in Brazil.

References

Aalim, H., Islam Shishir, M. R., Yosri, N., Arslan, M., Tahir, H. E., Hashim, S. B. H. et al. (2025). Systematic review of the digestive fate of rice phenolic compounds: Insights into bioavailability, influencing factors, encapsulation strategies, and health implications. Trends in Food Science and Technology 156(May 2024): 104833. https://doi.org/10.1016/j.tifs.2024.104833.

Akintunde, J. K., Akomolafe, V. O., Taiwo, O. A., Ahmad, I., Patel, H., Osifeso, A. et al. (2022). Antihypertensive activity of roasted cashew nut in mixed petroleum fractions-induced hypertension: An *in vivo* and *in silico* approaches. Heliyon 8(12): e12339. https://doi.org/10.1016/j.heliyon.2022.e12339.

Albergaria, E. T., Oliveira, A. F. M. and Albuquerque, U. P. (2020). The effect of water deficit stress on the composition of phenolic compounds in medicinal plants. South African Journal of Botany 131: 12–17. https://doi.org/10.1016/j.sajb.2020.02.002.

Alu'datt, M. H., Rababah, T., Alhamad, M. N., Al-Mahasneh, M. A., Almajwal, A., Gammoh, S. et al. (2017). A review of phenolic compounds in oil-bearing plants: Distribution, identification and occurrence of phenolic compounds. Food Chemistry 218: 99–106. https://doi.org/10.1016/j.foodchem.2016.09.057.

Alves-Santos, A. M., Fernandes, D. C. and Naves, M. M. V. (2021). Baru (Dipteryx alata Vog.) fruit as an option of nut and pulp with advantageous nutritional and functional properties: A comprehensive review. NFS Journal 24(May): 26–36. https://doi.org/10.1016/j.nfs.2021.07.001.

Apak, R., Özyürek, M., Güçlü, K. and Çapanoılu, E. (2016). Antioxidant activity/capacity measurement. 2. Hydrogen Atom Transfer (HAT)-Based, Mixed-Mode (Electron Transfer (ET)/HAT), and Lipid Peroxidation Assays. Journal of Agricultural and Food Chemistry 64(5): 1028–1045. https://doi.org/10.1021/acs.jafc.5b04743.

Asbaghi, O., Naeini, F., Ashtary-Larky, D., Kaviani, M., Rezaei Kelishadi, M., Eslampour, E. et al. (2021). Effects of chromium supplementation on blood pressure, body mass index, liver function enzymes and malondialdehyde in patients with type 2 diabetes: A systematic review and dose-response meta-analysis of randomized controlled trials. Complementary Therapies in Medicine 60: 102755. https://doi.org/10.1016/j.ctim.2021.102755.

Babaahmadifooladi, M., Jacxsens, L., Van de Wiele, T. and Laing, G. Du. (2020). Gap analysis of nickel bioaccessibility and bioavailability in different food matrices and its impact on the nickel exposure assessment. Food Research International 129(December 2019): 108866. https://doi.org/10.1016/j.foodres.2019.108866.

Baierle, M., Valentini, J., Paniz, C., Moro, A., Barbosa Junior, F. and Garcia, S. C. (2010). Possíveis efeitos do cobre sanguíneo sobre parâmetros hematológicos em idosas. Jornal Brasileiro de Patologia e Medicina Laboratorial 46(6): 463–470. https://doi.org/10.1590/s1676-24442010000600006.

Balci, Y. I., Karabulut, A., Gürses, D. and Çövüt, I. E. (2012). Prevalence and risk factors of anemia among adolescents in Denizli, Turkey. Iranian Journal of Pediatrics 22(1): 77–81.

Bazzano, M., McLean, A., Tesei, B., Gallina, E. and Laus, F. (2019). Selenium and vitamin E concentrations in a healthy donkey population in Central Italy. Journal of Equine Veterinary Science 78: 112–116. https://doi.org/10.1016/j.jevs.2019.04.003.

Bhattacharya, S. (2023). Edible nuts. pp. 173–192. In Snack Foods (Vol. 27). Elsevier. https://doi.org/10.1016/B978-0-12-819759-2.00006-9.

Bodoira, R. and Maestri, D. (2020). Phenolic compounds from nuts: extraction, chemical profiles, and bioactivity. Journal of Agricultural and Food Chemistry 68(4): 927–942. https://doi.org/10.1021/acs.jafc.9b07160.

Brown, K. and Arthur, J. (2007). Selenium, selenoproteins and human health: a review. Public Health Nutrition 4(2b): 593–599. https://doi.org/10.1079/phn2001143.

Brufau, G., Boatella, J. and Rafecas, M. (2006). Nuts: Source of energy and macronutrients. British Journal of Nutrition 96(SUPPL. 2): 24–28. https://doi.org/10.1017/BJN20061860.

Calvo, M. S., Moshfegh, A. J. and Tucker, K. L. (2014). Assessing the health impact of phosphorus in the food supply: Issues and considerations. Advances in Nutrition 5(1): 104–113. https://doi.org/10.3945/an.113.004861.

Campidelli, M. L. L., Carneiro, J. D. S., Souza, E. C., Magalhães, M. L., Nunes, E. E. C., Faria, P. B. et al. (2020). Effects of the drying process on the fatty acid content, phenolic profile, tocopherols and antioxidant activity of baru almonds (Dipteryx alata Vog.). Grasas y Aceites 71(1): 1–11. https://doi.org/10.3989/gya.1170182.

Campos, C. de M. F., Soares, A. K. de O., de ABREU, B. B., Morgano, M. A. and Moreira-Araújo, R. S. D. R. (2022). Development of functional cookies with Cerrado fruits and residues: sensory analysis, nutrients, and bioactive compounds. Food Science and Technology (Brazil) 42: 1–7. https://doi.org/10.1590/fst.17621.

Campos, C. de M. F., Soares, A. K. de O., Gonçalves, M. F. B., Hashimoto, J. M. and Moreira-Araújo, R. S. dos R. (2024). Nutritional and functional characteristics of whole-grain cookies with added Gurguéia Nut (Dipteryx Lacunifera Ducke) and Soursop Residue (Annona Muricata L.). Journal of Culinary Science and Technology 22(1): 110–124. https://doi.org/10.1080/15428052.2022.2034694.

Cao, B., Fang, L., Liu, C., Min, W. and Liu, J. (2018). Effects of high hydrostatic pressure on the functional and rheological properties of the protein fraction extracted from pine nuts. Food Science and Technology International 24(1): 53–66. https://doi.org/10.1177/1082013217726883.

Cardoso, B. R., Duarte, G. B. S., Reis, B. Z. and Cozzolino, S. M. F. (2017). Brazil nuts: Nutritional composition, health benefits and safety aspects. Food Research International 100(March): 9–18. https://doi.org/10.1016/j.foodres.2017.08.036.

Carvalho, A., Garcêz Maia, M., Maria Correia da Costa, J., Aurélio Barbosa de Souza, V. and Arraes Maia, G. (2008). Avaliação dos parâmetros físicos e nutricionais de amêndoas de chichá, sapucaia e castanha-do-gurguéia Evaluation of some physical and nutritional parameters of almonds of chichá, sapucaia and castanha-do-gurguéia. Revista Ciência Agronômica 39(4): 517–523. http://www.redalyc.org/articulo.oa?id=195317437006%0Awww.ccarevista.ufc.br.

Carvalho, M. G. de O., Brandão, M. L. L., Miranda, R. V. da S. L. de, Jacob, S. do C., Forsythe, S. J. and Sartori, A. V. (2024). Multi-mycotoxins and mycoflora in commercial Brazilian roasted baru nuts (Dipteryx alata vog). Food Control 168(August 2024). https://doi.org/10.1016/j.foodcont.2024.110967.

Cavalcante, Í. H. L., Osajima, J. A., Neto, R. F. and Silva, R. L. (2015). Gurguéia nut: A new and potential fruit crop. Comunicata Scientiae 6(1): 1–9.

Chamali, S., Bendaoud, H., Bouajila, J., Camy, S., Saadaoui, E., Condoret, J. S. et al. (2023). Optimization of accelerated solvent extraction of bioactive compounds from Eucalyptus intertexta using response surface methodology and evaluation of its phenolic composition and biological activities. Journal of Applied Research on Medicinal and Aromatic Plants 35(January): 100464. https://doi.org/10.1016/j.jarmap.2023.100464.

Chaves, M. H., Lustosa, M. C. G. and Citó, A. M. G. L. (2016). BIOMOLECULAS. 03(quadro 1).

Chaves, N., Santiago, A. and Alías, J. C. (2020). Quantification of the antioxidant activity of plant extracts: Analysis of sensitivity and hierarchization based on the method used. Antioxidants 9(1). https://doi.org/10.3390/antiox9010076.

Chung, Y. J., Luo, A., Park, K. C., Loonat, A. A., Lakhal-Littleton, S., Robbins, P. A. et al. (2019). Iron-deficiency anemia reduces cardiac contraction by downregulating RyR2 channels and suppressing SERCA pump activity. JCI Insight 4(7). https://doi.org/10.1172/jci.insight.125618.

Chunhieng, T., Pétritis, K., Elfakir, C., Brochier, J., Goli, T. and Montet, D. (2004). Study of selenium distribution in the protein fractions of the Brazil nut, Bertholletia excelsa. Journal of Agricultural and Food Chemistry 52(13): 4318–4322. https://doi.org/10.1021/jf049643e.

Costa, D. A., Chaves, M. H., Silva, W. C. S. and Cost, C. L. S. (2010). Constituintes químicos, fenóis totais e atividade antioxidante de Sterculia striata St. Hil. et Naudin. Acta Amazonica 40(1): 207–212. https://doi.org/10.1590/S0044-59672010000100026.

Cruz-Reina, L. J., Rodríguez-Cortina, J., Vaillant, F., Herrera-Orozco, I., Carazzone, C. and Sierra, R. (2024). Extraction of fermentable sugars and phenolic compounds from Colombian cashew (Anacardium occidentale) nut shells using subcritical water technology: Response surface methodology and chemical profiling. Chemical Engineering Journal Advances 20(October): 100661. https://doi.org/10.1016/j.ceja.2024.100661.

D'Elia, L. (2024). Potassium intake and human health. Nutrients 16(6): 14–18. https://doi.org/10.3390/nu16060833.

de Britto Policarpi, P., Turcatto, L., Demoliner, F., Ferrari, R. A., Bascuñan, V. L. A. F., Ramos, J. C. et al. (2018). Nutritional potential, chemical profile and antioxidant activity of Chichá (Sterculia striata) nuts and its by-products. Food Research International 106(December 2017): 736–744. https://doi.org/10.1016/j.foodres.2017.12.069.

Demoliner, F., de Britto Policarpi, P., Ramos, J. C., Bascuñan, V. L. A. F., Ferrari, R. A., Jachmanián, I. et al. (2018). Sapucaia nut (Lecythis pisonis Cambess) and its by-products: A promising and underutilized source of bioactive compounds. Part I: Nutritional composition and lipid profile. Food Research International 108(March): 27–34. https://doi.org/10.1016/j.foodres.2018.03.028.

Demoliner, F., de Carvalho, L. T., de Liz, G. R., Prudêncio, E. S., Ramos, J. C., Bascuñan, V. L. A. F. et al. (2020). Improving the nutritional and phytochemical compounds of a plant-based milk of sapucaia nut cake using block freeze concentration. International Journal of Food Science & Technology 55(8): 3031–3042. https://doi.org/10.1111/ijfs.14568.

Domej, W., Oettl, K. and Renner, W. (2014). Oxidative stress and free radicals in COPD-implications and relevance for treatment. International Journal of COPD 9: 1207–1224. https://doi.org/10.2147/COPD.S51226.

Elgadir, M. A., Chigurupati, S. and Mariod, A. A. (2023). Selected potential pharmaceutical and medical benefits of phenolic compounds: Recent advances. Functional Food Science 3(7): 108–128. https://doi.org/10.31989/ffs.v3i7.1118.

Fagundes, M. H. (2024). Maria Helena Fagundes – Técnica de Planejamento -TNS IV E. Conab 61: 1–8.

Faia, R., Lemos, L. M. de, Mourao-Júnior, C. A. and Andreazzi, A. E. (2016). Metabolismo do cobre e doença de Wilson: uma Revisão. Revista Médica de Minas Gerais 26(0): 5–8. http://www.rmmg.org/artigo/detalhes/2235.

Farapti, F., Buanasita, A., Atmaka, D. R., Setyaningtyas, S. W., Adriani, M., Rejeki, P. S. et al. (2022). Potassium intake is associated with nutritional quality and actual diet cost: A study at formulating a low sodium high potassium (LSHP) healthy diet. Journal of Nutritional Science 11(3): 1–9. https://doi.org/10.1017/jns.2021.104.

Feng, W., Mao, G., Li, Q., Wang, W., Chen, Y., Zhao, T. et al. (2015). Effects of chromium malate on glycometabolism, glycometabolism-related enzyme levels and lipid metabolism in type 2 diabetic rats: A dose-response and curative effects study. Journal of Diabetes Investigation 6(4): 396–407. https://doi.org/10.1111/jdi.12350.

Ferreira, C. P., de Lima, M. da C., da Silva, J. G. and Peixoto Araujo, N. M. (2024). Nutritional composition, phenolic compounds and biological activities of selected unconventional food plants. Food Research International 191(June): 114643. https://doi.org/10.1016/j.foodres.2024.114643.

Ferrera, T. S., Heldwein, A. B., Dos Santos, C. O., Somavilla, J. C. and Sautter, C. K. (2016). Substâncias fenólicas, flavonoides e capacidade antioxidante em erveiras sob diferentes coberturas do solo e sombreamentos. Revista Brasileira de Plantas Medicinais 18(2): 588–596. https://doi.org/10.1590/1983-084X/15_197.

Fonseca, F. L. da, Rover, O. J., Wadt, L. H. de O. and Cartaxo, C. B. da C. (2021). Sustainability indicators of the Brazil nut tree management. Revista Brasileira de Ciências Ambientais 56(4): 1–13. https://doi.org/10.5327/z21769478889.

Fontoura, B. H., Perin, E. C., Teixeira, S. D., Lima, V. A. de and Carpes, S. T. (2023). Multivariate and machine learning approaches for prediction of antioxidant potential in Bertholletia excelsa barks. Journal of King Saud University - Science 35(7): 102792. https://doi.org/10.1016/j.jksus.2023.102792.

Forouzesh, A., Forouzesh, F., Foroushani, S. S. and Forouzesh, A. (2024). A new method for calculating zinc content and determining appropriate zinc levels in foods. 09(5): 7352–7363.

French, M. R., Vernace-Inserra, F. and Hawker, G. A. (2013). A prospective study to identify factors affecting adherence to recommended daily calcium intake in women with low bone mineral density. Journal of the American College of Nutrition 27(1): 88–95. https://doi.org/10.1080/07315724.2008.10719679.

Gagnon, K. B. and Delpire, E. (2021). Sodium transporters in human health and disease. Frontiers in Physiology 11(February): 1–18. https://doi.org/10.3389/fphys.2020.588664.

Genchi, G., Lauria, G., Catalano, A., Sinicropi, M. S. and Carocci, A. (2023). Biological activity of selenium and its impact on human health. International Journal of Molecular Sciences 24(3). https://doi.org/10.3390/ijms24032633.

Gentile, L. (2020). Current Opinion in Colloid & Interface Science. Elsevier is not responsible for Protein-polysaccharide interactions and aggregates in food formulations. 48: 18–27.

Ghosh, G. C., Khan, M. J. H., Chakraborty, T. K., Zaman, S., Kabir, A. H. M. E. and Tanaka, H. (2020). Human health risk assessment of elevated and variable iron and manganese intake with arsenic-safe groundwater in Jashore, Bangladesh. Scientific Reports 10(1): 1–9. https://doi.org/10.1038/s41598-020-62187-5.

Goetz, B. M., Horst, E. A., Mayorga, E. J., Abeyta, M. A., Rodriguez-Jimenez, S., Carta, S. et al. (2023). Effects of cashew nut shell extract supplementation on production, rumen fermentation, metabolism, and inflammatory biomarkers in transition dairy cows. Journal of Dairy Science 106(12): 9843–9854. https://doi.org/10.3168/jds.2023-23563.

Goveia, G. R. (2007). Manual do Profissional. Manual de Nutrição, 1–8. https://crn5.org.br/wp-content/uploads/2013/05/Manual-Calorias-Macronutrientes-e-Micronutrientes.pdf.

Greco, M., Fuertes-Rabanal, M., Frey, C., Grosso, C. Del, Coculo, D., Moretti, P. et al. (2024). Phenolic compounds-enriched extract recovered from two-phase olive pomace serves as plant immunostimulants and broad-spectrum antimicrobials against phytopathogens including Xylella fastidiosa. Plant Stress 14(August). https://doi.org/10.1016/j.stress.2024.100655.

Hou, L., Rashid, M., Chhabra, M., Chandrasekhar, B., Amirthalingam, P., Ray, S. et al. (2021). The effect of Bertholletia excelsa on body weight, cholestrol, and c-reactive protein: A systematic review and meta-analysis of randomized controlled trials. Complementary Therapies in Medicine 57(50): 102636. https://doi.org/10.1016/j.ctim.2020.102636.

John, J. A. and Shahidi, F. (2010). Phenolic compounds and antioxidant activity of Brazil nut (Bertholletia excelsa). Journal of Functional Foods 2(3): 196–209. https://doi.org/10.1016/j.jff.2010.04.008.

Jurowski, K., Szewczyk, B., Nowak, G. and Piekoszewski, W. (2014). Biological consequences of zinc deficiency in the pathomechanisms of selected diseases. Journal of Biological Inorganic Chemistry 19(7): 1069–1079. https://doi.org/10.1007/s00775-014-1139-0.

Lemos, M. R. B., Siqueira, E. M. de A., Arruda, S. F. and Zambiazi, R. C. (2012). The effect of roasting on the phenolic compounds and antioxidant potential of baru nuts [Dipteryx alata Vog.]. Food Research International 48(2): 592–597. https://doi.org/10.1016/j.foodres.2012.05.027.

Lima, D. S., Egea, M. B., Cabassa, I. de C. C., Almeida, A. B. de, Sousa, T. L. de, Lima, T. M. de et al. (2021). Technological quality and sensory acceptability of nutritive bars produced with Brazil nut and baru almond coproducts. Lwt 137(September 2020): 110467. https://doi.org/10.1016/j.lwt.2020.110467.

Lucas, K. R. G., Caldarelli, C. E. and Ventura, M. U. (2023). Agriculture and biodiversity damage: A prospective evaluation of the impact of Brazilian agriculture on its ecoregions through life cycle assessment methodology. Science of the Total Environment 899(July). https://doi.org/10.1016/j.scitotenv.2023.165762.

Macêdo, É. M. C. de, Amorim, M. A. F., Silva, A. C. S. da and Castro, C. M. M. B. de. (2010). Efeitos da deficiência de cobre, zinco e magnésio sobre o sistema imune de crianças com desnutrição grave. Revista Paulista de Pediatria 28(3): 329–336. https://doi.org/10.1590/S0103-05822010000300012.

Maldonado, A. S. S., Fernández, M. I., Alemán, S. R., Fuentes, A. M. J. and Ferreira, da C. C. M. (2020). Determination of total phenolic compounds, antioxidant activity and nutrients in Brazil nuts (Bertholletia excelsa H. B. K.). Journal of Medicinal Plants Research 14(8): 373–376. https://doi.org/10.5897/jmpr2020.6953.

Marques, F., Pinho, M., Guerra, I. M. S., Conde, T. A., Silva, J., Cardoso, H. et al. (2024). Unlocking functional lipid ingredients from algae by food-grade biosolvents and ultrasound-assisted extraction for nutritional applications. Lwt 200(May). https://doi.org/10.1016/j.lwt.2024.116136.

Medicine, I. of. (2002). Dietary reference intakes for vitamin A, vitamin K, arsenic, boron, chromium, copper, iodine, iron, manganese, molybdenum, nickel, silicon, vanadium, and zinc : a report of the Panel on Micronutrients ... [et al.], Standing Committee on the Scientific Eva.

Mohamed, R. S., Fouda, K., Salama, A. and Akl, E. M. (2024). Peanut meal-derived bioactive compounds: Extraction, co-extrusion encapsulation and neuroprotection against aluminum-induced Alzheimer's disease via *in silico* and *in vivo* studies. Phytomedicine Plus 4(3). https://doi.org/10.1016/j.phyplu.2024.100588.

Mohammadnezhad, P., Valdés, A., Barrientos, R. E., Ibáñez, E., Block, J. M. and Cifuentes, A. (2023). A comprehensive study on the chemical characterization and neuroprotective evaluation of pracaxi nuts extracts obtained by a sustainable approach. Foods 12(20). https://doi.org/10.3390/foods12203879.

Morvaridzadeh, M., Estêvão, M. D., Qorbani, M., Heydari, H., Hosseini, A. sadat, Fazelian, S. et al. (2022). The effect of chromium intake on oxidative stress parameters: A systematic review and meta-analysis. Journal of Trace Elements in Medicine and Biology 69(March 2021). https://doi.org/10.1016/j.jtemb.2021.126879.

Nascimento, K. S. do. (2016). Compostos Fenólicos, Capacidade Antioxidante e Propriedades Físico-Químicas de méis de Apia mellifera do Estado do Rio Grande do Sul. In Dissertação.

Nwosu, N. B., Okoronkwo, N. E., Onwuka, O. M. and Osuchukwu, T. U. (2023). Phytochemical and nutritional compositions of two varieties of Anacardium occidentale L. World Journal of Advanced Research and Reviews 19(2): 966–977. https://doi.org/10.30574/wjarr.2023.19.2.1629.

Oliveira-Alves, S. C., Pereira, R. S., Pereira, A. B., Ferreira, A., Mecha, E., Silva, A. B. et al. (2020). Identification of functional compounds in baru (Dipteryx alata Vog.) nuts: Nutritional value, volatile and phenolic composition, antioxidant activity and antiproliferative effect. Food Research International 131(October 2019). https://doi.org/10.1016/j.foodres.2020.109026.

Organization, W. H. (2021). Iron deficiency anaemia. In Bmj (Vol. 314, Issue 7096, p. 1759). https://doi.org/10.1136/bmj.314.7096.1759.

Paiva, E. L. de, Lima, S. G., Waegeneers, N., Andjelkovic, M., Faustino de Almeida Marques, R. E., Morgano, M. A. et al. (2023a). Aluminium intake through the consumption of selected baby foods and risk characterization in a population of Brazilian infants aged 0 to 36 months. Journal of Food Composition and Analysis 115(June 2022). https://doi.org/10.1016/j.jfca.2022.105013.

Paiva, L. C. de M., Guedes, M. C., Ferreira, D. Q., Rodrigues, E. G., Machado, F. P. and Fernandes, C. P. (2023b). Preparation of nut-based milk alternative: evaluation of an Amazonian nut (Bertholletia excelsa) beverage with annatto nanodispersion. Food Chemistry Advances 3(October): 100506. https://doi.org/10.1016/j.focha.2023.100506.

Polmann, G. (2023). Valorização da Castanha do Gurguéia (Dipteryx Lacunifera Ducke) e seus Coprodutos Através da Obtenção e Caracterização de Extratos Obtidos por Tecnologias Limpas. 1–192.

Polmann, G., Badia, V., Danielski, R., Ferreira, S. R. S. and Block, J. M. (2021). Non-conventional nuts: An overview of reported composition and bioactivity and new approaches for its consumption and valorization of co-products. Future Foods 4(November): 100099. https://doi.org/10.1016/j.fufo.2021.100099.

Polmann, G., Teixeira, G. L., Santos, P. H., Rivera, G. Á., Ibañez, E., Cifuentes, A. et al. (2024). Chemical characterization of gurguéia nut (Dipteryx lacunifera Ducke) and press cake oil obtained by hydraulic pressing and supercritical extraction. Biomass Conversion and Biorefinery 14(16): 19065–19080. https://doi.org/10.1007/s13399-023-04042-x.

Queiroga-Neto, V., Bora, P. S., Diniz, Z. N., Cavalheiro, J. M. D. O. and Souza, P. A. S. (2009). Partial evaluation of Dipteryx lacunifera seed kernel as a nutritional food: Evaluación parcial de la almendra del fruto de Dipteryx lacunifera como alimento nutritivo. CYTA - Journal of Food 7(1): 23–29. https://doi.org/10.1080/11358120902850594.

Quesada, H. B., De Araújo, T. P., Cusioli, L. F., De Barros, M. A. S. D., Gomes, R. G. and Bergamasco, R. (2021). Evaluation of novel activated carbons from chichá-do-cerrado (Sterculia striata St. Hil. et Naud) fruit shells on metformin adsorption and treatment of a synthetic mixture. Journal of Environmental Chemical Engineering 9(1). https://doi.org/10.1016/j.jece.2020.104914.

Rico, R., Bulló, M. and Salas-Salvadó, J. (2016). Nutritional composition of raw fresh cashew (Anacardium occidentale L.) kernels from different origin. Food Science and Nutrition 4(2): 329–338. https://doi.org/10.1002/fsn3.294.

Roohani, N., Hurrell, R., Kelishadi, R. and Schulin, R. (2013). Zinc and its importance for human health: An integrative review. Journal of Research in Medical Sciences 18(2): 144–157.

Rosa, T. L. M., Araujo, C. P. de, Kamke, C., Ferreira, A., Ferreira, M. F. da S., Oliveira, J. P. B. de et al. (2020). Sapucaia nut: Morphophysiology, minerals content, methodological validation in image analysis, phenotypic and molecular diversity in Lecythis pisonis Cambess. Food Research International 137(February): 109383. https://doi.org/10.1016/j.foodres.2020.109383.

Ruskovska, T., Maksimova, V. and Milenkovic, D. (2020). Polyphenols in human nutrition: From the *in vitro* antioxidant capacity to the beneficial effects on cardiometabolic health and related inter-individual variability - An overview and perspective. British Journal of Nutrition 123(3): 241–254. https://doi.org/10.1017/S0007114519002733.

Sant´Ana, L. S. (2004). Mecanismos bioquímicos envolvidos na digestão, absorção e metabolismo dos acidos graxos omega. Revista Brasileira Em Promoção Da Saúde 211–216. https://doi.org/10.5020/18061230.2004.p211.

Santos, L. R. (2016). CONTRIBUIÇÃO AO CONHECIMENTO QUÍMICO E BIOLÓGICO DAS FOLHAS DE Senna acuruensis (Benth.) Irwin & Barneby (FABACEAE).

Santos, M. A., Guilherme, E. P. X., Carvalho, L. M., Brandi, I. V., Carvalho, G. G. P., Cota, J. et al. (2020). Fatty acid and proteomic analysis of Sterculia striata nut. Food Science and Technology 40(suppl 2): 489–495. https://doi.org/10.1590/fst.31819.

Semporé, J. N., Songré-Ouattara, L. T., Tarpaga, W. V., Bationo, F. and Dicko, M. H. (2021). Morphological characterization and quality assessment of cashew (Anacardium occidentale L.) nuts from 53 accessions of Burkina Faso. Journal of Agriculture and Food Research 6(September): 0–7. https://doi.org/10.1016/j.jafr.2021.100219.

Serna, J. and Bergwitz, C. (2020). Importance of dietary phosphorus for bone metabolism and healthy aging. Nutrients 12(10): 1–43. https://doi.org/10.3390/nu12103001.

Shokunbi, O. S., Adepoju, O. T., Ramaite, I. D. I., Shokunbi, O. S., Mojapelo, P. E. L. and Akinyele, I. O. (2023). Potassium, sodium, calcium and magnesium levels of commonly consumed foods and estimates of dietary intakes of selected Nigerian adults. Heliyon 9(3): e13729. https://doi.org/10.1016/j.heliyon.2023.e13729.

Silva, A. C. C., Frota, K. de M. G. and Arêas, J. A. G. (2012). Funções Plenamente Reconhecidas de Nutrientes Proteína. 20: 1–16.

Silva, A. G. de M. e and Fernandes, K. flávia. (2011). Composição química e antinutrientes presentes nas amêndoas cruas e torradas de chicha. Revista de Nutrição 24(2): 305–314.

Silva, A. S. L. da, Silva, A. de J., Latif, A. L. O., Santos, A. de F. J. and Benevides, C. M. de J. (2022). Uso de metodologias analíticas para determinação de compostos fenólicos em alimentos no Brasil: avanços e fragilidades. Research, Society and Development 11(2): e1311225193. https://doi.org/10.33448/rsd-v11i2.25193.

Silva, D. A. da, Rodrigues, A. M. da C., Santos, A. O. dos, Salvador-Reyes, R. and Silva, L. H. M. da. (2023). Physicochemical and technological properties of pracaxi oil, cupuassu fat and palm stearin blends enzymatically interesterified for food applications. Lwt 184(February): 114961. https://doi.org/10.1016/j.lwt.2023.114961.

Sousa, C. M. D. M., Silva, H. R. E., Vieira-Jr., G. M., Ayres, M. C. C., Costa, C. L. S. da, Araújo, D. S. et al. (2007). Fenóis totais e atividade antioxidante de cinco plantas medicinais. Química Nova 30(2): 351–355. https://doi.org/10.1590/S0100-40422007000200021.

Souza, I. S. de. (2018). Avaliação da capacidade antioxidante e compostos fenólicos em três sementes oleaginosas: castanha-do-brasil, castanha de caju e noz pecã. 80. http://hdl.handle.net/11449/154766.

Sruthi, P., Roopavathi, C. and Madhava Naidu, M. (2023). Profiling of phenolics in cashew nut (Anacardium occidentale L.) testa and evaluation of their antioxidant and antimicrobial properties. Food Bioscience 51(November 2022): 102246. https://doi.org/10.1016/j.fbio.2022.102246.

Teixeira, G. L., Ávila, S., Hornung, P. S., Barbi, R. C. T. and Ribani, R. H. (2018). Sapucaia nut (Lecythis pisonis Cambess.) flour as a new industrial ingredient: Physicochemical, thermal, and functional properties. Food Research International 109(March): 572–582. https://doi.org/10.1016/j.foodres.2018.04.071.

Teixeira, G. L., Maciel, L. G., Mazzutti, S., Gonçalves, C. B., Ferreira, S. R. S. and Block, J. M. (2020). Composition, thermal behavior and antioxidant activity of pracaxi (Pentaclethra macroloba) seed oil obtained by supercritical CO_2. Biocatalysis and Agricultural Biotechnology 24(January). https://doi.org/10.1016/j.bcab.2020.101521.

Vasconcelos, S. M. L., Goulart, M. O. F., Moura, J. B. D. F., Manfredini, V., Benfato, M. D. S. and Kubota, L. T. (2007). Reactive oxygen and nitrogen species, antioxidants and markers of oxidative damage in human blood: Main analytical methods for their determination. Quimica Nova 30(5): 1323–1338. https://doi.org/10.1590/s0100-40422007000500046.

Vizzotto, E. (2017). Radicais livres e mecanismos de proteção antioxidante. Disciplina de Fundamentos Bioquímicos dos Transtornos Metabólicos. Programa de Pós-Graduação Em Ciências Veterinárias, 10.

Vorland, C. J., Stremke, E. R., Moorthi, R. N. and Hill Gallant, K. M. (2017). Effects of excessive dietary phosphorus intake on bone health. Current Osteoporosis Reports 15(5): 473–482. https://doi.org/10.1007/s11914-017-0398-4.

Wadolowska, L., Ulewicz, N., Sobas, K., Wuenstel, J. W., Slowinska, M. A., Niedzwiedzka, E. et al. (2018). Dairy-related dietary patterns, dietary calcium, body weight and composition: A study of obesity in polish mothers and daughters, the MODAF project. Nutrients 10(1): 1–20. https://doi.org/10.3390/nu10010090.

Wang, C., Yolitz, J., Alberico, T., Laslo, M., Sun, Y., Wheeler, C. T. et al. (2014). Cranberry interacts with dietary macronutrients to promote healthy aging in drosophila. Journals of Gerontology - Series A Biological Sciences and Medical Sciences 69(8): 945–954. https://doi.org/10.1093/gerona/glt161.

Willians, M., Todd, G. and Ronney, N. (2022). Toxicological profile for manganese. National Library of Medicine, September. https://doi.org/10.1201/9781420061888_ch4.

Xu, X., Zeng, L., Jha, V., Cobb, L. K., Shibuya, K., Appel, L. J. et al. (2024). Potassium-enriched salt substitutes: a review of recommendations in clinical management guidelines. Hypertension 81(3): 400–414. https://doi.org/10.1161/HYPERTENSIONAHA.123.21343.

Zafeer, M. K. and Bhat, K. S. (2023). Valorisation of agro-waste cashew nut husk (Testa) for different value-added products. Sustainable Chemistry for Climate Action 2(October 2022): 100014. https://doi.org/10.1016/j.scca.2023.100014.

Zambeli, B. and Ciurli, S. (2013). Nickel in human health and disease. In Metal Ions in Life Sciences (Vol. 13). https://doi.org/10.1007/978-94-007-7500-8.

Zhai, J., Zhuang, Y., Sun, L., Gu, Y. and Fan, X. (2025). Nutritional health aspects and functional properties of nut yogurt: Future perspectives. Food Chemistry: X 25(December 2024): 102102. https://doi.org/10.1016/j.fochx.2024.102102.

Chapter 5

Bioactive Peptides from Fruits and Their Health-Promoting Properties

Sabrina Nunes da Silva,[1] *Mikael Kélvin de Albuquerque Mendes,*[1] *Carla Mariana da Silva Medeiros,*[1] *Sâmya Danielle Lima de Freitas,*[1] *Clecio Dantas,*[2] *Edivan Carvalho Vieira*[1] and *Cícero Alves Lopes Júnior*[1,*]

5.1 Introduction

Considering the great global escalation of health-related problems, diet has been recognized as a pivotal factor in preventing nutrition-associated ailments, reducing metabolic syndromes, and promoting human well-being (Zaky et al. 2022). The population's awareness of the impact of diet on health has gained importance and guided food choices. In this term, plant foods like fruits are rich sources of basic dietary nutrients and are cholesterol-free, an advantage over animal foods (Zaky et al. 2025, Sharma et al. 2023). Furthermore, bioactive compounds have also been explored in fruits, highlighting bioactive peptides (BAPs). In 1950, these compounds were identified for the first time in food sources (Sosalagere et al. 2022). Nowadays, ca. 7000 naturally occurring peptides have their structures known. These compounds are small biomolecules with a molecular weight of less than 10 kDa and composed of up to 50 amino acid residues linked by amide or peptide bonds (Sharma et al. 2023, Jakubczyk et al. 2020).

[1] Departamento de Química, Universidade Federal do Piauí – UFPI, P.O. Box 64049-550, Teresina, PI, Brasil.

[2] Departamento de Química, Universidade Estadual do Maranhão – UEMA, P.O. Box 65604-380, Caxias, MA, Brasil.

* Corresponding author: ciceroalj@ufpi.edu.br

DOI: 10.1201/9781003668169-7

In fruits and other foods, bioactive peptides are found as independent entities or encrypted in the primary sequence of native proteins, which are released through proteolysis by processes such as cooking, ripening, fermentation, and enzymatic hydrolysis (Xavier et al. 2024). Several biological activities have been reported for bioactive peptides, including antimicrobial, anticholesterolemic, antioxidative, antihypersensitive, antithrombotic, anticancer, antidiabetic, anti-inflammatory, and immunomodulatory (Sosalagere et al. 2022). Studies on BAPs from fruits are still restricted to a few species, e.g., moringa, walnut, jujube, coffee, and pepper (Sosalagere et al. 2022). Antioxidant, antihypersensitive, and anticancer activities have been the most investigated (Lath et al. 2023, Sosalagere et al. 2022, Olivares-Galván et al. 2020). Currently, fruit by-products have also been included in research focused on finding bioactive peptides, and the findings are promising (Panda et al. 2025). In fact, these biomolecules have specificity, tolerability, and efficacy. A key point is that the biological activity of BAPs depends on their physicochemical properties such as charge, hydrophobicity, length, and composition of the amino acid sequence (Chourasia et al. 2023, Sharma et al. 2023). So, the structural characterization approach applied to these bioactive compounds has a crucial role in exploring their potential as functional foods and nutraceuticals (Mendes et al. 2024, Ulug et al. 2021).

Although bioactive peptides from fruits have demonstrated several beneficial health properties as well as promising therapeutic candidates, some important aspects limit their use. The low yield of enzymatic proteolysis of fruit proteins in the gastrointestinal tract, combined with the low bioavailability of the BAPs produced, impairs their absorption in the human intestine (Sharma et al. 2023). Some BAPs have shown toxicity or allergenic effects, especially in individuals with celiac disease. In general, the bitter taste of bioactive peptides has low acceptability among consumers, hindering their incorporation into functional foods (Chourasia et al. 2023). The proteome of most fruits is unknown, limiting the selection of parental proteins that are sources of peptides with target biological activity (Jakubczyk et al. 2020).

Therefore, this chapter aims to discuss several aspects related to bioactive peptides from fruits, highlighting structural characteristics, production methods, analytical techniques for structural characterization, mechanism of action, and biological activities. Also, the challenges of converting the produced bioactive peptides into functional foods and nutraceuticals are mentioned here.

5.2 Bioactive Peptide: Definition, Classification, Production and Source Fruit

Bioactive peptides (BAPs) or cryptides are protein fragments encoded in the primary sequences of proteins. These compounds consist of 2 to 50 amino acid residues and have several significant pharmacological effects (Bidram and Ganjalikhany 2024, Pei et al. 2022). The field of peptide research began in 1902 when Starling and Bayliss discovered a substance secreted by the intestinal mucosa that had the potential to stimulate the secretion of digestive enzymes by the pancreas (Sosalagere et al. 2022).

This substance was later named secretin, identified as a peptide, and its amino acid sequence determined (Sosalagere et al. 2022).

However, it was only in 1950 that Mellender first observed a bioactive peptide from a food source (Sosalagere et al. 2022). Since then, several studies have been conducted to isolate and purify bioactive peptides from various matrices (Yang et al. 2021). As a result, the global market for bioactive peptides is expected to double from $48.6 billion in 2020 to $95.7 billion in 2028 (Du et al. 2023).

In general, BAPs are classified as exogenous or endogenous according to their different sources (Yang et al. 2021). Exogenous peptides are compounds present in parental proteins from external sources in the human body and are obtained by methods such as enzymatic hydrolysis and microbial fermentation, whereas endogenous peptides are produced by proteases in biological or plant organisms, synthesized by different types of cells in the human body, including neural and immune cells, or can be produced in glands such as the pituitary and adrenal glands (Zhang et al. 2024a, Akbarian et al. 2022).

Exogenous peptides have received more attention in research because they are more abundant as they are derived from proteins and can be obtained from a variety of sources, including dietary supplements, drugs, and diet (e.g., fruit) (Akbarian et al. 2022). Recently, the identification of exogenous peptides found in fruits has been gaining ground in research, as knowledge of the chemical composition, such as the presence of bioactive compounds associated with beneficial properties for the human body, strongly suggests the potential to obtain and isolate peptides from these matrices, enhancing the beneficial effects on health and adding value to the fruit (Sosalagere et al. 2022).

In addition to their phytochemical constituents, peptides of plant origin are considered to be non-toxic, have emulsifying properties, and are cholesterol-free, making them beneficial for maintaining the cardiovascular system and controlling the glycemic profile for an adequate supply of minerals, vitamins, and fiber (Sharma et al. 2023, Sosalagere et al. 2022). Furthermore, these plant-derived nutraceuticals are associated with greater accessibility and bioavailability in the body, which enhances the biological effect, making them more promising than animal-derived peptides (Sharma et al. 2023).

Naturally, bioactive peptides are inactive in the primary structure of proteins, but they exhibit biological properties—such as antioxidant, anticarcinogenic, antithrombotic, antihypertensive, and immunomodulatory activity—after being released through fermentation processes, enzymatic hydrolysis, simulated gastrointestinal digestion, and so on (Bidram and Ganjalikhany 2024, Zhu et al. 2024, Kehinde and Sharma 2018).

The release of bioactive peptides from microbial fermentation processes is well documented in the literature, as microbial fermentation improves the quality and nutritional profile of food by converting dietary components into valuable products such as bioactive peptides (Rizwan et al. 2023). This process involves the use of microorganisms capable of producing proteolytic enzymes with the function of hydrolyzing proteins into shorter peptides (Cruz-Casas et al. 2021). Proteases produced by microorganisms such as *Aspergillus niger*, *A. oryzae*, *Bacillus subtilis,* and *Lactobacillus* are commonly used (Zhu et al. 2024).

The principle of microbial fermentation lies in handling the appropriate substrate and microorganism and the ideal environmental conditions such as pH, temperature, and humidity to produce peptides with better bioactivity (Cruz-Casas et al. 2021). Microbial fermentation can be carried out using different methods, of which submerged fermentation and solid-state fermentation are the most important. Submerged fermentation involves the culture of microorganisms in a liquid medium, is a system restricted to microorganisms with high water activities, such as bacteria, and offers the advantage that the bioactive peptides produced are easy to purify (Cruz-Casas et al. 2021).

While solid-state fermentation consists of microbial growth on nutrient-rich solid substrates, it offers the use of substrate protein as a nutrient for strain growth and as a target for degradation during the fermentation process, which can reduce waste discharge (Zhu et al. 2024, Cruz-Casas et al. 2021). It also has the advantage of controlled nutrient release, making it suitable for fungi and microorganisms with lower moisture requirements.

In the work by Fideler et al. (2019), the authors used lactic acid bacteria to promote microbial fermentation in cucumbers. The authors isolated and identified five new peptides not previously reported in the literature: IPP, LPP, VPP, KP, and RY. These results show that fermentation using lactic acid bacteria can increase the content of bioactive peptides in vegetables compared to other methods, e.g., acid hydrolysis.

Enzymatic hydrolysis is the most commonly used method to obtain bioactive peptides and involves the incorporation of commercial enzymes derived from plants, animals, or microorganisms to break the peptide bonds of the parent protein, thus releasing the encoded peptide (Zhu et al. 2024, Cruz-Casas et al. 2021). To achieve this, it is essential that the enzyme first binds to the substrate, forming temporary bonds, and then proceeds with enzymatic catalysis so that when the enzyme-substrate complex is in a specific conformation, it can ensure the hydrolysis of the protein (Cruz-Casas et al. 2021).

Enzymatic hydrolysis of food proteins was developed in the 1990s as a biological method using proteases as an alternative to acid or alkaline hydrolysis (Habinshuti et al. 2023). This method is preferable because it produces minimal residual organic solvents or toxic chemicals, and the hydrolysis conditions, such as pH, time, and temperature, can be easily controlled (Marciniak et al. 2018). It is necessary to monitor these parameters during the reaction for an optimal degree of hydrolysis, as this factor is directly related to the size and composition of amino acids and, consequently, to the bioactivity of the peptides generated (Marciniak et al. 2018).

In this sense, different enzymes are used to break peptide bonds, such as pepsin, chymotrypsin, and pancreatin, as well as the commercial proteases Alcalase™, Protamex™, and Flavourzyme™ (Sosalagere et al. 2022). In the work by Kirac et al. (2024), the authors obtained bioactive peptides from white mulberry by enzymatic hydrolysis using pepsin, trypsin, chymotrypsin, and alcalase. The highest hydrolysis efficiency was observed when pepsin was used, as the degree of hydrolysis was higher compared to tests with the other enzymes.

Another method for obtaining bioactive peptides is simulated gastrointestinal digestion (SGD), which is considered an alternative and effective method for generating new bioactive peptides (Ni et al. 2020). This is because one of the difficulties in using BAPs is that many of these compounds, especially those derived from food sources, are susceptible to degradation by the digestive enzyme system, which reduces their potency and bioavailability (Liu and Pischetsrieder 2017, Mohan et al. 2015). Thus, SGD is an alternative that mimics the physiological conditions of the human body, making it possible to understand both the mechanism of absorption of these metabolites by the gut and the effects suffered until they reach the target sites (Ahmed et al. 2022, Mora et al. 2017).

In vitro assessment of the digestive stability of bioactive compounds is carried out using two- and three-phase systems, in which a gastric phase is modeled with pepsin at pH 3.0, while a subsequent intestinal phase involves incubation with pancreatic and brush border proteases at pH 7.0, providing a more critical proteolytic environment of the gastrointestinal tract and allowing the biostability of BAPs to be assessed (Ahmed et al. 2022). Although this technique is widely used to obtain peptides in different matrices, information on the application of SGD to obtain bioactive peptides in fruit is still scarce in the literature.

On the other hand, proteins are not always easily accessible, and the use of strategies to achieve higher yields is necessary (Mora and Toldrá 2023). Therefore, an effective and efficient protein extraction method is essential for the production of new BAPs (Mora and Toldrá 2023). The methods commonly used to extract these metabolites include maceration with alkaline solutions using the reagents NaOH and KOH, which have several disadvantages such as longer treatment time, higher solvent consumption, and lower extraction efficiency (Karki et al. 2025).

Therefore, new technologies, including methods of physical, chemical, and biological modification of the sample, are increasingly being used to modify the technical, functional, and structural properties of the protein, thereby increasing extraction efficiency (Karki et al. 2025). The most widely used current strategies include ultrasound-assisted extraction (UAE), pressurized liquid extraction, deep eutectic solvent extraction, microwave-assisted extraction (MAE), pulsed electric field (PEF) and high-voltage electric discharge (HEVD) extraction, nanomaterial extraction, subcritical fluid extraction (S_bFE), and supercritical fluid extraction (SFE) (Karki et al. 2025, Liu et al. 2025a, Sánchez-Elvira et al. 2024, Mora and Toldrá 2023).

In addition to these factors, obtaining the protein and releasing the peptide in the primary structure of the protein, the Design of Experiments (DoE) approach is used during *in silico* determinations. DoE has been used to study the effect of different hydrolysis parameters and to optimize the release of bioactive peptides by combining it with response surface methodology (Mora and Toldrá 2023). This makes it possible to achieve the optimal hydrolysis condition with the minimum number of experiments, i.e., to investigate the combination of factors that gives the desired response (Mendes et al. 2024).

Table 1. Bioactive peptides in different fruits.

Fruits	**Amino acid sequence***	**Health effects**	**References**
Tiger nut (*Cyperus esculentus*)	QPPSQPQPML, FLDPLIPPI and DEQPRPIPF	Antioxidant activity	Quan et al. 2025
Corn (*Zea mays*)	EWYF and EWFY	Antioxidant activity	Wayal et al. 2025
Hazelnut (*Corylus heterophylla*)	DNIQGITKPAIR, SDNAFGGR and YYLLVR	Antihypertensive activity	Song et al. 2025
Pepper (*Litsea cubeba*)	LSPRDAVGH, NAQRDAGIV, EQRPTYSN, EGDVFAVPR, NAQRDAGLL, HQEGPFVR, QEGPFVRV, DAGIVEGPK, ALVVDGK, ESEREVQ, LAGRDNIY, QEGPFVR, NAQRDAGIVEGPK, AQRDAGLL, AQRDAGIVEGPKGS, RDAGIVEGPKGS, EEVVHPY, GFDERVL, DVPPPRGPL and AGIVEGPKG	Antioxidant activity	Li et al. 2024
Walnut (*Juglans regia*)	LPFA, VPFWA and WGLP	Hypoglycemic and antioxidant activities	Mu et al. 2024
Moringa (*Moringa oleífera*)	KETTTIVR	Hypoglycemic activity	Mu et al. 2024
Green coffee (*Coffea robusta*)	IKK, LY, AH, EL, PHW, PHY, RHQ, VKL, KD, YVL, IR, LK, TY, TFE, VY and FLPE	Antioxidant activity	Ribeiro et al. 2021
Jackfruit (*Artocarpus heterophyllus*)	SSNSLGNR and VGPWQK	Antioxidant activity	Chai et al. 2021
Cowpea (*Vigna unguiculata*)	IAF, QGF e QDF	Hypercholesterolemic activity	Silva et al. 2021
Chickpea (*Cicer arietinum* L.)	FEI, FEL, FIE, FKN, FGKG, and MEE	Antioxidant activity	Quintero-soto et al. 2021
Potato (*Solanum tuberosum*)	IF	Antioxidant and antihypertensive activities	Tsai et al. 2020

Table 1 contd. ...

...Table 1 contd.

Fruits	Amino acid sequence*	Health effects	References
Raw olives (*Manzanilla*)	FDGEVK, VPLSPT, FAVV, KALM, KGAL, SSPL, and TLVY	Hypolipidemic effect	Prados et al. 2020
Jujube (*Zizyphus jujuba*)	IER and IGK	Antihypertensive activity	Memarpoor-Yazdi et al. 2020
Palm kernel (*Elaeis guineensis*)	QFGSSQESPFQSSRR, PRRADVFNPR, QLDRSHRQFLLAGR, NGLQVLRPS, QMGSSQESVRRAFVMR, NNQRTWLAGR, QRTWLAGR, TFVLPSDRK, MPALFTPR and YPVHSYNLVMK	Antifungal activity	Asri et al. 2020
Chinese quince (*Chaenomeles speciosa*)	RHAKF and NYRRE	Antioxidant activity	Deng et al. 2020
Pickling cucumbers (*Cucumis sativus*)	IPP, LPP and VPP	Anti-inflammatory activity	Fideler et al. 2019
Peach (*Prunus persica*)	IFSPR, IYTPH, VAIP and IYSPH	Antihypertensive activity	Vásquez-Villanueva et al. 2019

* A: Alanine; C: Cysteine; D: Aspartic acid; E: Glutamic acid; F: Phenylalanine; G: Glycine; H: Histidine; I: Isoleucine; K: Lysine; L: Leucine; M: Methionine; N: Asparagine; P: Proline; Q: Glutamine; R: Arginine; S: Serine; T: Threonine; V: Valine; W: Tryptophane; Y: Tyrosine

Table 1 summarizes the papers published in the last six years (2019–2025) that focus on producing bioactive peptides from fruit using various protein proteolysis protocols combined with extraction strategies. Although fruits are promising sources of BAPs, studies are still limited to the species most consumed in the diet of different countries, highlighting potato, walnut, jujube, coffee, and pepper. Furthermore, few studies have adopted a comprehensive approach, including production, purification, structural characterization, and investigation of the bioactive potential of the peptides using *in vivo* and *in vitro* assays, to pave the way for the transformation of these compounds into functional foods and nutraceuticals.

5.3 Characterization of Bioactive Peptides from Fruits

Peptides from fruits obtained by extraction and proteolysis are characterized through the application of analytical techniques to identify their sequences using mainly High-Performance Liquid Chromatography (HPLC) coupled with Tandem Mass Spectrometry (MS) (Colombo et al. 2024, Olivares-Galván et al. 2020). This peptidomics-based analysis makes it possible to have an effective and suitable method to obtain the mass and sequence of peptides in several different matrices (Colombo et al. 2024). The bioactivity of peptides depends on knowledge of their compositions and the sequencing of the amino acids of the structure (Chelliah et al. 2021).

The identification of peptides in databases using results obtained by mass spectrometry is indispensable and of utmost importance. Usually, parent proteins are not completely sequenced and unavailable in databases and, in this situation, it is necessary to determine the primary amino acid sequence of a given peptide based exclusively on the data obtained by MS, that is, without consulting databases (*de novo* sequencing) (Kuptawach et al. 2024, Ng et al. 2023, Olivares-Galván et al. 2020, Hernández-Corroto et al. 2020, 2019, González-García et al. 2015, 2014). In other cases, when the protein from fruits is present in databases, a homology-based method enables the identification of the peptide sequence (Olivares-Galván et al. 2020, Meshginfar et al. 2018).

MS analysis provides valuable information about the mass of a specific peptide and, in addition, can generate amino acid sequence information from tandem mass spectra. Peptide tracking by MS is viable using electrospray ionization (ESI) and matrix-assisted laser desorption/ionization (MALDI), since soft ionization maintains the polypeptide structure and allows the transfer of ions to the gas phase without internal fragmentation, being performed in positive mode (Olivares-Galván et al. 2020). The TOF (time-of-flight) analyzer and its configurations (e.g., Q-TOF) are the most widely used in peptidomics approaches because they present high precision and sensitivity (Colombo et al. 2024).

Four peptides with molecular masses of 1267.672 Da (EHIMPLGQNGR), 1682.694 Da (QPSPQDYLNAHNAAR), 1925.806 Da (YGENIAWSSGDLSGTAAVK), and 2778.353 Da (SEVGVPNLPWDDTVAAYAQNYANQR) were identified in the most antioxidant fraction obtained from *Geoffroea decorticans* seeds. MALDI-TOF evaluated the resulting peptide mixture obtained by tryptic digestion–TOF MS/MS to determine their amino acid sequences (Cotabarren et al. 2021). MALDI ionization preferentially results in monocharged ions and is much less prone to ionization suppression effects than ESI (Puchalska et al. 2015).

A complete characterization of bioactive peptides usually requires bidimensional chromatographic methods, for example, in the rigorous analysis of peach seed hydrolysates by LC-Q-q-TOF-MS/MS using hydrophilic interaction chromatography (HILIC) and reverse phase, RP-HPLC-MS/MS and HILIC-MS/MS or both modes in series, which allowed the identification of 33 bioactive compounds (Colombo et al. 2024, Vásquez-Villanueva et al. 2019, 2018). The separation of analytes by HILIC is based on the interaction with a hydrophilic stationary phase. However, HILIC uses water-miscible solvents (e.g., acetonitrile), and elution is acquired by a water gradient that makes this technique compatible with coupling with MS (Olivares-Galván et al. 2020).

The investigation of tomato seed peptides in the most active fractions was performed using a nano-liquid chromatography system coupled to a Q-TOF equipped with a nanoelectrospray ionization (nano-ESI) source. NanoLC–MS/MS analysis revealed peptides with molecular masses in the range of 400 and 3600 Da, with 27% of the identified peptides being below 600 Da, with 6 or fewer amino acids in their sequences (Moayedi et al. 2018).

Recently, Ferrara et al. (2025) studied grape seeds during the fermentation process, highlighting their richness in bioactive peptides from the analysis using an instrumental setup with an ultra-high-performance liquid chromatography system coupled to the Q-Exactive Orbitrap mass spectrometer (UHPLC-nanoESI-Orbitrap). The samples showed about 80% of the peptides deamidated. The data of the identified peptides were investigated using the Biopep algorithm.

Li et al. (2024b) employed an instrument Q Exactive™ Hybrid Quadrupole-Orbitrap™ for the identification of antioxidant peptides from the fraction of greatest activity obtained from *Litsea cubeba* fruits. MS-collected raw files were investigated by the PEAKS Studio 10.6 software database, and the most abundant active peptides were selected. Experimental data indicated 89.64% of peptides with MW < 1000 Da and 9.45% of the polypeptide fraction with MW > 1000 Da.

5.4 Mechanism of Action of Bioactive Peptides

Bioactive peptides have commonly shown potent antioxidant activity against free radicals and other reactive species (Akbarian et al. 2022). Excessive production of reactive oxygen species (ROS) and problems in their removal can lead to the onset of oxidative stress, which is responsible for several serious diseases, such as heart disease, cancer, obesity, diabetes, as well as severe neurological disorders, such as Alzheimer's syndrome (Colombo et al. 2024, Yang and Lian 2020). Thus, BAPs provide defense to cells against the harmful effects of ROS, thereby decreasing susceptibility to chronic illnesses (Sadiq 2021). As discussed above and based on studies of antioxidant peptides (Colombo et al. 2024, Zhang et al. 2024b, Sadiq 2021, Yang and Lian 2020), a proposal of possible mechanisms of the antioxidant action of peptides against the formation of ROS radicals was made, as described in Figure 1.

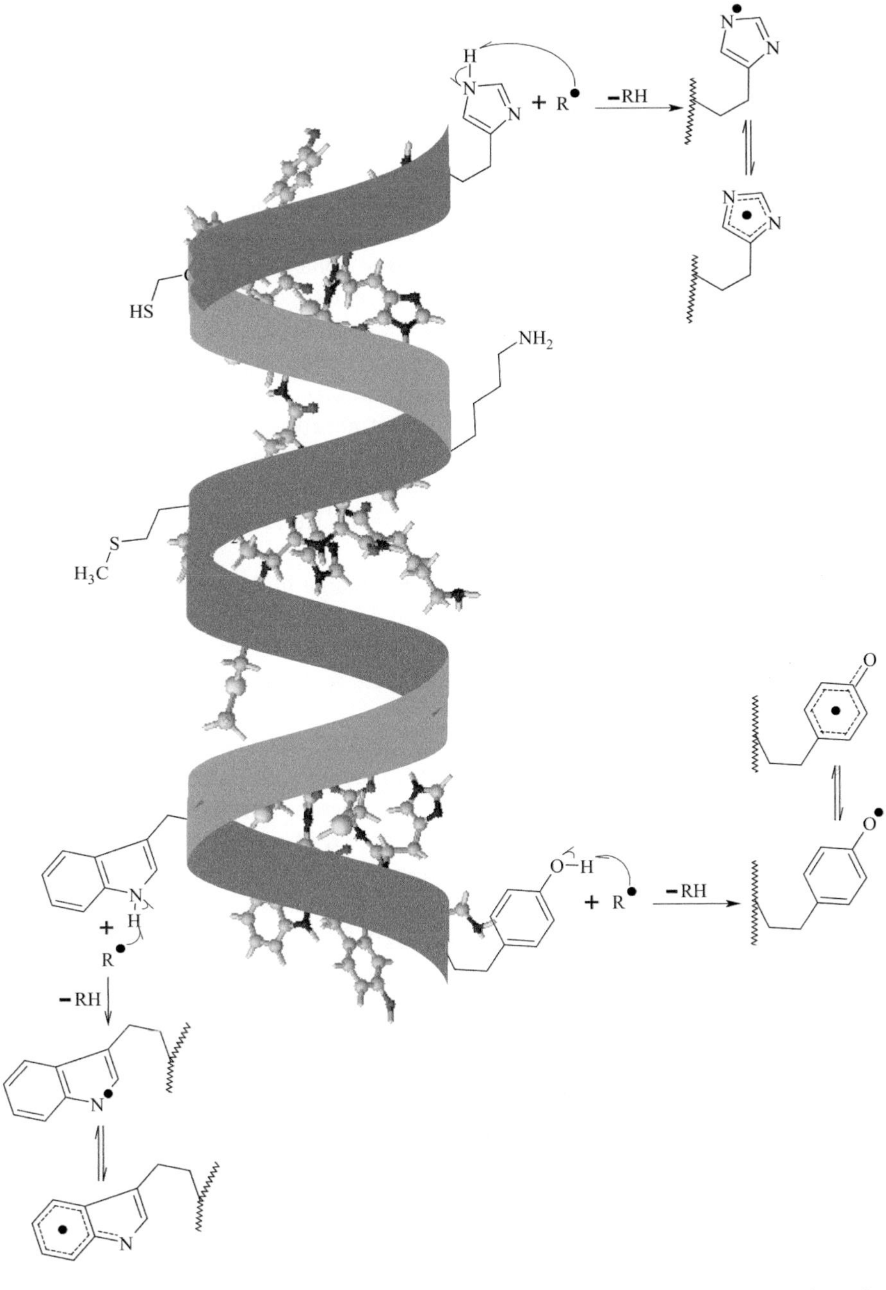

Figure 1. Possible mechanisms of action of the antioxidant potential of peptides against the formation of ROS radicals. R˙: includes a range of oxygen radical species, for example, $O_2^{\cdot -}$, HO˙, RO˙, and ROO˙.

Studies show that the prevention of enzymatic and nonenzymatic oxidation is efficient when it comes to the action of hydrolyzed peptides and proteins, through the removal of free radicals and chelation of metal ions, although the mechanism by which these compounds exert their antioxidant effects has not been fully elucidated (Akbarian et al. 2022). Antioxidant peptide action includes mechanisms such as scavenging free radicals, supplying protons and/or electrons, inhibiting lipid peroxidation, chelation of metal ions, and activating the antioxidant defense system of the body (Sharma et al. 2023, Sosalagere et al. 2022).

Histidine, cysteine, lysine, methionine, tryptophan, and tyrosine are efficient in defending the human body against reactive oxygen species, as well as in eliminating radicals (Maestri et al. 2016, Ryan et al. 2011). Aromatic amino acids, such as tyrosine, histidine, tryptophan, and phenylalanine, have been associated with antioxidant properties due to their ability to donate electrons to free radicals, turning them into stable molecules (Olivares-Galván et al. 2020, Dash and Ghosh 2017). Hydrophobic amino acids, such as valine and leucine, increase hydrogen transfer and capture of lipid peroxyl radicals, favoring their passage through cell membranes and promoting their accessibility to hydrophobic targets (Hernández-Corroto et al. 2018, Dash and Ghosh 2017).

Research shows that peptides containing histidine in their composition can act by capturing lipid peroxyl radicals, donating hydrogen, and chelating metal ions due to the imidazole group, while the sulfhydryl group in cysteine can interact directly with free radicals and implement its antioxidant action (Fan et al. 2022a, Brasili et al. 2016).

The GQVPP peptide identified in tomato seeds showed excellent DPPH radical scavenging activity, reaching 97% at 0.4 mM (Moayedi et al. 2018). The aromatic and hydrophobic amino acid residues present in the structure contributed significantly to such an antioxidant effect (Girgih et al. 2014). The DPPH radical scavenging activity and the greatest ferric reduction power of the GQVPP peptide could be associated with the presence of Gln, Val, and Pro (Moayedi et al. 2018).

Implementation of anticancer peptides (ACPs) in therapeutic strategies can promote large advances and non-resistant treatments, offering new alternatives regarding the prevention and cure of diseases and disorders (Ahmed et al. 2023, Lath et al. 2023, Blasio 2021). Some characteristics such as high selectivity, viability for modification, and efficient penetration in cancer cells make ACPs very promising (Chiangjong et al. 2020).

Peptides can act as anticancer agents through different mechanisms, by destroying the cell membrane, inducing premalignant and malignant cells to undergo apoptosis, inhibiting angiogenesis, and regulating immune responses (Colombo et al. 2024, Lath et al. 2023). The efficiency of ACPs in targeting cancer cells is due to various aspects such as orientation, reduced size, net positive charge, as well as amphipathic nature and hydrophobicity (Lath et al. 2023).

The bioactive peptides are readily found in fruits and vegetables, enabling the future development of novel bioactive peptides with anticancer properties (Bidram and Ganjalikhany 2024, Lath et al. 2023, Sosalagere et al. 2022). Studies show the ability of peptides to stimulate apoptosis and inhibit angiogenesis, which are the main factors affecting metastasis of tumors (Mejia and Dia 2010). The anticancer peptide (BG-4) obtained from *Momordica charantia* was evaluated for toxicity against ovarian cancer cells (COV318 and A27801AP) and showed a significant reduction of 20% in the viable cell count of COV318 and 65% of A27801AP (Lath et al. 2023, Bloom 2018). The mechanism involved in the decrease in viable cell count was related to higher apoptotic induction (from 7% to 24%) in the A27801AP cell line with a BG-4 concentration of 250 μg/mL (Bloom 2018). These findings support the hypothesis that there is anticancer potential in bioactive peptides from fruits, although the mechanisms of inhibition of tumor cell proliferation need to be studied in detail.

5.5 Diverse Biological Activities of Bioactive Peptides

Bioactive peptides, specific protein fragments, are characterized by their positive impact on disease prevention and the maintenance of quality of life and are beneficial to human health (Espitia et al. 2012). These molecules can enter the cell where they exert their biological functions, including antioxidant, antihypertensive, hypoglycemic, hypolipidemic, and hypocholesterolemic activities, as well as antifungal, antibacterial, and anticarcinogenic activities (Figure 2) (Jahandideh et al. 2022). The potential of these compounds is related to peptide sequence, amino acid distribution, and hydrophobicity/hydrophilicity or peptide conformation, L or D configuration—factors that can directly influence the bioactivity of BAPs (Colombo et al. 2024). In view of this, various studies have been carried out to understand the role of bioactive peptides in biological processes in the human body. Table 1

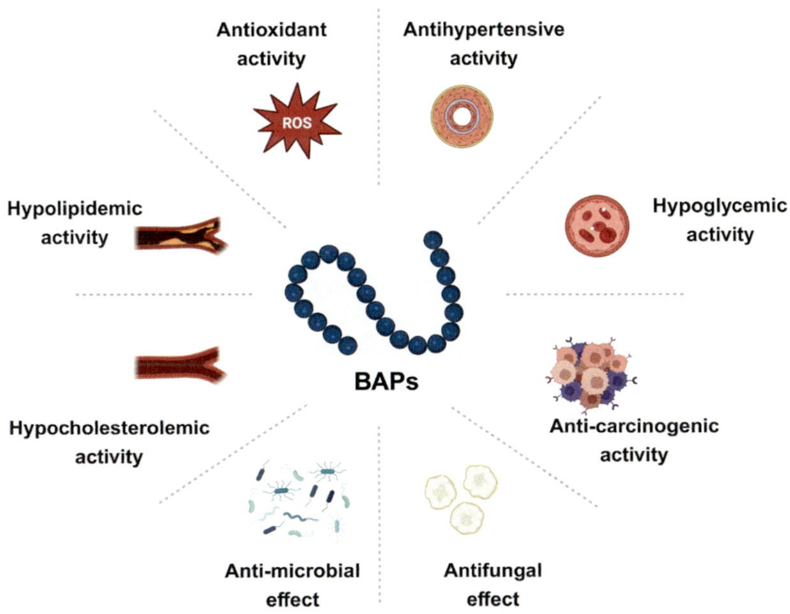

Figure 2. Physiological functions related to bioactive peptides (BAPs).

summarizes information on peptides identified in different fruits and their bioactive potential.

5.5.1 Antioxidant Activity

Cellular oxidation, induced by reactive oxygen species (ROS), is a harmful factor associated with the development of several human diseases and is considered detrimental to health (Guo et al. 2020, Espitia et al. 2012). These reactive species are naturally generated during respiration and also arise through secondary reactions, disrupting homeostasis in the body and negatively affecting human health (Sarmadi and Ismail 2010).

Therefore, protection against free radical-induced oxidation is a key factor in disease prevention. Currently, treatment strategies involve the use of synthetic antioxidants. However, these compounds are subject to regulatory oversight due to potential health risks and toxic effects (e.g., allergenic potential) (Guo et al. 2020). As a result, there has been growing research interest aimed at replacing synthetic antioxidants with natural alternatives (Guo et al. 2020).

In recent years, bioactive peptides with antioxidant potential have attracted significant research interest due to their ability to combat reactive oxygen species (Guo et al. 2020). These bioactive compounds, typically low-molecular-weight peptides composed of 5 to 16 amino acid residues, are considered safe, healthy, cost-effective, highly active, and easily absorbed (Sarmadi and Ismail 2010). In addition, bioactive peptides offer several advantages over enzymatic antioxidants, including a simpler structure, greater stability under various human physiological conditions, and a lower risk of triggering immune reactions (Tonolo et al. 2023, Sarmadi and Ismail 2010).

The antioxidant activity of peptides can be exerted through three main mechanisms: hydrogen atom transfer (HAT), single electron transfer (SET), and transition metal chelation (Tonolo et al. 2023). In the HAT mechanism, the antioxidant peptide donates a hydrogen atom to stabilize free radicals, whereas in the SET mechanism, it donates an electron to the radical (Tonolo et al. 2023, Aguilar-Toalá and Liceaga 2021, Sarmadi and Ismail 2010).

In addition to transfer mechanisms, antioxidant peptides can also act by chelating pro-oxidant transition metals, forming stable complexes with transition metals such as Fe^{2+} and Cu^{2+}. This limits their participation in hydroxyl radical generation via the Fenton reaction, thereby reducing oxidative stress (Tonolo et al. 2023). Figure 3 illustrates the three antioxidant reaction mechanisms involving peptides.

In the study conducted by Quan et al. (2025), the authors isolated and identified bioactive peptides derived from tiger nut (*Cyperus esculentus*). The peptides QPPSQPQPML, FLDPLIPPI, and DEQPRPIPF exhibited promising antioxidant potential in electron transfer-based assays, including 2,2-diphenyl-1-picrylhydrazyl (DPPH), 2,2′-azino-bis(3-ethylbenzothiazoline-6-sulfonic acid) (ABTS), and Ferric Reducing Antioxidant Power (FRAP). The authors also assessed the antioxidant potential of the peptides through their metal-chelating activity with Fe^{2+}. In all assays, the peptides demonstrated excellent antioxidant capacity, with QPPSQPQPML

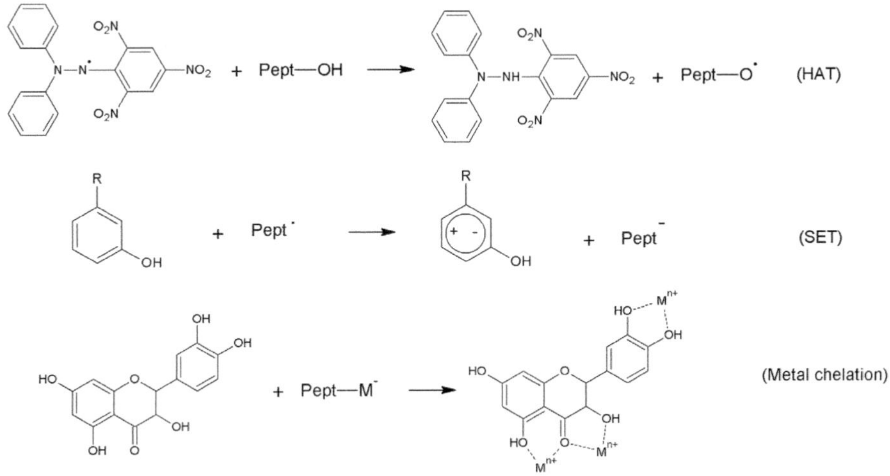

Figure 3. Reaction scheme of antioxidant peptides (Pept) by hydrogen transfer (HAT), single electron transfer (SET), and metal chelation.

standing out for exhibiting greater antioxidant activity than the positive control, glutathione.

In the work by Tsai and colleagues (2020), the authors isolated and identified a bioactive peptide, IF, in potatoes (*Solanum tuberosum*). In an *in vivo* model using rats, the authors investigated whether the bioactive peptide was able to protect against oxidative stress in kidneys damaged by hypertension. The results showed that most of the proteins that act against oxidative stress, including heme oxygenase, glutathione peroxidase 4, superoxide dismutase 1 and 2, and peroxiredoxin, were increased in the kidney tissue of the IF group compared to the control group. Therefore, the authors concluded that the IF peptide protects kidney cells against hypertension-associated oxidative stress by increasing the expression of proteins that act against ROS to improve kidney cell survival in a hypertensive rat model.

5.5.2 Antihypertensive Activity

Systemic arterial hypertension, commonly known as high blood pressure, is a chronic condition that directly contributes to the development of cardiovascular diseases such as arteriosclerosis, stroke, and myocardial injury (Shori and Baba 2015). In this context, the angiotensin-converting enzyme (ACE) plays a key role in blood pressure regulation by catalyzing the conversion of angiotensin I, an inactive decapeptide, into angiotensin II, an active octapeptide with strong vasoconstrictive properties (Kaprasob et al. 2022). Moreover, angiotensin II inhibits bradykinin, a potent vasodilator, further contributing to elevated blood pressure (Liu et al. 2024, Kaprasob et al. 2022, Zheng et al. 2017).

ACE-inhibitory peptides have been explored as an alternative to antihypertensive drugs due to the high cost and undesirable side effects commonly associated with synthetic medications, such as cough, headache, throat irritation, and elevated potassium levels (Fontana et al. 2025, Liu et al. 2024, Shori and Baba 2015). Antihypertensive peptides exert their effect by inhibiting ACE, thereby promoting the relaxation and dilation of blood vessels and improving blood flow (Aiemratchanee et al. 2021). In fact, ACE inhibition by natural compounds is one of the most widely accepted mechanisms for managing hypertension, offering fewer side effects compared to synthetic drugs (Kaprasob et al. 2022, Shori and Baba 2015).

In the study conducted by Kaprasob et al. (2022), novel bioactive peptides were isolated and identified from a protein hydrolysate of king bolete mushroom (*Boletus edulis*) using bromelain as the hydrolyzing enzyme. The researchers identified six peptides with high ACE-inhibitory activity, with peptide KBMPHF4 standing out and exhibiting an IC50 of 0.12 mg mL^{-1}. The authors suggest that these peptides may be promising candidates for use as ACE-inhibitory agents in the food and nutraceutical industries.

In the work by Memarpoor-Yazdi et al. (2020), the authors isolated two bioactive peptides, IER and IGK, from the fruit of the jujube (*Zizyphus jujuba*) using the enzymes trypsin, papain, and their combination. The resulting peptides were purified and evaluated for their ACE inhibitory activity. The results showed that the two tripeptides, IER and IGK, have the ability to inhibit ACE with IC50 values equal to 0.060 and 0.072 mg mL^{-1}, respectively.

5.5.3 Hypoglycemic Activity

Diabetes mellitus is a chronic metabolic disease with a high prevalence rate and has emerged as a significant public health concern, as it is associated with the formation of free radicals and the complication of various diseases such as diabetic ketoacidosis, hyperglycemia, and cardiovascular disorders (Wang et al. 2024a, Ademiluyi et al. 2015). Characterized by persistent hyperglycemia, clinical treatments for managing diabetes mellitus include the use of insulin and oral hypoglycemic agents such as glibenclamide, metformin, sulfonylureas, and inhibitors of α-glucosidase, α-amylase, and dipeptidyl peptidase-4 (Gong et al. 2024, Jiang-Peng et al. 2024).

However, prolonged use of pharmacological agents is associated with adverse effects, including gastrointestinal discomfort, hypoglycemia, and weight gain (Jiang-Peng et al. 2024). According to the 2023 Diabetes Atlas published by the International Diabetes Federation (IDF), the number of people living with diabetes—both type 1 (T1DM) and type 2 (T2DM)—is projected to reach approximately 700 million by 2045 (Pavkov and Miyamoto 2023).

Therefore, there is a growing need for the development of diabetes treatments with minimal side effects. In this regard, bioactive peptides with hypoglycemic potential have been widely investigated as a promising alternative for the treatment of diabetes mellitus, as they have demonstrated the ability to reduce blood glucose levels without the need for synthetic drugs (Yu et al. 2023).

The antidiabetic potential of peptides is related to the hydrophobic and hydrophilic properties of their side chains (R-groups), which enable them to interact with and inhibit key enzymes such as α-amylase, α-glucosidase, and dipeptidyl peptidase-4 (DPP4) (Quintero-Soto et al. 2021). These enzymes play an essential role in the digestion and absorption of carbohydrates and stimulate insulin secretion in the human body (Quintero-Soto et al. 2021). By reducing the activity of these enzymes in the small intestine, peptides can delay or inhibit glucose absorption, effectively lowering blood glucose concentrations (Lu et al. 2023).

The oral administration of bioactive peptides with hypoglycemic potential is considered a safe, accessible, and effective treatment for diabetes. However, the bioavailability of these compounds must be carefully evaluated to ensure their therapeutic efficacy (Peng et al. 2023). Studies indicate that hypoglycemic peptides may be more effective than synthetic drugs due to their high specificity and safety (Lu et al. 2023). Furthermore, they offer superior clinical value, demonstrating improved therapeutic effects and fewer adverse reactions (Peng et al. 2023).

In the study conducted by Quintero-Soto et al. (2021), the authors evaluated the antidiabetic potential of chickpea (*Cicer arietinum*) protein hydrolysates obtained through enzymatic digestion with alcalase, using albumin and globulin fractions. The isolated and purified peptide FGKV exhibited strong inhibition of glucose-elevating enzymes, reducing α-amylase activity and α-glucosidase activity by 54% and 56%, respectively. The authors concluded that this peptide may be a viable candidate for use as a nutraceutical supplement in the treatment of diabetes mellitus.

In the study by Mu et al. (2024), the authors identified 117 peptide sequences in walnuts and evaluated the dipeptidyl peptidase IV (DPP-IV) inhibitory potential, molecular docking, absorption transport of the peptides in a Caco-2 cell model, and performed oral glucose tolerance tests. Considering the 117 peptides identified, three were found to be highly active DPP-IV inhibitors. The BAPs were identified as LPFA, VPFWA, and WGLP, and molecular docking showed that these three peptides bind to DPP-IV mainly via hydrogen bonds and van der Waals forces. The cell assay showed that the peptides can cross the intact Caco-2 cell monolayer with *in situ* IC(50s) of 267.9 ± 7.2 μM, 325.0 ± 8.4 μM and 350.9 ± 8.3 μM, respectively. Oral glucose tolerance tests showed that the three inhibitory peptides significantly improved glucose metabolism in normal ICR mice.

5.5.4 Hypolipidemic and Hypocholesterolemic Activities

Hypercholesterolemia is a metabolic disorder associated with a high risk of developing cardiovascular diseases (Chai et al. 2020). This condition is characterized by elevated cholesterol levels in the bloodstream and, although commonly treated with synthetic medications, the continuous use of such drugs may lead to undesirable side effects, including digestive disturbances, nausea, increased hepatic transaminases and creatine kinase levels, as well as muscle weakness and headaches (Prados et al. 2023).

In light of these drawbacks, the search for natural alternative compounds for the treatment or prevention of hypercholesterolemia has become increasingly relevant. Among the approaches explored in recent years, bioactive peptides have stood out due to their ability to lower lipid and lipoprotein levels in the blood, thereby preventing excessive cholesterol accumulation in blood vessels by inhibiting lipase activity (Chai et al. 2020). As a result, these metabolites with hypolipidemic and hypocholesterolemic potential are being investigated as effective alternatives with fewer adverse effects for the management of hypercholesterolemia.

Bioactive peptides that act in the control of dyslipidemia perform their functions through various mechanisms, contributing to the reduction of exogenous cholesterol or modulating endogenous levels via metabolic pathways (Li et al. 2024a). The regulation of exogenous cholesterol occurs mainly through the inhibition of dietary cholesterol absorption in the intestine by interfering with the enterohepatic circulation of bile acids and reducing the micellar solubility of cholesterol in the small intestine (Li et al. 2024a).

In addition, bioactive peptides can regulate cholesterol biosynthesis, primarily by inhibiting the enzyme 3-hydroxy-3-methylglutaryl-CoA reductase (HMG-CoAR), one of the main regulators of the mevalonate pathway, which is responsible for endogenous cholesterol synthesis. The modulation of endogenous cholesterol levels depends on the absorption of these peptides into the bloodstream and their bioavailability in target organs (Li et al. 2024b). Studies have shown that many fruit-derived peptides exhibit hypocholesterolemic activity by inhibiting HMG-CoAR activity, making them promising candidates for the development of natural therapies for hypercholesterolemia (Sánchez et al. 2024, Silva et al. 2021).

In the study conducted by Silva et al. (2020), the authors isolated and identified 13 peptides from cowpea (*Vigna unguiculata*) and evaluated their inhibitory activity against HMG-CoA reductase. Among the peptides identified, only three – QGF, IAF, and QDF – demonstrated potential for cholesterol regulation. The authors concluded that these peptides may reduce cholesterol synthesis through a mechanism of regulation similar to that of statins.

5.5.5 Antimicrobial Activity

Antimicrobial peptides are bioactive substances that play a crucial role in the immune system of plants, acting as the first line of defense against infections caused by exogenous microorganisms (Tavares et al. 2024). Their therapeutic potential has been widely studied, particularly due to characteristics that make them especially promising, such as low molecular weight, rapid antimicrobial activity, and a low likelihood of inducing resistance in target microorganisms (Zupin et al. 2022).

Among the main properties of antimicrobial peptides are their broad-spectrum efficacy against various pathogens, including superbugs, their ability to inhibit biofilm formation – a key mechanism of bacterial resistance – and a primary mode of action that is less prone to triggering antibiotic resistance (Wu et al. 2018). In addition, some studies have shown that antimicrobial peptides also exhibit beneficial secondary effects, such as modulating the immune response, enhancing wound healing, and promoting angiogenesis (Corrêa et al. 2023). The wide range of biological activities attributed to antimicrobial peptides is illustrated in Figure 4, which is adapted from Sharma et al. (2025).

Currently, the literature reports the identification of over 2,700 types of antimicrobial peptides, of which approximately 1,000 exhibit antifungal activity (Sharma et al. 2025). Peptides with antifungal potential are particularly attractive,

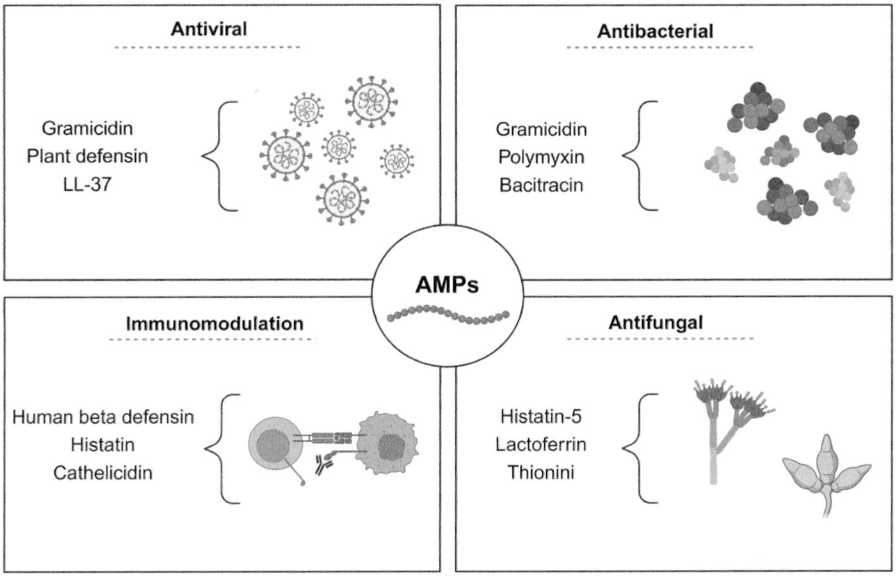

Figure 4. Activities carried out by antimicrobial peptides (AMPs) (modified from Sharma et al. 2025).

as fungi are among the most prevalent contaminants in the food industry, posing a threat to human health due to the damage caused by mycotoxins (Dong et al. 2024). In addition to their importance in food safety, these metabolites play a key role in food preservation by extending shelf life through the inhibition of yeast, mold, and spoilage bacteria (Han et al. 2019).

The growing demand for antifungal agents has intensified due to the limited number of effective options currently available and the emergence of fungal strains resistant to conventional treatments. In the study conducted by Asri et al. (2020), the authors identified ten peptides with antifungal activity, generated from palm kernel cake through a lactofermentation process. These compounds were characterized by the following amino acid sequences: QFGSSQESPFQSSRR, PRRADVFNPR, QLDRSHRQFLLAGR, NGLQVLRPS, QMGSSQESVRRAFVMR, NNQRTWLAGR, QRTWLAGR, TFVLPSDRK, MPALFTPR, and YPVHSYNLVMK. The antifungal potential of these peptides was evaluated in whole wheat bread inoculated with *Aspergillus flavus*, *A. niger*, *Fusarium* sp., and *Penicillium* sp. The results revealed a significant reduction in conidial counts in the treated bread samples, with inhibitory activity ranging from 69% to 88%. Moreover, the peptides effectively delayed fungal growth and extended the shelf life of the bread by up to 10 days.

5.5.6 Anticarcinogenic Activity

Cancer is one of the most lethal diseases worldwide, involving multiple stages, mutational changes, and uncontrolled cell proliferation (Song et al. 2024). Due to its high incidence and mortality rate, numerous studies have sought alternative approaches for cancer prevention and treatment, prioritizing more natural, less aggressive, and cost-effective methods that minimize damage to healthy cells (Sah et al. 2015). Among the promising strategies is the use of bioactive peptides, natural compounds that may exhibit potential anticancer activity (Görgüç et al. 2020).

Over the past decades, numerous studies have been conducted to develop protein hydrolysates or peptides with anticancer properties derived from various food protein sources (Li and Yu 2015). The anticarcinogenic activity of peptides is closely related to their structural and physicochemical characteristics, including amino acid composition, electric charge, hydrophobicity, and spatial conformation. Research indicates that amino acids such as tryptophan, arginine, lysine, and cysteine play a crucial role in peptide interactions with tumor cell membranes (Chalamaiah et al. 2018).

Moreover, hydrophobic residues located in the inner core of the parent proteins facilitate membrane insertion and the modulation of intracellular pathways (Chalamaiah et al. 2018, Li and Yu 2015). As a result, these bioactive compounds from natural sources, such as fruits, induce apoptosis, inhibit cell proliferation, and block tumor angiogenesis, making them promising candidates for the development of novel anticancer therapies (Chalamaiah et al. 2018, Ma et al. 2015).

In the study developed by Ma et al. (2015), the authors isolated a novel biopeptide, CTLEW, and evaluated its anticarcinogenic potential in human breast adenocarcinoma cells (MCF-7). The biopeptide induced apoptosis and autophagy in MCF-7 cells, significantly inhibited the growth of cancer cells such as Caco-2 and HeLa, and showed cytotoxic activity against non-cancerous IEC-6 cells.

5.5.7 Immunomodulatory Activity and Anti-inflammatory Activities

Immunity is a vital physiological function of the human body. It involves a complex defense mechanism that protects against pathogens and external stimuli, thereby maintaining physiological balance (Xu et al. 2025). Dysfunction or impairment of the immune system can increase susceptibility to bacterial and viral infections (Shen et al. 2025). This disorder may be exacerbated by chronic diseases, microbial infections, stress, and poor dietary habits, necessitating the use of immune regulators to modulate the immune response in the human body (Xu et al. 2025).

Immunomodulatory peptides are bioactive fragments consisting of 2–10 amino acid residues with relatively small molecular masses (Yang et al. 2025). They directly influence the immune system, enhancing the body's defense against external pathogens and regulating cytokine and antibody production. They also stimulate lymphocyte proliferation and increase macrophage phagocytosis (Jiang et al. 2025, Yan et al. 2024). The immunomodulatory potential of peptides is typically evaluated based on their ability to produce cytokines and other inflammatory mediators (Yan et al. 2024). BAPs with immunomodulatory activity, especially those associated with anti-inflammatory effects, have gained prominence (Yan et al. 2024).

The inflammatory response in the body involves the recruitment and activation of leukocytes, as well as the release of interleukin (IL)-1, IL-6, and tumor necrosis factor (TNF)-α. This also includes the production of prostaglandin E2 (PGE2) and the activation of important enzymes such as cyclooxygenase (COX)-2 and inducible nitric oxide synthase (iNOS) (Lv et al. 2024, Medeiros et al. 2022). Peptides with anti-inflammatory potential target leukotriene and platelet-activating factor receptors, thereby inhibiting TNF-α secretion and simultaneously inhibiting lipoxygenase (LOX) and COX activity, while reducing interleukin levels (Fan et al. 2024, Lv et al. 2024, Medeiros et al. 2022). Therefore, BAPs with anti-inflammatory properties act by reducing the release of inflammatory cytokines in the body.

In their study, Li et al. (2021) evaluated the anti-inflammatory potential of a peptide fraction (IV2) in mice subjected to a high-fat diet (HFD). This fraction was obtained from *Litsea cubeba* fruit flour using the two-step sequential hydrolysis method with the enzymes alcalase and papain. The authors observed that the peptide fraction modulated the inflammatory response in the mice, reducing the levels of TNF-α and IL-6. Consequently, they confirmed that the IV2 fraction had a restorative effect on the inflammatory lesion induced by the HFD.

5.6 Challenges Associated with Bioactive Peptides as Health-promoting Agents

As has been discussed, bioactive peptides derived from fruits represent an emerging class of health-promoting compounds with immense therapeutic potential. They are particularly attractive due to their natural origin, biodegradability, and multifunctional properties. Nevertheless, the translation of these peptides from bench to market is hindered by several scientific, technological, and regulatory challenges. The success of bioactive peptides in functional foods or nutraceuticals will ultimately depend on overcoming issues related to bioavailability, toxicity, allergenic potential, and sensory acceptability (Figure 5), which are deeply interconnected and rooted in peptide chemistry, host biology, and processing technologies (Patil et al. 2022, Daliri et al. 2017).

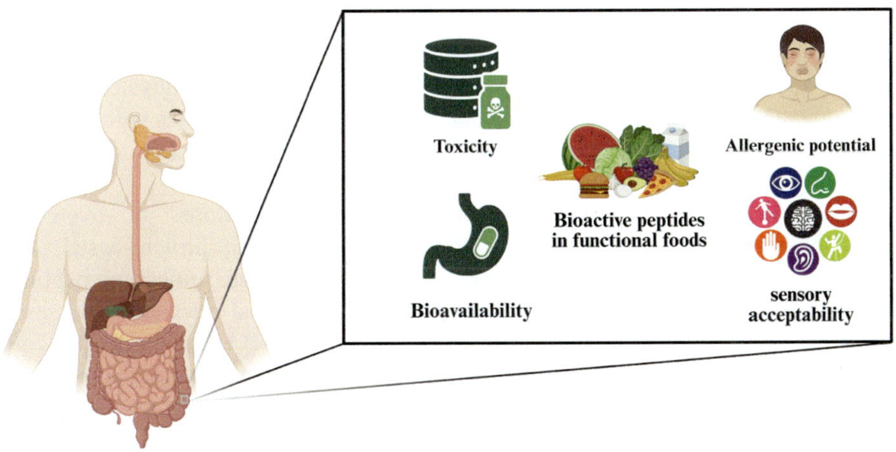

Figure 5. Crucial factors for the success of bioactive peptides as functional food raw materials.

5.6.1 Bioavailability and Bioaccessibility

The successful application of fruit-derived bioactive peptides (BAPs) in functional foods or therapeutic strategies critically depends on two interconnected parameters: bioaccessibility—the fraction of peptides released from the food matrix during digestion—and bioavailability, which refers to the fraction that is absorbed and reaches systemic circulation in biologically active form. Although numerous *in vitro* studies have demonstrated promising biological effects, such as antioxidant and antihypertensive activities, the real challenge lies in preserving their activity throughout the complex digestive and metabolic journey in the human body (Rizzello et al. 2016).

In the gastrointestinal tract, BAPs face immediate degradation by proteolytic enzymes including pepsin, trypsin, and chymotrypsin, often breaking down into smaller, biologically inactive fragments (Pimentel et al. 2020). This enzymatic instability is a well-known issue for many peptides, such as β-casomorphin-7 and GLP-1 analogues, which are quickly hydrolyzed by human peptidases or dipeptidyl peptidase-IV in the plasma, thus limiting their systemic effectiveness (Pascale et al. 2024, Malm-Erjefält et al. 2010). The degradation profile is heavily dependent on peptide size, sequence, and specific residues—particularly those that influence interaction with hydrolases or promote aggregation.

Even when partially protected against enzymatic hydrolysis, most BAPs show poor intestinal permeability. Only small peptides, mainly di- and tripeptides, are efficiently absorbed through PepT1 transporters, which display a strong preference for proline-rich or hydrophobic N-terminal sequences (Xu et al. 2019). However, this mechanism alone is insufficient for many larger or charged peptides. Furthermore, secondary structure plays a crucial role in determining uptake: while amphipathic α-helices may promote membrane interaction, they may also lead to self-aggregation, hindering translocation across the intestinal epithelium (Belwal et al. 2021, Ablan et al. 2016).

Compounding these challenges are the complex interactions between BAPs and the fruit matrix itself. Fruits are naturally rich in dietary fibers, polyphenols, and organic acids, all of which can bind peptides or modulate the physicochemical environment of digestion. Certain polyphenols form stable complexes with peptides, which can either inhibit enzymatic degradation or sequester the peptides, limiting their availability for absorption (Lin et al. 2023, Dias et al. 2019).

Nanoencapsulation strategies have been widely explored to overcome these physiological and biochemical hurdles. Lipid-based, protein-based, and polysaccharide-based nanocarriers have proven effective in protecting peptides from gastrointestinal enzymes and facilitating their targeted release in the small intestine. Encapsulation of short-chain bioactive peptides (BAPs) within lipid-core nanoparticles has been shown to enhance their stability against gastrointestinal degradation and markedly improve their bioaccessibility and absorption in simulated oral delivery models. However, the design of such carriers must be highly specific to the peptide's charge, hydrophobicity, and conformational flexibility to ensure efficiency and food-grade compliance (Atma et al. 2024).

The liver, as well as the kidney, lung, blood, nasal epithelial cells, placenta, and skin, is capable of metabolizing and degrading bioactive peptides, especially those that are not modified to acquire greater stability through the processes of N- and/or C-terminal substitution, D-amino acid substitution, or unnatural amino acids. These unmodified peptides often have a prohibitively short *in vivo* half-life due to rapid proteolytic degradation mainly in the blood, liver, and kidneys by the action of endogenous proteases (Evans et al. 2020, Yao et al. 2018).

Given the multifaceted nature of these challenges, computational and experimental integration has become increasingly relevant. Wang et al. (2024b) emphasized the importance of *in silico* modeling tools combined with dynamic digestion simulations for predicting peptide behavior in complex food systems. These models allow for the evaluation of peptide solubility, digestibility, and permeability before conducting labor-intensive *in vitro* or *in vivo* assays.

In conclusion, the bioavailability and bioaccessibility of fruit-derived BAPs are constrained by digestive instability, low epithelial permeability, matrix-binding interactions, and hepatic clearance. Addressing these issues requires an interdisciplinary strategy combining rational peptide sequence design, protective formulation technologies (such as nanocarriers), realistic digestive simulations, and eventual clinical validation. Only through such multifactorial approaches will the *in vivo* potential of fruit BAPs align with their *in vitro* promise.

5.6.2 Toxicity

While bioactive peptides are generally perceived as safe due to their natural origin, this assumption is misleading without empirical toxicological validation. Peptides may exert unintended bioactivities at pharmacological doses, including cytotoxic, hemolytic, or genotoxic effects, especially when derived from novel enzymatic hydrolysis pathways or samples of plant origin (Pitchakarn et al. 2024).

The cytotoxicity of bioactive peptides has shown promising potential in the search for new therapeutic strategies against cancer. Studies have demonstrated that peptide fractions extracted from Flathead by-products exhibit significant antioxidant activity and can inhibit the growth of colon cancer cells (HT-29) by up to 91% (Nurdiani et al. 2017). However, this activity was found to be non-selective, highlighting the need for further refinement to ensure safety in clinical applications. In contrast, peptides isolated from *Salvia hispanica* seeds—such as KLKKNL, MLKSKR, and KKYRVF—displayed selective cytotoxic effects against various human cancer cell lines, including MCF-7, Caco2, HepG2, DU145, and HeLa. The cell death mechanisms induced by these peptides varied, with KLKKNL primarily triggering apoptosis, while KKYRVF induced necrosis, emphasizing their potential as specific anticancer agents (Bojórquez and Campos 2024). These findings underscore the importance of evaluating selective cytotoxicity as a critical criterion in assessing the efficacy and safety of bioactive peptides for therapeutic applications.

Peptides, particularly epirubicin-loaded Boc-FF dipeptides, demonstrate notable cytotoxic effects against non-small cell lung cancer (NSCLC) cells (A549) by significantly reducing cell viability, enhancing oxidative stress, and promoting DNA damage and apoptosis (Altuntas 2023). Dermaseptin S4 and its N-terminal analogs exhibit marked cytotoxicity against colon cancer cell lines while preserving low hemolytic activity (Altuntas 2023, Belaid et al. 2021). Hemolytic activity is a crucial parameter in the characterization of bioactive peptides, as it reflects their membrane selectivity and potential cytotoxicity. The comparative study of peptides P6 and P6.2 demonstrated that affinity for anionic lipids, such as those found in bacterial membranes (PG), is associated with strong antimicrobial activity and low cytotoxicity, whereas affinity for zwitterionic lipids, typical of erythrocyte membranes (PC), leads to increased hemolytic activity. Therefore, a peptide's ability to preferentially target bacteria over mammalian membranes is key to its therapeutic potential (Maturana et al. 2017). Some antimicrobial peptides (AMPs) are designed to minimize hemolytic activity while maintaining antimicrobial efficacy, as seen in studies of lipid selectivity and peptide design (Maturana et al. 2017, Frecer et al. 2004).

The genotoxicity assessment of bioactive peptides is a critical aspect of ensuring their safety, especially as their use in functional foods and biotherapeutics becomes more widespread. Despite their promising bioactivities, peptides—particularly those derived from meat by-products or modified with non-natural amino acids or chemical linkers—require thorough toxicological evaluation. *In vitro* genotoxicity testing and ADME (absorption, distribution, metabolism, and excretion) studies are considered essential prerequisites before advancing to clinical applications or commercialization. While peptides composed solely of natural amino acids typically do not raise genotoxic concerns, modifications such as conjugation with linkers or incorporation of non-natural residues may introduce mutagenic risks. In such cases, *in vivo* testing is often more appropriate to assess potential genotoxic metabolites. Regulatory guidance, including ICH S6 (R1) and ICH M7, emphasizes a case-by-case approach, considering peptide structure, route of exposure, and likelihood of systemic distribution. The current lack of harmonized global regulations and limited genotoxicity data remains a challenge, underscoring the need for further research to establish safety standards and facilitate international trade of peptide-based products (Madhu et al. 2022, Thybaud et al. 2016).

Conventional toxicity assessments, including acute and subchronic animal studies, *in vitro* cytotoxicity assays, and computational toxicology tools, are indispensable (Singh et al. 2023, Ye et al. 2022). However, they are seldom applied to fruit-derived peptides. This raises regulatory concerns, especially given the growing popularity of "clean label" functional foods that incorporate hydrolysates or extracts with unknown peptide content.

Moreover, the degree of purification significantly influences toxicity profiles. Crude peptide mixtures may contain residual enzymes, phenolics, or fermentation by-products that confound toxicity results. Thus, standardized isolation, purification, and characterization protocols are needed to reliably assess the safety of fruit peptides across different stages of development.

5.6.3 Allergenicity and Bitterness

Bioactive peptides derived from enzymatic hydrolysis of food proteins have aroused great interest in the development of functional foods due to their proven health benefits, such as antioxidant, antihypertensive, antimicrobial, and immunomodulatory activities (Admassu et al. 2018, Maturana et al. 2017). However, despite their promising bioactivity, two main limitations hinder their application: allergenic potential and bitter taste. These factors are related to both safety and sensory acceptance and must be addressed to ensure the successful incorporation of bioactive peptides into food matrices intended for broad consumer use.

There is evidence in the literature that various cytotoxic compounds act selectively on malignant cells, suggesting a possible protective effect against cancer by certain proteins and bioactive peptides. Furthermore, all known food constituents with allergenic potential are protein in nature. Some products resulting from protein degradation, such as peptide fragments, may retain the allergenicity of the original protein, either fully or partially, and may therefore also act as allergens (Sánchez and Vásquez 2017).

The allergenic potential of bioactive peptides is particularly concerning when they are derived from known allergenic sources, such as cow's milk proteins. Peptides can retain or even expose new IgE-binding epitopes depending on the degree and type of hydrolysis. In this context, enzymatic hydrolysis has emerged as an effective strategy to reduce allergenicity by breaking down conformational and linear epitopes. Pang et al. (2025a) demonstrated that using combinations of endopeptidases (such as bromelain and papain) with exopeptidases (such as ProteAXH) on whey protein isolates significantly reduced IgE-binding capacity—by up to 51.9%—while simultaneously enhancing emulsifying and antioxidant properties. This dual benefit of hypoallergenic and functional improvement was attributed to increased peptide diversity and structural disruption caused by enzymatic action.

To further enhance the efficiency of hydrolysis, physical methods such as ultrasound pretreatment have been employed. Ultrasound alters protein conformation, increasing enzyme accessibility to hydrolysis sites. Pang et al. (2025b) reported that sonication prior to hydrolysis led to deeper fragmentation of whey proteins, producing hydrolysates with lower allergenicity (55.6% IgE-binding), higher antioxidant capacity, and better functional performance, such as foaming and emulsifying properties. These findings suggest that combining enzymatic and physical approaches is a promising path for reducing allergenic risks while preserving or even enhancing functional benefits.

Another effective method to reduce allergenicity is microbial fermentation. Liu et al. (2025a) studied *Lactiplantibacillus plantarum* AHQ-14, a strain capable of degrading allergenic proteins such as β-lactoglobulin and α-casein. Peptidomic and genomic analyses confirmed the destruction of known IgE-binding epitopes and the presence of functional genes related to proteolytic systems and peptide transport. These features not only contribute to allergenicity reduction but also enhance the bioactivity and probiotic potential of the fermented product.

In addition to enzymatic and microbial strategies, selecting naturally hypoallergenic sources, such as donkey milk, is a viable alternative. Cunsolo et al. (2017) conducted proteomic analyses showing that donkey milk proteins share higher similarity with human milk proteins and exhibit fewer known allergenic epitopes compared to bovine proteins. These intrinsic characteristics explain its low immunogenicity and support its use for populations with cow's milk protein allergy.

The study by Liu et al. (2020) addresses the safety considerations of food protein-derived bioactive peptides, analyzing the potential formation of allergenic and toxic peptides from parental proteins. The authors demonstrate that some potentially toxic and allergenic peptides—often exhibiting biological activities—can be released during the hydrolysis of food proteins. One example includes amatoxins and phallotoxins, two naturally occurring toxic bioactive peptides found in mushrooms. Both exhibit potent hepatotoxicity, primarily through the inhibition of RNA polymerase II, which impairs protein synthesis, leads to cell necrosis, acute liver failure, and, in severe cases, death.

Bioactive peptides derived from plant sources often exhibit a bitter taste, which remains a significant sensory challenge for their application in functional foods and nutraceuticals. This bitterness is primarily attributed to the presence of hydrophobic amino acids located at the C-terminal ends of peptide chains generated during protein hydrolysis. Bitterness mainly arises from the presence of short-chain peptides rich in hydrophobic amino acids—such as leucine, isoleucine, valine, phenylalanine, and tyrosine—which strongly interact with TAS2R bitter taste receptors in the human tongue, resulting in a pronounced and often lingering bitterness that negatively impacts consumer acceptance (Tagliamonte et al. 2024).

To overcome this sensory limitation, enzymatic approaches used to reduce allergenicity can also be optimized to mitigate bitterness. Pang et al. (2025a) noted that the use of exopeptidases in combination with endopeptidases helped cleave terminal hydrophobic residues, thereby reducing bitterness without affecting bioactivity. Similarly, Zhang et al. (2021) demonstrated that enzymatic cross-linking using transglutaminase after Alcalase hydrolysis of soybean protein resulted in structural rearrangements that masked bitter side chains, significantly reducing the perception of bitterness while improving techno-functional properties.

Ultrasound, as previously mentioned for allergenicity mitigation, also shows potential in managing bitterness. Pang et al. (2025b) observed that ultrasound-assisted hydrolysis promoted the release of peptides with lower hydrophobicity and bitterness intensity, while improving their solubility and antioxidant properties. Other strategies to reduce flavor include activated carbon treatments, alcohol extraction, isoelectric precipitation, chromatographic methods, and chemical modifications such as amination, deamination, acetylation, or cross-linking (Mirzapour-Kouhdasht et al. 2023). Furthermore, encapsulation has emerged as a promising technique, offering effective taste masking, preserving biological activity, improving stability, and enabling targeted administration (Aguilar-Toala et al. 2022).

Overall, the integrated application of targeted enzymatic hydrolysis, physical pretreatments, microbial fermentation, and source selection provides a comprehensive solution to overcome both allergenic and sensory barriers. This multifaceted approach is essential for the effective development of safe, bioactive, and palatable peptide ingredients for functional food applications. By aligning food safety and consumer satisfaction, these strategies pave the way for broader use of bioactive peptides in the health-oriented food industry.

5.7 Conclusion and Perspectives

Protein sources like fruits have attracted more attention nowadays because of their sustainability and the trend of plant-based foods. Bioactive peptides derived from fruits have shown great potential for promoting health and developing functional foods and nutraceuticals due to their diverse biological activities, but their practical application is still limited by challenges such as low bioavailability, possible toxicity, bitter taste, and the limited knowledge of fruit proteomes.

To overcome these challenges, future research should focus on utilizing innovative processing technologies and artificial intelligence (AI) as advanced *in silico* methods. This approach will enhance the discovery of bioactive peptides by optimizing peptide design and more efficiently predicting their bioactive profiles. Additionally, understanding peptide–receptor interactions, applying omics techniques in mechanistic studies, conducting clinical trials, and developing and commercializing bioactive peptides from diverse tropical fruits are important steps forward.

By addressing these challenges, fruit-derived bioactive peptides may play a transformative role in preventive healthcare, offering safe and effective alternatives to synthetic drugs while contributing to environmental sustainability and economic growth in the agri-food sector.

References

Ablan, F. D. O., Spaller, B. L., Abdo, K. I. and Almeida, P. F. (2016). Charge distribution fine-tunes the translocation of α-helical amphipathic peptides across membranes. Biophysical Journal 111(8): 1738–1749. https://doi.org/10.1016/j.bpj.2016.08.047.

Ademiluyi, A. O., Oboh, G., Boligon, A. A. and Athayde, M. L. (2015). Dietary supplementation with fermented legumes modulate hyperglycemia and acetylcholinesterase activities in Streptozotocin-induced diabetes. Pathophysiology 22(4): 195–201. https://doi.org/10.1016/j.pathophys.2015.08.003.

Admassu, H., Gasmalla, M. A. A., Yang, R. and Zhao, W. (2018). Bioactive peptides derived from seaweed protein and their health benefits: antihypertensive, antioxidant, and antidiabetic properties. Journal of Food Science 83(1): 6–16. https://doi.org/10.1111/1750-3841.14011.

Aguilar-Toalá, J. E. and Liceaga, A. M. (2021). Cellular antioxidant effect of bioactive peptides and molecular mechanisms underlying: beyond chemical properties. International Journal of Food Science and Technology 56(5): 2193–2204. https://doi.org/10.1111/ijfs.14855.

Aguilar-Toala, J. E., Quintanar-Guerrero, D., Liceaga, A. M. and Zambrano- Zaragoza, M. L. (2022). Encapsulation of bioactive peptides: A strategy to improve the stability, protect the nutraceutical bioactivity and support their food applications. RSC Advances 12(11): 6449–6458. https://doi.org/10.1039/d1ra08590e.

Ahmed, S. A., Taie, H. A. A. and Abdel Wahab, W. A. (2023). Antioxidant capacity and antitumor activity of the bioactive protein prepared from orange peel residues as a by-product using fungal protease. International Journal of Biological Macromolecules 234(October 2022): 123578. https://doi.org/10.1016/j.ijbiomac.2023.123578.

Ahmed, T., Sun, X. and Udenigwe, C. C. (2022). Role of structural properties of bioactive peptides in their stability during simulated gastrointestinal digestion: A systematic review. Trends in Food Science and Technology 120(December 2021): 265–273. https://doi.org/10.1016/j.tifs.2022.01.008.

Aiemratchanee, P., Panyawechamontri, K., Phaophu, P., Reamtong, O. and Panbangred, W. (2021). *In vitro* antihypertensive activity of bioactive peptides derived from porcine blood corpuscle and plasma proteins. International Journal of Food Science & Technology 56(5): 2315–2324. https://doi.org/10.1111/ijfs.14853.

Akbarian, M., Khani, A., Eghbalpour, S. and Uversky, V. N. (2022). Bioactive peptides: synthesis, sources, applications, and proposed mechanisms of action. International Journal of Molecular Sciences 23(3). https://doi.org/10.3390/ijms23031445.

Altuntas, S. (2023). The effects of epirubicin-loaded Boc-L-Diphenylalanine peptide nanoparticles on cytotoxicity, genotoxicity, oxidative stress, and apoptosis in non-small cell lung cancer cells. Food and Chemical Toxicology 175(February): 113690. https://doi.org/10.1016/j.fct.2023.113690.

Atma, Y., Murray, B. S., Sadeghpour, A. and Goycoolea, F. M. (2024). Encapsulation of short-chain bioactive peptides (BAPs) for gastrointestinal delivery: a review. Food and Function 15(8): 3959–3979. https://doi.org/10.1039/d3fo04195f.

Bakr Shori, A. and Salihin Baba, A. (2015). Fermented milk derives bioactive peptides with antihypertensive effects. Integrative Food, Nutrition and Metabolism 2(3): 180–183. https://doi.org/10.15761/IFNM.1000126.

Blasio, B. F. (2021). Infectious disease threats. Meeting the Challenges of Existential Threats Through Educational Innovation: A Proposal for an Expanded Curriculum (1st ed.). Routledge. https://doi.org/10.4324/9781003019480.

Belaid, A., Braiek, A., Alibi, S., Hassen, W., Beltifa, A., Nefzi, A. et al. (2021). Evaluating the effect of dermaseptin S4 and its derivatives on multidrug-resistant bacterial strains and on the colon cancer cell line SW620. Environmental Science and Pollution Research 28(30): 40908–40916. https://doi.org/10.1007/s11356-021-13683-2.

Belwal, V. K., Vijayakumar, A. and Chaudhary, N. (2021). Inhibition of huntingtin aggregation by its N-terminal 17-residue peptide and its analogs. Archives of Biochemistry and Biophysics 712(April): 109033. https://doi.org/10.1016/j.abb.2021.109033.

Bidram, M. and Ganjalikhany, M. R. (2024). Bioactive peptides from food science to pharmaceutical industries: Their mechanism of action, potential role in cancer treatment and available resources. Heliyon 10(23): e40563. https://doi.org/10.1016/j.heliyon.2024.e40563.

Bloom, A. D. (2018). BG-4, a bioactive peptide from *Momordica charantia*, promotes BG-4, a bioactive peptide from Momordica charantia, promotes apoptosis in ovarian cancer cells apoptosis in ovarian cancer cells. 3–4. https://trace.tennessee.edu/utk_chanhonoproj/2178.

Bojórquez, N. del C. Q. and Campos, M. R. S. (2024). Novel peptides derived from *S. hispanica* seeds induce selective cytotoxicity on human cancer cells. Food Chemistry 460(April). https://doi.org/10.1016/j.foodchem.2024.140470.

Brasili, D., Watly, J., Simonovsky, E., Guerrini, R., Barbosa, N. A., Wieczorek, R. et al. (2016). The unusual metal ion binding ability of histidyl tags and their mutated derivatives. Dalton Transactions 45(13): 5629–5639. https://doi.org/10.1039/c5dt04747a.

Chai, K. F., Voo, A. Y. H. and Chen, W. N. (2020). Bioactive peptides from food fermentation: A comprehensive review of their sources, bioactivities, applications, and future development. Comprehensive Reviews in Food Science and Food Safety September: 3825–3885. https://doi.org/10.1111/1541-4337.12651.

Chai, T. T., Xiao, J., Mohana Dass, S., Teoh, J. Y., Ee, K. Y., Ng, W. J. et al. (2021). Identification of antioxidant peptides derived from tropical jackfruit seed and investigation of the stability profiles. Food Chemistry 340(August 2020): 127876. https://doi.org/10.1016/j.foodchem.2020.127876.

Chalamaiah, M., Yu, W. and Wu, J. (2018). Immunomodulatory and anticancer protein hydrolysates (peptides) from food proteins: A review. Food Chemistry 245(September 2017): 205–222. https://doi.org/10.1016/j.foodchem.2017.10.087.

Chelliah, R., Wei, S., Daliri, E. B. M., Elahi, F., Yeon, S. J., Tyagi, A. et al. (2021). The role of bioactive peptides in diabetes and obesity. Foods 10(9). https://doi.org/10.3390/foods10092220.

Chiangjong, W., Chutipongtanate, S. and Hongeng, S. (2020). Anticancer peptide: Physicochemical property, functional aspect and trend in clinical application (Review). International Journal of Oncology 57(3): 678–696. https://doi.org/10.3892/ijo.2020.5099.

Chourasia, R., Chiring Phukon, L., Abedin, M. M., Padhi, S., Singh, S. P. and Rai, A. K. (2023). Bioactive peptides in fermented foods and their application: a critical review. Systems Microbiology and Biomanufacturing 3(1): 88–109. https://doi.org/10.1007/s43393-022-00125-4.

Colombo, R., Pellicorio, V., Barberis, M., Frosi, I. and Papetti, A. (2024). Recent advances in the valorization of seed wastes as source of bioactive peptides with multifunctional properties. Trends in Food Science and Technology 144(August 2023): 104322. https://doi.org/10.1016/j.tifs.2023.104322.

Corrêa, J. A. F., de Melo Nazareth, T., Rocha, G. F. da and Luciano, F. B. (2023). Bioactive antimicrobial peptides from food proteins: perspectives and challenges for controlling foodborne pathogens. Pathogens 12(3). https://doi.org/10.3390/pathogens12030477.

Cotabarren, J., Ozón, B., Claver, S., Garcia-Pardo, J. and Obregón, W. D. (2021). Purification and identification of novel antioxidant peptides isolated from *Geoffroea decorticans* seeds with anticoagulant activity. Pharmaceutics 13(8). https://doi.org/10.3390/pharmaceutics13081153.

Cruz-Casas, D. E., Aguilar, C. N., Ascacio-Valdés, J. A., Rodríguez-Herrera, R., Chávez-González, M. L. and Flores-Gallegos, A. C. (2021). Enzymatic hydrolysis and microbial fermentation: The most favorable biotechnological methods for the release of bioactive peptides. Food Chemistry: Molecular Sciences 3. https://doi.org/10.1016/j.fochms.2021.100047.

Cunsolo, V., Salete, R., Muccilli, V., Gallina, S., Francesco, A. D. and Foti, S. (2017). Proteins and bioactive peptides from donkey milk: The molecular basis for its reduced allergenic properties. Food Research International 99: 41–57. https://doi.org/10.1016/j.foodres.2017.07.002.

Daliri, E. B. M., Oh, D. H. and Lee, B. H. (2017). Bioactive peptides. Foods 6(5): 1–21. https://doi.org/10.3390/foods6050032.

Dash, P. and Ghosh, G. (2017). Proteolytic and antioxidant activity of protein fractions of seeds of *Cucurbita moschata*. Food Bioscience 18(December 2016): 1–8. https://doi.org/10.1016/j.fbio.2016.12.004.

Mendes, M. K. D. A., Oliveira, C. B. D. S., Medeiros, C. M. D. S., Dantas, C., Carrilho, E., Nogueira, A. R. D. A. et al. (2024). Application of experimental design as a statistical approach to recover bioactive peptides from different food sources. Food Science and Biotechnology 33(7): 1559–1583. https://doi.org/10.1007/s10068-024-01540-0.

Pascale, S.D., Picariello, G., Troise, A.D., Caira, S., Pinto, G., Marino, F. et al. (2024). *Ex vivo* degradation of *β*-Casomorphin-7 by human plasma peptidases: Potential implications for peptide systemic effects. Journal of Functional Foods 113(November 2023): 106004. https://doi.org/10.1016/j.jff.2024.106004.

Deng, Y., Huang, L., Zhang, C., Xie, P., Cheng, J., Wang, X. et al. (2020). Skin-care functions of peptides prepared from *Chinese quince* seed protein: Sequences analysis, tyrosinase inhibition and molecular docking study. Industrial Crops and Products 148(January): 112331. https://doi.org/10.1016/j.indcrop.2020.112331.

Dias, R., Brás, N. F., Pérez-Gregorio, M., Fernandes, I., Mateus, N. and Freitas, V. (2019). A multi-spectroscopic study on the interaction of food polyphenols with a bioactive gluten peptide: From chemistry to biological implications. Food Chemistry 299(June): 125051. https://doi.org/10.1016/j.foodchem.2019.125051.

Dong, J., Zhu, L., Sun, Z., Mandlaa and Chen, Z. (2024). Purification and structural characterization of APTs, a novel antifungal peptide produced by *Lactobacillus paracasei* ALAC-4, and its inhibitory effect on *Candida albicans*. Lwt 199(January): 116077. https://doi.org/10.1016/j.lwt.2024.116077.

Du, Z., Ding, X., Xu, Y. and Li, Y. (2023). UniDL4BioPep: A universal deep learning architecture for binary classification in peptide bioactivity. 24(April): 1–10.

Espitia, P. J. P., Soares, N. de F. F., Coimbra, J. S. dos R., Andrade, N. J. de, Cruz, R. S. and Medeiros, E. A. A. (2012). Bioactive peptides: synthesis, properties, and applications in the packaging and preservation of food. Comprehensive Reviews in Food Science and Food Safety 11(2): 187–204. https://doi.org/10.1111/j.1541-4337.2011.00179.x.

Evans, B. J., King, A. T., Katsifis, A., Matesic, L. and Jamie, J. F. (2020). Methods to enhance the metabolic stability of peptide-based PET radiopharmaceuticals. Molecules 25(10). https://doi.org/10.3390/molecules25102314.

Fan, H., Liu, H., Zhang, Y., Zhang, S., Liu, T. and Wang, D. (2022). Review on plant-derived bioactive peptides: biological activities, mechanism of action and utilizations in food development. Journal of Future Foods 2(2): 143–159. https://doi.org/10.1016/j.jfutfo.2022.03.003.

Fan, S., Li, J., Zhang, X., Xu, D., Liu, X., Dias, A. C. P. et al. (2022). A study on the identification, quantification, and biological activity of compounds from *Cornus officinalis* before and after *in vitro* gastrointestinal digestion and simulated colonic fermentation. Journal of Functional Foods 98(September): 105272. https://doi.org/10.1016/j.jff.2022.105272.

Fan, X., Zhang, L., Sun, Y., Zhou, C. and Xia, Q. (2024). Food science and human wellness identification and the molecular mechanism of novel duck liver-derived. Food Science and Human Wellness 13(6): 3595–3605. https://doi.org/10.26599/FSHW.2023.9250041.

Ferrara, A., D'Auria, G., Barile, D., Baller, M. I., Nitride, C., Mamone, G. et al. (2025). The valorization of grape pomace from Montepulciano winemaking: A new source of functional ingredients for sustainable food industry. Food Research International 200: 115443. https://doi.org/10.1016/j.foodres.2024.115443.

Fideler, J., Johanningsmeier, S. D., Ekelöf, M. and Muddiman, D. C. (2019). Discovery and quantification of bioactive peptides in fermented cucumber by direct analysis IR-MALDESI mass spectrometry and LC-QQQ-MS. Food Chemistry 271(March 2018): 715–723. https://doi.org/10.1016/j.foodchem.2018.07.187.

Fontana, A., Knuf, F., Monasterio, R. and Schieber, A. (2025). Screening of wine industry by-products as a source of bioactive peptides: Fractionation, *in vitro* antihypertensive activity and peptidomics analysis. Food Chemistry 476(October 2024): 143478. https://doi.org/10.1016/j.foodchem.2025.143478.

Frecer, V., Ho, B. and Ding, J. L. (2004). *De novo* design of potent antimicrobial peptides. Antimicrobial Agents and Chemotherapy 48(9): 3349–3357. https://doi.org/10.1128/AAC.48.9.3349-3357.2004.

Girgih, A. T., He, R., Malomo, S., Offengenden, M., Wu, J. and Aluko, R. E. (2014). Structural and functional characterization of hemp seed (*Cannabis sativa* L.) protein-derived antioxidant and antihypertensive peptides. Journal of Functional Foods 6(1): 384–394. https://doi.org/10.1016/j.jff.2013.11.005.

Gong, P., Ke, Y., Wang, X., Long, H., Yang, W., Wang, J. et al. (2024). Structural characterization and hypoglycemic activity of *Cordyceps sinensis* polysaccharide (CSP-3). Food Bioscience 61(August): 104999. https://doi.org/10.1016/j.fbio.2024.104999.

González-García, E., Marina, M. L. and García, M. C. (2014). Plum (*Prunus domestica* L.) by-product as a new and cheap source of bioactive peptides: Extraction method and peptides characterization. Journal of Functional Foods 11(C): 428–437. https://doi.org/10.1016/j.jff.2014.10.020.

González-García, E., Puchalska, P., Marina, M. L. and García, M. C. (2015). Fractionation and identification of antioxidant and angiotensin-converting enzyme-inhibitory peptides obtained from plum (*Prunus domestica* L.) stones. Journal of Functional Foods 19: 376–384. https://doi.org/10.1016/j.jff.2015.08.033.

Görgüç, A., Gençdağ, E. and Yılmaz, F. M. (2020). Bioactive peptides derived from plant origin by-products: Biological activities and techno-functional utilizations in food developments – A review. Food Research International 136(July): 109504. https://doi.org/10.1016/j.foodres.2020.109504.

Guo, H., Guo, S. and Liu, H. (2020). Antioxidant activity and inhibition of ultraviolet radiation-induced skin damage of Selenium-rich peptide fraction from selenium-rich yeast protein hydrolysate. Bioorganic Chemistry 105(October): 104431. https://doi.org/10.1016/j.bioorg.2020.104431.

Habinshuti, I., Nsengumuremyi, D., Muhoza, B., Ebenezer, F., Yinka, A. and Antoine, M. (2023). Recent and novel processing technologies coupled with enzymatic hydrolysis to enhance the production of antioxidant peptides from food proteins: A review. Food Chemistry 423(November 2022): 136313. https://doi.org/10.1016/j.foodchem.2023.136313.

Han, Y., Zhao, J., Zhang, B., Shen, Q., Shang, Q. and Li, P. (2019). Effect of a novel antifungal peptide P852 on cell morphology and membrane permeability of *Fusarium oxysporum*. Biochimica et Biophysica Acta - Biomembranes 1861(2): 532–539. https://doi.org/10.1016/j.bbamem.2018.10.018.

Hernández-Corroto, E., Marina, M. L. and García, M. C. (2018). Multiple protective effect of peptides released from *Olea europaea* and *Prunus persica* seeds against oxidative damage and cancer cell proliferation. Food Research International 106(January): 458–467. https://doi.org/10.1016/j.foodres.2018.01.015.

Hernández-Corroto, E., Marina, M. L. and García, M. C. (2019). Extraction and identification by high resolution mass spectrometry of bioactive substances in different extracts obtained from pomegranate peel. Journal of Chromatography A 1594: 82–92. https://doi.org/10.1016/j.chroma.2019.02.018.

Hernández-Corroto, E., Plaza, M., Marina, M. L. and García, M. C. (2020). Sustainable extraction of proteins and bioactive substances from pomegranate peel (*Punica granatum* L.) using pressurized liquids and deep eutectic solvents. Innovative Food Science and Emerging Technologies 60(December 2019): 102314. https://doi.org/10.1016/j.ifset.2020.102314.

Jahandideh, F., Bourque, S. L. and Wu, J. (2022). A comprehensive review on the glucoregulatory properties of food-derived bioactive peptides. Food Chemistry: X 13(January): 100222. https://doi.org/10.1016/j.fochx.2022.100222.

Jakubczyk, A., Karas, M., Rybczynska-Tkaczyk, K., Zielinska, E. and Zielinski, D. (2020). Current trends of bioactive peptides - New sources and therapeutic effect. Foods 9(7). https://doi.org/10.3390/foods9070846.

Jiang-Peng, S., Jia-Qin, F., Chuang, L., Shou-Ping, L., Chun, C., Chin-Ping, T. et al. (2024). Advances in structure-hypoglycemic activity relationship and mechanisms of berry polysaccharides. Food Bioscience 62(September): 105472. https://doi.org/10.1016/j.fbio.2024.105472.

Jiang, Y., Li, S., Jiang, L., Mu, G. and Jiang, S. (2025). Immunomodulatory activity and molecular mechanisms of action of peptides derived from casein hydrolysate by alcalase and flavourzyme based on virtual screening. Journal of Dairy Science 108(3): 2152–2168. https://doi.org/10.3168/jds.2024-25224.

Kaprasob, R., Khongdetch, J., Laohakunjit, N., Selamassakul, O. and Kaisangsri, N. (2022). Isolation and characterization, antioxidant, and antihypertensive activity of novel bioactive peptides derived from hydrolysis of King Boletus mushroom. Lwt 160(February): 113287. https://doi.org/10.1016/j.lwt.2022.113287.

Karki, S., Prathumpai, W. and Anal, A. K. (2025). Microwave-assisted protein extraction from foxtail millet: Optimization, structural characterization, techno-functional properties, and bioactivity of peptides. International Journal of Biological Macromolecules 293(October 2024): 139312. https://doi.org/10.1016/j.ijbiomac.2024.139312.

Kehinde, B. A. and Sharma, P. (2018). Recently isolated antidiabetic hydrolysates and peptides from multiple food sources: a review. Critical Reviews in Food Science and Nutrition 60(2): 322–340. https://doi.org/10.1080/10408398.2018.1528206.

Kuptawach, K., Noitung, S., Buakeaw, A., Puthong, S., Sawangkeaw, R., Sangtanoo, P. et al. (2024). Lemon basil seed-derived peptide: Hydrolysis, purification, and its role as a pancreatic lipase inhibitor that reduces adipogenesis by downregulating SREBP-1c and PPAR-γ in 3T3-L1 adipocytes. Plos One 19(5). https://doi.org/10.1371/journal.pone.0301966.

Lath, A., Santal, A. R., Kaur, N., Kumari, P. and Singh, N. P. (2023). Anti-cancer peptides: their current trends in the development of peptide-based therapy and anti-tumor drugs. Biotechnology and Genetic Engineering Reviews 39(1): 45–84. https://doi.org/10.1080/02648725.2022.2082157.

Li, J., Bollati, C., d'Adduzio, L., Fanzaga, M., Cruz-Chamorro, I., Arnoldi, A. et al. (2024). Food-derived peptides with hypocholesterolemic activity: Production, transepithelial transport and cellular mechanisms. Trends in Food Science and Technology 143(October 2023): 104279. https://doi.org/10.1016/j.tifs.2023.104279.

Li, L., Wang, Y. M., Zeng, X. Y., Hu, Y., Zhang, J., Wang, B. et al. (2024). Bioactive proteins and antioxidant peptides from *Litsea cubeba* fruit meal: Preparation, characterization and ameliorating function on high-fat diet-induced NAFLD through regulating lipid metabolism, oxidative stress and inflammatory response. International Journal of Biological Macromolecules 280(July). https://doi.org/10.1016/j.ijbiomac.2024.136186.

Li, Y. and Yu, J. (2015). Research progress in structure-activity relationship of bioactive peptides. Journal of Medicinal Food 18(2): 147–156. https://doi.org/10.1089/jmf.2014.0028.

Lin, D., Sun, L. C., Huo, W. Sen, Zhang, L. J., Chen, Y. L., Miao, S. et al. (2023). Improved functionality and safety of peptides by the formation of peptide-polyphenol complexes. Trends in Food Science and Technology 141(September): 104193. https://doi.org/10.1016/j.tifs.2023.104193.

Liu, R., Chen, Z., Li, W., Xu, Q., Yang, M. and Wang, Y. (2025a). Exploring proteolytic activity of *Lactiplantibacillus plantarum* AHQ-14 reducing the allergenicity of milk protein and its probiotic potential based on peptidomics and genomics. Journal of Dairy Science. https://doi.org/10.3168/jds.2024-26245.

Liu, L., Li, S., Zheng, J., Bu, T., He, G. and Wu, J. (2020). Safety considerations on food protein-derived bioactive peptides. Trends in Food Science & Technology 96: 199–207. https://doi.org/10.1016/j.tifs.2019.12.022.

Liu, C., Wen, C., Olatunji, O. J., Suttikhana, I. and Ashaolu, T. J. (2024). Biologically active peptides from soy: Updates on antihypertensive action and gut microbiota modulation. Journal of Functional Foods 123(November): 106592. https://doi.org/10.1016/j.jff.2024.106592.

Liu, Yixuan, Berrada, H., Castagnini, J. M. and Barba, F. J. (2025b). Supercritical fluid extraction prior pulsed electric fields to improve high-added-value compounds recovery from salmon side streams: Evaluation of protein, bioactive peptides and minerals. Food Chemistry 467(November 2024): 142283. https://doi.org/10.1016/j.foodchem.2024.142283.

Liu, Yufang and Pischetsrieder, M. (2017). Identification and relative quantification of bioactive peptides sequentially released during simulated gastrointestinal digestion of commercial kefir. Journal of Agricultural and Food Chemistry 65(9): 1865–1873. https://doi.org/10.1021/acs.jafc.6b05385.

Lu, H., Xie, T., Wu, Q., Hu, Z., Luo, Y. and Luo, F. (2023). Alpha-glucosidase inhibitory peptides: sources, preparations, identifications, and action mechanisms. Nutrients 1–24.

Lv, R., Sun, N., Mao, C., Zheng, Z. and Lin, S. (2024). Prevention and potential repair of colitis: Beneficial effects and regulatory mechanisms of food-derived anti- inflammatory peptides. Critical Reviews in Food Science and Nutrition 64(23): 8184–8202. https://doi.org/10.1080/10408398.2023.2197068.

Ma, S., Huang, D., Zhai, M., Yang, L., Peng, S., Chen, C. et al. (2015). Isolation of a novel bio-peptide from walnut residual protein inducing apoptosis and autophagy on cancer cells. BMC Complementary and Alternative Medicine 15(1): 1–14. https://doi.org/10.1186/s12906-015-0940-9.

Madhu, M., Kumar, D., Sirohi, R., Tarafdar, A., Dhewa, T., Aluko, R. E. et al. (2022). Bioactive peptides from meat: Current status on production, biological activity, safety, and regulatory framework. Chemosphere 307(P1): 135650. https://doi.org/10.1016/j.chemosphere.2022.135650.

Maestri, E., Marmiroli, M. and Marmiroli, N. (2016). Bioactive peptides in plant-derived foodstuffs. Journal of Proteomics 147: 140–155. https://doi.org/10.1016/j.jprot.2016.03.048.

Marciniak, A., Suwal, S., Naderi, N., Pouliot, Y. and Doyen, A. (2018). Enhancing enzymatic hydrolysis of food proteins and production of bioactive peptides using high hydrostatic pressure technology. Trends in Food Science & Technology 80(February): 187–198. https://doi.org/10.1016/j.tifs.2018.08.013.

Malm-Erjefält, M., Bjørnsdottir, I., Vanggaard, J., Helleberg, H., Larsen, U., Oosterhuis, B. et al. (2010). Metabolism and excretion of the once-daily human glucagon-like peptide-1 analog liraglutide in healthy male subjects and its *in vitro* degradation by dipeptidyl peptidase IV and neutral endopeptidase. Drug Metabolism and Disposition 38(11): 1944–1953. https://doi.org/10.1124/dmd.110.034066.

Maturana, P., Martinez, M., Noguera, M. E., Santos, N. C., Disalvo, E. A., Semorile, L. et al. (2017). Lipid selectivity in novel antimicrobial peptides: Implication on antimicrobial and hemolytic activity. Colloids and Surfaces B: Biointerfaces 153: 152–159. https://doi.org/10.1016/j.colsurfb.2017.02.003.

Medeiros, A. F. De, Luana, J., Queiroz, C. De, Leal, B. and Maciel, L. (2022). Hydrolyzed proteins and vegetable peptides: anti-inflammatory mechanisms in obesity and potential. Nutrients 14. https://doi.org/10.3390/nu14030690.

Mejia, E. G. De and Dia, V. P. (2010). The role of nutraceutical proteins and peptides in apoptosis, angiogenesis, and metastasis of cancer cells. Cancer and Metastasis Reviews 29(3): 511–528. https://doi.org/10.1007/s10555-010-9241-4.

Memarpoor-Yazdi, M., Zare-Zardini, H., Mogharrab, N. and Navapour, L. (2020). Purification, characterization and mechanistic evaluation of angiotensin converting enzyme inhibitory peptides derived from *Zizyphus jujuba* fruit. Scientific Reports 10(1): 3976. https://doi.org/10.1038/s41598-020-60972-w.

Meshginfar, N., Sadeghi Mahoonak, A., Hosseinian, F., Ghorbani, M. and Tsopmo, A. (2018). Production of antioxidant peptide fractions from a by-product of tomato processing: mass spectrometry identification of peptides and stability to gastrointestinal digestion. Journal of Food Science and Technology 55(9): 3498–3507. https://doi.org/10.1007/s13197-018-3274-z.

Mirzapour-Kouhdasht, A., McClements, D. J., Taghizadeh, M. S., Niazi, A. and Garcia-Vaquero, M. (2023). Strategies for oral delivery of bioactive peptides with focus on debittering and masking. Npj Science of Food 7(1). https://doi.org/10.1038/s41538-023-00198-y.

Moayedi, A., Mora, L., Aristoy, M. C., Safari, M., Hashemi, M. and Toldrá, F. (2018). Peptidomic analysis of antioxidant and ACE-inhibitory peptides obtained from tomato waste proteins fermented using *Bacillus subtilis*. Food Chemistry 250(January): 180–187. https://doi.org/10.1016/j.foodchem.2018.01.033.

Mohamad Asri, N., Muhialdin, B. J., Zarei, M. and Saari, N. (2020). Low molecular weight peptides generated from palm kernel cake via solid state lacto-fermentation extend the shelf life of bread. LWT 134(September): 110206. https://doi.org/10.1016/j.lwt.2020.110206.

Mohan, A., Rajendran, S. R. C. K., He, Q. S., Bazinet, L. and Udenigwe, C. C. (2015). Encapsulation of food protein hydrolysates and peptides: A review. RSC Advances 5(97): 79270–79278. https://doi.org/10.1039/c5ra13419f.

Mora, L., Bolumar, T., Heres, A. and Toldrá, F. (2017). Effect of cooking and simulated gastrointestinal digestion on the activity of generated bioactive peptides in aged beef meat. Food and Function 8(12): 4347–4355. https://doi.org/10.1039/c7fo01148b.

Mora, L. and Toldrá, F. (2023). Advanced enzymatic hydrolysis of food proteins for the production of bioactive peptides. Current Opinion in Food Science 49: 100973. https://doi.org/10.1016/j.cofs.2022.100973.

Mu, X., Li, D., Xiao, R., Guan, K., Ma, Y., Wang, R. et al. (2024). Discovery of novel dipeptidyl peptidase-IV inhibitory peptides derived from walnut protein and their bioactivities *in vivo* and *in vitro*. Current Research in Food Science 9(September): 100893. https://doi.org/10.1016/j.crfs.2024.100893.

Ni, X., Zhai, Z., Xin, Z., Tan, C., Chen, Y., Deng, J. et al. (2020). Antioxidant properties and transepithelial transportation of di-/tripeptides derived from simulated gastrointestinal digestion of pig blood cells hydrolysates. Food 1(3): 254–269. https://doi.org/10.2991/efood.k.200422.001.

Nurdiani, R., Vasiljevic, T., Yeager, T., Singh, T. K. and Donkor, O. N. (2017). Bioactive peptides with radical scavenging and cancer cell cytotoxic activities derived from Flathead (*Platycephalus fuscus*) by-products. European Food Research and Technology 243(4): 627–637. https://doi.org/10.1007/s00217-016-2776-z.

Olivares-Galván, S., Marina, M. L. and García, M. C. (2020). Extraction and characterization of antioxidant. Foods; Basel 9(8): 1018.

Panda, J., Amrit, R., Mishra, A. K., Chakraborty, A., Rustagi, S., Nath, P. C. et al. (2025). Sustainable valorization of fruit and vegetable waste for bioactive compounds: advancing functional food and wellness. Waste and Biomass Valorization 0123456789. https://doi.org/10.1007/s12649-025-02937-6.

Pang, L., Huang, Z., Li, R., Su, Y., Shi, J., Yang, X. et al. (2025a). Reducing the allergenicity of whey proteins while improving their functional properties and bioactivity using combined enzymes. Food Chemistry 481: 144052. https://doi.org/10.1016/j.foodchem.2025.144052.

Pang, L., Zhang, W., Chen, C., Li, R., Huang, Z., Liu, S. et al. (2025b). Effect of enzymatic hydrolysis and ultrasound pretreatment on allergenicity, functional properties and bioactivity of whey protein isolates. LWT - Food Science and Technology 210: 116891. https://doi.org/10.1016/j.lwt.2024.116891.

Patil, P. J., Usman, M., Zhang, C., Mehmood, A., Zhou, M., Teng, C. et al. (2022). An updated review on food-derived bioactive peptides: Focus on the regulatory requirements, safety, and bioavailability. Comprehensive Reviews in Food Science and Food Safety 21(2): 1732–1776. https://doi.org/10.1111/1541-4337.12911.

Pavkov, M. E. and Miyamoto, Y. (2023). Diabetes and kidney disease. International Diabetes Federation.

Pei, J., Gao, X., Pan, D., Hua, Y., He, J., Liu, Z. et al. (2022). Advances in the stability challenges of bioactive peptides and improvement strategies. Current Research in Food Science 5(November): 2162–2170. https://doi.org/10.1016/j.crfs.2022.10.031.

Peng, T., Shao, X., Long, L., Liu, H., Song, W., Hou, J. et al. (2023). Nano today rational design of oral delivery nanosystems for hypoglycemic peptides. Nano Today 53(September): 102031. https://doi.org/10.1016/j.nantod.2023.102031.

Pimentel, F. B., Cermeño, M., Kleekayai, T., Harnedy, P. A., FitzGerald, R. J., Alves, R. C. et al. (2020). Effect of *in vitro* simulated gastrointestinal digestion on the antioxidant activity of the red seaweed

Porphyra dioica. Food Research International 136(228): 109309. https://doi.org/10.1016/j.foodres.2020.109309.

Pitchakarn, P., Buacheen, P., Taya, S., Karinchai, J., Temviriyanukul, P., Inthachat, W. et al. (2024). Anti-inflammatory, cytotoxic, and genotoxic effects of soybean oligopeptides conjugated with mannose. Foods 13(16). https://doi.org/10.3390/foods13162558.

Prados, I. M., Dominguez-vega, E., Marina, M. L. and Concepci, M. (2023). Maximizing hypocholesterolemic peptides from an olive byproduct by enzymatic hydrolysis. LWT 187(September). https://doi.org/10.1016/j.lwt.2023.115371.

Prados, I. M., Orellana, J. M., Marina, M. L. and García, M. C. (2020). Identification of peptides potentially responsible for *in vivo* hypolipidemic activity of a hydrolysate from olive seeds. Journal of Agricultural and Food Chemistry 68(14): 4237–4244. https://doi.org/10.1021/acs.jafc.0c01280.

Puchalska, C., Esteve, M. L., Marina, M. C. and García, M. C. (2015). Peptides. pp. 329–355. In Handbook of Food Analysis; CRC Press:Boca Raton, FL, USA, ISBN 978-1-4665-5654-6.

Quan, Y., Ren, H., Liu, S., Zhao, X. and Hao, J. (2025). Identification and molecular mechanism of novel antioxidant peptides from tiger nut (*Cyperus esculentus* L.). Food Bioscience 63(September 2024): 105738. https://doi.org/10.1016/j.fbio.2024.105738.

Quintero-soto, M. F., Chávez-Ontiveros, J., Garzón-Tiznado, J., Salazar-Salas, N. Y., Pineda-Hidalgo, K. V, Delgado-Vargas, F. et al. (2021). Characterization of peptides with antioxidant activity and antidiabetic potential obtained from chickpea (*Cicer arietinum* L.) protein hydrolyzates. Process Biochemistry April: 2962–2977. https://doi.org/10.1111/1750-3841.15778.

Ribeiro, E., Rocha, T. de S. and Prudencio, S. H. (2021). Potential of green and roasted coffee beans and spent coffee grounds to provide bioactive peptides. Food Chemistry 348(June 2020): 129061. https://doi.org/10.1016/j.foodchem.2021.129061.

Rizwan, D., Masoodi, F. A., Wani, S. M. and Mir, S. A. (2023). Bioactive peptides from fermented foods and their relevance in COVID-19 mitigation. Food Production, Processing and Nutrition 5(1). https://doi.org/10.1186/s43014-023-00165-w.

Rizzello, C. G., Tagliazucchi, D., Babini, E., Sefora Rutella, G., Taneyo Saa, D. L. and Gianotti, A. (2016). Bioactive peptides from vegetable food matrices: Research trends and novel biotechnologies for synthesis and recovery. Journal of Functional Foods 27: 549–569. https://doi.org/10.1016/j.jff.2016.09.023.

Ryan, J. T., Ross, R. P., Bolton, D., Fitzgerald, G. F. and Stanton, C. (2011). Bioactive peptides from muscle sources: Meat and fish. Nutrients 3(9): 765–791. https://doi.org/10.3390/nu3090765.

Sadiq, I. Z. (2021). Free radicals and oxidative stress: signaling mechanisms, redox basis for human diseases, and cell cycle regulation. Current Molecular Medicine 23(1): 13–35. https://doi.org/10.2174/1566524022666211222161637.

Sah, B. N. P., Vasiljevic, T., Mckechnie, S. and Donkor, O. N. (2015). Identification of anticancer peptides from bovine milk proteins and their potential roles in management of cancer: A critical review. Comprehensive Reviews in Food Science and Food Safety 14(2): 123–138. https://doi.org/10.1111/1541-4337.12126.

Sánchez-Elvira, A., Hernández-Corroto, E., García, M. C., Castro-Puyana, M. and Marina, M. L. (2024). Sustainable extraction of proteins from lime peels using ultrasound, deep eutectic solvents, and pressurized liquids, as a source of bioactive peptides. Food Chemistry 458(May). https://doi.org/10.1016/j.foodchem.2024.140139.

Sánchez, A. and Vásquez, A. (2017). Bioactive peptides: A review. International Journal Bioautomation 15(4): 223–250. https://doi.org/10.1093/fqs/fyx006.

Sánchez, G. S., López, A. I. Á., España, E. P., Lardone, P. J., Vico, A. C. and Chamorro, I. C. (2024). Food-derived peptides with inhibitory capacity for HMG-CoA reductase activity: a potential nutraceutical for hypercholesterolemia. Food Science and Human Wellness 13(6): 3083–3094. https://doi.org/10.26599/fshw.2023.9250001.

Sarmadi, B. H. and Ismail, A. (2010). Antioxidative peptides from food proteins: A review. Peptides 31(10): 1949–1956. https://doi.org/10.1016/j.peptides.2010.06.020.

Sharma, A., Singh, G., Bhatti, J. S., Gill, S. K. and Arya, S. K. (2025). Antifungal peptides: Therapeutic potential and challenges before their commercial success. International Journal of Biological Macromolecules 284(P1): 137957. https://doi.org/10.1016/j.ijbiomac.2024.137957.

Sharma, V., Gupta, P. and Sharma, P. (2023). Recent advancements in novel bioactive peptides and protein hydrolysates isolated from different medicinal plants along with their applications in food and pharmaceutical industries. International Journal of Peptide Research and Therapeutics (Vol. 29, Issue 6). Springer Netherlands. https://doi.org/10.1007/s10989-023-10565-8.

Shen, W., Tu, Y., Li, J., Wu, Q., Huang, H., Lin, H. et al. (2025). Immunomodulatory mechanisms of low molecular weight peptides from Pinctada martensii: A network pharmacology and *in vivo* study on gut microbiota and immune function. Journal of Functional Foods 129(April): 106874. https://doi.org/10.1016/j.jff.2025.106874.

Silva, M., Philadelpho, B., Santos, J., Souza, V., Souza, C., Santiago, V. et al. (2021). IAF, QGF, and QDF peptides exhibit cholesterol-lowering activity through a statin-like HMG-CoA reductase regulation mechanism: *In silico* and *in vitro* approach. International Journal of Molecular Sciences 22(20). https://doi.org/10.3390/ijms222011067.

Singh, P. P., Gupta, V. and Prakash, B. (2023). Recent advancement in functional properties and toxicity assessment of plant-derived bioactive peptides using bioinformatic approaches. Critical Reviews in Food Science and Nutrition 63(20): 4503–4521. https://doi.org/10.1080/10408398.2021.2002807.

Song, W., Huang, H., Liu, J., Zhao, X., Fang, L., Liu, C. et al. (2025). Network pharmacology and cell experiment demonstrate the neuroprotective potential on hypertension-induced nerve injury of hazelnut-derived peptide YYLLVR. Food Bioscience 65(February): 106089. https://doi.org/10.1016/j.fbio.2025.106089.

Song, Y., Wang, Z., Ji, H., Jiang, Z., Li, X., Du, Z. et al. (2024). Fatty acid modification of casein bioactive peptides nano-assemblies, synthesis, characterization and anticarcinogenic effect. International Journal of Biological Macromolecules 254(P1): 127718. https://doi.org/10.1016/j.ijbiomac.2023.127718.

Sosalagere, C., Adesegun Kehinde, B. and Sharma, P. (2022). Isolation and functionalities of bioactive peptides from fruits and vegetables: A reviews. Food Chemistry 366(March 2021): 130494. https://doi.org/10.1016/j.foodchem.2021.130494.

Tagliamonte, S., Oliviero, V. and Vitaglione, P. (2024). Food bioactive peptides: functionality beyond bitterness. Nutrition Reviews 83(2): 369–381. https://doi.org/10.1093/nutrit/nuae008.

Tavares, T. D., Ribeiro, A. R. M., Silva, C., Antunes, J. C. and Felgueiras, H. P. (2024). Combinatory effect of nisin antimicrobial peptide with bioactive molecules: A review. Journal of Drug Delivery Science and Technology 91(November 2023). https://doi.org/10.1016/j.jddst.2023.105246.

Thybaud, V., Kasper, P., Sobol, Z., Elhajouji, A., Fellows, M., Guerard, M. et al. (2016). Genotoxicity assessment of peptide/protein-related biotherapeutics: Points to consider before testing. Mutagenesis 31(4): 375–384. https://doi.org/10.1093/mutage/gew013.

Tonolo, F., Grinzato, A., Bindoli, A. and Rigobello, M. P. (2023). From *in silico* to a cellular model: molecular docking approach to evaluate antioxidant bioactive peptides. Antioxidants 12(3): 665. https://doi.org/10.3390/antiox12030665.

Tsai, B. C. K., Hsieh, D. J. Y., Lin, W. T., Tamilselvi, S., Day, C. H., Ho, T. J. et al. (2020). Functional potato bioactive peptide intensifies Nrf2-dependent antioxidant defense against renal damage in hypertensive rats. Food Research International 129(December 2019): 108862. https://doi.org/10.1016/j.foodres.2019.108862.

Ulug, S. K., Jahandideh, F. and Wu, J. (2021). Novel technologies for the production of bioactive peptides. Trends in Food Science and Technology 108(March 2020): 27–39. https://doi.org/10.1016/j.tifs.2020.12.002.

Vásquez-Villanueva, R., Muñoz-Moreno, L., José Carmena, M., Luisa Marina, M. and Concepción García, M. (2018). *In vitro* antitumor and hypotensive activity of peptides from olive seeds. Journal of Functional Foods 42(January): 177–184. https://doi.org/10.1016/j.jff.2017.12.062.

Vásquez-Villanueva, R., Orellana, J. M., Marina, M. L. and García, M. C. (2019). Isolation and characterization of angiotensin converting enzyme inhibitory peptides from peach seed hydrolysates: *in vivo* assessment of antihypertensive activity. Journal of Agricultural and Food Chemistry 67(37): 10313–10320. https://doi.org/10.1021/acs.jafc.9b02213.

Wang, T., Bo, N., Sha, G., Guan, Y., Yang, D., Shan, X. et al. (2024a). Identification and molecular mechanism of novel hypoglycemic peptide in ripened pu-erh tea: Molecular docking, dynamic simulation, and cell experiments. Food Research International 194(August): 114930. https://doi.org/10.1016/j.foodres.2024.114930.

Wang, C., Udenigwe, C. C., Zheng, L. and Zhao, M. (2024b). Bioactivities, preparation, and bioavailability of food-derived proline-containing peptides. Trends in Food Science & Technology 104776. https://doi.org/10.1016/j.tifs.2024.104776.

Wayal, V., Wang, S. Der and Hsieh, C. C. (2025). Novel bioactive peptides alleviate Western diet-induced MAFLD in C57BL/6J mice by inhibiting NLRP3 inflammasome activation and pyroptosis via TLR4/NF-κB and Keap1/Nrf2/HO-1 signaling pathways. International Immunopharmacology 148(January): 114177. https://doi.org/10.1016/j.intimp.2025.114177.

Wu, D., Gao, Y., Tan, Y., Liu, Y., Wang, L., Zhou, M. et al. (2018). Discovery of distinctin-like-peptide-PH (DLP-PH) from the skin secretion of *Phyllomedusa hypochondrialis*, a prototype of a novel family of antimicrobial peptide. Frontiers in Microbiology 9(MAR): 1–11. https://doi.org/10.3389/fmicb.2018.00541.

Xavier, J. R., Sanjay, B. S., Gupta, D., Mehta, S. and Chauhan, O. P. (2024). Bioactive compounds of foods: Phytochemicals and peptides. Food and Humanity 3(April): 100354. https://doi.org/10.1016/j.foohum.2024.100354.

Xu, Q., Hong, H., Wu, J. and Yan, X. (2019). Bioavailability of bioactive peptides derived from food proteins across the intestinal epithelial membrane: A review. Trends in Food Science and Technology 86(February): 399–411. https://doi.org/10.1016/j.tifs.2019.02.050.

Xu, Y., Yi, W., Zhao, Q., Zhao, R., Jiang, S. and Tang, Y. (2025). Exploring immunomodulatory peptides from monkfish (*Lophius litulon*) roe: Identification, *in vitro* activity, and molecular docking analysis. International Immunopharmacology 156(December 2024): 114698. https://doi.org/10.1016/j.intimp.2025.114698.

Yang, F. J., Chen, X., Huang, M. C., Yang, Q., Cai, X. X., Chen, X. et al. (2021). Molecular characteristics and structure–activity relationships of food-derived bioactive peptides. Journal of Integrative Agriculture 20(9): 2313–2332. https://doi.org/10.1016/S2095-3119(20)63463-3.

Yang, T., Meng, Y., Li, S., Zhao, X. and Hou, H. (2025). Preparation, unique structural characteristics and immunomodulatory effects of peptides from Antarctic krill (*Euphausia superba*). Food Research International 116592. https://doi.org/10.1016/j.foodres.2025.116592.

Yan, Z., Gui, Y., Liu, C., Zhang, X., Wen, C., Olatunji, O. J. et al. (2024). Gastrointestinal digestion of food proteins: Anticancer, antihypertensive, anti-obesity, and immunomodulatory mechanisms of the derived peptides. Food Research International 189(May): 114573. https://doi.org/10.1016/j.foodres.2024.114573.

Yao, J. F., Yang, H., Zhao, Y. Z. and Xue, M. (2018). Metabolism of peptide drugs and strategies to improve their metabolic stability. Current Drug Metabolism 19(11): 892–901. https://doi.org/10.2174/1389200219666180628171531.

Yang, S. and Lian, G. (2020). ROS and diseases: role in metabolism and energy supply. Molecular and Cellular Biochemistry 467(1–2): 1–12. https://doi.org/10.1007/s11010-019-03667-9.

Ye, H., Tao, X., Zhang, W., Chen, Y., Yu, Q. and Xie, J. (2022). Food-derived bioactive peptides: production, biological activities, opportunities and challenges. Journal of Future Foods 2(4): 294–306. https://doi.org/10.1016/j.jfutfo.2022.08.002.

Yu, S., Wang, W., Bu, T., Zhao, R., Niu, R., Liu, L. et al. (2023). Digestion, absorption, and transport properties of soy-fermented douchi hypoglycemic peptides VY and SFLLR under simulated gastrointestinal digestion and Caco-2 cell monolayers. Food Research International 164(July 2022): 112340. https://doi.org/10.1016/j.foodres.2022.112340.

Zaky, A. A., Simal-Gandara, J., Eun, J. B., Shim, J. H. and Abd El-Aty, A. M. (2022). Bioactivities, applications, safety, and health benefits of bioactive peptides from food and by-products: a review. Frontiers in Nutrition 8(January): 1–18. https://doi.org/10.3389/fnut.2021.815640.

Zaky, A. A., Witrowa-Rajchert, D. and Nowacka, M. (2025). Insights into plant-origin bioactive peptides: extraction, bioactivities, *in silico* approaches, and applications. In International Journal of Peptide Research and Therapeutics 31(2). https://doi.org/10.1007/s10989-024-10684-w.

Zhang, Q., Cheng, Z., Wang, Y., Zheng, S., Wang, Y. and Fu, L. (2021). Combining Alcalase hydrolysis and transglutaminase-cross-linking improved bitterness and techno-functional properties of hypoallergenic soybean protein hydrolysates through structural modifications. Lwt 151: 112096. https://doi.org/10.1016/j.lwt.2021.112096.

Zhang, Y., Li, H., Liu, P., Chen, K., Ma, S. and Cai, W. (2024a). Extensive characterization and antioxidant function prediction of endogenous peptides derived from *Polygonatum cyrtonema* Hua. Microchemical Journal 203(February): 110872. https://doi.org/10.1016/j.microc.2024.110872.

Zhang, Y., Li, Y., Quan, Z., Xiao, P. and Duan, J. A. (2024b). New insights into antioxidant peptides: an overview of efficient screening, evaluation models, molecular mechanisms, and applications. Antioxidants 13(2). https://doi.org/10.3390/antiox13020203.

Zheng, Y., Li, Y., Zhang, Y., Ruan, X. and Zhang, R. (2017). Purification, characterization, synthesis, *in vitro* ACE inhibition and *in vivo* antihypertensive activity of bioactive peptides derived from oil palm kernel glutelin-2 hydrolysates. Journal of Functional Foods 28: 48–58. https://doi.org/10.1016/j.jff.2016.11.021.

Zhu, M., Li, F., Zhang, Y., Yu, J., Wei, Y. and Gao, X. (2024). Preparation, bioactivities, and food industry applications of tuber and tuberous roots peptides: A review. Food Chemistry 456(June): 140027. https://doi.org/10.1016/j.foodchem.2024.140027.

Zupin, L., Santos-Silva, C. A. dos, Al Mughrbi, A. R. H., Vilela, L. M. B., Benko-Iseppon, A. M. and Crovella, S. (2022). Bioactive antimicrobial peptides: a new weapon to counteract zoonosis. Microorganisms 10(8). https://doi.org/10.3390/microorganisms10081591.

Chapter 6

Polysaccharides in Fruits and Their Bioactivities

Francisca Maria Macêdo Pereira,[1,2] *Iago Cavalcante Oliveira,*[1,2] *Shellda Sousa Ribeiro,*[1] *Bruno Prado Moreira,*[1] *Edson Cavalcanti da Silva Filho*[1,2,3] and *Jurandy do Nascimento Silva*[1,2,*]

6.1 Introduction

Polysaccharides are macromolecules composed of monosaccharide units (simple sugars) linked together by glycosidic bonds, forming a polymeric chain (Wu et al. 2016). These compounds are carbohydrates and have a general formula $[C(H_2O)]n$, from which their name derives; they can also be known as carbon hydrates. This polymer results from the repetition of monosaccharides and is called a homopolysaccharide if its units are identical, or a heteropolysaccharide when there are different types of monomers (Nelson et al. 2006).

They are considered "life fuels" in living beings and are responsible for energy storage, in the form of starch (plants) and glycogen (animals), and for the release of this energy through metabolic chains when converted into other compounds, such as in the breakdown of glucose during cellular respiration for ATP synthesis. They are

[1] Laboratory of Ecobiomaterials, Federal University of Piauí, Piauí, Brazil.
[2] Postgraduate Program in Chemistry, Federal University of Piauí, Piauí, Brazil.
[3] Postgraduate Program in Materials Science, Federal University of Piauí, Piauí, Brazil.
* Corresponding author: jurandy@ufpi.edu.br

DOI: 10.1201/9781003668169-8

also the main products generated in the process of photosynthesis, playing structural and defense functions in the cell walls of bacteria, fungi, and plants, and acting as signaling molecules in cellular activities and functions (Pomin and Mourão 2006).

Carbohydrates represent the main class of compounds present in fruits. As a tropical country with a vast territorial extension and diverse biomes, Brazil offers favorable climatic conditions for the continuous production of various fruits throughout the year. This diversity also directly influences the composition and abundance of available polysaccharides. Insufficient fruit consumption is one of the main factors contributing to the development of chronic noncommunicable diseases (NCDs), such as obesity, hypertension, and cardiovascular diseases (Madsen et al. 2023). Among the impacts resulting from carbohydrate deficiency, decreased vigor stands out, since carbohydrates are the main source of energy. A diet deficient in carbohydrates also impacts the absorption of nutrients necessary to maintain a healthy body (Clemente-Suárez et al. 2022).

Thus, this chapter aims to present the main classes, extraction, purification, and characterization methods of fruit polysaccharides, highlighting their bioactive properties and functional and technological applications.

6.2 Main Polysaccharides Present in Fruits

6.2.1 Pectin

The main polysaccharides found in fruit cell walls belong to the pectin group, which are heteropolysaccharides with a main structure that consists of a linear chain of α-D-galacturonic acid and its O-methylated derivatives (Figure 1). Galacturonic acid is usually esterified, and the presence of α-L-rhamnose residues can promote

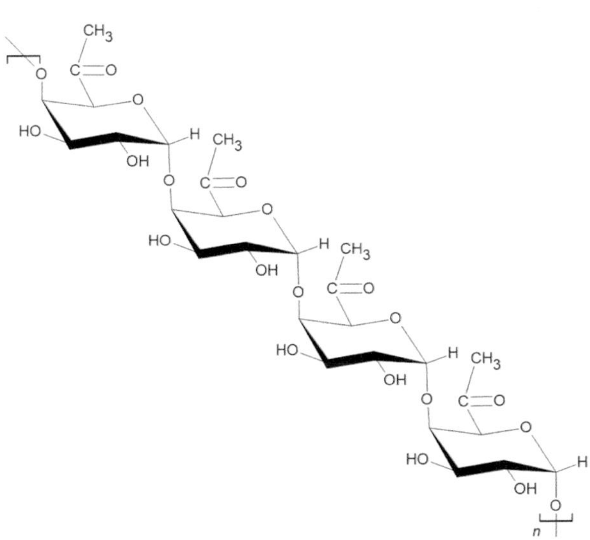

Figure 1. Chemical structure of pectin.

breaks along this chain. The degree of esterification and the number and distribution of rhamnose residues are determining factors in the solubility and gel-forming capacity of these polysaccharides. Regardless of the pectin's origin, the galacturonic acid region has between 72 and 100 residues. In regions rich in rhamnose residues, there are branches with neutral sugars, predominantly arabinose, galactose, and arabinogalactans (Cunha et al. 2009).

Pectins can be found in all parts of plants, with higher concentrations in the peels of citrus fruits like lemon and orange. They are included in the group of soluble dietary fibers; they have a high fluid retention capacity, promote a feeling of satiety, and exhibit significant hypocholesterolemic and hypoglycemic effects (Seyfried et al. 2016).

6.2.2 Cellulose and Hemicelluloses

Cellulose is the main component of plant cell walls and consists of a linear chain polymer formed exclusively by D-glucose units linked together by β-1,4-glycosidic bonds (Figure 2). The anomeric carbon at the C1 position adopts the β configuration, and each glucose unit is rotated approximately 180° relative to its adjacent unit, resulting in a repeating disaccharide structure (Nishiyama et al. 2003). This formation is highly resistant to traction forces and makes up between 20% and 30% of the dry mass of the primary cell wall, potentially occupying up to 15% of its volume. In secondary cell walls, this proportion can range from 40% to 90% of the total biomass weight (Seyfried et al. 2016).

Hemicelluloses are heteropolysaccharides whose main chain is made of glucose (D-Glcp), mannose (D-Manp), or xylose (D-Xylp) units, connected by β-1,4-type bonding, with the C1 and C4 carbon atoms linked in an equatorial position (Figure 3). This polysaccharide can be classified based on the type of predominant monosaccharide in its main chain, with the most frequent being xylans, arabinoxylans, galactomannans, glucomannans, xyloglucans, mannans, and β-(1-3, 1-4)-glucans. In addition to composing the cell wall matrix, they also act as signaling molecules, precursors, and reserve substances (Seyfried et al. 2016).

Integrating the hemicellulose class, arabinoxylans have a linear structure of D-xylose residues linked by β-1,4 with frequent substitutions for L-arabinose. Phenolic acids such as ferulic acid can be linked to arabinose units, forming cross-links that influence the structure and properties (Botticella et al. 2021). They are part of the main components of cereal grains, including wheat, rice, rye, corn,

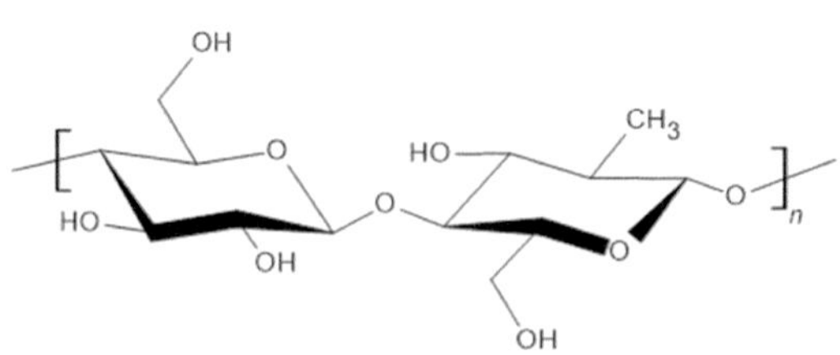

Figure 2. Basic chemical structure of cellulose.

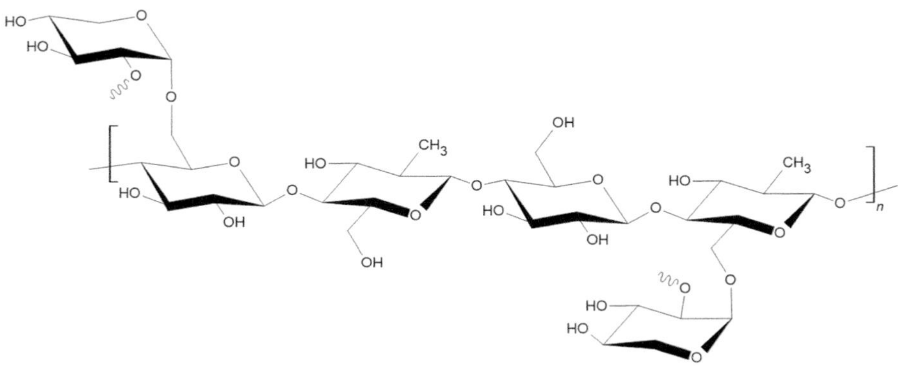

Figure 3. Chemical structure of hemicelluloses.

sorghum, and oats, and are non-digestible fibers. On the other hand, they can also be found in fruit peels such as apple peel and pineapple (Bernaud and Rodrigues 2013).

Also composing the hemicellulose class, galactans are galactose polymers, and their chain is most frequently formed by β-(1→4) linkages with L-arabinofuranose branches every 16–21 residues of the main chain, and can be formed by β-(1→3) and (1→6) bonding (Buckeridge et al. 2000). When they exhibit arabinose residues in their branches, arabinogalactans, they act as cell wall reserve polysaccharides, controlling cell expansion. They are often found in various fruits, such as apple, pear, watermelon, mango, plums, avocado, and cherry (Martins et al. 2020).

6.2.3 Starch

Starch is a reserve carbohydrate widely distributed in various plant species, present in different parts of the plant, which can include immature or green fruits (Lajolo and Menezes 2006). The conformation of starch can be linear (amylose) or branched (amylopectin). The type of structure directly influences the physical and physiological properties of starch. Amylose is a linearly structured macromolecule composed of D-glucose units linked by α-1,4-type bonds (Figure 4). However, some amylose molecules may have a few branches joined by α-1,6-type bonds. Amylopectins, on the other hand, contain about 4 to 5% branches, also formed by D-glucose, with α-1,4-bonds in the main chain and α-1,6-bonds at the branching points. Due to its chemical structure, amylopectin crystallizes easily and is directly involved in the formation of starch granules (Bertoft et al. 2008).

According to studies by Englyst et al. (1992), starch can be classified for nutritional purposes into three categories: rapidly digestible starch (RDS), slowly digestible starch (SDS), and resistant starch (RS). This classification was determined based on the speed and extent (where the starch will be metabolized) of *in vitro* hydrolysis. Of these, resistant starch is classified by the AACC (American Association of Cereal Chemists) and the Food and Nutrition Board of the IOM (Institute of Medicine of the National Academies) as a type of dietary fiber. When fermented in the large intestine, it produces gases and short-chain fatty acids that are associated

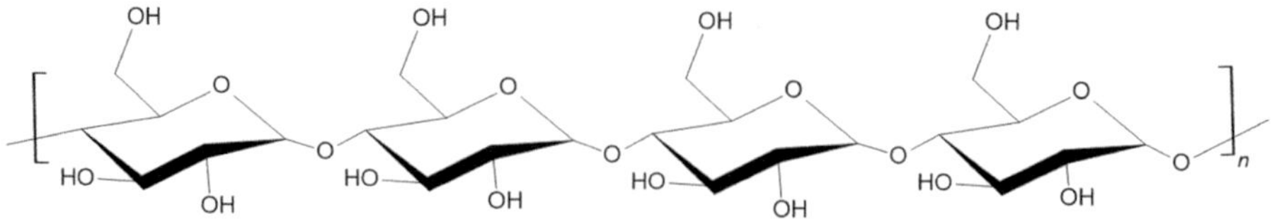

Figure 4. Chemical structure of starch.

with preventing diet-related diseases, cardiovascular issues, and aiding in weight loss (Koakuzu et al. 2015).

6.2.4 Fructans

Fructans are reserve polysaccharides composed mainly of fructose units (Pollock and Cairns 1991). They exhibit characteristics like those of dietary fibers and can be used as a sucrose substitute in functional foods, in addition to being applied in dairy products, confectionery, and baked goods. The name "fructan" is used to refer to non-reducing oligo- and polysaccharides naturally occurring in plants. These compounds have linear or branched chains, formed by β-2,1 or β-2,6 glycosidic linkages between fructose units, with both linkages potentially found within the same molecule (Figure 5) (Fernández-Lainez 2023).

This class of polysaccharides exhibits a degree of polymerization ranging from 2 to 60 fructose molecules, rarely exceeding 150 units in plants, but potentially reaching up to 100,000 units in microorganisms. The simplest form, with one linked fructose molecule, is called monofructosyl sucrose, while linear compounds with a degree of polymerization between 2 and 10 fructose molecules are known as oligofructose or fructo-oligosaccharides (Fernández-Lainez 2023).

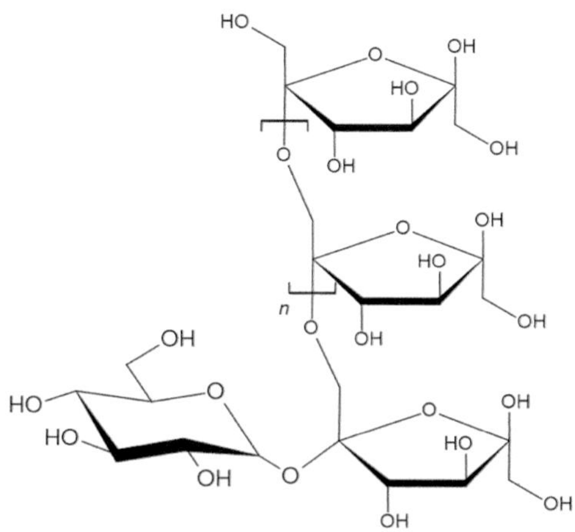

Figure 5. Chemical structure of fructan.

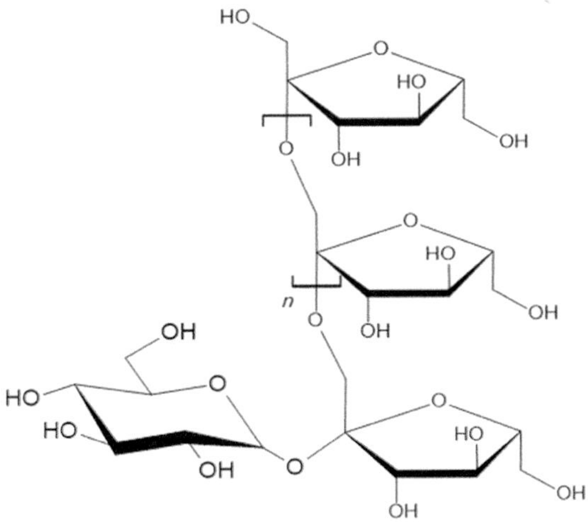

Figure 6. Chemical structure of inulin.

Inulin is an example of fructo-oligosaccharides found in fruits, belonging to the fructan class, with a wide variety of uses in the food and pharmaceutical sectors. They are known as prebiotics because they contain carbohydrates that are not digested by the body, and strong evidence indicates that prebiotics, like inulin, can influence the regulation of immune functions (Amarante et al. 2018). Its molecule (Figure 6) is water-soluble, with the degree of solubility dependent on temperature. For this reason, when inulin is chilled or frozen, precipitation of fructo-oligosaccharides occurs, caused by thc temperature reduction (Toneli et al. 2008).

Cruz-Cárdenas et al. (2015) reported the presence of this polysaccharide in banana fruits, and techniques to boost future work aimed at improving the fructan content in banana plants (banana fructosyltransferases), as well as enhancing the nutritional properties of this fruit.

6.2.5 Other Relevant Polysaccharides

Beyond the most commonly found polysaccharides in fruits, mentioned previously, others can be found, albeit in smaller quantities. These include glycoproteins like arabinogalactans, which are macromolecules with a high content of hydroxyproline and a large number of glycosides, potentially containing up to 95% carbohydrates in their composition. These compounds play structural and functional roles, especially in the ripening of fruits such as apple, grape, pear, and tomato (Leszczuk et al. 2020).

Furthermore, in some fruits rich in phenolic compounds, such as grapes, these are bound to carbohydrates. These compounds bind to polysaccharides present in the cell wall. This binding occurs through hydrogen bonds and hydrophobic interactions with polyphenols. Consequently, extracting the polysaccharides is a way to release the polyphenols that are trapped within the cell wall structure (Beres et al. 2015).

6.3 Extraction, Purification, and Characterization Methods

Having in mind their existence as metabolites in different plant species, it's safe to say that different polysaccharides will exhibit different chemical compositions. These structural divergences imply varying responsiveness to certain extraction methods, as well as highlighting the relevance of the characterization methods employed in elucidating the structures of these sugars.

6.3.1 Extractive Methods

6.3.1.1 Aqueous Extraction (Conventional Method)

This is the most common method for extracting water-soluble polysaccharides, primarily used when the sugars of interest are neutral. However, acidified water can be employed to extract compounds with reduced solubility (Huang et al. 2022a). Generally, the first step involves preparing the plant sample by drying and grinding to maximize the material's contact surface with the extraction liquid. However, depending on the material, high drying temperatures, especially for long durations, may be discouraged due to the possibility of thermodegradation (Sant'anna et al. 2021). The obtained powder is then subjected to extraction in organic solvents so that the lipid content is removed, and subsequently, the residue is immersed in water to obtain the soluble content.

It's worth noting that if the water temperature during the extraction process is too high for an extended period, it can also lead to the breakage of the saccharide molecular chain, exposing reactive functional groups and causing rearrangements that reduce the molecular weight of the obtained compounds (Lu 2023). Additionally, during extraction, enzymes like pectinase can be added to the medium to facilitate cell wall breakdown and easier solubilization of neutral polysaccharides (Zhai et al. 2017). Subsequently, the aqueous extract is commonly centrifuged to separate the supernatant, which is then subjected to vacuum concentration. Finally, the polysaccharide is precipitated using ethanol. This yields a polysaccharide-rich mass that can then undergo purification methods.

6.3.1.2 Ultrasound-Assisted Extraction

As described by Wang et al. (2023), the use of ultrasound for polysaccharide extraction offers two main advantages. Firstly, the application of these waves can generate cavitations in plant cell walls, destroying them and efficiently releasing the sugars into the extraction medium. Additionally, the vibrations caused by ultrasound promote faster and more efficient solubilization of polysaccharides. It's relevant to note that increasing ultrasound power can enhance the extraction rate until the extractive medium becomes saturated. Beyond this point, further intensification of power will not increase the extraction rate. A prolonged extraction period in the presence of ultrasound, as well as excessive power, can even result in a smaller quantity of products (yield) due to the chemical degradation of polysaccharides (Ying et al. 2011).

Due to these advantages, the application of ultrasound waves constitutes a method capable of obtaining polysaccharides with higher molecular weight when compared to hot aqueous extraction (Xu et al. 2015). However, relevant parameters such as extraction time, water mass, ultrasound power, and extraction temperature must be controlled.

6.3.1.3 Enzyme-Assisted Extraction

Enzymes are proteinaceous biomolecules capable of acting as biocatalysts in a wide variety of chemical reactions, and therefore can be employed to increase the yield of polysaccharide extractions (Bhotmange et al. 2017). This application generally occurs due to the enzymes' ability to degrade complex carbohydrates and plant cell walls, facilitating the dissolution of the polysaccharides of interest.

The efficiency of this methodology depends on the enzyme to be used for each extraction, as well as the carbohydrates aimed for extraction. Pectinase enzyme, for example, has been shown to degrade pectin present in lychee peel, increasing the extraction rate (Wang et al. 2024). Certain precautions must be taken when applying these catalysts, as factors such as temperature, pH, and the presence of ultrasound can affect enzyme stability. As an example, cellulase can be activated by ultrasound, catalyzing the breakdown of plant cell walls; however, if the ultrasound power is too high, the enzyme in question will agglomerate, and its activity will be reduced (Yun et al. 2023).

Thus, it can be concluded that the application of enzymes for obtaining polysaccharides is possible and, in many cases, beneficial. However, the influences of other factors present during extraction on enzymatic activity must be studied to ensure that the use of these catalysts will bring full benefits.

6.3.1.4 Microwave-Assisted Extraction

Microwave frequencies are capable of accelerating the transfer of molecules of interest from the plant material to the extractive medium (Yang et al. 2021). This occurs because polar materials can align with the electromagnetic field, generating friction and altering their molecular conformation (Oyeyinka et al. 2021). These modifications cause tissue damage to the plant material, releasing the glycosidic content, which then dissolves in the extractive medium.

However, it's necessary to note that prolonged microwave usage can result in the cleavage of glycosidic bonds, leading to the obtainment of polysaccharides with lower molecular weights. These cleavages can result in a higher content of obtained monosaccharides, such as arabinose, xylose, and glucose (Lu 2023).

Thus, the use of microwaves presents itself as an effective alternative for obtaining polysaccharides, as it promotes the rapid and effective release of these constituents. Nevertheless, the exposure time must be carefully controlled to avoid polysaccharide degradation. Therefore, careful monitoring of the process, as well as optimization of the adopted methodology, is essential to ensure the integrity of the compounds of interest and the efficiency of the extraction.

6.3.2 Purification Methods

6.3.2.1 Fractionation with Solvents or Salts

When an amount of inorganic salts is added, the solubility of dissolved polysaccharides is reduced, leading to the initial precipitation of higher molecular weight carbohydrates (Huang et al. 2022a). The presence of salts in the precipitate, however, can be an issue, as a subsequent purification step will be necessary to remove them. Nevertheless, due to the low cost of the salts used, it remains a relevant method.

It's also possible to induce carbohydrate precipitation through the use of organic solvents. Due to its low toxicity and cost, ethanol is usually the most employed solvent for this purpose (Xu et al. 2024).

The advantage of both these methods lies in the possibility of obtaining fractions with carbohydrates of different molecular weights (as those with higher molecular weight tend to precipitate at lower salt or ethanol concentrations) and their low cost. However, the production of residues is a possibility, requiring the application of subsequent purification techniques.

Thus, this chapter aims to present the main classes, extraction, purification, and characterization methods of fruit polysaccharides, highlighting their bioactive properties and functional and technological applications.

6.3.2.2 Membrane Filtration

Using absolute or diluted ethanol to effect polysaccharide precipitation, while effective, creates a new problem: the formation of residue (Kang et al. 2015). To mitigate this issue, filtration through a membrane, positioned transversely to the flow of the solution containing the polysaccharide of interest, can be employed.

Typically, the sample needs to be diluted to 0.5% (w/w), and the chosen membrane should have a molecular weight cut-off of 3 kDa (Huang et al. 2022a). Sequential filtrations can be performed to isolate different fractions containing molecules with varying molecular weights.

This method allows for obtaining a purified solution without the application of toxic organic solvents. Subsequently, the obtained fraction is commonly concentrated by lyophilization.

6.3.2.3 Chromatographic Techniques

Chromatography is a separation technique that employs a mobile phase and a stationary phase, with the components of a mixture separated due to their differing affinities for the stationary phase. Different types of high-performance liquid chromatography (HPLC) can be used to separate polysaccharides of interest.

One of the most established methods for separating biopolymers, such as proteins, nucleic acids, and polysaccharides, is ion-exchange chromatography (IEC) (Zheng et al. 2023). The principle of this chromatography relies on the presence of a positively or negatively charged stationary phase. When the mixture is injected into the column at a controlled pH, ionized acidic groups, which confer a negative charge to carbohydrates, will be attracted and retained by a positively charged column.

Similarly, basic compounds can be retained by a column containing a negative charge (Fekete et al. 2015). After ion exchange, the polysaccharide retained by the stationary phase is neutralized and re-elutes through the system, later being detected.

A second commonly employed chromatographic variation is size-exclusion chromatography (SEC), also known as "gel filtration chromatography" or "gel exclusion chromatography." The columns used in this case have porosities capable of retaining the mixture's components. Larger carbohydrates, due to their size, are not retained in the smaller pores and, therefore, elute with greater fluidity toward the detector (Pitkänen 2025). This technique is widely recognized for its ability to purify antigenic carbohydrates present in the cell membranes of microorganisms, which is essential for determining new targets for vaccines (Mihooliya et al. 2025).

Certain polysaccharides can reversibly bind to specific molecules, such as lectins (Shi 2016). This affinity between chemical species, common in biological systems, is utilized in affinity chromatography (AC). In this method, polysaccharides capable of binding to the stationary phase components are temporarily retained and, therefore, elute at a slower rate through the column. Despite its high efficiency, the selection of specific ligands for the polysaccharide of interest must be careful and is often time-consuming (Ahmad et al. 2022).

6.3.2.4 Ammonium Salts-Induced Precipitation

Since quaternary ammonium salts, due to their permanently positively charged nitrogen atom, are capable of forming relatively strong interactions with some acidic polysaccharides, it's possible to use them in the precipitation of these carbohydrates (Shi 2016). Starting from a previously deproteinized polysaccharide solution, salts like cetyltrimethylammonium bromide (CTAB), its conjugate base (CTA-OH), or cetylpyridinium chloride (CPC) are added, which interact with the acidic polysaccharides. Subsequently, the precipitation of the formed complex is induced by adding a sodium chloride solution (3 to 4 mol L^{-1}). Finally, the addition of ethanol promotes the selective precipitation of the polysaccharides, while the ammonium salts remain in solution (Huang et al. 2022a).

This is a simple alternative that can yield satisfactory results. However, it's important to note that the polysaccharides' ability to interact with ammonium salts depends on their acidity, as well as the pH of the medium. If the goal is to precipitate only the most acidic carbohydrates, an excessively alkaline reaction medium should be avoided.

6.3.3 Characterization Methods

6.3.3.1 Determination of Monosaccharide Composition

The identification of the constituent monosaccharides of the biopolymer of interest is carried out by comparison with standards using High-Performance Anion-Exchange Chromatography with Pulsed Amperometric Detection (HPAEC-PAD) (Huang et al. 2022a). Initially, a standard curve is constructed from pure monosaccharides. Next, the target polysaccharide is subjected to acid hydrolysis using concentrated sulfuric acid and subsequently diluted. The hydrolyzed sample is then injected into the chromatographic system. Under standardized analytical conditions, the monosaccharide composition can be determined by comparing the retention times and peak areas of the standards, allowing for the estimation of the relative percentage of each monosaccharide present in the analyzed polysaccharide.

6.3.3.2 Zeta Potential

Zeta potential, which arises from the charge present on particle surfaces, is extensively studied due to its relevance in diverse fields (pharmaceutical, food, medicinal, and materials engineering) (Mohammadi-Jam et al. 2022). This physical property indicates the tendency of particles to repel or agglomerate in an aqueous medium and is generally determined by zeta potential measured through electrophoresis. However, it's crucial to highlight that the repulsive force between these particles is determined by the ionization of functional groups in the polysaccharide, which can be influenced by the medium. Therefore, it is important to maintain a constant solution pH during sample preparation.

6.3.3.3 Fourier-Transform Infrared Spectroscopy (FTIR)

Despite its usefulness, the qualitative analysis of polysaccharides using infrared spectroscopy is challenging, mainly due to the frequent band overlap that often masks important characteristics of the analyte (Hong et al. 2021). Nevertheless, this technique remains an interesting alternative for identifying some specific functional groups, the presence of substituent groups, and the configuration of anomeric carbons.

6.4 Bioactive and Technological Properties of Fruit Polysaccharides

Polysaccharides are widely present in various living organisms and exhibit complex and diverse biological activities due to differences in their composition and structure (Xu and Zhang 2025). In fruits, they display a wide range of bioactive properties, such as antifungal, antioxidant, anticancer, immunomodulatory, neuroprotective, and prebiotic activities, among others, which play a fundamental role in maintaining the health and life of living beings (Alves et al. 2022).

In recent years, interest in the properties of these natural bioactive compounds has grown considerably, driven by the search for safer and more sustainable therapeutic alternatives compared to traditional synthetic drugs. The focus is on finding polysaccharides that demonstrate low toxicity, easy obtainment and manipulation, in addition to high biocompatibility, aiming for applications in the food, pharmaceutical, and biomedical industries (Alves et al. 2022, Ferreira et al. 2022).

6.4.1 Antioxidant Activity

Numerous studies have demonstrated that polysaccharides exhibit immunomodulatory, antitumor, anti-inflammatory, anti-aging, and other properties, with a large part of these effects being associated with their antioxidant activity (Huang et al. 2017).

These antioxidant properties refer to the ability of certain substances to neutralize free radicals, protecting cells against damage caused by oxidative reactions. Thus, they play an essential role in defending against oxidative stress, a condition related to the development of various diseases (Menezes et al. 2023).

Studies investigating these antioxidant activities in polysaccharides have indicated that some of them have scavenging effects on reactive oxygen species (ROS), originating from physical, chemical, and biological sources (Huang et al. 2017). Thus, these compounds have been described as a safe and natural approach to prevent and/or mitigate damage induced by oxidative stress (Liu and Sun 2020).

The mechanism of antioxidant action of polysaccharides is primarily suggested through the activation of the endogenous Nrf2/ARE antioxidant pathway, which regulates the expression of antioxidant enzymes. These enzymes are capable of interrupting the free radical chain reaction, reducing their generation. Additionally, these compounds can act by inhibiting the expression of iNOS mRNA and reducing nitric oxide (NO) production, thereby substantially increasing antioxidant capacity and attenuating damage caused by oxidative stress (Mu et al. 2021).

In a recent study by Ghosh and Abdullah (2024), a polysaccharide fraction extracted from the fruits of the plant species *Neolamarckia cadamba*, a plant originating from the tropical regions of South and Southeast Asia, showed significant antioxidant potential, suggesting broad applications for controlling oxidative stress-related disorders.

The investigation of the antioxidant activities of polysaccharides and their derivatives holds high practical relevance, aiming for the development of more effective natural antioxidants and the progressive replacement of synthetic antioxidants, which may present toxicity and potential carcinogenicity (Mei et al. 2017).

6.4.2 Prebiotic Effect

Prebiotics are non-digestible food components that impact the composition and activity of the gut microbiota, which, in turn, leads to improved health conditions (Vazquez-Olivo et al. 2019). The introduction of probiotics into the human diet is beneficial for the gut microbiota. They can be consumed in the form of raw vegetables and fruits, fermented pickles, or dairy products (Markowiak and Śliżewska 2017).

Foods that exhibit such activity stand out because they are not digested by the gastrointestinal tract. Thus, various studies aim to understand and optimize these processes by comprehending prebiotic activity, which refers to the ability of certain compounds, like polysaccharides present in fruits, to selectively stimulate the growth and/or bactericidal activity beneficial to the gut, resulting in health benefits for the individual. These compounds serve as a substrate, providing a balanced and functional gut microbiota (Monteiro et al. 2021, Leite et al. 2023a).

The mechanisms of action and modulation of the gut microbiota by prebiotics occur through the selective stimulation of beneficial bacteria via fruit polysaccharides through fermentation, and the production of short-chain fatty acids (SCFAs), also via fermentation, but with the capacity to reduce intestinal pH and promote an environment less favorable for pathogen development and, concurrently, more favorable for beneficial bacteria (Corrêa et al. 2022, Leite et al. 2023a, b, Torres et al. 2024).

Studies on the modulation of gut microbiota through fruit consumption, specifically the use of orange juice enriched with oligosaccharides and dextran using the enzyme dextran-sucrase, promoted a digestive pathway suitable for the microbiota to consume added compounds, resulting in the production of short-chain fatty acids, such as lactate and acetate. Furthermore, other tests also showed an increase in the abundance of beneficial bacteria, such as *Lactobacillus ruminis* and *Bifidobacterium adolescentis* (Leite et al. 2023b, Corrêa et al. 2022).

Pectin is an example of dietary polysaccharides present in fruits, in the form of compounds like fructooligosaccharides (FOS) and inulin, which act as selective substrates for beneficial bacteria, favoring their growth and increasing the production of short-chain fatty acids (SCFAs), including acetate, propionate, and butyrate. The function of these metabolites improves intestinal homeostasis, reducing inflammatory processes and strengthening the epithelial barrier, as well as exerting systemic effects, influencing and aiding in immune and metabolic regulation (Monteiro et al. 2021, Leite et al. 2023b).

Inulin is a water-soluble polysaccharide used as an energy reserve, classified as a non-digestible carbohydrate belonging to the fructan group. It is commonly used as a prebiotic and for the development of functional foods due to its beneficial role in gastric health, being employed as a functional ingredient and in the development of new products (Shoaib et al. 2016).

Additionally, the fermentation of these polysaccharides contributes to the reduction of intestinal pH, creating an unfavorable environment for pathogens and promoting a beneficial microbial balance. Thus, the regular inclusion of polysaccharide-rich fruits in the diet represents an effective strategy to improve gut health and prevent gastrointestinal disorders (Lee et al. 2023, Leite et al. 2023b).

6.4.3 Immunomodulatory Properties

For example, in the study by Salman et al. (2008), citrus pectins, such as those obtained from fruits, were shown to impact the production of inflammatory and anti-inflammatory cytokines, depending on the chemical structure (degree of esterification), indicating a potential to modulate the inflammatory response through alteration in cytokine release by human mononuclear cells, with no significant change in TNF-α or IL-6.

In addition to guava, studies indicate that polysaccharides isolated from byproducts of the processing of acerola (*Malpighia emarginata*), cashew (*Anacardium occidentale* L.), and pineapple (*Ananas comosus* L.) can increase the activity of the enzymes beta-glutamylcysteine synthetase and glutathione reductase, both associated with elevated levels of reduced glutathione (GSH). This increase was observed in the epithelial tissue of the stomach after alcohol consumption, suggesting that these carbohydrates have the potential to reduce the damage caused by free radical-forming agents (Sabino et al. 2020).

Studies have also identified immunomodulatory effects in polysaccharides extracted from fruits such as chayote (*Sechium edule*). There is evidence that extracts of this plant, especially hybrid variants, demonstrate antioxidant and anti-inflammatory activity. A study with hybrid *Sechium* reduced pro-inflammatory cytokines (TNF-α, IFN-γ, IL-6) and increased IL-10 in mice. This profile indicates a possible immunomodulatory effect, i.e., both pro-inflammatory (via ROS, NO, TNFs) and anti-inflammatory potential, dependent on the dose, type of extract, and biological context (Aguiñiga-Sánchez et al. 2020).

Still in the immunomodulatory context, polysaccharides isolated from edible fruits of the Solanaceae family, such as bell pepper (*Capsicum annuum*), show the capacity to influence cytokine secretion by macrophages. A pectic fraction extracted from bell pepper was shown to stimulate the production of TNF-α, IL-1β, and IL-10 by THP-1 cells differentiated into macrophages. After certain structural modifications, the fraction showed changes in the induction of these cytokines, indicating that differences in the degree of methyl-esterification, molecular weight, and presence of side chains can modulate the immune response (Nascimento et al. 2017).

These results highlight the potential of fruit polysaccharides as immunomodulatory agents, demonstrating how their ability to influence the body's inflammatory response and the understanding of their effects can contribute to the development of nutritional and therapeutic strategies aimed at modulating the immune system and promoting health (Aguiñiga-Sánchez et al. 2020).

6.4.4 Glycemic Regulation

Regulating blood glucose is crucial for maintaining metabolic health, preventing serious diseases like type 2 diabetes and dyslipidemias, and it's one of the most studied areas for developing dietary solutions (Ferreira et al. 2025).

The incorporation of polysaccharides into the diet has shown significant beneficial effects on blood glucose and lipid metabolism, contributing to improved metabolic health. Fructooligosaccharides (FOS), not being digested in the small intestine, impact glycemic response modulation. They are fermented in the colon, leading to the production of SCFAs, which stimulate the release of intestinal hormones like GLP-1 and PYY, both involved in glucose regulation (Ferreira et al. 2025). Studies conducted in animal models have demonstrated that FOS administration resulted in significant improvements in blood glucose levels and insulin sensitivity. This effect is attributed to the absorption of bile acids and the production of short-chain fatty acids (SCFAs) by bacterial fermentation in the colon, which inhibit hepatic cholesterol synthesis and increase LDL clearance (Nurkolis et al. 2025, Figueroa et al. 2023).

Understanding the molecular mechanisms that regulate blood glucose is essential for developing effective therapeutic strategies in managing this metabolic disorder. Modulating insulin signaling pathways, activating PPARs, and controlling lipotoxicity emerge as promising targets for interventions aimed at restoring metabolic homeostasis and preventing associated complications. This highlights the importance of dietary interventions focused on macronutrient balance to preserve metabolic homeostasis and prevent diseases associated with glycemic and lipid disorders (Nurkolis et al. 2025).

Inulin is a polysaccharide that can be used as a sugar substitute, applied in the development of functional foods aimed at improving health, especially for people needing glycemic control. It can be used in the formulation of new products, becoming an important component in developing foods with better sensory and nutritional properties (Shoaib et al. 2016).

6.4.5 *Other Biological Activities*

In addition to the previously recognized nutritional properties, polysaccharides extracted from fruits have exhibited a wide range of additional biological activities.

Antinociceptive activity refers to the ability to inhibit or decrease the sensitivity of nociceptors to stimuli, especially regarding inflammation. This property was observed in arabinoxylan (hemicellulose), a non-starchy polysaccharide purified from tomato (*Solanum lycopersicum*) mucilage, which showed a significant reduction in acetic acid-induced abdominal writhing and in the inflammatory phase of formalin-induced nociception when the aforementioned fraction was administered intraperitoneally to mice, indicating an analgesic effect mediated by anti-inflammatory mechanisms (Nascimento et al. 2016).

Studies conducted by Prado et al. (2019) demonstrated that a soluble pectin fraction extracted from papaya (*Carica papaya*), present at an intermediate stage of ripening, inhibits galectin-3 and can reduce the viability of colorectal cancer cells (HCT116 and HT29 cell lines). Low molecular weight pectins (typically 3–30 kDa) reduced the viability of SW-480 (colorectal) cells and decreased galectin-3 levels *in vitro*, indicating a correlation between low molecular weight, presence of galactans, and antitumor activities (Pedrosa et al. 2023).

Although most research on polysaccharides with neuroprotective activity has focused on sources such as seaweed, it is plausible that fruit polysaccharides also have similar properties, given the presence of bioactive compounds with antioxidant and anti-inflammatory potential that contribute to the preservation and restoration of neural activity. A study by Liu et al. (2018) demonstrated that selenium-containing polysaccharides extracted from the fruit of Chinese pomegranate (*Rosa laevigata*) efficiently protected neuronal cells (SH-SY5Y line) against induced oxidative stress. A similar effect was observed with polysaccharides derived from soursop (*Annona muricata*), which protected HT22 hippocampal cells exposed to H_2O_2 (Xu et al. 2022).

6.4.6 *Technological and Functional Applications*

6.4.6.1 Functional Food Formulation

Functional foods are those that, besides providing essential nutrients, offer additional health benefits, contributing to disease prevention and improving quality of life. These foods contain bioactives, such as polysaccharides extracted from fruits, that play a crucial role in the food industry. These compounds not only offer beneficial health properties but also improve the technological characteristics of products.

Chemical modifications of polysaccharides for food and non-food applications have been practiced for years to improve their functional properties (Spotti and Campanella 2017). Due to their high molecular weight and the abundance of intramolecular hydrogen bonds, these can be modified chemically and biochemically with relative ease to increase the biostability of foods. For this reason, some studies have evaluated the use of different polysaccharide matrices in the assembly and protection of food ingredients for the formulation of functional foods (Lu et al. 2019).

Functional foods with the presence of fruit polysaccharides have been gaining significant prominence and becoming more popular due to their health benefits and potential for innovation in the industry, such as the development of "clean label" products (Spotti and Campanella 2017). A study about yellow passion fruit flour (*Passiflora edulis* f. *flavicarpa*) indicates that the presence of pectin in this food contributes to intestinal health, promoting the balance of bacterial flora, improving the absorption of minerals such as calcium and magnesium, and preventing diarrhea, besides being indicated for the control of type II diabetes mellitus. In the digestive tract, pectins bind to bile salts and fat, facilitating the excretion of these compounds and contributing to the reduction of LDL cholesterol levels (Manzoli et al. 2021, Antonini 2023).

The presence of pectin in yellow passion fruit flour helps to delay gastric emptying and intestinal transit; this occurs due to pectin's ability to form gels, which can have a positive impact on various bodily functions, in addition to promoting a feeling of satiety for longer (Manzoli et al. 2021).

Another example of functional foods with the presence of fruit polysaccharides is fermented beverages based on quinoa and amaranth. This beverage stands out as a nutritious and viable alternative for people with dietary restrictions, especially due to the addition of polysaccharides derived from quinoa and amaranth, such as inulin and fructooligosaccharides (FOS), which are essential for intestinal health and contribute to the balance of the microbiome (Oliveira et al. 2020). Inulin and FOS are known for their prebiotic properties, which favor the growth of beneficial bacteria in the intestine, contributing to the improvement of digestive well-being and strengthening the immune system (Bezerra et al. 2020, Chagas et al. 2020).

In the study by Phirom-on and Apiraksakorn (2021), prebiotic fiber based on cellulose from banana peel obtained by enzymatic hydrolysis showed a positive prebiotic value, being higher than that of inulin, a commercially used prebiotic fiber.

6.4.6.2 Applications in the Pharmaceutical and Cosmetic Industries

6.4.6.2.1 Pharmaceutical Excipients

Polysaccharides derived from fruits have gained prominence as pharmaceutical excipients due to their properties that favor the controlled release of drugs, contributing to improved bioavailability and stability of active compounds.

Research has explored the use of fruit-derived polysaccharides, such as pectin, in pharmaceutical systems for controlled drug release. One example is the development of films composed of pectin and starch, which have proven effective in the gradual release of drugs. Variations in the proportion of these biopolymers directly influenced properties such as disintegration time, highlighting the potential of these materials in mucoadhesive systems that ensure controlled and precise release of active compounds (Murata et al. 2019).

Another important example proposed the use of enhanced hydrogels for the delivery and controlled release of pilocarpine in ocular treatment. Biological activity increased, and the preliminary results presented indicate that the use of a pilocarpine complex in combination with hydrogels can result in prolonged activity and greater bioavailability of the active ingredient (Burgalassi 1996).

Cellulose and starch are among the most widely used biomaterials in biomedical applications, due to their biocompatibility, biodegradability, and lack of toxicity. These polysaccharides, along with their derivatives, such as dextrins, ethylcellulose, methylcellulose, hydroxypropyl methylcellulose, and carboxymethylcellulose, among others, have broad applications in pharmaceutical technology as excipients in solid, semi-solid, and liquid dosage forms (Lukova et al. 2023).

6.4.6.2.2 Cosmetics Production

Fruits are a rich source of polysaccharides, and a growing number of studies have shown that these have a wide range of biological functions and applications, among which we can highlight their use in the development of sustainable cosmetics. These also align with the growing demand for eco-friendly and effective products (Shi et al. 2024).

Arabinogalactan, a highly branched polysaccharide primarily composed of galactose and arabinose, found in various fruits, demonstrated anti-aging action in work by Huang et al. (2022b). This suggests its potential use in the formulation of cosmetic products for dermal application.

A recent study indicates that, while the mechanisms by which polysaccharides act in cosmetics have been widely investigated, the correlation between their chemical structure and cosmetic activity remains poorly elucidated. This gap limits the use of these biopolymers in the sector and, therefore, represents a promising technological and scientific window (Wu et al. 2024).

6.4.6.2.3 Thickeners, Emulsifiers, and Encapsulating Agents

Polysaccharides extracted from fruits have gained prominence in the food and pharmaceutical industries due to their functional properties as thickeners, emulsifiers, and encapsulating agents. These natural compounds contribute to improving the quality and stability of food products.

Thickeners are substances that increase the viscosity of foods, enhancing their texture and consistency. They are water-soluble and hydrophilic, used to disperse, stabilize, or prevent the sedimentation of suspended substances. Studies show that using fructooligosaccharides as thickeners in the industry is highly effective, as it influences the rheological properties of foods, increases viscosity, and improves the texture of various food products without altering the flavor, thereby maintaining sensory characteristics (Mecedo et al. 2020).

Emulsifiers are characterized as mixtures of two immiscible liquids that act by reducing the surface tension between substances and stabilizing the mixture. The combination of these polysaccharides as emulsifiers, such as açaí oil, guarantees the stability and homogeneity of the system, facilitating the incorporation of the oil into water-based food products, with possible applications in the food and pharmaceutical industries (McClements and Jafari 2018, Rabelo et al. 2018). Protein/polysaccharide complexes are considered good alternatives to synthetic emulsifiers and stabilizers for formulating stable emulsion-based foods, with improved stability due to enhanced water-holding capacity through conjugation (Seo and Yoo 2022).

Encapsulating agents are substances that form capsules or coatings to encapsulate other ingredients, such as oils, acids, vitamins, or fragrances, protecting them and ensuring their stability and controlled release during consumption or use. The use of pectin as an encapsulating material offers advantages like biodegradability, safety for consumption, and the possibility of modulating the release of active substances. This is due to its gel-forming capacity, as well as its biocompatibility and non-toxicity. Encapsulating drugs in pectin nanoparticles improves drug stability, facilitating the gradual release of the active compound (Han et al. 2022, Rosales and Fabi 2023).

The wide variety of polysaccharides, their properties, and the vast possibility of modifying these compounds make them polymers of great interest for future and promising studies (Maugeri Filho and Goldbeck 2019).

6.5 Conclusion and Perspectives

The structural complexity and compositional variability of polysaccharides extracted from fruits bestow upon them functional and bioactive properties of great relevance to human health, the food industry, and the pharmaceutical sector. Extraction and purification techniques, ranging from aqueous extraction to ultrasound-, microwave-, and enzyme-assisted technologies, followed by chromatography and membrane filtration purification methods, allow for the isolation of specific polysaccharide fractions with distinct physicochemical and functional characteristics.

In terms of bioactivity, these polysaccharides demonstrate a diverse set of benefits, acting as antioxidants, protecting against oxidative stress; exerting prebiotic effects, promoting the growth of beneficial bacteria and the production of short-chain fatty acids; showing immunomodulatory action through selective cytokine induction and activation of immune cells; aiding glycemic regulation by influencing hormone secretion and insulin sensitivity; and displaying other activities, such as antinociception, antitumor effects, and neuroprotective potential. These bioactive effects reinforce the potential of fruit polysaccharides as safe, sustainable, and multifunctional alternatives to synthetic drugs and artificial additives.

In the field of technological applications, fruit-extracted polysaccharides are being employed in the development of functional foods, improving the texture, stability, and nutritional profile of "clean label" products, as well as in the formulation of pharmaceutical excipients and cosmetics, where their gelling properties, mucoadhesion, and controlled active encapsulation are especially valued.

The structural and functional richness of fruit polysaccharides makes them polymers of high scientific and technological value. Continuous investigation into their extraction methods, purification, characterization, and mechanisms of bioactive action will not only deepen the understanding of their properties but also expand their applications in the health, nutrition, and industrial sectors.

Thus, interdisciplinary research focused on the structure, functionality, and mechanisms of action of fruit polysaccharides will be fundamental to transforming their potential into concrete solutions that meet the growing demands for innovation, health, and sustainability.

References

Aguiñiga-Sánchez, I., Soto-Hernández, M., Cadena-Iñiguez, J., Suwalsky, M., Colina, J. R., Castillo, I. et al. (2020). Phytochemical analysis and antioxidant and anti-inflammatory capacity of the extracts of fruits of the sechium hybrid. Molecules 25(20): 4637. https://doi.org/10.3390/molecules25204637.

Ahmad, M., Shahid, A., Iqbal, Y., Hussain, A. I., Khan, I. and Xie, F. (2022). Recent trends in extraction, purification, and antioxidant activity evaluation of plant leaf-extract polysaccharides. Biotechnology and Bioprocess Engineering 16(6): 1820–1848. https://doi.org/10.1002/bbb.2405.

Alves, A. S. S., Costa, A. M. T., Silva, A. V., Simões, C. V., Souza, I. A., Costa, R. M. C. E. et al. (2022). Revestimentos comestíveis à base de polissacarídeos em frutas: uma revisão narrativa. pp. 388–402. *In*: Verruck, S. (Org.), Avanços em Ciência e Tecnologia de Alimentos. Cientifica Digital. https:// 10.37885/220308060.

Amarante, M. K., Ariza, C. B., Oliveira, C. E. C. D., Gualtieri, K. D. A., Oda, J. M. M. and Watanabe, M. A. E. (2018). Inulin induces IL-10 secretion and increased FOXP3 gene expression in human peripheral blood mononuclear cells. Brazilian Archives of Biology and Technology 61: e18160591. https://doi.org/10.1590/1678-4324-2018160591.

Antonini, L. M. (2023). Compostos bioativos em alimentos funcionais e seus benefícios à saúde: uma revisão. Editora Publicar 156–172. http://10.47402/ed.ep.c2023526014938.

Beres, C., Mellinger-Silva, C., Freitas, S. P. and Cabral, L. M. C. (2015). Extração de polissacarídeos do bagaço de uva com aplicação tecnológica. Blucher Chemical Engineering Proceedings 1(2): 4019–4026. http://dx.doi.org/10.1016/chemeng-cobeq2014-0872-22988-178601.

Bernaud, F. S. R. and Rodrigues, T. C. (2013). Fibra alimentar: ingestão adequada e efeitos sobre a saúde do metabolismo. Arquivos Brasileiros de Endocrinologia & Metabologia 57(6): 397–405. https://doi.org/10.1590/S0004-27302013000600001.

Bertoft, E., Piyachomkwan, K., Chatakanonda, P. and Sriroth, K. (2008). Internal unit chain composition in amylopectins. Carbohydrate Polymers 74(3): 527–543. https://doi.org/10.1016/j.carbpol.2008.04.011.

Bezerra, N. M., Oliveira, G. L., Victorino, J. L., Ferreira, T. N. and Fabri, S. G. (2020). Barra de cereal proteica à base de quinoa e amendoim enriquecido com frutooligossacarídeo. Revista Principia 49: 106–113. https://doi.org/10.18265/1517-03062015v1n49p106-113.

Bhotmange, D. U., Wallenius, J. H., Singhal, R. S. and Shamekh, S. S. (2017). Enzymatic extraction and characterization of polysaccharide from Tuber aestivum. Bioactive Carbohydrates and Dietary Fibre 10: 1–9. https://doi.org/10.1016/j.bcdf.2017.02.001.

Botticella, E., Savatin, D. V. and Sestili, F. (2021). The triple jags of dietary fibers in cereals: how biotechnology is longing for high fibergrains. Frontiers in Plant Science 12: 745579. https://doi.org/10.3389/fpls.2021.745579.

Buckeridge, M. S., Tiné, M. A. S., Santos, H. P. and Lima, D. U. (2000). Polissacarídeos de reserva de parede celular em sementes. Estrutura, metabolismo, funções e aspectos ecológicos. Revista Brasileira de Fisiologia Vegetal 12: 137–162.

Burgalassi, S., Chetoni, P. and Saettone, M. F. (1996). Hydrogels for ocular delivery of pilocarpine: Preliminary evaluation in rabbits of the influence of viscosity and of drug solubility. European Journal of Pharmaceutics and Biopharmaceutics 42: 385–392.

Chagas, A. A., Silva, L. R., Oliveira, C. A. and Rodrigues, M. B. (2020). Bioactive compounds of interest to the food industry: properties, applications and perspectives for the consumer market. Research, Society and Development 9(10): e3469108094. https://doi.org/10.33448/rsd-v9i10.8094.

Clemente-Suárez, V. J., Mielgo-Ayuso, J., Martín-Rodríguez, A., Ramos-Campo, D. J., Redondo-Flórez, L. and Tornero-Aguilera, J. F. (2022). The burden of carbohydrates in health and disease. Nutrients 14(18): 3809. https://doi.org/10.3390/nu14183809.

Corrêa, A. R. A., Souza, C. L., Nunes, G. Z. and Ferreira, J. C. S. (2022). O mecanismo de ação dos bióticos na disbiose e a relação com a obesidade e ansiedade. Revista Ibero-Americana de Humanidades, Ciências e Educação 8(4): 975–1001. https://doi.org/10.51891/rease.v8i4.5104.

Cruz-Cárdenas, C. I., Miranda-Ham, M. L., Castro-Concha, L. A., Ku-Cauich, J. R., Vergauwen, R., Reijnders, T. et al. (2015). Fructans and other water soluble carbohydrates in vegetative organs and fruits of different Musa spp. accessions. Frontiers in Plant Science 6: 1–10. https://doi.org/10.3389/fpls.2015.00395.

Cunha, P. L. R. D., Paula, R. C. M. D. and Feitosa, J. (2009). Polysaccharides from Brazilian biodiversity: an opportunity to change knowledge into economic value. Química Nova 32(3): 649–660. https://doi.org/10.1590/S0100-40422009000300009.

Englyst, H. N., Kingman, S. M. and Cummings, J. H. (1992). Classification and measurement of nutritionally important starch fractions. European Journal of Clinical Nutrition 46(2): S33–S50.

Fekete, S., Beck, A., Veuthey, J. L. and Guillarme, D. (2015). Ion-exchange chromatography for the characterization of biopharmaceuticals. Journal of Pharmaceutical and Biomedical Analysis 113: 43–55. https://doi.org/10.1016/j.jpba.2015.02.037.

Fernández-Lainez, C., Stegge, M. A., Silva-Lagos, L. A., López-Velázquez, G. and Vos, P. (2023). β(2 → 1)-β(2 → 6) and β(2 → 1) fructans protect from impairment of intestinal tight junction's gene expression and attenuate human dendritic cell responses in a fructan-dependent fashion. Carbohydrate Polymers 320: 121259. https://10.1016/j.carbpol.2023.121259.

Ferreira, G. A., Silva, R. N. and Monteiro, V. N. (2022). Estudo da dinâmica de polissacarídeos da parede celular e das enzimas hidrolíticas de espécies de trichoderma durante o micoparasitismo. Revista Sapiência: sociedade, saberes e práticas educacionais 11(2): 277–290. https://doi.org/10.31668/revsap.v12i1.13938.

Ferreira, T. H., Maximiano, P. and Simões, P. N. (2025). Effect of fructooligosaccharides in full-hydrated lactic acid bacteria membrane models during thermal stress: A molecular simulation study. Food Research International 200: 115475. https://doi.org/10.1016/j.foodres.2024.115475.

Figueroa, G. M., Costa, D. N., Miranda, A. C. dos A., Valasques Júnior, G. L., Mendes, T. P. S. and Pires Cedro, P. É. P. (2023). Polissacarídeos fúngicos e seu potencial hipoglicemiante para atuar como adjuvante no tratamento do diabetes mellitus. Evidência 23(1): 47–62. https://doi.org/10.18593/evid.32561.

Gavlighi, H. A., Tabarsa, M., You, S., Surayot, U. and Ghaderi-Ghahfarokhi, M. (2018). Extraction, characterization and immunomodulatory property of pectic polysaccharide from pomegranate peels: Enzymatic vs conventional approach. International Journal of Biological Macromolecules 116: 698–706. https://doi.org/10.1016/j.ijbiomac.2018.05.083.

Ghosh, S. and Abdullah, M. F. (2024). Extraction of polysaccharide fraction from cadamba (*Neolamarckia cadamba*) fruits and evaluation of its *in vitro* and *in vivo* antioxidant activities. International Journal of Biological Macromolecules 279(4): 135564. https://doi.org/10.1016/j.ijbiomac.2024.135564.

Han, S. S., Ji, S. M., Park, M. J., Suneetha, M. and Uthappa, U. T. (2022). Pectin based hydrogels for drug delivery applications: A mini review. Gels 8(12): 834. https://doi.org/10.3390/gels8120834.

Hong, T., Yin, J. Y., Nie, S. P. and Xie, M. Y. (2021). Applications of infrared spectroscopy in polysaccharide structural analysis: Progress, challenge and perspective. Food Chemistry 12: 100168. https://doi.org/10.1016/j.fochx.2021.100168.

Huang. G., Mei, X. and Hu, J. (2017). The antioxidant activities of natural polysaccharides. Current Drug Targets 18(11): 1296–1300. https://10.2174/1389450118666170123145357.

Huang, X. J., Ai, C., Yao, H., Zhao, C., Xiang, C., Hong, T. et al. (2022a). Guideline for the extraction, isolation, purification, and structural characterization of polysaccharides from natural resources. eFood 3(6): 1–16. https://doi.org/10.1002/efd2.37.

Huang, W., Zhao, M., Wang, X., Tian, Y., Wang, C., Sun, J. et al. (2022b). Revisiting the structure of arabinogalactan from Lycium barbarum and the impact of its side chain on antiageing activity. Carbohydrate Polymers 286: 119282. https://doi.org/10.1016/j.carbpol.2022.119282.

Kang, J., Hua, X., Yang, R., Chen, Y. and Yang, H. (2015). Characterization of natural low-methoxyl pectin from sunflower head extracted by sodium citrate and purified by ultrafiltration. Food Chemistry 180: 98–105. https://doi.org/10.1016/j.foodchem.2015.02.037.

Koakuzu, S., Araújo, E. J., Bassinello, P. Z., Carvalho, R. N. and Teixeira, M. C. (2015). Procedimento para determinação de amido resistente, amido não resistente (solúvel) e total (Comunicado técnico, No. 228). Embrapa Arroz e Feijão.

Lajolo, F. M. and Menezes, E. W. (2006). Carbohidratos en alimentos regionales Iberoamericanos. EDUSP. https://doi.org/10.1590/S1516-93322006000400018.

Lee, Y. R., Lee, H. B., Kim, Y., Shin, K. S. and Park, H. Y. (2023). Prebiotic and anti-adipogenic effects of radish green polysaccharide. Microorganisms 11(7): 1862. https://10.3390/microorganisms11071862.

Leite, L. T. E., Santos, R. P., Tecchio, J. V. S. and Cardoso, M. R. S. (2023a). Potencial da curcumina na modulação da microbiota intestinal e seus efeitos na saúde gastrointestinal. Revista Ibero-Americana de Humanidades, Ciências e Educação 9(11): 1818–1834. https://doi.org/10.51891/rease.v9i11.12612.

Leite, A. K. F., Fonteles, T. V., Alves Filho, E. G., Oliveira, F. A. S. and Rodrigues, S. (2023b). Impact of orange juice containing potentially prebiotic ingredients on human gut microbiota composition and its metabolites. Food Chemistry 405(Part B): 134706. https://doi.org/10.1016/j.foodchem.2022.134706.

Leszczuk, A., Kalaitzis, P., Blazakis, K. N. and Zdunek, A. (2020). The role of arabinogalactan proteins (AGPs) in fruit ripening—a review. Horticulture Research 7(176): 1–12. https://doi.org/10.1038/s41438-020-00397-8.

Liu, Z. and Sun, X. (2020). A critical review of the abilities, determinants, and possible molecular mechanisms of seaweed polysaccharides antioxidants. International Journal of Molecular Sciences 21(20): 7774. https://doi: 10.3390/ijms21207774.

Liu, X., Gao, Y., Li, D., Liu, C., Jin, M., Bian, J. et al. (2018). The neuroprotective and antioxidant profiles of selenium-containing polysaccharides from the fruit of Rosa laevigata. Food & Function 9(3): 1800–1808. https://doi.org/10.1039/c7fo01725a.

Lu, X., Chen, J., Guo, Z., Zheng, Y., Rea, M. C., Su, H. et al. (2019). Using polysaccharides for the enhancement of functionality of foods: A review. Trends in Food Science & Technology 86: 311–327. https://doi.org/10.1016/j.tifs.2019.02.024.

Lu, X. (2023). Changes in the structure of polysaccharides under different extraction methods. eFood 4(2): 1–7. https://10.1002/efd2.82.

Lukova, P., Katsarov, P. and Pilicheva, B. (2023). Application of starch, cellulose, and their derivatives in the development of microparticle drug-delivery systems. Polymers 15(17): 3615. https://10.3390/polym15173615.

Macedo, L. L., Vimercati, W. C. and Araújo, C. S. (2020). Fruto-oligossacarídeos: aspectos nutricionais, tecnológicos e sensoriais. Brazilian Journal of Food Technology 23: e2019080. https://doi.org/10.1590/1981-6723.08019.

Madsen, H., Sen, A. and Aune, D. (2023). Fruit and vegetable consumption and the risk of hypertension: A systematic review and meta-analysis of prospective studies. European Journal of Nutrition 62(5): 1941–1955. https://doi.org/10.1007/s00394-023-03145-5.

Manzoli, A. L., Araújo, D. L., Julião, G. B. and Eduardo, A. M. L. N. (2021). Farinha de maracujá-amarelo – Passiflora edulis f. flavicarpa: Aplicabilidade no tratamento da Diabetes Mellitus Tipo II. Revista Ibero-Americana de Humanidades, Ciências e Educação 7(5): 462–479. https://doi.org/10.51891/rease.v7i5.1201.

Markowiak, P. and Śliżewska, K. (2017). Effects of probiotics, prebiotics, and synbiotics on human health. Nutrients 9(9): 1021. https://10.3390/nu9091021.

Martins, A., Pequito, A. and Baltazar, A. L. (2020). Disbiose Intestinal e Síndrome do Intestino Irritável: efeito de uma dieta baixa em FODMAPs. Acta Portuguesa de Nutrição 22: 38–41. https://dx.doi.org/10.21011/apn.2020.2207.

Maugeri Filho, F. and Goldbeck, R. (2019). Produção de polissacarídeos. *In*: Alterthum, F. et al. (eds.). Biotecnologia Industrial: Processos fermentativos e enzimáticos (p. 760). Blucher.

McClements, D. J. and Jafari, S. M. (2018). Improving emulsion formation, stability and performance using mixed emulsifiers: A review. Advances in Colloid and Interface Science 251: 55–79. https://doi.org/10.1016/j.cis.2017.12.001.

Mei, X., Yi, C. and Huang, G. (2017). Preparation methods and antioxidant activities of polysaccharides and their derivatives. Mini-Reviews in Medicinal Chemistry 17(10): 863–868. https://10.2174/1389557517666170116114657.

Menezes, D. P., Monção, F. E. S., Vieira, J. G. M., Ferreira, L. P. F., Batista, C. L., Acha, B. T. et al. (2023). Atividade antioxidante de extratos de *Libidibia ferrea*: revisão sistemática. Anais do Simpósio de Plantas Medicinais do Brasil. https://resumos.sbpmed.org.br/index.php/spmb/article/view/217.

Mihooliya, K. N., Kumar, A., Heiss, C., Kumari, A., Boniche-Alfaro, C., Azadi, P. et al. (2025). Evaluating size exclusion chromatography for nucleic acid removal in Klebsiella pneumoniae cell surface polysaccharide purification. Carbohydrate Polymers 358: 123531. https://doi.org/10.1016/j.carbpol.2025.123531.

Mohammadi-Jam, S., Waters, K. E. and Greenwood, R. W. (2022). A review of zeta potential measurements using electroacoustics. Advances in Colloid and Interface Science 309: 102778. https://doi.org/10.1016/j.cis.2022.102778.

Monteiro, R. G., Silva, E. C., Martins, A. L. C., Porto, A. L. F. and Soares, M. T. C. V. (2021). The effect of exopolysaccharides obtained from lactic acid bacteria as a prebiotic: a systematic review. Research, Society and Development 10(15): e194101522547. https://doi.org/10.33448/rsd-v10i15.22547.

Mu, S., Yang, W. and Huang, G. (2021). Antioxidant activities and mechanisms of polysaccharides. Chemical Biology & Drug Design 97(3): 628–632. https://10.1111/cbdd.13798.

Murata, Y., Maida, C. and Kofuji, K. (2019). Drug release profiles and disintegration properties of pectin films. Materials 12(3): 355. https://doi.org/10.3390/ma12030355.

Nascimento, G. E., Baggio, C. H., Werner, M. F. P., Iacomini, M. and Cordeiro, L. M. C. (2016). Arabinoxylan from mucilage of tomatoes (*Solanum lycopersicum* L.): Structure and antinociceptive effect in mouse models. Journal of Agricultural and Food Chemistry 64(6): 1239–1244. https://doi.org/10.1021/acs.jafc.5b05134.

Nascimento, G. E., Winnischofer, S. M. B., Ramirez, M. I., Iacomini, M. and Cordeiro, L. M. C. (2017). The influence of sweet pepper pectin structural characteristics on cytokine secretion by THP-1 macrophages. Food Research International 102: 588–594. https://doi.org/10.1016/j.foodres.2017.09.037.

Nelson, D. L. and Cox, M. M. (2006). Lehninger: Princípios de bioquímica (pp. 626–631). Editora Sarvier.

Nishiyama, Y., Sugiyama, J., Chanzy, H. and Langan, P. (2003). Crystal structure and hydrogen bonding system in cellulose I(α) from synchrotron X-ray and neutron fiber diffraction. Journal of the American Chemical Society 125(47): 14300–14306. https://doi.org/10.1021/ja037055w.

Nurkolis, F., Harbuwono, D. S., Taslim, N. A., Soegondo, S., Suastika, K., Sparringa, R. A. et al. (2025). New insight on dietary strategies to increase insulin sensitivity and reduce diabetes prevalence: an expert perspective and recommendation. Discover Food 5(136): 1–32. https://doi.org/10.1007/s44187-025-00422-6.

Oliveira, C., Silva, M. R., Santos, L. A. and Oliveira, A. A. (2020). Estudo comparativo: uso de extrato de quinoa e amaranto para produção de bebida fermentada. Revista Engenho 12(1): 84–104. https://revistas.anchieta.br/index.php/RevistaEngenho/article/view/1698.

Oyeyinka, S. A., Akintayo, O. A., Adebo, O. A., Kayitesi, E. and Njobeh, P. B. (2021). A review on the physicochemical properties of starches modified by microwave alone and in combination with other methods. International Journal of Biological Macromolecules 176: 87–95. https://doi.org/10.1016/j.ijbiomac.2021.02.066.

Pedrosa, L. d. F., Nascimento, K. R., Soares, C. G., Oliveira, D. P. d., de Vos, P. and Fabi, J. P. (2023). Unveiling plant-based pectins: exploring the interplay of direct effects, fermentation, and technological applications in clinical research with a focus on the chemical structure. Plants 12(14): 2750. https://doi.org/10.3390/plants12142750.

Phirom-on, K. and Apiraksakorn, J. (2021). Development of cellulose-based prebiotic fiber from banana peel by enzymatic hydrolysis. Food Bioscience 41: 101083. https://doi.org/10.1016/j.fbio.2021.101083.

Pitkänen, L. (2025). Potential of size-exclusion chromatography and asymmetric flow fieldflow fractionation in separation and characterization of plant polysaccharides. Journal of Chromatography A 1748: 465862. https://doi.org/10.1016/j.chroma.2025.465862.

Pollock, C. J. and Cairns, A. J. (1991). Fructan metabolism in grasses and cereals. Annual Review of Plant Physiology and Plant Molecular Biology 42(1): 77–101. https://doi.org/10.1146/annurev.pp.42.060191.000453.

Pomin, V. H. and Mourão, P. A. S. (2006). Carboidratos. Ciência Hoje 35: 24–35.

Prado, S. B. R. D., Santos, G. R. C., Mourão, P. A. S. and Fabi, J. P. (2019). Chelate-soluble pectin fraction from papaya pulp interacts with galectin-3 and inhibits colon cancer cell proliferation. International Journal of Biological Macromolecules 126: 170–178. https://doi.org/10.1016/j.ijbiomac.2018.12.191.

Rabelo, C. A. S., Taarji, N., Khalid, N., Kobayashi, I., Nakajima, M. and Neves, M. A. (2018). Formulation and characterization of water-in-oil nanoemulsions loaded with açaí berry anthocyanins: Insights of degradation kinetics and stability evaluation of anthocyanins and nanoemulsions. Food Research International 106: 542–548. https://doi.org/10.1016/j.foodres.2018.01.017.

Rosales, T. K. O. and Fabi, J. P. (2023). Pectin-based nanoencapsulation strategy to improve the bioavailability of bioactive compounds. International Journal of Biological Macromolecules 229: 11–21. https://doi.org/10.1016/j.ijbiomac.2022.12.292.

Sabino, L. B. de S., Gonzaga, M. L. C., Oliveira, L. de S., Duarte, A. S. G., Silva, L. M. A., Brito, E. S. de et al. (2020). Polysaccharides from acerola, cashew apple, pineapple, mango and passion fruit co-products: Structure, cytotoxicity and gastroprotective effects. Bioactive Carbohydrates and Dietary Fibre 24: 100228. https://doi.org/10.1016/j.bcdf.2020.100228.

Salman, H., Bergman, M., Djaldetti, M., Orlin, J. and Bessler, H. (2008). Citrus pectin affects cytokine production by human peripheral blood mononuclear cells. Biomedicine & Pharmacotherapy 62(9): 579–582. https://doi.org/10.1016/j.biopha.2008.07.058.

Sant'anna, V. B. R., Santos, A. G. D., Brandelli, A. and Veras, F. F. (2021). Análise multivariada do efeito de diferentes processos de secagem de bagaço de uvas de vinho sobre compostos bioativos e propriedades físico-químicas. Revista Brasileira de Meio Ambiente & Sustentabilidade 1(1): 84–106. https://rbmaes.emnuvens.com.br/revista/article/view/11.

Seo, C. W. and Yoo, B. (2022). Effect of milk protein isolate/κ-carrageenan conjugates on rheological and physical properties of whipping cream: A comparative study of Maillard conjugates and electrostatic complexes. Food Science of Animal Resources 42(5): 889–902. https://doi.org/10.5851/kosfa.2022.e42.

Seyfried, M., Soldera-Silva, A., Bovo, F., Stevan-Hancke, F. R., Maurer, J. B. B. and Zawadzki-Baggio, S. F. (2016). Pectins of medicinal plants: structural characteristics and immunomodulatory activities. Revista Brasileira de Plantas Medicinais 18(1): 201–214. https://doi.org/10.1590/1983-084X/15_078.

Shi, L. (2016). Bioactivities, isolation and purification methods of polysaccharides from natural products: A review. International Journal of Biological Macromolecules 92: 37–48. https://doi.org/10.1016/j.ijbiomac.2016.06.100.

Shi, L., He, Q., Li, J., Liu, Y., Cao, Y., Liu, Y. et al. (2024). Polysaccharides in fruits: Biological activities, structures, and structure-activity relationships and influencing factors—a review. Food Chemistry 451: 139408. https://doi.org/10.1016/j.foodchem.2024.139408.

Shoaib, M., Shehzad, A., Omar, M., Rakha, A., Raza, H., Sharif, H. R. et al. (2016). Inulin: Properties, health benefits and food applications. Carbohydrate Polymers 147: 444–454. https://doi.org/10.1016/j.carbpol.2016.04.020.

Spotti, M. J. and Campanella, O. H. (2017). Functional modifications by physical treatments of dietary fibers used in food formulations. Current Opinion in Food Science 15: 70–78. https://doi.org/10.1016/j.cofs.2017.10.003.

Toneli, J. T. C. L., Park, K. J., Murr, F. E. X. and Negreiros, A. A. (2008). Effect of moisture on the microstructure of inulin powder. Food Science and Technology 28(1): 122–131. https://doi.org/10.1590/S0101-20612008000100018.

Torres, R. C., Silva, L. F., Santos, M. L. and Oliveira, F. C. (2024). Modulação da microbiota intestinal como estratégia para o tratamento da obesidade: efeitos de prebióticos e probióticos no perfil metabólico. Journal of Medical and Biosciences Research 1(3): 1076–1081. https://doi.org/10.70164/jmbr.v1i3.161.

Vazquez-Olivo, G., Gutiérrez-Grijalva, E. P. and Heredia, J. B. (2019). Prebiotic compounds from agro-industrial by-products. Journal of Food Biochemistry 43(6): e12711. https://10.1111/jfbc.12711.

Wang, Y., Xiong, X. and Huang, G. (2023). Ultrasound-assisted extraction and analysis of maidenhairtree polysaccharides. Ultrasonics Sonochemistry 95: 106395. https://doi.org/10.1016/j.ultsonch.2023.106395.

Wang, Y., Huang, G. and Huang, H. (2024). Ultrasonic/enzymatic extraction, characteristics and comparison of leechee peel polysaccharide. Ultrasonics Sonochemistry 108: 106948. https://doi.org/10.1016/j.ultsonch.2024.106948.

Wu, J., Shi, S., Wang, H. and Wang, S. (2016). Mechanisms underlying the effect of polysaccharides in the treatment of type 2 diabetes: A review. Carbohydrate Polymers 144: 474–494. https://doi.org/10.1016/j.carbpol.2016.02.040.

Xu, J., Li, M. F. and Sun, R. C. (2015). Identifying the impact of ultrasound-assisted extraction on polysaccharides and natural antioxidants from *Eucommia ulmoides* Oliver. 50(3): 473–481. https://doi.org/10.1016/j.procbio.2014.12.021.

Xu, X. L., Li, S., Zhang, R. and Le, W. D. (2022). Neuroprotective effects of naturally sourced bioactive polysaccharides: An update. Neural Regeneration Research 17(9): 1907–1912. https://doi.org/10.4103/1673-5374.335142.

Xu, M., Ren, J., Jiang, Z., Zhou, S., Wang, E., Li, H. et al. (2024). Structural characterization and immunostimulant activities of polysaccharides fractionated by gradient ethanol precipitation method from Panax ginseng C. A. Meyer. Frontiers in Pharmacology 15: 1388206. https://10.3389/fphar.2024.1388206.

Xu, Q. and Zhang, Y. (2025). Research progress on the bioactivity of compound polysaccharides: A review. International Journal of Biological Macromolecules 306(3): 141693. https://10.1016/j.ijbiomac.2025.141693.

Yang, Y., Zhao, M., Bai, J., Bu, X., Tang, Y., Han, W. et al. (2021). Microwave-assisted aqueous two-phase extraction of phenolic compounds from *Ribes nigrum* L. and its antibacterial effect on foodborne pathogens. Food Control 119: 107449. https://doi.org/10.1016/j.foodcont.2020.107449.

Ying, Z., Han, X. and Li, J. (2011). Ultrasound-assisted extraction of polysaccharides from mulberry leaves. Food Chemistry 127(3): 1273–1279. https://doi.org/10.1016/j.foodchem.2011.01.083.

Yun, C., Ji, X., Chen, Y., Zhao, Z., Gao, Y., Gu, L. et al. (2023). Ultrasound-assisted enzymatic extraction of Scutellaria baicalensis root polysaccharide and its hypoglycemic and immunomodulatory activities. International Journal of Biological Macromolecules 227: 134–145. https://doi.org/10.1016/j.ijbiomac.2022.12.115.

Zhai, X., Zhu, C., Li, Y., Zhang, Y., Duan, Z. and Yang, X. (2017). Optimization for pectinaseassisted extraction of polysaccharides from pomegranate peel with chemical composition and antioxidant activity. International Journal of Biological Macromolecules 109: 244–253. https://doi.org/10.1016/j.ijbiomac.2017.12.064.

Zheng, Y., Yan, J., Cao, C., Liu, Y., Yu, D. and Liang, X. (2023). Application of chromatography in purification and structural analysis of natural polysaccharides: A review. Journal of Separation Science 46: e2300368. https://doi.org/10.1002/jssc.202300368.

Chapter 7

Carotenoids from Fruits

Chemistry, Occurrence, Bioavailability, and Biological Activities

Irakerley Alves Fernandes,[1,*] *Lucivania Rodrigues dos Santos,*[2] *Lidiane da Silva Araújo,*[1] *Kerlane Alves Fernandes*[3] and *Edivan Carvalho Vieira*[2]

7.1 Introduction

Carotenoids and their derivatives form parts of the isoprenoid family and are critical for both plants and animals (Mein et al. 2008). Among them, at the cellular and molecular levels, their importance is widely acknowledged, especially for their antioxidant properties and provitamin A functions in human nutrition. Furthermore, post-harvest treatments and processing techniques influence the composition and bioavailability of carotenoids in food products (Rao and Rao 2007). This chapter covers theoretical perspectives on structural properties, biosynthesis, composition, bioavailability, and effects on human health of the major carotenoids in fruit.

Carotenoids, found in fruit and vegetables, have been widely studied for their functional properties, health benefits, and potential to prevent several important chronic diseases (Viñas-Ospino et al. 2023). They are natural pigments metabolized by plants, algae, and photosynthetic bacteria, and are responsible for the yellow, orange, and red colors of various fruits and vegetables (Weber and Grune 2023).

Based on their functional groups, carotenoids can be classified into two groups: xanthophylls, which contain oxygen as a functional group, including lutein and

[1] Federal Institute of Education, Science and Technology of Maranhão, Campus Codó, 65400-000, Maranhão, Brazil.
[2] Federal University of Piauí, Campus Ministro Petrônio Portela, 64049-550, Piauí, Brazil.
[3] Federal University of Maranhão, Campus São Bernardo, 65550-000, Maranhão, Brazil.
* Corresponding author: irakerley.alves@ifma.edu.br

DOI: 10.1201/9781003668169-9

zeaxanthin, and carotenes, which have only the original hydrocarbon chain, such as α-carotene, β-carotene, and lycopene. The addition of polar groups (epoxy, hydroxyl, and keto) alters the polarity of carotenoids and influences their biological functions (Rodriguez-Amaya 2001). Currently, more than 600 types of carotenoids have been identified in nature (Rodriguez-Amaya 2016).

Carotenoids are found within subcellular organelles, called plastids, such as chloroplasts and chromoplasts. In chloroplasts, these pigments primarily bind to proteins and function as accessory components of photosynthesis. Conversely, in chromoplasts, they exist either as crystals or as oily droplets. Some xanthophylls, for example, participate in energy transfer in the presence of chlorophyll in plants and are involved in photosynthesis (Cazzonelli and Pogson 2010).

Fruits are the main source of carotenoids and therefore have an essential function in the diet as a result of vitamin A activity. Carotenoids are also important for their antioxidant, anti-inflammatory, and immune-support activities (Rao and Rao 2007). Studies suggest that diets high in carotenoids are associated with a reduced risk of cancer, cardiovascular diseases, age-related macular degeneration, and the development of cataracts (Sathasivam and Ki 2018).

The data published in the literature is highly useful to food researchers, nutritionists, and health professionals, helping promote appropriate diets to minimize vitamin A deficiency and maintain healthy eating habits. This chapter explores the theoretical foundations of carotenoids, including their structural properties, biosynthesis in plants, composition in fruits, bioavailability, and their impact on human health, with an emphasis on their biological activities.

7.2 Carotenoids: An Overview

Carotenoids are a class of fat-soluble pigments of great biological importance in plants, acting mainly in photoprotection and photosynthesis. Chemically, carotenoids are tetraterpenoids, containing 40 carbon atoms, derived from the union of eight isoprene units (Fraser and Bramley 2004). This structure allows the formation of long conjugated chains of double bonds, which confer characteristic optical and antioxidant properties (Rodriguez-Amaya 2016).

The presence of an extensively conjugated system of double bonds is responsible for light absorption in the visible spectrum, contributing to the yellow, orange, and red coloration observed in many plant tissues (Rodriguez-Amaya 2016). The system of conjugated double bonds determines both the color and the chemical reactivity of carotenoids, and is also mainly responsible for their instability to light, oxygen, and heat (Britton Functions 2008).

Carotenoids are divided into two large groups: carotenes and xanthophylls, where carotenes, such as β-carotene and lycopene, are pure hydrocarbons, while xanthophylls contain oxygen in functional groups such as hydroxyls or epoxides (Figure 1) (Khoo et al. 2011). This chemical difference impacts both the solubility and the intracellular localization of the compounds.

In plants, these pigments are mainly located in chloroplasts (in leaves) and chromoplasts (in fruits and flowers), where they associate with proteins and lipid membranes. The spatial configuration, predominantly *all-trans*, can isomerize into

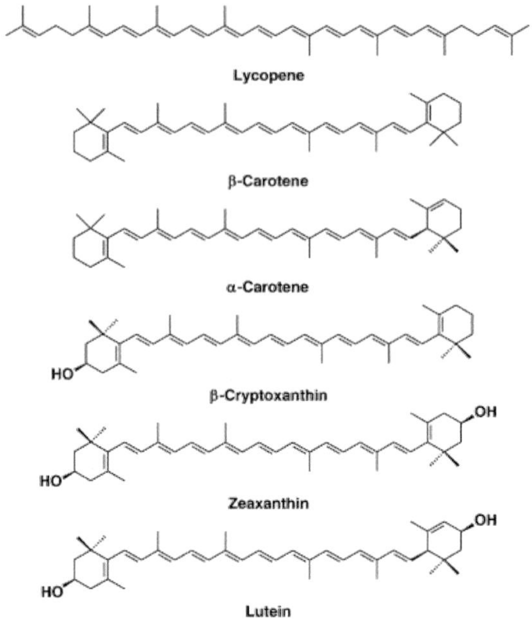

Figure 1. Illustration representing the structure of carotenoids.

cis forms under the influence of environmental factors (Rodriguez-Amaya 2016). The isomerization of carotenoids can be induced by light, heat, or acids, altering their biochemical and functional properties (Khoo et al. 2011).

7.3 Biosynthesis of Carotenoids in Plants

The biosynthesis of carotenoids in plants occurs in plastids, especially in chloroplasts and chromoplasts, via the methylerythritol phosphate (MEP) pathway in the stroma (Fraser and Bramley 2004). The carotenoid biosynthesis pathway in plants is summarized in Figure 2.

The process involves several enzymatic steps starting from the 20-carbon isoprenoid intermediate geranylgeranyl diphosphate (GGPP) (Cazzonelli and Pogson 2010). The first specific step in carotenoid biosynthesis is the condensation of two GGPP molecules to form 15-Z-phytoene, catalyzed by phytoene synthase (PSY) (Fraser and Bramley 2004). This reaction is considered the control point of the pathway and is subject to transcriptional and post-transcriptional regulation (Cazzonelli and Pogson 2010).

The process involves several enzymatic steps starting from the 20-carbon isoprenoid intermediate geranylgeranyl diphosphate (GGPP) (Cazzonelli and Pogson 2010). The first specific step in carotenoid biosynthesis is the condensation of two GGPP molecules to form 15-Z-phytoene, catalyzed by phytoene synthase (PSY) (Fraser and Bramley 2004). This reaction is considered the control point of the pathway and is subject to transcriptional and post-transcriptional regulation (Cazzonelli and Pogson 2010).

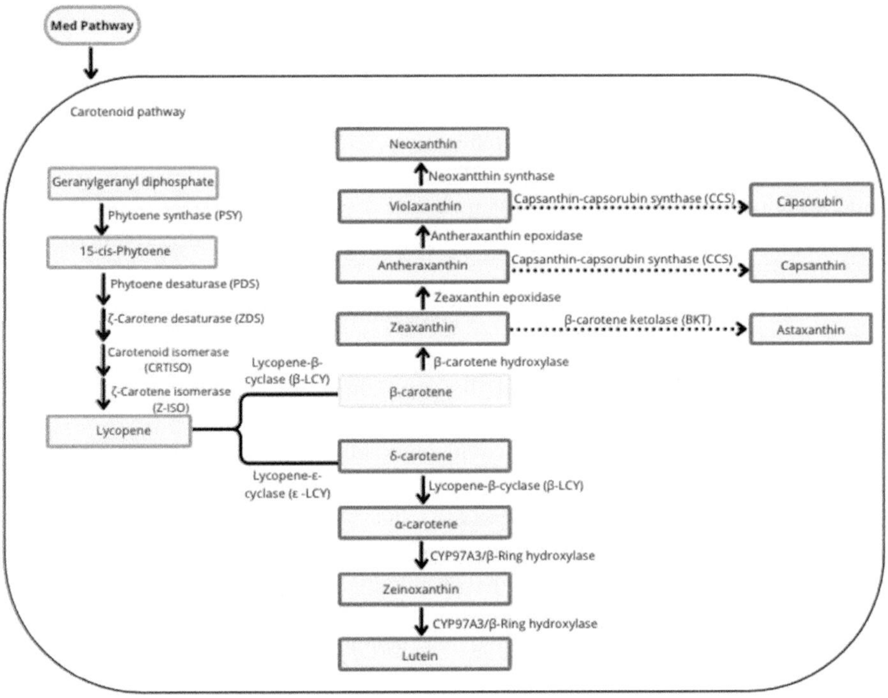

Figure 2. Biosynthesis of carotenoids in plants.

From lycopene, cyclization reactions occur that lead to the formation of α-carotene and β-carotene, mediated by the enzymes lycopene β-cyclase (LCYB) and lycopene ε-cyclase (LCYE); β-carotene is formed when two LCYB molecules act on lycopene, while the combined action of LCYB and LCYE gives rise to α-carotene (Giuliano et al. 2008).

The cyclization of lycopene with lycopene ε-cyclases (ε-LCY) and lycopene β-cyclases (β-LCY) to form δ-carotene and γ-carotene, respectively, is the fundamental branching point in carotenoid biosynthesis. In one branch, a single enzyme, β-LCY, introduces a β-ionone ring at both ends of lycopene, forming β-carotene. However, in the other branch, which leads to lutein, ε-LCY and β-LCY are required to introduce the β-ring and the ε-ionone ring into lycopene to form α-carotene. The β-ionone ring is required to provide provitamin A activity (Send and Sundholm 2007).

α-carotene metabolism entails its conversion into zeinoxanthin by β-ring hydroxylase. This product is hydroxylated by an ε-ring hydroxylase to lutein. Then, a similar metabolic pathway proceeds as follows: β-carotene is hydroxylated by β-carotene hydroxylase to yield zeaxanthin. Zeaxanthin can then be epoxidized to antheraxanthin and violaxanthin by zeaxanthin epoxidase and antheraxanthin epoxidase, respectively. Lastly, violaxanthin is converted into neoxanthin by neoxanthin synthase (Giuliano et al. 2008).

In addition to the enzymes involved, the regulation of carotenoid biosynthesis is highly influenced by environmental factors such as light, temperature, and nutrient availability. Light, for example, not only stimulates gene expression of biosynthetic enzymes but also affects the cell's energy demand and redox state, which are fundamental to metabolic activity. The interaction between hormonal signaling, such as abscisic acid (ABA) and ethylene, also plays a significant role in regulating carotenoid levels during fruit development and ripening (Othman et al. 2014).

Research into the biosynthesis of carotenoids in plants not only advances understanding of metabolic mechanisms but also opens avenues for biotechnology, enabling the development of cultivars enriched in carotenoids (Hannoufa and Hossain 2012).

7.4 Carotenoids in Fruits

Carotenoids can be linked to chlorophylls, resulting in diverse colors in plants, fruits, and vegetables. However, as fruits mature, chlorophyll levels decrease, resulting in the prominence of colored carotenoid pigments (Khoo et al. 2011). Therefore, the type and availability of carotenoids in fruits can be predicted by their color; for example, yellow-orange fruits are generally rich in β-carotene and α-carotene, and orange fruits contain α-cryptoxanthin and zeaxanthin (Krinsky and Johnson 2005, Meléndez-Martínez 2019). In this chapter, comprehensive data on the composition of the main carotenoids in fruit are presented in Table 1.

Knowledge of the composition of carotenoids in different edible parts and cultivars will be useful for nutrition experts in selecting nutrient-rich plants for appropriate diet recommendations. The total carotenoid content ranged from 0.54 µg/g in banana to 513.87 µg/g in buriti, with β-carotene and lutein as the main carotenoids present in the fruit, at 0.23 to 372.32 µg/g and 0.06 to 172.00 µg/g, respectively.

7.4.1 Bioavailability of Carotenoids from Fruits

Carotenoids are fat-soluble pigments widely distributed in fruits and vegetables; however, the bioavailability of these compounds, i.e., the fraction that is absorbed and used by the body, can vary significantly depending on several factors, such as the food matrix, processing, type of carotenoid, interactions with other nutrients, and individual consumer characteristics (Castenmiller and West 1998, Reboul 2013).

In fruit, the most abundant carotenoids include β-carotene, α-carotene, lutein, and zeaxanthin, as shown in Table 1, whose concentrations vary according to the species, variety, degree of ripeness, and growing conditions (Rodriguez-Amaya 2001). Although fruits such as mango, papaya, and buriti have high carotenoid concentrations, their bioavailability is not directly proportional to the total content in the food (Van Het Hof et al. 2000).

The fruit's food matrix directly influences the release of carotenoids during digestion. In fresh fruit, these compounds are typically stored in crystals or bound to proteins, hindering their release in the gastrointestinal tract (Reboul 2019). Thermal processing (such as cooking or homogenization) can break down cell structures and increase the availability of carotenoids, as demonstrated in studies with purees and

Table 1. Carotenoid content in different fruits.

Common name	Species/cultivars	Carotenoid contents (µg/ g, dry weight)							References
		Total carotenoids	β-carotene	α-carotene	β-cryptoxanthin	zeaxanthin	lycopene	lutein	
Acerola	*Malpighia glabra*	-	12.0	1.1	1.1	-	-	1.3	(Rodriguez-Amaya et al. 2008)
Apple green	*Malus domestica*	-	1.4	-	-	0.02	-	0.71	(Delgado-Pelayo et al. 2014)
Apple red	*Malus domestica*	-	0.17	-	-	0.01	-	0.06	
Apricot	*Prunus armeniaca*	173.00	170.00	-	0.73	2.7	-	0.09	(Campbell and Padilla-Zakour 2013)
Peach	*Prunus persica*	12.0	9.3	-	1.6	1.1	-	-	
Mango	*Mangifera indica*	-	-	-	-	1.5	-	31.7	
Banana	*Musa paradisiaca*	1.24	0.66	0.18	-	-	-	0.4	(Lokesh et al. 2014)
		0.54	0.26	0.32	-	-	-	0.06	
		19.7	8.38	8.62	-	-	-	0.9	
Grapes	*Vitis vinifera*		0.23	-	-	-	-	0.47	(Bunea et al. 2012)
Jambolão fruit	*Syzygium cumini*	-	0.23	-	0.003	0.02	-	0.39	(Faria et al. 2011)
Mango	*Mangifera indica*	-	-	-	-	1.5	-	31.7	(Khonsarn and Lawan n.d.)
Papaya	*Carica papaya*	-	-	-	-	14.1	-	237	
Peels of peach palm	*Bactris gasipaes* Kunth	69.88	14.78	0.61	-	-	-	1.19	(Menezes Silva et al. 2023)

Table 1 contd. ...

...Table 1 contd.

Common name	Species/cultivars	Carotenoid contents (µg/ g, dry weight)							References
		Total carotenoids	β-carotene	α-carotene	β-cryptoxanthin	zeaxanthin	lycopene	lutein	
Pineapple	*Ananas comosus*	-	9.9	-	-	-	-	-	(Freitas et al. 2015)
Watermelon	*Curcubita pepo Subsp. pepo*	185	-	-	-	-	7.0	172	(Martínez-Valdivieso et al. 2014)
Buriti	*Mauritia flexuosa*	513.87	372.32	3.23	1.28	-	-	0.03	(De Rosso and Mercadante 2007)
Mamey	*Mammea americana*	62.53	20.37	0.65	1.71	-	-	-	
Marimari	*Geoffrola striata*	37.98	23.21	0.17	0.23	0.17	-	0.21	
Peach palm	*Bactrys gasipaes*	197.66	55.51	1.78	0.12	-	-	-	
Physalis	*Physalis angulata*	80.89	62.23	1.25	1.87	0.40	-	-	
Tucuma	*Astrocaryum aculeatum*	62.65	47.36	1.68	1.64	1.02	0.04	0.79	

juices (Tang et al. 2009). However, heat can also degrade sensitive carotenoids such as lutein, reducing their total content (D'evoli et al. 2013).

The presence of lipids in the diet is essential for the efficient absorption of carotenoids, as these compounds are solubilized in intestinal lipid micelles. Studies have shown that consuming fruits rich in carotenoids with a fat source significantly increases the absorption of these nutrients (Unlu et al. 2005). For example, the addition of olive oil to the consumption of salad containing papaya and mango increases the plasma concentration of β-carotene (Brown et al. 2004).

In addition, individual genetic and physiological factors, such as intestinal transporter expression, nutritional status, and gastrointestinal health, also affect bioavailability (Borel et al. 2021). This implies that strategies to optimize bioavailability must consider not only the food, but also consumer characteristics.

7.5 Benefits of Carotenoids for Human Health

Carotenoids are natural pigments widely distributed in fruits and vegetables and are responsible for vibrant colors ranging from yellow to red (Zia-Ul-Haq et al. 2021). In addition to their aesthetic function, they play crucial roles in human health, such as antioxidant activity, prevention of chronic diseases, and, in some cases, as sources of provitamin A (Amorim-Carrilho et al. 2014, Eggersdorfer and Wyss 2018). However, humans cannot synthesize valuable pigments; therefore, to benefit from carotenoids in their diet, they must obtain them from food or supplements (Lafarga et al. 2021). Fruits are rich sources of carotenoids and are easily accessible to the population.

7.5.1 Provitamin A Activity

The amount of vitamin A in plant foods is determined by the efficiency with which the body converts provitamin A carotenoids into vitamin A (retinol), a process mediated by the enzyme carotenoid dioxygenase (Von Lintig and Vogt 2000). For a carotenoid to act as provitamin A, it must contain at least one unmodified β-ionone ring in its structure (Send and Sundholm 2007). Thus, β-carotene, α-carotene, γ-carotene, and β-cryptoxanthin are the carotenoids with provitamin A activity. However, the conversion capacity varies: β-carotene, because it has two β-ionone rings, has 100% activity, while the others offer approximately 50% (Von Lintig and Vogt 2000).

In particular, β-carotene's conversion rate into vitamin A from plant-based diets is lower than from animal-based sources and vitamin A supplements (Haskell et al. 2004). For example, scientific studies on humans have shown that dietary β-carotene is efficient at producing vitamin A, with 10 to 28 grams of β-carotene producing 1 gram of vitamin A (Tang et al. 2005). Using this data, the Food and Nutrition Board revised the estimated conversion efficiency of dietary β-carotene to vitamin A from 6:1 (weight ratio) to 12:1 (weight ratio). For α-carotene and β-cryptoxanthin (mainly derived from mixed plant foods in the U.S. diet), conversion efficiency is similarly modified, now reaching 24:1 (*Food and Nutrition Board, Institute of Medicine* 2001).

Thus, the percentage absorption and conversion of carotenoids are calculated using the ratio 1 RE = 12 μg of β-carotene and 24 μg of other carotenoids, assuming 50% biological activity.

7.5.2 Biological Activities

Carotenoids perform biological activities in animals, which are assessed by the conversion of carotenoids into retinol equivalents (1 RE = 1 μg of retinol); the RDI for carotenoids is 1000 RE for adult men, 800 RE for adult and pregnant women, and 1300 RE for lactating women (*Food and Nutrition Board, Institute of Medicine* 2001).

The main biological activities of carotenoids in humans are shown in Figure 3, and Table 2 presents evidence of the biological activities of the main carotenoids found in fruit.

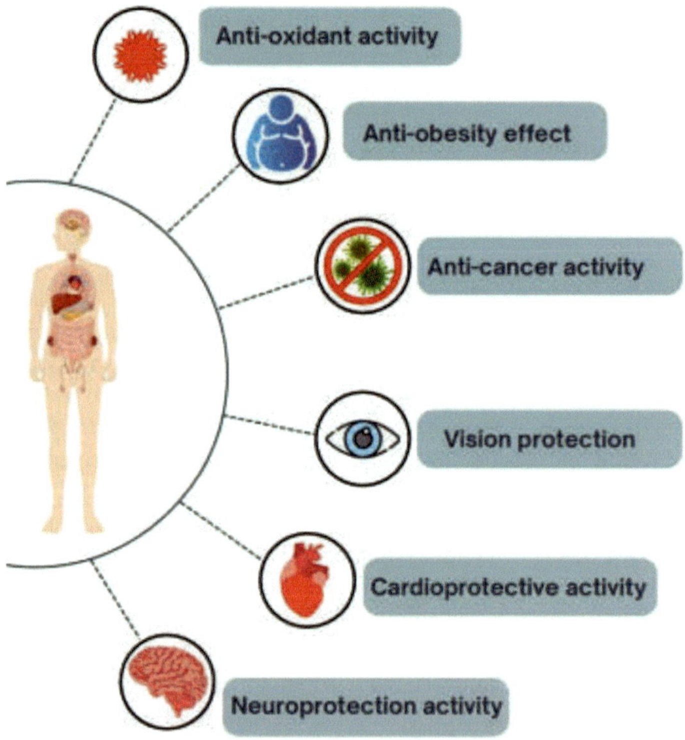

Figure 3. Bioactive functions of carotenoids.

Table 2. Proven biological activities of the major carotenoids found in fruits (β-carotene, α-carotene, β-cryptoxanthin, zeaxanthin, lycopene, and lutein).

Health benefits	Carotenoids	Experimental procedure	Reference
Antioxidant activity	β-carotene	*In vivo* - Oxygen radical absorbance capacity (ORAC), 2,2-diphenyl-1-picrylhydrazyl (DPPH), and 2,2′-azino-bis(3-ethylbenzothiazoline-6-sulfonic acid (ABTS).	(Teow et al. 2007)
	α-carotene, β-carotene, zeaxanthin, β-cryptoxanthin, and lutein	Quantum mechanical calculations should allow the description of quantitative structure activity relationships (QSARs) for the antioxidant activity of carotenoids.	(Soffers et al. 1999)
	Lutein and zeaxanthin	*In vitro* and *in vivo* - Inhibition of lipid peroxidation, scavenging and reducing activity: DPPH, ferric reducing antioxidant power (FRAP), nitric oxide radical, and ABTS.	(Sindhu 2010)
	Lycopene	*In vitro* - Scavenging and reducing activity: DPPH, FRAP, and β-carotene assays	(Naz et al. 2013)
Anti-inflammatory	Lycopene, lutein, β-carotene	*In vivo* (mouse) - Mouse model of sterile peritonitis.	(Hadad and Levy 2012)
Anti-cancer activity	lycopene	*In vivo* (mice) - Human prostate cancer (PC-3, LNCaP, and DU145 cell).	(Jiang et al. 2019)
	β-carotene	*In vitro* - Prostate cancer (PC-3 cell).	(Jayappriyan et al. 2013)
Cardioprotective activity	β-carotene	*In vivo* - Eastern Finnish men. Elevated serum concentrations of β-carotene.	(Karppi et al. 2012)
Anti-obesity effect	β-carotene, β-cryptoxanthin, lutein, and zeaxanthin	*In vitro* - 3T3-L1 cells, and *in vivo*- high-fat diet (HFD) induced obese mice.	(Kim et al. 2020)
Anti-diabetic activity	β-carotene and α-carotene	The European Prospective Study on Cancer and Nutrition - Netherlands Cohort (EPIC-NL) investigated the associations and incidence of type 2 diabetes.	(Sluijs et al. 2015)
Vision protection	Lycopene	*In vitro* and *in vivo* (rat lenses). Cataract prevention	(Gupta et al. 2003)
	Lutein and zeaxanthin	*In vitro* - Diabetic retinopathy	(Neelam et al. 2017)
Skin Protection	Lycopene and lutein	*In vivo* - UV radiation human skin.	(Grether-Beck et al. 2017)
	β-carotene	*In vitro* - UV-irradiated on human skin fibroblasts.	(Fernández et al. 2015)
Neuroprotection activity	Lutein	*In vivo* (mice)	(Sun et al. 2014)

7.6 Conclusion and Perspectives

Fruits are one of the main sources of carotenoids in the human diet, with buriti being the fruit with the highest total content of these compounds. Among the carotenoids identified and quantified are lutein, β-carotene, lycopene, α-carotene, β-cryptoxanthin, and zeaxanthin. The most abundant carotenoids in fruit include β-carotene, α-carotene, lutein, and zeaxanthin. However, their bioavailability is not directly proportional to the total content present in the food. The food matrix of the fruit influences the release of carotenoids during digestion. In fresh fruits, these compounds are stored in crystals or bound to proteins, hindering their release in the gastrointestinal tract. Thermal processing increases the availability of carotenoids, but can degrade the more sensitive ones. The presence of lipids in the diet is essential for the efficient absorption of carotenoids.

The regulation of carotenoid biosynthesis is highly influenced by environmental factors such as light, temperature, and nutrient availability. Light, for example, not only stimulates gene expression of biosynthetic enzymes but also affects the cell's energy demand and redox state, which are fundamental to metabolic activity.

The carotenoids present in fruit have various beneficial properties for human health, such as antioxidant, anti-angiogenic, anti-diabetic, anti-obesity, anti-inflammatory, skin protection, neuroprotection, vision protection, and cardioprotection. Knowledge of the composition of carotenoids across different edible parts and cultivars will be useful to nutritional experts in selecting nutrient-rich plants and recommending an appropriate diet.

References

Amorim-Carrilho, K. T., Cepeda, A., Fente, C. and Regal, P. (2014). Review of methods for analysis of carotenoids. In TrAC - Trends in Analytical Chemistry 56: 49–73. https://doi.org/10.1016/j.trac.2013.12.011.

Borel, P., Desmarchelier, C., Nowicki, M. and Bott, R. (2021). Lycopene bioavailability is associated with a combination of genetic variants. Free Radical Biology & Medicine 83: 238–244. https://doi.org/10.1016/j.freeradbiomed.2015.02.033.

Britton Functions, G. A. (2008). Carotenoids: Natural Functions. Functions of Intact Carotenoids. Birkhäuser Verlag Basel 4: 189–212.

Brown, M. J., Ferruzzi, M. G., Nguyen, M. L., Cooper, D. A., Eldridge, A. L., Schwartz, S. J. et al. (2004). Carotenoid bioavailability is higher from salads ingested with full-fat than with fat-reduced salad dressings as measured with electrochemical detection 1–3. The American Journal of Clinical Nutrition 80: 396–403.

Bunea, C. I., Pop, N., Babeş, A. C., Matea, C., Dulf, F. V. and Bunea, A. (2012). Carotenoids, total polyphenols, and antioxidant activity of grapes (Vitis vinifera) cultivated in organic and conventional systems. Chemistry Central Journal 6: 1–9. https://doi.org/10.1186/1752-153X-6-66.

Campbell, O. E. and Padilla-Zakour, O. I. (2013). Phenolic and carotenoid composition of canned peaches (Prunus persica) and apricots (Prunus armeniaca) as affected by variety and peeling. Food Research International 54(1): 448–455. https://doi.org/10.1016/j.foodres.2013.07.016.

Castenmiller, J. J. M. and West, C. E. (1998). Bioavailability and bioconversion of carotenoids. Annu. Rev. Nutr. 18: 19–32. https://doi.org/10.1146/annurev.nutr.18.1.19.

Cazzonelli, C. I. and Pogson, B. J. (2010). Source to sink: regulation of carotenoid biosynthesis in plants. Trends in Plant Science 15(Issue 5): 266–274. https://doi.org/10.1016/j.tplants.2010.02.003.

Eggersdorfer, M. and Wyss, A. (2018). Carotenoids in human nutrition and health. In Archives of Biochemistry and Biophysics 652: 18–26. https://doi.org/10.1016/j.abb.2018.06.001.

Food and Nutrition Board, Institute of Medicine. (2001). Dietary Reference Intakes for Vitamin A, Vitamin K, Arsenic, Boron, Chromium, Copper, Iodine, Iron, Manganese, Molybdenum, Nickel, Silicon, Vanadium, and Zinc. http://www.nap.edu/catalog/10026.html.

Fraser, P. D. and Bramley, P. M. (2004). The biosynthesis and nutritional uses of carotenoids. Progress in Lipid Research 43(3): 228–265. https://doi.org/10.1016/j.plipres.2003.10.002.

Giuliano, G., Tavazza, R., Diretto, G., Beyer, P. and Taylor, M. A. (2008). Metabolic engineering of carotenoid biosynthesis in plants. Trends in Biotechnology 26(3): 139–145. https://doi.org/10.1016/j.tibtech.2007.12.003.

Hacke, A., Marques, M. C., Rebellato, A. P., Rodrigues, D. B., Pallone, J. A. L. and Mariutti, L. R. B. (2023). How does the food matrix affect the provitamin A carotenoid *in vitro* bioaccessibility in fruit-based baby food? Journal of Food Composition and Analysis 123: 1–11. https://doi.org/10.1016/j.jfca.2023.105538.

Hannoufa, A. and Hossain, Z. (2012). Regulation of carotenoid accumulation in plants. In Biocatalysis and Agricultural Biotechnology 1(3): 198–202. https://doi.org/10.1016/j.bcab.2012.03.004.

Haskell, M. J., Jamil, K. M., Hassan, F., Peerson, J. M., Hossain, I., Fuchs, G. J. et al. (2004). Daily consumption of Indian spinach (Basella alba) or sweet potatoes has a positive effect on total-body vitamin A stores in Bangladeshi men 1–3. The American Journal of Clinical Nutrition 80: 705–714. https://doi.org/10.1093/ajcn/80.3.705.

Khoo, H. E., Prasad, K. N., Kong, K. W., Jiang, Y. and Ismail, A. (2011). Carotenoids and their isomers: Color pigments in fruits and vegetables. Molecules 16(2): 1710–1738. https://doi.org/10.3390/molecules16021710.

Krinsky, N. I. and Johnson, E. J. (2005). Carotenoid actions and their relation to health and disease. Molecular Aspects of Medicine 26(6): 459–516. https://doi.org/10.1016/j.mam.2005.10.001.

Lafarga, T., Sánchez-Zurano, A., Morillas-España, A. and Acién-Fernández, F. G. (2021). Extremophile microalgae as feedstock for high-value carotenoids: A review. International Journal of Food Science and Technology 56(10): 4934–4941. https://doi.org/10.1111/ijfs.15069.

Mein, J. R., Lian, F. and Wang, X. D. (2008). Biological activity of lycopene metabolites: Implications for cancer prevention. Nutrition Reviews 66(12): 667–683. https://doi.org/10.1111/j.1753-4887.2008.00120.x.

Meléndez-Martínez, A. J. (2019). An overview of carotenoids, apocarotenoids, and vitamin A in agro-food, nutrition, health, and disease. Molecular Nutrition and Food Research 63(15): 1–11. https://doi.org/10.1002/mnfr.201801045.

Othman, R., Mohd Zaifuddin, F. A. and Hassan, N. M. (2014). Carotenoid biosynthesis regulatory mechanisms in plants. Journal of Oleo Science 63(8): 753–760. https://doi.org/10.5650/jos.ess13183.

Rao, A. V. and Rao, L. G. (2007). Carotenoids and human health. Pharmacological Research 55(3): 207–216. https://doi.org/10.1016/j.phrs.2007.01.012.

Reboul, E. (2013). Absorption of vitamin A and carotenoids by the enterocyte: Focus on transport proteins. Nutrients 5(9): 3563–3581. https://doi.org/10.3390/nu5093563.

Reboul, E. (2019). Mechanisms of carotenoid intestinal absorption: Where do we stand? Nutrients 11(838): 1–12. https://doi.org/10.3390/nu11040838.

Rodriguez-Amaya, D. B. (2016). Natural food pigments and colorants. Current Opinion in Food Science 7: 20–26. https://doi.org/10.1016/j.cofs.2015.08.004.

Rodriguez-Amaya, D. B. (2001). A Guide to Carotenoid Analysis in Foods. ILSI Press, Washington, DC, 64: 1–60.

Sathasivam, R. and Ki, J. S. (2018). A review of the biological activities of microalgal carotenoids and their potential use in healthcare and cosmetic industries. Marine Drugs 16: 1–31. https://doi.org/10.3390/md16010026.

Send, R. and Sundholm, D. (2007). The role of the β-ionone ring in the photochemical reaction of rhodopsin. Journal of Physical Chemistry A 111(1): 27–33. https://doi.org/10.1021/jp065510f.

Tang, G., Jian, Q., Dolnikowski, G. G., Russell, R. M. and Grusak, M. A. (2009). Golden rice is an effective source of vitamin A. American Journal of Clinical Nutrition 89(6): 1776–1783. https://doi.org/10.3945/ajcn.2008.27119.

Tang, G., Qin, J., Dolnikowski, G. G., Russell, R. M. and Grusak, M. A. (2005). Spinach or carrots can supply significant amounts of vitamin A as assessed by feeding with intrinsically deuterated vegetables. The American Journal of Clinical Nutrition 82: 821–828. https://doi.org/10.1093/ajcn/82.4.821.

Unlu, N. Z., Bohn, T., Clinton, S. K. and Schwartz, S. J. (2005). The addition of avocado or avocado oil enhances human nutrition and metabolism, and carotenoid absorption from salad and salsa. The Journal of Nutrition 135(3): 431–436. https://doi.org/10.1093/jn/135.3.431.

Van Het Hof, K. H., West, C. E., Weststrate, J. A. and Hautvast, J. G. A. J. (2000). Recent advances in nutritional sciences: dietary factors that affect the bioavailability of carotenoids. The Journal of Nutrition 130(3): 503–506. https://doi.org/10.1093/jn/130.3.503.

Viñas-Ospino, A., López-Malo, D., Esteve, M. J., Frígola, A. and Blesa, J. (2023). Green solvents: emerging alternatives for carotenoid extraction from fruit and vegetable by-products. Foods 12(4): 1–19. https://doi.org/10.3390/foods12040863.

Von Lintig, J. and Vogt, K. (2000). Filling the gap in vitamin A research. Molecular identification of an enzyme cleaving β-carotene to retinal. Journal of Biological Chemistry 275(16): 11915–11920. https://doi.org/10.1074/jbc.275.16.11915.

Weber, D. and Grune, T. (2023). Carotenoid basics: from food to skin. Redox Experimental Medicine 2023(1): 1–8. https://doi.org/10.1530/rem-23-0019.

Zia-Ul-Haq, M., Dewanjee, S. and Riaz, M. (2021). Carotenoids: structure and function in the human body. Springer Nature 862. https://doi.org/10.1007/978-3-030-46459-2.

Chapter 8
Functional Foods
Fruits as Disease Preventers

*Shanta Karki** and *Govinda Rizal*

8.1 Introduction

Although cereals form the basic staple food for most humans, a healthy diet, technically referred to as a balanced diet, comprises fresh fruits and vegetables in a significant portion of each meal. A balanced diet is essential to prevent both communicable and non-communicable diseases as well, and it helps in faster curing of several diseases. Fruits are considered functional foods because they provide various health-benefiting factors in addition to the basic nutrients. Fruits are rich in vitamins, minerals, anthocyanins, proanthocyanins, fiber, fructose, and amino acids (Table 1). Fruit and vegetable intake was associated with reduced risk of cardiovascular diseases, cancer, and premature mortality (Aune et al. 2017). Similarly, nuts, which are fruits consisting of a hard shell protecting edible kernels, such as almond, cashew, hazel, macadamia, pecan, and walnut, have a high nutritional and economic value, but are low in volume and weight and require less labor for cultivation; that's why these are considered high-value fruit crops. They are easily dried, storable for months, and hence form a good commodity for trade.

Walnuts have a rich, distinctive flavor and are a source of dietary fats, protein, vitamins, and minerals. Walnut farming has emerged to be a very stable business, attributable to the health benefits of the walnut kernels. Seeds are a highly dense source of nutrients, particularly proteins, fiber, and essential fatty acids (Tan et al. 2022). Walnuts improve cardiovascular function, can be used to treat type 2 diabetes, are protective against metabolic syndrome, have anticancer benefits, and have

Central Agricultural Laboratory, Harihar Bhawan, Lalitpur, Nepal.
* Corresponding author: shantakyoto@gmail.com

DOI: 10.1201/9781003668169-10

Table 1. Nutrient contents of some of the most popular fruits.

S.N.	Scientific name	Common name	Major nutrient contents	Diseases prevented	References
1	*Prunus amygdalus*	Almond	Healthy fats, Protein, Vitamin E, Calcium, Magnesium	Alzheimer's disease, cancer, diabetes, heart disease, and inflammation	Wang et al. 2020
2	*Malus pumila*	Apple	Iron, Zinc, Vitamin C, E, Polyphenols	Heart attack, atherosclerosis, asthma, and allergies	Bao et al. 2013
3	*Persea americana*	Avocado	Unsaturated fat, Vitamins C, K, E; Fiber	Cardiovascular diseases (including stroke, high blood pressure), and diabetes	Pacheco et al. 2022
4	*Musa paradisiaca*	Banana	Potassium, Vitamins B6 and C, Antioxidants	Heart diseases (including blood pressure, heart attack, stroke), constipation, and colorectal cancer	Jyothirmayi and Rao 2015
5	*Vaccinium* spp.	Blueberry	Anthocyanins, Vitamin C and K, Fiber, Manganese	Heart diseases, cancer, oxidative stress, type 2 diabetes, and cognitive decline	Arif et al. 2022, Kalt et al. 2020
6	*Cocos nucifera*	Coconut	Potassium, Manganese, Copper, Iron, Selenium, Proteins and Vitamin B, C	High blood pressure, chronic diseases, and diarrhea	DebMandal and Mandal 2011
7	*Vitis vinifera*	Grape	Vitamin C and K, Calcium, Magnesium, Resveratrol	Heart diseases, cancers, oxidative stress, Alzheimer's and Parkinson's diseases	Joseph et al. 2009
8	*Actinidia deliciosa*	Kiwi	Vitamins C, K, E; Fiber, and Antioxidants, Folate, Potassium, Copper, Magnesium	Heart diseases (including high blood pressure), cancer, oxidative stress, and eye diseases	Denise et al. 2016
9	*Citrus aurantiifolia*	Acid lime	Vitamin C, Citrate, Potassium	Scurvy, kidney stones, heart diseases, and respiratory diseases	Richa et al. 2023
10	*Mangifera indica*	Mango	Vitamins A, B6, C, E, K; Folate (B9), Copper, Magnesium	Cancer, heart diseases, eye and skin diseases	Lauricella et al. 2017
11	*Citrus reticulata*	Mandarin orange	Antioxidants, Vitamin C, Beta-cryptoxanthin, Fiber, Calcium, Potassium	Cancer, heart diseases (including stroke, high blood cholesterol, high blood pressure), skin and gum diseases	Richa et al. 2023

12	*Carica papaya*	Papaya	Vitamin A, C, Folate, Potassium, Magnesium, Lycopene, Beta-carotene	Heart diseases, diabetes, cancer, age-related eye diseases, and dengue fever	Ali et al. 2012
13	*Punica granatum*	Pomegranate	Folate, Vitamin C and K, Potassium, Copper	Heart diseases (including high blood pressure), cancer, eye and skin diseases	Zarfeshany et al. 2014
14	*Fragaria ananassa*	Strawberry	Vitamin C and B9, Manganese, Potassium, Calcium, Ellagic acid and Anthocyanins	Blood sugar, heart diseases (including high blood pressure), cancer, and constipation	Cianciosi et al. 2021
15	*Juglans regia*	Walnut	Alpha-linolenic acid (ALA), a plant-based omega-3 fatty acid, Antioxidants	Cardiovascular diseases, cancer, oxidative stress, type 2 diabetes, and inflammation	Tan et al. 2022

anti-inflammatory properties (Tan et al. 2022). These nuts are a rich source of healthy oils that provide energy and health-benefiting nutrients, minerals, antioxidants, and vitamins that are essential for optimum health. According to a randomized controlled trial, consumption of two to three servings of walnuts per day has been found to consistently decrease total cholesterol and LDL cholesterol, and it has also been shown to improve endothelial function (EF) in individuals with hypercholesterolemia and type 2 diabetes, suggesting a potential reduction in overall cardiac risk (Ma et al. 2009).

Fruits are crucial in our daily diets as they provide essential vitamins, minerals, fiber, and antioxidants, which contribute to overall health by supporting digestion, boosting immunity, maintaining heart health, and potentially reducing the risk of chronic diseases like cancer and diabetes (Table 1). Most health guidelines recommend consuming at least 1.5–2 cups of fruit per day, which is equivalent to 240–320 grams, aiming for a variety of different types to maximize varied nutrient availability and intake.

8.2 Fruits in Chronic Disease Management

Chronic diseases are diseases whose conditions last one year or more and require continuous medical attention or otherwise limit daily activities or both. Heart diseases, stroke, cancer, and diabetes are some of the major chronic diseases worldwide (Hacker 2024). In a meta-analysis of 95 studies, it was found that an intake of 800 g/day of fruits and vegetables reduced the risk of cardiovascular diseases (Aune et al. 2017). In a review by Boeing et al. (2012), it was found that for hypertension, coronary heart diseases, and stroke, there is convincing evidence that increasing the consumption of vegetables and fruit reduces the risk of these diseases.

Key points about the importance of fruits in chronic diseases can be summarized as follows.

8.2.1 Dense in Nutrients

Fruits are packed with essential vitamins like vitamin C, vitamin A, vitamin K, potassium, folate, and various phytonutrients, which are vital for several types of disease prevention, chronic disease management, and thus promoting healthy bodily functions (Table 1). Because fruits are dense in nutrients, several bioactive compounds are extracted from them (Cho et al. 2009, Dhalaria et al. 2020), which have been used in the management of different types of chronic diseases.

8.2.2 Good Source of Fiber

Dietary fiber in fruits aids in digestion, promotes regular bowel movements, regulates blood sugar, and helps with satiety, thereby helping in weight control. Fibers in fruits can be soluble and insoluble, both of which help in preventing and managing various chronic diseases.

8.2.3 Fruit as a Powerful Antioxidant

Many fruits like apples, berries, citrus, grapes, and apricots contain antioxidants such as flavonoids, which neutralize free radicals, thereby combating free radical damage to the body and potentially reducing the risk of chronic diseases (Table 1).

8.2.4 Hydration

Fresh fruits in general contain 70 to 95% water by weight, which makes them highly perishable. The high water content in fruits contributes to overall hydration of the body, which is important to prevent dehydration and help in nutrient uptake, as water is imperative for water-soluble nutrients.

8.2.5 Fruits Help with Body Weight Management

Including fruits in the diet helps to manage healthy body weight by promoting fullness and reducing unhealthy snacking, which prevents obesity and helps maintain appropriate body weight (Table 1).

8.2.6 General Daily Requirement of Fruits

Although general recommendations of fruits vary from region to region or nation to nation, most health organizations recommend consuming 1.5–2 cups of fruit per day, which can be split into different servings throughout the day (Wallace et al. 2019). Even though fruits are healthy, consuming excessive amounts can still contribute to excess calories, so maintaining the appropriate portion is more beneficial. Recognizing the importance of fruits as an essential component of a healthy diet, the World Health Organization (WHO) and the Food and Agriculture Organization (FAO) recommend a minimum consumption of 400 g of fruits and vegetables per person per day, and this diet would help prevent chronic diseases as well as prevent and alleviate several micronutrient deficiencies (Caprile and Rossi 2021).

8.2.7 Nutritional Diversity in Fruits

For maximum health benefits, one must eat a diverse range of fruits to obtain a wider spectrum of nutrients. Adding variety to the daily fruit intake will multiply the health benefits. Even citrus has many varieties, such as sweet orange, mandarin orange, pummelo, acid lime, sweet lime, etc. (Figure 1), and a combination of fruits from different groups would add even more variety.

8.2.8 Consumption of Whole Fruits

It is better to consume whole fruits (fresh fruits) over fruit juices or derivative processed products whenever possible, as they provide more fiber and nutrients in their natural forms. Processed products contain additives and preservatives that are not good for health.

Figure 1. Citrus fruit diversity.

8.3 Functional Foods Targeting Specific Health Conditions

Certain fruits are better for specific conditions. Viral diseases have no medical cure, and they can be cured by our indigenous immune system, enhanced by the consumption of various fruits and functional foods targeted for specific health conditions (Finnegan et al. 2023). Some viral diseases, like dengue fever, can be dangerous, and appropriate nutrition plays an important role in recovery. Including fruits rich in vitamins, minerals, fiber, and other essential nutrients will help fight the virus (Finnegan et al. 2023). Consuming pomegranate is beneficial as it is high in iron content, which supports maintaining a healthy blood platelet count (Zarfeshany et al. 2014). Similarly, kiwi contains copper, which helps in producing healthy red blood cells and enhances resistance to infection (Denise et al. 2016). Kiwi also contains vitamin C, which boosts immunity. Oranges are a rich source of antioxidants (limonoids, flavonoids, carotenoids) and vitamins, which, when consumed daily, significantly strengthen the immune system and fight fatigue. Bananas are great for dengue as they are rich in potassium and vitamins B6 and C, which help replenish electrolytes as well as aid digestion and thus act as boosters in recovering from the disease (Jyothirmayi and Rao 2015). Moreover, anticarcinogenic properties of fruits due to high levels of some micronutrients, fiber, and dietary bioactive compounds like polyphenols have the capacity to block or attenuate initiation or suppress the stages of promotion and progression of carcinogenesis by controlling proliferation, apoptosis, cell cycle progression, and inflammation (Wallace et al. 2019).

Fruits like grapes, strawberries, blueberries, cherries, and oranges are rich in antioxidants (Table 1). Antioxidants help stop the proliferation of cancerous cells by neutralizing harmful free radicals, thus reducing DNA damage and potentially preventing cancer development. Antioxidants help maintain a healthy balance of oxidative and antioxidative processes in cells, which supports cells to function properly and resist transformation into cancerous cells, making them less susceptible to becoming cancerous (Kalt et al. 2020).

Fruits play a significant role in mitigating the effects of aging and age-related disorders due to their rich content of antioxidants and bioactive compounds. These compounds, like polyphenols and flavonoids, help combat oxidative stress, a key factor in aging and age-related disease development (Arif et al. 2022). Specifically, fruits like blueberries, apples, grapes, and berries have been linked to improved cognitive function, reduced risk of cardiovascular diseases, and even potential longevity benefits (Arif et al. 2022).

A study conducted by Bao et al. (2013) in mice showed that apple polyphenol protects against cigarette smoke–induced acute lung injury, indicating that apple polyphenols can be a potential dietary nutrient supplement that can be used to improve the quality of life of people suffering from chronic obstructive pulmonary disease. Similarly, neuroinflammation plays a significant role in the aging process, which also causes neurodegenerative diseases such as Alzheimer's disease (Arif et al. 2022). Therefore, certain therapeutic nutrition that has anti-inflammatory effects would be beneficial to brain health, thereby preventing such diseases. Many such therapeutic nutrients have been derived from fruits.

Walnut kernels contain large quantities of unsaturated fatty acids, peptides, and phenolic compounds that have anti-inflammatory effects (Tan et al. 2022). The long-term intake of walnuts has been found to improve cognitive function and memory in rats and humans (Tan et al. 2022). More clinical studies are required to detect the potential influence and identify main regulating mechanisms of walnuts and their active ingredients on neuroinflammation, including the regulation of microglia activation induced by amyloid β or lipopolysaccharides, inhibition of peripheral inflammation mediated by macrophages, reduction in oxidative stress by decreasing free radical levels and boosting antioxidant defenses, and control of gut microbes to maintain homeostasis (Dhalaria et al. 2020). Such studies would shed new insights into the neuroinflammation-regulating mechanisms and natural active ingredients of walnuts and lead to the development of walnut-based functional foods for the alleviation of neurodegenerative diseases (Zhang et al. 2023).

Oxidative reactions of many kinds occur inside the human body due to exposure to environmental toxins, which cause several diseases and disorders (Tan et al. 2022). Toxins like cigarette smoke and exhaust from automobiles are in the air we breathe, the water we drink contains charged ions, and the food we eat may contain chemical food additives such as preservatives, artificial colors, flavors, emulsifiers, lubricants, bleaching agents, flavor enhancers, and synthetic sweeteners. These toxins trigger oxidative reactions in the body, leading to several types of adverse effects (Bagchi et al. 2000). Proanthocyanidins (PAs) are an important therapeutic class of flavonoids that help prevent oxidative damage in the body, fight free radicals, and even help repair cell damage that has already occurred. PAs have a broad range of pharmacologic activity through increasing vitamin C levels (Yamakoshi et al. 2004), decreasing capillary permeability and fragility (Bagchi et al. 2000), scavenging free radicals and inhibiting destruction of collagen, lowering cholesterol levels, and reducing arterial cholesterol deposits (Cho et al. 2009). Nandakumar et al. (2008) summarized and highlighted the latest developments on anti-carcinogenic activities of PAs from different sources, specifically from grape seeds.

Cardiovascular diseases (CVDs) are defined as disorders of the heart and vessels and include coronary heart disease (CHD) and stroke. Diet represents the most important modifiable factor to prevent CVDs. There is evidence that plant-based dietary components are associated with a lower risk of CVDs, among which fruits constitute the most important components with a key role in preventing CVDs (Zhao et al. 2017). Furthermore, fruits such as grapes, blueberries, pomegranate, apple, hawthorn, and avocados have been widely studied and have shown strong cardiovascular protective effects (Table 1). Moreover, the effectiveness of fruit intake in the primary prevention of CVDs has been revealed by growing clinical data in patients with high metabolic risk factors (Zhao et al. 2017). In this way, fruits in general and specific fruits in particular have been proven to have a crucial function to prevent and even cure specific health conditions.

8.4 Consumer Trends and Future Research in Fruits

Globally, there's an increasing trend of fruit consumption due to improved perception toward healthy eating, with consumers increasingly choosing fresh fruits as a key portion of their daily diets, which is reflected by growing demand for various fruits and has led to the increased production of various fruits across different regions (FAO 2024). Similarly, the trade of fruits, both in fresh form and processed product forms, is observed to increase. This is driven by consumers' increased awareness of the health benefits of fruits and hence a desire for a balanced diet with a focus on overall well-being.

8.4.1 Overall Growth of the Fruit Sector

The global fruit and vegetable markets are experiencing significant growth due to increased consumption, with sales value and volume increasing over the past few years. Global fruit production has steadily increased over the past decade, with a notable growth of 30% between 2010 and 2023 (FAO 2024). In 2023, total fruit production reached 952 million tons, with bananas being the most produced fruit at 139 million tons. Other major fruits include watermelons (105 million tons), apples (97 million tons), grapes (72 million tons), and oranges (70 million tons) (FAO 2024). Emphasis on fruit consumption has increased research on various aspects of the fruit sector so that production and productivity can be increased as per the growing demand and recommended daily doses can be met in all parts of the world.

Not only the production, but also the consumption of fruits has increased greatly over the years. The global average per capita fruit intake was 102.86 g in 1961, which has reached 240.62 g in 2022, an increase of 134% (Ritchie et al. 2023). The foremost factor contributing to increased fruit consumption can be attributed to healthy eating habits emphasized by government programs across the world. Consumers are prioritizing healthy diets and choosing fruits and vegetables for their nutritional benefits, including vitamins, minerals, fiber, and antioxidants.

A significant portion of consumers are seeking out inclusive dietary options, including vegetarian and vegan choices, which are being met by fruits with a wider range of product offerings. Fruit and vegetable consumption levels are often tied to individuals' health behaviors and awareness, including factors like exercise, smoking, and knowledge of healthy eating recommendations.

8.4.2 Future Research on Fruits

Fruits have become a very important component of daily human diets, thereby directly affecting the food and nutrition security of the world. They are an integral part of each meal, and their demand is continuously increasing due to various factors, including medical/nutritional recommendations and increased affordability and accessibility. Therefore, more work on fruit research is crucial for food security as it supports both nutritional security and stable food systems as well as improved health. By improving fruit cultivation practices, post-harvest management, and overall food processing, research contributes to increased yields, reduced losses (both pre-harvest and post-harvest), and healthier diets, ultimately bolstering food availability and access.

Future fruit research must prioritize varietal improvement through modern breeding techniques for adaptation traits related to changing environmental and climatic conditions for sustainable high production, tolerating drought, heat stress, and other extreme weather conditions. Fruit breeding and research programs also should focus on improving traits like taste, bioavailability of nutrients, reducing anti-nutritional factors and/or toxins, appearance, prolonged shelf life, disease and pest resistance. Fruits are quite expensive, so maximizing yields and minimizing the cost of production through cheaper cultivation practices, ensuring sustainable production through advancements in technology, precision agricultural practices including safe food production, and appropriate post-harvest management are crucial to make fruits affordable and available for consumption by all groups of people irrespective of their economic conditions. Due to climate change, rainfall patterns have been changing globally, which is affecting fruit production and quality. Therefore, future fruit research also should focus on different aspects of irrigation, which has an important impact on fruit quality. For successful fruit cultivation, appropriate irrigation management technologies need to be developed so that fruit trees can be optimally irrigated, which can improve the fruit production and fruit yield as well as ensure quality throughout the productive years. By studying the advantages and disadvantages of irrigation water treatment technology (particularly in the case of water purification and fertigation), different irrigation methods, different irrigation timing, and different irrigation amounts, suitable methods for each fruit crop can be selected to improve fruit quality (Ni et al. 2024). Additionally, further research can be conducted on the relationship between irrigation (water availability) and fruit nutrient components, such as water content and nutrient elements, and their impact on fruit nutrient components. By finding the optimum amount of water required per quality fruit production or per productive tree for each of the major specific fruit crops, valuable fresh water can be conserved by avoiding overwatering, and providing the required amount of water during critical times can save the fruit trees from prolonged drought. Most fruit crops, being perennial in nature, require long-term research, which should be considered during planning so that multi-year projects are designed with realistic action plans, skilled human resource development, and continuity should be ensured.

8.5 Conclusion

Fruits are indispensable components of daily human diets. Fruits play a crucial role in preventing various diseases due to their nutritional contents, including but not limited to various vitamins, minerals, antioxidants, fiber, and phytochemicals. Most fruits can control metabolic risk factors such as hypertension, dyslipidemia, diabetes, and overweight, and inhibit atherosclerosis, which is the key pathological process of heart attacks and strokes. Fruits as functional foods have been proven by experiments and clinically; hence, their balanced intake should be promoted through increased production, reduced cost, proper storage, post-harvest, and transportation to make them accessible to all throughout the year. Therefore, prioritizing fruit research at a global level to meet the consumption at recommended daily intake amount by the growing population is crucial to prevent life-threatening diseases.

References

Ali, A., Devarajan, S., Waly, I. M., Essa, M. M. and Rahman, M. S. (2012). Nutritional and medicinal values of papaya (*Carica papaya* L.). *In*: Essa, M. M., Manickavasagan, A. and Sukumar, E. (eds.). Natural Products and Their Active Compounds on Disease Prevention (Food Science and Technology). Nova Science Pub Inc.

Arif, U. M., Khan, K. I. M., Riaz, S., Nazir, A., Maan, A. A., Amin, U. et al. (2022). Role of fruits in aging and age-related disorders. Experimental Gerontology 162: 111763. https://doi.org/10.1016/j.exger.2022.111763.

Aune, D., Giovannucci, E., Boffetta, P., Fadnes, T. L., Keum, N., Norat, T. et al. (2017). Fruit and vegetable intake and the risk of cardiovascular disease, total cancer and all-cause mortality-a systematic review and dose-response meta-analysis of prospective studies. Int. J. Epidemiol. 46(3): 1029–1056. doi: 10.1093/ije/dyw319.

Bagchi, D., Bagchi, M., Stohs, S. J., Das, D. K., Ray, S. D., Kuszynski, C. A. et al. (2000). Free radicals and grape seed proanthocyanidin extract: importance in human health and disease prevention. Toxicol. 148: 187–97.

Bao, M. J., Shen, J., Jia, Y. L., Li, F. F., Ma, W. J., Shen, H. J. et al. (2013). Apple polyphenol protects against cigarette smoke-induced acute lung injury. Nutrition 29: 235–243.

Boeing, H., Bechthold, A., Bub, A., Ellinger, S., Haller, D., Kroke, A. et al. (2012). Critical review: vegetables and fruit in the prevention of chronic diseases. Eur J. Nutr. 51(6): 637–663.

Caprile, A. and Rossi, R. (2021). At a glance: 2021 International year of fruits and vegetables. European Parliamentary Research Service, PE 689.367.

Cho, M. L., Heo, Y. J., Park, M. K., Oh, H. J., Park, J. S., Woo, Y. J. et al. (2009). Grape seed proanthocyanidin extract (GSPE) attenuates collagen-induced arthritis. Immunology Letters 124: 102–110.

Cianciosi, D., Ansary, J., Regolo, L., Giampieri, F. and Forbes, T. (2021). The preventive effect of strawberries against the onset of common human diseases induced by oxidative stress. Acta Horticulturae 1309: 917–924. doi: 10.17660/ActaHortic.2021.1309.131.

DebMandal, M. and Mandal, S. (2011). Coconut (*Cocos nucifera* L.: Arecaceae): In health promotion and disease prevention. Asian Pacific Journal of Tropical Medicine 4: 241–247. https://doi.org/10.1016/51995-7645(11)60078-3.

Dhalaria, R., Verma, R., Kumar, D., Puri, S., Tapwal, A., Kumar, V. et al. (2020). Bioactive compounds of edible fruits with their anti-aging properties: A comprehensive review to prolong human life. Antioxidants 9(11): 1123. doi: 10.3390/antiox9111123.

FAO. (2024). FAOSTAT: Production: Crops and Livestock Products.

Finnegan, D., Tocmo, R. and Loscher, C. (2023). Targeted application of functional foods as immune fitness boosters in the defense against viral infection. Nutrients 15(15): 3371.

Hacker, K. (2024). The burden of chronic disease. Mayo Clinic Proceedings: Innovations, Quality & Outcomes 8: 112–119.

Joseph, A. J., Shukitt-Hale, B. and Willis, M. L. (2009). Grape juice, berries, and walnuts affect brain aging and behavior. J. Nutr. 139(9): 1813S–7S.

Jyothirmayi, N. and Rao, N. M. (2015). Banana medicinal uses. Jour of Med Sc & Tech 4(2): 152–160.

Kalt, W., Cassidy, A., Howard, R. L., Krikorian, R., Stupp, J. A., Tremblay, F. et al. (2020). Recent research on the health benefits of blueberries and their anthocyanins. Adv. Nutr. 11(2): 224–236. doi: 10.1093/advances/nmz065.

Lauricella, M., Emanuele, S., Calvaruso, G., Giuliano, M. and D'Anneo, A. (2017). Multifaceted health benefits of Mangifera indica L. (Mango): The inestimable value of orchards recently planted in Sicilian rural areas. Nutrients 9(5): 525. doi: 10.3390/nu9050525.

Ma, Y., Njike, Y. V., Millet, J., Dutta, S., Doughty, K., Treu A. J. et al. (2009). Effects of walnut consumption on endothelial function in type 2 diabetic subjects: a randomized controlled crossover trial. Diabetes Care 2010. doi: 10.2337/dc09-1156.

Nandakumar, V., Singh, T. and Katiyar, S. K. (2008). Multi-targeted prevention and therapy of cancer by proanthocyanidins. Cancer Letters 269: 378–387.

Ni, F., Li, W. and Huang, J. (2024). Worldwide productivity and research trend on fruit quality: a biometric study. Frontiers in Plant Science 14: 1294989.

Pacheco, S. L., Li, Y., Rimm, B. E., Manson, E. J., Sun, Q., Rexrode, K. et al. (2022). Avocado consumption and risk of cardiovascular disease in US adults. Journal of the American Heart Association 11: e024014. doi:10.1161/JAHA.121.024014.

Richa, R., Kohli, D., Vishwakarma, D., Mishra, A., Kabdal, B., Kothakota, A. et al. (2023). Citrus fruit: Classification, value addition, nutritional and medicinal values, and relation with pandemic and hidden hunger. Journal of Agriculture and Food Research 14: 100718. https://doi.org/10.1016/j.jafr.2023.100718.

Ritchie, H., Rosado, P. and Roser, M. (2023). Diet Compositions. OurWorldinData.org. Retrieved from: 'https://ourworldindata.org/diet-compositions'.

Tan, B., Wang, Y., Zhang, X. and Sun, X. (2022). Recent studies on protective effects of walnuts against neuroinflammation. Nutrients 14(20): 4360.

Wallace, T. C., Bailey, R. L., Blumberg, J. B., Burton-Freeman, B., Chen, C. O., Crowe-White, K. M. et al. (2019). Fruits, vegetables, and health: A comprehensive narrative, umbrella review of the science and recommendations for enhanced public policy to improve intake. Critical Reviews in Food Science and Nutrition 60(13): 2174–2211.

Wang, J., Bravatti, A. M., Johnson, J. E. and Raman, G. (2020). Daily almond consumption in cardiovascular disease prevention via LDL-C change in the U. S. population: a cost-effective analysis. MBC Public Health 20: 558. doi: 10.1186/s12889-020-08642-4.

Yamakoshi, J., Sano, A., Tokutake, S., Saito, M., Kikuchi, M., Kubota, Y. et al. (2004). Oral intake of proanthocyanidin-rich extract from grape seeds improves chloasma. Phytother. Res. 18: 895–899.

Zarfeshany, A., Asgary, S. and Javanmard, H. S. (2014). Potent health effects of pomegranate. Adv Biomed Res. 3: 100. doi: 10.4103/2277-9175.129371.

Zhao, C. N., Meng, X., Li, Y., Li, S., Liu, Q., Tang, G. Y. et al. (2017). Fruits for prevention and treatment of cardiovascular diseases. Nutrients 9: 598.

Zhang, W., Xiao, D., Mao, Q. and Xia, H. (2023). Role of neuroinflammation in neurodegeneration development. Nature 267.

Chapter 9

Nutritional and Therapeutic Value of Wild Edible Fruits

Sujan Chaudhary,[1] *Hari Achhami,*[1] *Saroj Babu Koirala,*[2] *Bishwa Bandhu Pachhai*[1] and *Lok Ranjan Bhatt*[1,*]

9.1 Introduction

The usage of wild edible fruits (WEFs) dates back to the dawn of human civilization. The term "WEFs" describes the fruits harvested from uncultivated plant species that are used by local inhabitants or collected by them around their environment to use in their food system (Ju et al. 2013). While domestic and cultivated plants are used as the primary source for food requirements for the rural population, they are still not sufficient to fulfill the yearly food demands (Asfaw et al. 2023, Semwal et al. 2023). Widely occurring fruits provide essential food diversity that both prevents nutritional deficiencies while offering valuable alternative dietary options, especially where food shortages occur (Bhatt 2021). The therapeutic and nutritional properties of WEFs have been considered as an essential part of the human diet and demonstrate the need for a comprehensive understanding of their role in creating sustainable food systems (FAO 2019).

According to the current trends, rural and food-insecure groups rely on wild food-producing plants more frequently and extensively compared to urban areas (Sadia et al. 2022). These plants can be the best alternatives for food even during drought periods as they are highly resistant to environmental changes. Generally,

[1] Biological Resource Unit, Faculty of Science, Nepal Academy of Science and Technology, Lalitpur, Nepal.
[2] Faculty of Technology, Nepal Academy of Science and Technology, Lalitpur, Nepal.
* Corresponding author: lokranjan2000@yahoo.com

DOI: 10.1201/9781003668169-11

wild edible fruits, due to their delicious taste and flavor are eaten raw, without cooking. As a result, the local communities have these fruits to enhance their health and nutritional balance. They are also an excellent source of macro- and micronutrients (Bhatt 2021, Borelli et al. 2022). These plants are rich in vitamins, phenols, flavonoids, antioxidants, microelements, and fiber compared to cultivated crops (Motti 2022). Wild plants are often seen as a healthier option compared to cultivated vegetables, because cultivated vegetables may contain pesticides and other chemicals. WEFs could hold significant potential as a source of unique colors, flavors, bioactive compounds, and dietary supplements. Some prominent wild edible fruits such as *Aegle marmelos*, *Berberis asiatica*, *Diploknema butyracea*, *Elaeagnus infundibularis*, *Myrica esculenta*, *Phoenix loureiroi* var. *pedunculata*, *Phyllanthus emblica*, and *Rubus ellipticus* consumed across the different geographical locations (Chaudhary et al. 2025) have been provided in Figure 1 A–H.

In today's world, consumers are aware of their diet, health, and appearance. Consumers now seek food that not only meets their nutritional needs but also offers additional health benefits. Since time immemorial, people have been using plants as medicine for their health care and benefits. From the use of teas to the extraction of morphine from opium in the early 19th century has gone through significant evolution, and therefore, administration methods have changed exceptionally (Newman et al. 2000). This is possible due to the presence of bioactive compounds in them that can be directly or indirectly used against human ailments and disorders (Riaz et al. 2023). These bioactive compounds can be isolated from plants, fungi, bacteria, and marine organisms (Ferreira et al. 2017). Such compounds have contributed to 70% of the total drug formulations (Cechinel-Filho 2012). The bioactive compounds isolated from plants are called phyto-bioactive compounds, which include several secondary metabolites and other bioactive compounds (Riaz et al. 2023). Bioactive natural products have always played a crucial role as innovative therapeutic agents in human health and the medicinal sector. They possess vast medicinal properties that have a direct or indirect contribution to improving human health (Karasawa and Mohan 2018). Wild edible fruits possesses a variety of bioactive compounds, including flavonoids, stilbenes, carotenoids, and so on (Banwo et al. 2021), that have contributed to improving human health. As a result, these plants have been evaluated for their pharmacological properties in many parts of the world (Punchay et al. 2020, Shin et al. 2018, Vitalini et al. 2013). Wild food species are much healthier and better alternatives to processed food products and medications. WEFs have provided a varied diet and serve as a valuable source of essential macronutrients and micronutrients, helping to decrease vulnerability to diseases (Garekae and Shackleton 2020). Wild fruits play a significant role in managing a range of health conditions, including cardiovascular diseases, stroke, diabetes, hypertension, and other chronic illnesses (Patil et al. 2012). The consumption of wild fruits has been linked to improved digestion (Kalt et al. 2020), skin health (Lau et al. 2009), and respiratory issues (Li et al. 2016). Moreover, the rise of interest in natural and plant-based remedies has further amplified the interest in wild fruits for promoting health benefits.

Figure 1. Some popular wild edible fruits (A = *Aegle marmelos* (L.) Corrêa; B = *Berberis asiatica* Roxb. ex DC.; C = *Diploknema butyracea* (Roxb.) H. J. Lam; D = *Elaeagnus infundibularis* Momiy.; E = *Myrica esculenta* Buch.-Ham. ex D.Don; F = *Phoenix loureiroi var. pedunculata* (Griff.) Govaerts; G = *Phyllanthus emblica* L.; H = *Rubus ellipticus* Sm.) consumed across the geographical regions.

The increasing awareness of the nutritional and therapeutic benefits of WEFs highlights their capacity to improve human nutrition, human health, and contribute to global food security. Taking WEFs as a contemporary diet not only boosts human health but also preserves traditional practices and knowledge passed down through generations. It is crucial to further explore the uncovered potential of these fruits and encourage the integration of these foods into the global food system. Despite the vast nutritional and medicinal benefits of wild edible fruits, they remain underutilized and under-researched compared to cultivated crops and medicinal plants. Therefore, this chapter has highlighted the nutritional and therapeutic potential of wild edible fruits, promoting their role in improving food security, human health, and preserving traditional knowledge.

9.2 Macro- and Micronutrients in WEFs

The natural growth of WEFs in diverse environmental conditions produces fruits with enriched and complex nutritional profiles (Mahapatra et al. 2012). Nutrients are generally categorized into two types, viz., macronutrients and micronutrients. Energy-providing nutrients required in large quantities are macronutrients such as carbohydrates, proteins, and fats. In contrast, micronutrients like vitamins and minerals are required in small amounts to support essential metabolic functions

(Morris and Mohiuddin 2020). WEFs are recognized as crucial dietary resources, which contain essential nutrients that consist of macronutrients like carbohydrates, protein, and fats, along with micronutrients such as vitamins and minerals (Shrestha et al. 2021). The natural ability of WEFs to adapt to a challenging environment boosts their resistance and directly enhances their nutritional value. The biochemical composition of WEFs is not only shaped by variations in their species and genetic makeup but also by ecological factors such as soil composition (Wyszkowska et al. 2024).

From a macronutrient perspective, WEFs contain fundamental substances that enable energy metabolism and support growth functions as well as fundamental biological processes. Carbohydrates function as the main macronutrient found in WEFs to provide an essential energy source that fuels body cells and supports physical movement (Singh et al. 2017). Proteins are present in different amounts, but their amino acids help tissues repair themselves and produce enzymes while aiding physical expansion (Wu 2009). Although WEFs contain fewer lipids than other foods, the lipid composition primarily consists of unsaturated fats, which enhance both cardiovascular health and cellular integrity (Maikhuri et al. 2021).

The micronutrient profile of WEFs is also remarkable, with high concentrations of vitamins and minerals that are essential for maintaining physiological balance and disease prevention (FAO 2019). Vitamins A, C, and E are found in most of the WEFs, which act as potent antioxidants, lowering oxidative stress and improving immunological functions (Alpsoy and Yalvac 2011, Lau et al. 2009, Singh et al. 2005). Additionally, several fruits are abundant in minerals like potassium, calcium, and iron, which help to improve bone health, maintain electrolyte balance, and support hemoglobin formation (Stadlmayr et al. 2013). Their natural growth without chemical intervention frequently results in less pollution and improved nutrient bioavailability when compared to conventionally grown products (Newman et al. 2000, Punchay et al. 2020, Rai 2024).

Recent studies draw attention to WEFs because they can help fight global malnutrition as well as ensure sustainable food system development (Padulosi et al. 2013). The existing research on wild fruits in Nepal demonstrates interesting possibilities in nutraceutical and food development, agro-biodiversity and conservation, pharmaceutical potential, economic opportunities, and rural development due to their nutritional richness and adaptation to the local ecological conditions (Bhatt 2021, Bhandari et al. 2025). The nutrient compositions of WEFs show regional variability according to geographical region, which indicates environmental factors directly influence on their dietary value (Karki et al. 2024). WEFs possess ample nutritional elements and versatility, which enables them to add value to conventional agriculture practices while expanding food options (Sahoo et al. 2021). Integrating these plants into commercial food systems has the potential to enhance global dietary nutrition while preserving ecological balance and traditional cultural practices (Gairola et al. 2021).

WEFs serve as important sustainable nutrition sources because they connect traditional food habits to modern dietary requirements (Roberts and Mattoo 2018). The evaluation of WEFs' macronutrient composition, along with their micronutrient contents and their environmental and cultural aspects, leads to complete potential discovery. Systematic research and policy attention focused on WEFs holds the potential to increase their daily life use, which will enhance the global lifestyle of rural and urban dwellers (FAO 2019).

9.2.1 Macronutrients in WEFs

9.2.1.1 Carbohydrates

Carbohydrates are mainly derived from sugar and fiber, providing a quick energy source and promoting digestive health (Lunn and Buttriss 2007). It is a significant component of plant tissues and is classified into several types, including free sugars, starch, pectin, water-soluble polysaccharides, gums and mucus, as well as hemicelluloses and cellulose. During photosynthesis, carbohydrates are produced in the leaves in response to sunlight. The sugar solutions in their transport forms, such as sucrose and sorbitol, spread from the leaves across the phloem to the other sections of the plant, where they participate in metabolic processes (Komor 2000).

The carbohydrate in WEFs is composed of monosaccharides (such as glucose, fructose), disaccharides (such as sucrose, pentose), starches, and dietary fibers. Among these, simple sugars like glucose, fructose, sucrose, and pentose found in wild fruits act as a source of quick energy (Cowan 2017, Vincent et al. 2022). WEFs provide both immediate and sustained sources of energy due to their varied carbohydrate profiles. Simple carbohydrates, such as glucose, fructose, and sucrose, are prevalent in fruits like *Phoenix dactylifera* (El-Sohaimy and Hafez 2010), *Ficus carica* (Vemmos et al. 2013), *Pyracantha crenulata* (Rana et al. 2018), *Eugenia rothii* (Mahapatra et al. 2012), and *Prunus cerasoides* (Bhandari et al. 2025, Sundriyal and Sundriyal 2001), delivering quick energy and offering an immediate fuel source for physical activity. The bloodstream quickly absorbs these sugars, which makes them ideal energy sources for quick activities and higher energy requirements. The energy from starch and resistant carbohydrates in fruits is released slowly in the bodies of *Castanea* spp. and *Adansonia digitata* (Camisón et al. 2020, Gurashi et al. 2016).

Furthermore, dietary fibers found in wild fruits such as Baobab (Gurashi et al. 2016), *Ficus capensis* (Uzoekwe and Mohammed 2015), and *Carissa macrocarpa* (Souilem et al. 2019) aid in digestion and help to regulate blood sugar levels, improve bowel movement, and prevent constipation (Mehmood et al. 2014). WEFs contain two types of dietary fibers, namely soluble and insoluble, which function to manage blood sugar levels, promote digestive health and reduce the risk of type 2 diabetes and cardiovascular diseases (Fernandes et al. 2023). Drinkable fibers such as pectin present in wild apple (*Malus sylvestris*) and wild guava (*Psidium guajava*) help to lower the cholesterol level and serve as prebiotics (Efimtseva and Chelpanova 2020, Hui 2020). Antioxidative insoluble fibers in *Adansonia digitata* and wild chestnut fruits increase stool volume, promoting regular bowel movements and lowering the risk of constipation (Molino et al. 2025, Silva et al. 2023). Additionally, the fiber content in wild fruits helps improve glucose regulation, making them beneficial for individuals with diabetes or those at an increased risk of developing diabetes (Dreher 2018).

9.2.1.2 Protein

The protein composition in WEFs varies considerably among species (Levey et al. 2000). Wild fruits such as *Benthamidia capitata* (Wall.) (Rana et al. 2018), *Elaeagnus umbellata* Thunb (Sibiya et al. 2021), *Cucumis melo* Linn. (Sundriyal and Sundriyal 2001), *Phoenix reclinata* Jacq. (Aderinola et al. 2018), and *Eugenia stipitata* Mc Vaugh (Rogez et al. 2004) are the common fruits which have a significant amount of protein content. Other wild fruits like jackfruit (*Artocarpus heterophyllus*) exhibit higher variability with protein content (0.51–18.12%) (Abedin et al. 2012). In African Baobab fruit (*Adansonia digitata*), it ranged between 1.86 –2.24 g 100 g^{-1} dw (Muthai et al. 2017). Wild apricots (*Prunus armeniaca*) and wild dates (*Phoenix dactylifera*) also contain a moderate amount of protein, ranging between 1–3% of dry weight (El-Sohaimy and Hafez 2010, Nawaz et al. 2015). It is largely influenced by various factors such as specific species, geographical location, and ecological conditions such as soil, temperature, humidity, fertility, etc. (Achhami et al. 2025).

The protein present in WEFs contributes significantly to human health by supporting physiological growth, facilitating tissue repair, and improving overall physiological functions (Bayang et al. 2021). These protein molecules are composed of essential amino acids that maintain dietary balance in the human body (Sibiya et al. 2021). In particular, amino acids present in WEFs play a crucial role in immune defense and enzyme activities (Wu 2009). The regular consumption of WEFs boosts nutritional intake of that vital amino acids which might be often limited in staple foods (Lara-Arevalo 2023). The amino acid profile supports optimal protein synthesis in the body, improving the nutritional status of populations reliant on plant-based diets. For instance, fruits like *Prunus armeniaca* (wild apricots) and *Tamarindus indica* (Imli) contain two essential amino acids, methionine and lysine, that are often limited in cereal-based diets (Sibiya et al. 2021). This feature is particularly important in areas where protein malnutrition remains a concern, as WEFs can act as a valuable complementary protein source (Mahapatra et al. 2012).

The high protein levels in wild fruits such as *Tamarindus indica* support against protein-energy malnutrition, a significant issue in many developing countries (Ishaku et al. 2016). WEFs become more biologically absorbable through protein when combined with locally available grains, legumes, and tubers (Anwar et al. 2024, Jiru et al. 2023, Singh et al. 2017). The integration of wild edible fruits with locally available food sources improves protein utilization and helps address nutrient deficiencies such as Kwashiorkor and Marasmus, particularly in contexts where access to animal-based protein is limited (Chakravarty et al. 2016, Duguma 2020).

9.2.1.3 Fats

Fats provide essential fatty acids that support cell structure and hormone production in the body (Calder 2015). The human body requires fats for energy storage and cellular structure maintenance, together with fat-soluble vitamins A, D, E, and K. Although wild edible fruits have lower fat content compared to other nutrients, their unsaturated fatty acids provide health benefits for heart wellness and brain performance (Bajramova and Spégel 2022).

Wild edible fruits contain both saturated and unsaturated fats. Both of these benefit human body by minimizing the risks of chronic diseases (Kapoor et al. 2021). Various wild fruits such as *Persea americana* var. wild cultivar (Dreher 2018), *Dacryodes edulis* (Ajayi and Adesanwo 2009), *Olea europaea* (Guo et al. 2018), and *Hippophae rhamnoides* (Yang and Kallio 2001) contain elevated levels of advantageous fats. WEFs also contain omega-3 and omega-6 essential fatty acids in addition to being a source of healthy unsaturated fats that support heart health, minimize inflammation, and promote neurological well-being (Kapoor et al. 2021). The major fat components found in wild edible fruits include monounsaturated and polyunsaturated fatty acids (Bajramova and Spégel 2022). Monounsaturated fatty acids (MUFA) operate to diminish dangerous cholesterol while protecting against heart disease (Kris-Etherton 1999). The seeds from wild olives (e.g., *Olea europaea*), along with multiple other WEFs, contain healthy fats like oleic acid (Guo et al. 2017). Wild avocados (e.g., *Persea americana*) along with wild olives are rich sources of polyunsaturated fatty acids (PUFA), including essential omega-3 and omega-6 fatty acids that support proper brain health and cardiovascular functions (Dreher 2018). Some tropical fruits, including coconut (*Cocos nucifera*) and palm (*Elaeis guineensis*), have high saturated fat content that provides energy, yet people need to consume them only in moderate amount (Choudhary and Grover 2019, Hewlings 2020).

Persea americana contains 13.38% fat, the majority of which consists of monounsaturated fats (Surukite et al. 2013). It also possesses hepatoprotective activity (Stephen and Radhakrishnan 2022). The fatty substance in coconut (*Cocos nucifera*) consists primarily of medium-chain triglycerides (MCTs), which process differently than other fats and might provide health advantages for metabolic fat breakdown and management of weight (Hewlings 2020). Baobab fruit pulp contains omega-6 fatty acids and other unsaturated fatty acids, which enhance the nutritional value of the fruit (Muthai et al. 2017). The seeds from wild mango (*Mangifera indica*) contains 3.7–13.7 crude fat, mostly of unsaturated fatty acid (Solís-Fuentes et al. 2020). The beneficial fats found in these fruits help build cell structures while delivering long-lasting energy (Lebaka et al. 2021). The list of some wild fruits with their carbohydrate, protein, and fat content is provided in Table 1.

Table 1. Macronutrients present in some wild fruits across different countries.

SN	Plant scientific name	Carbohydrate (%)	Protein (%)	Fat (%)	Country	References
1	*Ancylobotrys capensis* (Oliv.)	33.1	1	0.2	South Africa	(Ngemakwe et al. 2017)
2	*Annona senegalensis* (Pers.)	25.3	8.8	24	South Africa	(Wilson and Downs 2012)
3	*Antidesma ghaesembilla* Gaertn.	48.96	7.79	2.4	Pakistan	(Sangsopha et al. 2023)
4	*Ardisia macrocarpa* Wall.	77.75	2.9	0.1	India	(Sundriyal and Sundriyal 2001)
5	*Azanza garckeana* (F. Hoffm.)	13.54	12	1.1	Ethiopia	(Duguma 2020)
6	*Baccaurea sapida* Muell. Arg.	86.41	5.43	1.24	India	(Pandey et al. 2018)
7	*Benthamidia capitata* (Wall.)	78.51	8.51	0.71	Nepal	(Rana et al. 2018)
8	*Berberis aristata* DC.	19.92	3.48	-	India	(Singh et al. 2023)
9	*Bridelia tomentosa* Blume	16.26	3.17	0.7	India	(Mahapatra et al. 2012)
10	*Calamus guruba* Buch.-Ham.	6.45	1.73	-	India	(Nayak and Basak 2015)
11	*Careya arborea* Roxb.	11.8	0.9	-	India	(Nayak and Basak 2015)
12	*Carissa macrocarpa* (Eckl.) A.DC.	21.57	1.96	0.87	South Africa	(Sibiya et al. 2021)
13	*Carissa macrocarpa* (Eckl.) A.DC.	-	0.8	-	South Africa	(Moodley et al. 2012)
14	*Carissa spinarum* L.	25.94	1.24	4.94	Ethiopia	(Siyum and Meresa 2021)
15	*Carmona retusa* (Vahl) Masam.	4.33	4.11	-	India	(Nayak and Basak 2015)
16	*Carpobrotus edulis* (L.) N.E.Br.	10.97	3.45	0.21	South Africa	(Sibiya et al. 2021)
17	*Citrullus lanatus* (Thunb.)	7.19	0.44	0.15	Nigeria	(Anhwange et al. 2010)
18	*Coccinia sessilifolia* (Sond.)	13	2.1	0.2	South Africa	(Sibiya et al. 2021)
19	*Cordyla africana* Lour.	19.94	0.004	3.55	South Africa	(Sibiya et al. 2021)
20	*Cordyla Africana* Lour.	75.6	-	-	South Africa	(Ngemakwe et al. 2017)
21	*Cucumis melo* Linn.	32.53	10.32	9.33	India	(Sundriyal and Sundriyal 2001)
22	*Diospyros mespiliformis* Hochst. ex A.DC.	4.6	2.6	0.11	Namibia	(Ngemakwe et al. 2017)

Table 1 contd. ...

...Table 1 contd.

SN	Plant scientific name	Carbohydrate (%)	Protein (%)	Fat (%)	Country	References
23	*Diploknema butyracea* Roxb.	4.05	6.4	20.23	Nepal	(Dawadi et al. 2023)
24	*Dovyalis abyssinica* (A.Rich.)	78.27	3.01	1.46	Ethiopia	(Jiru et al. 2023)
25	*Dovyalis caffra* (Hook.fil. & Harv.)	27.58	0.4	0.4	Kenya	(Chaka and Osano 2019)
26	*Dovyalis caffra* (Hook.fil. & Harv.)	14.2	0.83	0.73	South Africa	(Sibiya et al. 2021)
27	*Dovyalis longispina* (Harv.) Warb.	30.91	1.94	1.49	South Africa	(Sibiya et al. 2021)
28	*Elaeagnus umbellata* Thunb	70.82	14.46	-	India	(Sibiya et al. 2021)
29	*Elaeocarpus sikkimensis* Mast.	88.17	6.93	1.02	India	(Pandey et al. 2018)
30	*Elaeocarpus sikkimensis* Roxb.	73.05	5.37	0.21	India	(Sundriyal and Sundriyal 2001)
31	*Elaeagnus latifolia* L.	74.06	7.8	0.52	India	(Sundriyal and Sundriyal 2001)
32	*Englerophytum magalismontanum* (Sond.)	3.1	0.83	0.31	South Africa	(Sibiya et al. 2021)
33	*Eriolobus indica* Schn.	71.73	1.75	0.35	India	(Sundriyal and Sundriyal 2001)
34	*Eugenia roxburghii* DC.	19.01	0.65	-	India	(Mahapatra et al. 2012)
35	*Eugenia stipitata* Mc Vaugh	49.2	11.9	-	Brazil	(Rogez et al. 2004)
36	*Ficus hispida* Roxb. ex Wall.	1.75	1.13	-	India	(Nayak and Basak 2015)
37	*Ficus mucuso* Welw. ex Ficalho	71.88	5.11	3.31	Ethiopia	(Jiru et al. 2023)
38	*Ficus sycomorus* L.	55.8	5.6	8.9	South Africa	(Ogunlaja et al. 2020)
39	*Flacourtia indica* (Burm. fil.) Merr.	89.72	3.17	0.53	India	(Perera et al. 2022)
40	*Garcinia livingstonei* T.Anderson	95.02	0.65	2.56	India	(Joseph et al. 2017)
41	*Garcinia livingstonei* T. Anderson	4.64	2.16	0.69	South Africa	(Sibiya et al. 2021)
42	*Gardenia erubescens* Stapf & Hutch.	77.88	4.22	1.4	Ethiopia	(Jiru et al. 2023)
43	*Halleria lucida* L.	11.41	6.83	1.44	South Africa	(Sibiya et al. 2021)

Table 1 contd. ...

...Table 1 contd.

SN	Plant scientific name	Carbohydrate (%)	Protein (%)	Fat (%)	Country	References
44	*Harpephyllum caffrum* Bernh.	9.1	0.7	0.2	Southern Africa	(Alfred 2019)
45	*Hippophae rhamnoides* D.Don	84.07	3.5	0.59	India	(Sundriyal and Sundriyal 2001)
46	*Hippophae salicifolia* D. Don	31.67	7.01	-	India	(Singh et al. 2023)
47	*Lannea edulis* (Sond.) Engl.	75.9	6.5	9.7		(Malaisse and Parent 1985)
48	*Machilus edulis* Roxb.	51.5	4.51	22.5	India	(Sundriyal and Sundriyal 2001)
49	*Malus baccata* Loisel	25.51	3.31	-	India	(Singh et al. 2023)
50	*Mangifera sylvatica* Roxb.	66	2.94	21.5	India	(Sundriyal and Sundriyal 2001)
51	*Manilkara mochisia* (Baker) Dubard	36.98	2.19	1.38	South Africa	(Sibiya et al. 2021)
52	*Melastoma malabathricum* L.	7.75	5.48	-	India	(Nayak and Basak 2015)
53	*Mimusops elengi* L.	18.15	0.61	-	India	(Mahapatra et al. 2012)
54	*Morus alba* Wall.	80.61	1.25	3.9	India	(Sundriyal and Sundriyal 2001)
55	*Pappea capensis* Eckl. & Zeyh.	25	4.33	5.11	South Africa	(Sibiya et al. 2021)
56	*Parinari curatellifolia* Planch.	21.18	2.61	0.25	South Africa	(Sibiya et al. 2021)
57	*Parinari curatellifolia* Planch.	21.4	1.1	1.7	Zimbabwe	(Benhura et al. 2012)
58	*Passiflora foetida* Vell.	3.6	0.15	13	China	(Song et al. 2018)
59	*Passiflora indica* Linn.	77.85	8.29	0.3	India	(Sundriyal and Sundriyal 2001)
60	*Phoenix dactylifera* L.	65.99	13.23	8.67	Nigeria	(Aderinola et al. 2018)
61	*Phoenix reclinata* Jacq.	60.44	10.39	15.5	Nigeria	(Aderinola et al. 2018)
62	*Phoenix reclinata* Jacq.	11.44	4.08	1.44	South Africa	(Sibiya et al. 2021)
63	*Phyllanthus acidus* (L.) Skeels	6.7	0.4	0.52	Indonesia	(Brooks et al. 2020)
64	*Platonia insignis* Mart	49.7	6.4	13.5	Brazil	(Rogez et al. 2004)
65	*Polyalthia suberosa* (Roxb.)	4.25	1.96	-	India	(Mahapatra et al. 2012)

Table 1 contd. ...

...Table 1 contd.

SN	Plant scientific name	Carbohydrate (%)	Protein (%)	Fat (%)	Country	References
66	*Prunus cerasoides* D.Don	87.55	5.5	0.21	India	(Sundriyal and Sundriyal 2001)
67	*Prunus cornuta* Wall.	78.4	8.46	-	India	(Singh et al. 2023)
68	*Prunus persica* (L.) Batsch	65.13	7.08	-	India	(Singh et al. 2023)
69	*Pyrus pashia* Buch.	69.2	1.62	-	India	(Singh et al. 2023)
70	*Ramaria botrytis* (Pers.) Bourdot	23	0.38	-	India	(Singh et al. 2023)
71	*Rosa webbiana* Wall.	65.08	3.44	-	India	(Singh et al. 2023)
72	*Rubus ellipticus* Smith	72.7	4	7.1	India	(Sundriyal and Sundriyal 2001)
73	*Salacia kraussii* (Harv.)	-	3.7	0.8	Mozambique	(Magaia et al. 2013)
74	*Solanum torvum* Sw.	11.9	1.46	-	India	(Mahapatra et al. 2012)
75	*Spondias axillaris* Roxb.	52.28	4.11	0.05	India	(Sundriyal and Sundriyal 2001)
76	*Streblus asper* Lour.	5.59	1.48	-	India	(Nayak and Basak 2015)
77	*Strychnos spinosa* Lam.	(37.39–42.24)	(2.85–9.19)	(0.95–2.67)	South Africa	(Mbhele et al. 2025)
78	*Syzygium cordatum* Hochst.	ND	10.1	4	India	(Sibiya et al. 2021)
79	*Syzygium cordatum* Hochst.	63.9	5.9	7.7	South Africa	(Maliehe 2015)
80	*Syzygium guineense* (Willd.) DC	1.01	1.66	7.74	South Africa	(Sibiya et al. 2021)
81	*Terminalia chebula* Retz.	80.81	2.32	0.56	India	(Sundriyal and Sundriyal 2001)
82	*Terminalia citrina* (Gaertn.) Roxb.	3.1	0.92	-	India	(Mahapatra et al. 2012)
83	*Theobroma grandiflorum* (Willd. ex Spreng.)	49.5	8.8	12.7	Brazil	(Rogez et al. 2004)
84	*Viburnum mullaha*	17.3	1.17	2.1	Nepal	(Dawadi et al. 2022)
85	*Vangueria infausta* Burch.	78.1	4.7	0.7	South Africa	(Mothapo et al. 2014)
86	*Viburnum cotinifolium* D.Don	63.33	7.34	-	India	(Singh et al. 2023)

Table 1 contd. ...

...*Table 1 contd.*

SN	Plant scientific name	Carbohydrate (%)	Protein (%)	Fat (%)	Country	References
87	*Ximenia caffra* Sond.	26.3	3.1	1.3	South Africa	(Goosen et al. 2018)
88	*Ziziphus mucronata* Willd.	23.61	35.1	27.4	Botswana	(Aganga and Mosase 2001)
89	*Ziziphus oenoplia* (L.) Mill.	17.13	0.7	-	India	(Mahapatra et al. 2012)
90	*Ziziphus spina-christi* (L.) Desf.	70.41	5.31	1.65	Ethiopia	(Jiru et al. 2023)

9.2.2 Micronutrients in WEFs

Micronutrients are vitamins and minerals required by the body in trace amounts. However, their importance to human health is crucial, and a lack of any of them can result in severe and even fatal illnesses. WEFs feature abundant micronutrients, which are critical to enhance human nutrition, especially when alternative food sources remain scarce. The amounts of micronutrients present in WEFs can vary based on the species as well as their ecological habitats and the parts used for consumption.

9.2.2.1 Occurrence of Vitamins in WEFs

9.2.2.1.1 Vitamin A

Vitamin A (all-trans-retinol and its esters) and provitamin A (β-carotene) are important nutrients that contain retinol. Both retinyl esters and β-carotene are metabolized into retinol. The retinol-binding proteins, on binding retinol, provide a way for solubilizing retinol for distribution to target tissues and controlling retinol plasma concentrations (Napoli 2016). Wild fruits such as *Hippophae* spp. (sea buckthorn berries) (Zielińska and Nowak 2017), *Adansonia digitata* (baobab) (Muthai et al. 2017), *Irvingia gabonensis* (African mango) (Mateus-Reguengo et al. 2020), *Rosa canina* (de Almeida Caetano et al. 2023), etc., contain substantial amounts of carotenoids, contributing to vitamin A intake.

9.2.2.1.2 Vitamin B

WEFs are notable sources of vitamin B, including niacin and folate (Sánchez-Mata et al. 2016). Wild fruits like *Morus indica*, *Meyna laxiflora*, *Solanum gilo*, *Solanum kurzii*, *Persea americana*, *Careya arborea*, and *Viburnum foetidum* contain significant amounts of thiamin (B1), riboflavin (B2), niacin (B3), and folate (B9), which are essential for producing energy as well as improving cell division (Seal and Chaudhuri 2017, Mmuo and Okoli 2022). In addition, wild bananas (*Musa* spp.) are good sources of pyridoxine (B6), which helps metabolism and brain function (Chandler 1995).

9.2.2.1.3 Vitamin C

Vitamin C is vital for humans, exhibiting pleiotropic effects due to its capacity to transfer electrons (Zhitkovich 2020). It is an effective antioxidant and cofactor for several biosynthetic and gene regulatory enzymes. Vitamin C aids immune defense by supporting the cellular functioning of both the innate and adaptive immune systems (Jabłońska-Ryś et al. 2009). Many WEFs, especially fruits from tropical and subtropical regions, are excellent sources of vitamin C. For instance, *Phyllanthus emblica* (Khomdram and Guruaribam 2010) and *Haematocarpus validus* (Rymbai et al. 2023) are rich in vitamin C.

9.2.2.1.4 Vitamin E

The fat-soluble antioxidant tocopherol (vitamin E) protects cellular structures from damage and strengthens immune defense mechanisms (Ncube et al. 2008). It functions as an antioxidant by donating a hydrogen atom to peroxyl radicals in unsaturated lipid molecules, resulting in the formation of a hydroperoxide and a tocopheroxyl radical that combine with peroxyl radicals to form more stable adducts (Morales et al. 2013). *Rosa canina*, *Fragaria vesca*, and *Rubus chamaemorus* supplement vitamin E through their fruits (Jaakkola et al. 2012). *Rubus ulmifolius* Schott contains a good amount of tocopherol (13.48 mg/100 g), which inhibits lipid peroxidation (Morales et al. 2013).

9.2.2.2 Minerals in WEFs

Wild edible fruits serve as rich sources of essential minerals, often exhibiting higher mineral content than cultivated crops, largely due to their natural growth in mineral-rich soils. Key minerals found in wild fruits include calcium, iron, magnesium, potassium, zinc, and phosphorus, which support metabolic function, bone health, and immunity (Deshmukh and Waghmode 2011). *Dialium guineense* (Ayessou et al. 2014) and *Fragaria vesca* (Jurgiel-Małecka et al. 2017) contain a significant amount of calcium, which plays a fundamental role in maintaining bones, muscles and electrical impulses in the nerves. Wild fruits, *Meyna laxiflora* and *Flacourtia indica*, are rich in iron content, which is good for blood circulation and overall health (Valvi and Rathod 2011). Magnesium helps in proper functioning of muscles and nerves, and it assists both energy metabolism and bone health. The combination of wild fruits, including *Grewia tiliifolia* and *Flacourtia indica*, provides an essential magnesium supply (Valvi and Rathod 2011). *Persea americana* (wild avocado) contains abundant magnesium that strengthens the cardiovascular system along with building bone health (Bhuyan et al. 2019).

Wild edible fruits are notable sources of potassium, an essential mineral that supports nerve function, muscle contraction, and fluid balance. Various studies have shown that the potassium content in these fruits can vary significantly (Ishaku et al. 2016, Jiru et al. 2023, Mahapatra et al. 2012, Stadlmayr et al. 2013). *Musa* spp. (wild bananas) and *Ficus carica* (wild figs) effectively fulfill the recommended potassium intake of the human body (Bayang et al. 2021). Moreover, wild fruits such as *Arbutus pavarii* that possess the high potassium content (349.33 mg/100 g FW) (Hegazy et al. 2013), can become good source of essential micronutrients. The list of some wild fruits with different micronutrient components has been provided in Table 2.

Table 2. Micronutrients present in some wild fruits across different countries.

SN	Scientific name	K (mg/100 g)	Na (mg/100 g)	Ca (mg/100 g)	Mg (mg/100 g)	Fe (mg/100 g)	P (mg/100 g)	Zn (mg/100 g)	Asc (mg/100 g)	BC (mg/100 g)	Nia (mg/100 g)	Rib (mg/100 g)	Country	Reference
1	*Adansonia digitata* L.	1730	14.3	275	232	6.2	51	1.36	-	-	-	-	South Africa	(Stadlmayr et al. 2013)
2	*Anacardium occidentale* L.	556.7	3.08	64.05	277.09	3.89	1101.04	4.98	-	-	-	-	Brazil	(de Almeida Caetano et al. 2023)
3	*Balanites aegyptiaca* (L.) Delile	131.6	1.43	10.58	166.81	0.34	55.23	2.02	-	-	-	-	Ghana	(Achaglinkame et al. 2019)
4	*Balanites aegyptiaca* (L.)	1110	-	164	44	13.8	51	1.77	-	-	-	-	South Africa	(Stadlmayr et al. 2013)
5	*Dacryodes edulis* (G.Don)	443	0.8	50	47	1.7	60	0.47	-	-	-	-	South Africa	(Stadlmayr et al. 2013)
6	*Dialium guineense* Willd.	260	390	47	300	4.1	42.9	2.5	6.2	362	0.52	0.06	Nigeria	(Adepoju 2009)
7	*Diospyros mespiliformis* Hochst.	129.4	1.25	8.81	162.98	1.46	64.78	1.7	-	-	-	-	Ghana	(Achaglinkame et al. 2019)
8	*Dovyalis abyssinica* (A. Rich.)	183.36	10.86	120.18	5.62	2.09	168	0.62	-	-	-	-	Ethiopia	(Jiru et al. 2023)
9	*erminalia citrina* Gaertn.	1460.72	18.65	44.28	-	3.11	-	41.93	-	-	-	-	India	(Mahapatra et al. 2012)
10	*Eugenia pyriformis* Cambess.	134	-	341.33	41	5.37	134	1.03	-	-	-	-	Brazil	(de Almeida Caetano et al. 2023)

Table 2 contd. ...

...Table 2 contd.

SN	Scientific name	K (mg/100 g)	Na (mg/100 g)	Ca (mg/100 g)	Mg (mg/100 g)	Fe (mg/100 g)	P (mg/100 g)	Zn (mg/100 g)	Asc (mg/100 g)	BC (mg/100 g)	Nia (mg/100 g)	Rib (mg/100 g)	Country	Reference
11	*Eugenia rothii* Panigrahi	2009.9	25.09	60.8	-	4.29	-	13.57	-	-	-	-	India	(Mahapatra et al. 2012)
12	*Ficus mucuso* Welw. ex Ficalho	165.84	4.88	190.18	56.55	20.96	168	0.62	-	-	-	-	Ethiopia	(Jiru et al. 2023)
13	*Gardenia erubescens* Stapf & Hutch.	107.54	7.21	98.89	3.44	15.04	79.15	6.23	-	-	-	-	Ethiopia	(Jiru et al. 2023)
14	*Gardenia erubescens* Stapf & Hutch.	133.41	2.64	6.48	168.07	0.48	46.37	0.81	-	-	-	-	Ghana	(Achaglinkame et al. 2019)
15	*Genipa americana* L.	-	-	45.82	-	0.8	33.5	-	-	-	-	-	Brazil	(de Almeida Caetano et al. 2023)
16	*Glycosmis pentaphylla* (Retz.)	675.78	31.01	88.87	-	5.28	-	2.95	-	-	-	-	India	(Mahapatra et al. 2012)
17	*Mimusops elengi* L.	808.02	52.97	1975.16	-	4.71	-	2.1	-	-	-	-	India	(Mahapatra et al. 2012)
18	*Mondia whitei* (Hook.fil.) Skeels	410	360	300	300	2	170	6.2	35.8	105	0.52	0.08	Nigeria	(Adepoju 2009)
19	*Morinda coreia* Buch.-Ham.	223.44	17.11	163.22	-	2.33	-	1.42	-	-	-	-	India	(Mahapatra et al. 2012)
20	*Polyalthia suberosa* Roxb.	845.76	25.09	62.72	-	4.19	-	2.25	-	-	-	-	India	(Mahapatra et al. 2012)

21	*Sclerocarya birrea* (A.Rich.)	325	-	35	19	3.4	31	0.31	-	-	-	-	South Africa	(Stadlmayr et al. 2013)
22	*Spondias mombin* Jacq.	270	400	31.8	465	3.2	37.1	0.2	34	85	0.5	0.07	Nigeria	(Adepoju 2009)
23	*Spondias mombin* Jacq.	288.27	0.55	11.03	15.09	0.32	32.84	-	-	-	-	-	Brazil	(de Almeida Caetano et al. 2023)
24	*Streblus taxoides* (B.Heyne ex Roth)	256.39	18.16	30.93	-	2.92	-	0.196	-	-	-	-	India	(Mahapatra et al. 2012)
25	*Syzygium guineense* (Willd.)	-	-	67	34	7.9	-	-	-	-	-	-	South Africa	(Stadlmayr et al. 2013)
26	*Tamarindus indica* L.	933	30.2	192	66	3.1	87	3.1	-	-	-	-	South Africa	(Stadlmayr et al. 2013)
27	*Toddalia asiatica* (L.) Lam.	599.09	27.47	66.48	-	3.36	-	1.35	-	-	-	-	India	(Mahapatra et al. 2012)
28	*Uapaca kirkiana* Müll.Arg.	375	10	17	39	11.8	15	-	-	-	-	-	South Africa	(Stadlmayr et al. 2013)
29	*Vitex doniana* Sweet	637	8	21	17	1.3	77	-	-	-	-	-	South Africa	(Stadlmayr et al. 2013)
30	*Ziziphus mauritiana* Lam	256	6	23	5	0.8	19	0.3	-	-	-	-	South Africa	(Stadlmayr et al. 2013)
31	*Ziziphus oenoplia* (L.) Mill.	720.38	26.15	94.76		4.28		2.57	-	-	-	-	India	(Mahapatra et al. 2012)
32	*Ziziphus spina-christi* (L.) Desf.	148.91	6.01	67.17	81.15	29.13	55.56	8.34	-	-	-	-	Ethiopia	(Jiru et al. 2023)

Note: K: Potassium; Na: Sodium; Ca = Calcium; Mg = Magnesium; Fe = Iron; P = Phosphorus; Zn = Zinc; Asc = Ascorbic acid; BC = β-carotene; Nia = Niacin; Rib = Riboflavin

9.3 Environmental Impact on Nutritional Composition

The ecological parameters and environmental changes have the potential to affect the nutritional quality of fruits and nuts. For instance, an increase or decrease in salinity, availability of moisture, carbon dioxide, and temperature can alter the yield and nutritional quality of fruits (Alae-Carew et al. 2020). In high-altitude regions, due to the stress from cold climates, there is an elevation in sugar levels of fruits such as *Prunus armeniaca* (apricots), resulting in a sweeter taste with greater carbohydrate content (Nicolétis et al. 2019). The tropical wild jackfruit *Artocarpus heterophyllus* possesses high starch, which is transformed into sugars when it matures, while delivering both fast energy alongside dietary fiber (Ranasinghe et al. 2019). Harsh, dry climates favor *Ziziphus* spp. (wild jujube), because they build up sugar reserves during times of water scarcity, which makes them an essential energy reserve for inhabitants of arid areas (Muhammad et al. 2022). Through ecological adaptations during stress, WEFs enable to maintain required nutrients, which includes multiple varieties that adjust carbohydrate content in various climatic zones (González-Zamorano et al. 2025).

Protein content in WEFs depends on different ecological factors, including altitudinal variations, soil types, and climatic conditions (Bvenura et al. 2017). Wild fruits growing in soils with low nutritional availability increase their protein content through this survival strategy (Achaglinkame et al. 2019). Higher protein content is found in tropical fruits (such as wild jackfruit and baobab) in comparison to temperate fruits due to ecological stress (Ranasinghe et al. 2019). The protein composition of fruits changes according to their maturity stage because unripe fruits typically possess higher levels compared to ripe counterparts due to sugar and starch levels (Singh et al. 2017). Similarly, the content of fat in WEFs depends significantly on environmental conditions, specifically climate, soil types, water availability, and ecological pressure (Achaglinkame et al. 2019). It affects the lipid metabolism process of the plant (Yang and Kallio 2001).

Micronutrient availability to plants is dependent on soil conditions and parent material (Riaz et al. 2023). Soil composition has a direct impact on mineral uptake (Tyler and Olsson 2001). Fruits of *Carissa* spp. in iron-rich soils contain five times more iron than those in poor soils (Bala 2021). The combination of hot, dry climates facilitates these plants to accumulate calcium and magnesium minerals that support their tolerance of environmental stress (Stadlmayr et al. 2013). *Careya arborea* (wild guavas) and *Adansonia digitata* (baobab) fruits grow naturally in a tropical climate, resulting in higher vitamin C content (Kashyap et al. 2022, Silva et al. 2023).

9.4 Therapeutic Properties of Wild Edible Fruits

WEFs hold immense potential as both nutraceuticals as well as in therapeutic applications (Semwal et al. 2023). The abundance of diverse bioactive compounds in notable amounts makes them promising therapeutic agents (Cristina and Jordao 2024). Traditionally, these fruits have been used to cure gastrointestinal, respiratory, reproductive, and other human diseases in different geographical regions (Mahapatra et al. 2012). In the last decade, the focus has shifted towards the scientific exploration and development of pharmaceuticals with antioxidant, antitumor, anti-inflammatory, and antimicrobial effects, among diverse bioactive effects (Li et al. 2016).

9.4.1 Ethnomedicinal Applications of Wild Fruits

WEFs have been traditionally used for medicinal purposes, such as treating respiratory, gastrointestinal, cardiovascular, reproductive, dermatological, ENT-related, neurological, muscular, and bone diseases (Abbasi et al. 2013, Hazarika et al. 2012). Most of these plants have been used both as foods and medicines, also referred to as medicinal foods or nutraceuticals (Deshmukh and Waghmode 2011). Several studies based on different countries, such as India (Bhatia et al. 2018, Mahapatra and Panda 2012), Nepal (Ojha Khatri et al. 2021, Panta et al. 2021, Dulal et al. 2022, Munankarmi et al. 2025), Indonesia (Mukhtar 2016), Ethiopia (Melaku and Ebrahim 2021), and Pakistan (Khan et al. 2015) have reported their traditional medicinal applications. The main driving factor behind the ethnomedicinal utilization of WEFs is socio-economic rather than climatic or environmental. The socio-economic factors, including cultural values, fads (temporal trends), and flavor preferences, were the major influencing factors in the selection of WEFs (Abbasi et al. 2013). Ethnomedicinal knowledge is found to be transferred orally without proper documentation through generations. Thus, a detailed scientific study considering its therapeutic effectiveness and safety is necessary to validate the traditional knowledge (Hazarika et al. 2012).

9.4.2 Bioactive Compounds and Phytochemical Profile

Plant secondary metabolites with beneficial pharmacological activities are known as bioactive compounds (Jain et al. 2019). Mostly, these metabolites include glycosides, flavonoids, tannins, terpenoids, alkaloids, and coumarins (Bernhoft 2010). WEFs are also abundant in flavonoids, phenolic acids, tannins, tocopherols, glycosides, steroids, carotenoids, malic acid, oleanolic acid, allicin, and glucosinolates, among others (Camara et al. 2016). These phytoconstituents lead to diverse pharmacological activities such as antimicrobial, antioxidant, anti-inflammatory, anticancer, neuroprotective, anti-diabetic, anti-hypolipidemic, anti-diarrheal, antiscorbutic, and aphrodisiac properties (Cámara et al. 2016, Pratap et al. 2023). For example, wild fruits from the Himalayan region like *Berberis asiatica*, *Myrica esculenta*, *Celtis australis*, *Morus alba*, *Phyllanthus emblica*, *Ficus palmata*, *Prunus armeniaca*, and *Fragaria indica* were also reported with similar patterns of phytochemicals and therapeutic values (Semwal et al. 2023). Antioxidant activity in wild fruits is primarily attributed to diverse polyphenols, which further contribute to anticancer, neuroprotective, and cardiovascular protective effects (Murillo et al. 2012).

9.4.3 Pharmacological Activities

9.4.3.1 Antioxidant Potential

Antioxidant properties in WEFs are strongly correlated with the abundance of secondary metabolites like flavonoids and phenolic compounds. Fruits of *Detarium microcarpum* with a robust amount of phenolic and flavonoid content were reported to have higher antioxidant activities among the studied WEFs (Lamien-Meda et al. 2008). These secondary metabolites directly affect bioactivity, including antioxidant potential (Kaurinovic and Vastag 2019). In addition, antioxidant activity in wild fruits is primarily attributed to diverse polyphenols, which further contribute to anticancer, neuroprotective, and cardiovascular protective ailments (Murillo et al. 2012). Notable phenolic compounds reported in wild fruits include gallic acid, vanillic acid, chlorogenic acid, caffeic acid, syringic acid, p-coumaric acid, ferulic acid, sinapic acid, salicylic acid, ellagic acid, and trans-cinnamic acid (Cosmulescu et al. 2017). Similarly, a significant amount of phenolic content and antioxidant activity was also observed in *Emblica officinalis*, *Spondias axillaris*, and *Baccaurea sapida*, in which compounds like caffeic acid, ellagic acid, and gallic acid were detected (Prakash et al. 2012). Another study on WEFs from the Indian Himalayan region, such as *Phyllanthus emblica, Morus alba, Terminalia chebula, Myrica esculenta, Berberis asiatica, Pyaracantha chebula*, and *Ficus palmata* were reported to contain phenolic acids, viz. gallic, catechin, chlorogenic, caffeic, and p-coumaric acid, which contributed to the antioxidant properties (Bhatt et al. 2017). Phenolic compounds are generally found in both free and bound forms. Both free and bound phenolics were evaluated among the 28 dry and fresh WEFs of Cameroon and were identified to exhibit antioxidant properties (Koubala et al. 2021). Notable antioxidant activity in DPPH, ABTS, and FRAP assays was recorded in the cases of *Vitex diversifolia, Ximenia americana, Carissa edulis, Hexabolus monopetalus, Haematostaphis barteri*, and *Cordia sinensis* (Koubala et al. 2021).

The solvent used for the extraction of phytochemicals also played a critical role in the antioxidant potential of a particular fruit. According to Lamien-Meda et al. (2008), maximum phenolic and flavonoid content was extracted in acetone compared to methanol. Similarly, a study conducted in the Himalayan region of Pakistan among 20 different WEFs revealed the highest amount of flavonoids and phenolic compounds in the acetone extract of *Prunus domestica* (113.55 ± 0.61 mg GAE/100 g) and *Rubus ellipticus* (200.06 ± 1.57 mg RtE/100 g). On the other hand, flavonoid content was highest in the water extract of *Rosa moschata* (194.82 ± 3.46 mg RtE/100 g). *Berberis lycium* acetone extract showed the maximum total antioxidant activity among the tested fruits (Shan et al. 2019).

Several bioactive compounds, including ascorbic acid, carotenoids, anthocyanins, tocopherols, tannins, and proanthocyanidins, have been detected in WEFs (Egea et al. 2010, Morales et al. 2013, Rymbai et al. 2023). Specifically, catechin, epicatechin, rutin, myricetin, and quercetin were the identified flavonoids with antioxidant activities (Cosmulescu et al. 2017). Prominent WEFs found in different geographical locations with antioxidant potentials are listed in Table 3.

Table 3. Some antioxidant potential WEFs from different geographical locations.

SN	Scientific names	Solvent Used	Assay	Studied Bioactive Antioxidant Compound	Country	References
1	*Arbutus pavarii* Pamp.	Methanol	DPPH, H_2O_2 scavenging assay	Tannins, anthocyanins, carotenoids	Saudi Arabia	(Hegazy et al. 2013)
2	*Chrysophyllum cainito* L.	Methanol	DPPH	Polyphenols	Panama	(Murillo et al. 2012)
3	*Citrus hystrix* DC.	Methanol	DPPH	Phenolics, flavonoids	India	(Biswas et al. 2021)
4	*Detarium microcarpum* Guill. & Perr.	Acetone, Methanol	DPPH, FRAP, ABTS	Phenolics, flavonoids	Burkina Faso	(Lamien-Meda et al. 2008)
5	*Fragaria nubicola* Lindl.	Methanol	ABTS, DPPH, FRAP, OH-	Phenolics, flavonoids, flavonols, tannins, proanthocyanidins	India	(Biswas et al. 2021) (Bahukhandi et al. 2020)
6	*Morella esculenta* (Buch.-Ham. ex D.Don) I.M.Turner	Acetone	DPPH	Phenolics, flavanol	India	(Seal 2011)
7	*Morella esculenta* (Buch.-Ham. ex D.Don) I.M.Turner	Methanol	ABTS, DPPH, FRAP	Gallic acid, catechin, chlorogenic acid, ρ-coumaric acid	India	(Rawat et al. 2011)
8	*Nageia nagi* (Thunb.) Kuntze	Methanol	DPPH, FRAP	Ascorbic acid, carotenoid, monomeric anthocyanin, flavonoids, phenolics, flavonol	India	(Rymbai et al. 2023)
9	*Potentilla indica* (Andrews) Th.Wolf, Rubus ellipticus Sm.	Acetone	DPPH, FRAP	Polyphenols	Pakistan	(Shan et al. 2018)
10	*Prunus padus* L.	Methanol	DPPH	Gallic acid, vanillic acid, chlorogenic acid, caffeic acid, syringic acid, p-coumaric acid, ferulic acid, sinapic acid, salycilic acid, elagic acid, trans-cinnamic acid, catechin, epicatechin, rutin, myricetin, quercetin	Romania	(Cosmulescu et al. 2016)

Table 3 contd. ...

...Table 3 contd.

SN	Scientific names	Solvent Used	Assay	Studied Bioactive Antioxidant Compound	Country	References
11	*Rosa canina* L.	-	FRAP, ABTS	Vitamin C, phenolics	Poland	(Jabłońska-Ryś et al. 2009)
12	*Rosa canina* L.	Phosphate buffer	ABTS	Phenolics, carotenoids, ascorbic acid	Spain	(Egea et al. 2010)
13	*Rubus uimifolius* Schott	Methanol	Lipid peroxidation inhibition assay (Beta-carotene, TBARS assay)	Tocopherols, vitamin C, organic acids	Spain	(Morales et al. 2013)
14	*Spondias pinnata* (L.fil.) Kurz	Methanol	DPPH	Phenolics	Nepal	(Chalise et al. 2010)
15	*Thespesia garckeana* F.Hoffm.	Methanol	FRAP, TAA, DPPH	Phenolics, flavonoids	Malawi	(Mwamatope et al. 2023)
16	*Rubus acuminatus* Sm.	Methanol	DPPH	Phenolics, flavonoids, carotenoids, vitamin C	Nepal	(Bhusal et al. 2020)
17	*Rubus ellipticus* Sm.	Methanol	DPPH	Phenolics, flavonoids, carotenoids	Nepal	(Bhusal et al. 2020)
18	*Diospyros malabarica*	Methanol	DPPH	Phenolics, flavonoids,	Nepal	(Shrestha et al. 2021)
19	*Viburnum mullaha*	Methanol	DPPH	Phenolics, flavonoids, vitamin C	Nepal	(Dawadi et al. 2022)

9.4.3.2 Antimicrobial Activity

The extracts of WEFs were reported to possess antibacterial, antifungal, and antiviral activities. Fruit pulp of *Myrica negi* exhibited antimicrobial activities against ten bacterial and three fungal species, including *Escherichia coli* and *Streptococcus pyogenes* (Saklani and Chandra 2012). *In vitro* antibacterial activities have been reported in both gram-positive and gram-negative bacterial strains (Anand et al. 2017). Gram-negative bacteria, such as *Escherichia coli*, *Pseudomonas aeruginosa*, *Salmonella typhi*, and *Klebsiella pneumoniae*, and gram-positive bacteria, including *Bacillus subtilis*, *Salmonella typhi*, and *Staphylococcus aureus*, were inhibited by the extract of *Catunaregam spinosa* (Anand et al. 2017), *Adansonia digitata*

(Tshikalange et al. 2017), *Eugenia operculata* (Islary et al. 2018), *Garcinia gummi-gutta*, and *Artocarpus lacucha* (Biswas et al. 2021).

Wild berries, namely *Vaccinium myrtillus*, *V. vitis-idaea*, *Fragaria vesca*, and *Rubus idaeus*, rich in anthocyanins, were found to inhibit the replication of influenza virus A/H3N2 and coxsackie virus B1 (Nikolaeva-Glomb et al. 2014). In another study, the crude fruit extract of *Ribes nigrum* was proven to demonstrate potential antiviral activity against influenza virus A and B (Knox et al. 2003).

Several secondary metabolites, including phenolics, flavonoids, terpenoids, saponins, and alkaloids, among others, play major roles in the antimicrobial activity of plant extracts (Patra 2012). Fruits of *Ficus auriculata*, rich in phytochemicals like flavonoids, phenolic compounds, tannins, glycosides, and resins, exhibited *in vitro* antimicrobial activity against *Shigella flexneri* (14 ± 1 mm), *E. coli* (13 ± 1 mm), and *Staphylococcus epidermidis* (12 ± 1 mm) (Saklani and Chandra 2012). A protein fraction rich in peptides, protease inhibitors, and enzymes extracted from the pulp of *Bromelia pinguin* was reported to have antibacterial effects against *E. coli* (MIC 0.3492 mg/mL) and *S. aureus* (0.6845 mg/mL). This extracted protein fraction might be utilized in therapeutic applications as it is not toxic to consume and has an equivalent mechanism of action to antibiotics in bacterial inhibition (Carlos Ruiz-Ruiz et al. 2017).

The growing trend of antibiotic resistance in microorganisms has raised concerns as bacteria utilize diverse mechanisms to escape drug effects, including biofilm formation. Wild fruits have been also reported to be beneficial for the inhibition of biofilm formation. A study by Compaoré and Kiendrebeogo (2018) on the anti-biofilm activity of six WEFs against *P. aeruginosa* showed maximum inhibition of about 72% with the extract of the wild fruit *Ziziphus mauritiana*. Antibiotic-resistant pathogens like MRSA (methicillin-resistant *Staphylococcus aureus*), VRE (vancomycin-resistant enterococci), and ESBL (Extended-spectrum β-lactamases), using the juices of *Docynia indica* and *Emblica officinalis*, exhibited inhibition against all three tested pathogens (Punjabi et al. 2017). In this study, organic acids and polyphenols were found to be mainly associated with antimicrobial activity, although the exact mechanism is not yet comprehensively defined.

9.4.3.3 Antiinflammatory Effects

Secondary metabolites from WEFs, including anthocyanins, proanthocyanidins, polyphenols, and ursolic acid, are reported to have anti-inflammatory effects (Li et al. 2016). In a Zimbabwe-based study of *Vanguria infausta*, three different formulations—crude extract, unsaponifiable fraction, and flavonoid fraction—showed anti-inflammatory effects in both NO radical scavenging and egg albumin inhibition assays (Gwatidzo et al. 2018). Among them, crude extract and unsaponifiable fraction exhibited better results and can be developed into potential anti-inflammatory drugs in the future (Gwatidzo et al. 2018). Similarly, wild fruits collected from India, namely *Canthium parviflorum* and *Limonia acidissima*, were reported to be rich in phytoconstituents like polyphenols, flavonoids, β-carotene, tocopherol, coumarins, tannins, tyramine, and ascorbic acid, and were also documented to exhibit anti-inflammatory properties (Pratap et al. 2023). Moreover, different species of wild blueberry, such as *Rubus rosifolius*, *Rubus racemosus*, and *Rubus jamaicensis*, were shown to inhibit cyclooxygenase (COX-1 and COX-2) enzymes, nitric oxide, and inducible nitric oxide synthase (Bowen-Forbes et al. 2010, Cuevas-Rodríguez et al. 2010).

9.4.3.4 Anticancer Activity

Several studies based on different WEFs have been reported with anticancer activities against several tumor cell lines (Li et al. 2016). Brazilian wild fruits, including *Campomanesia cambessedeana*, *Pouteria guardneriana*, and *Byrsonoma verbascifolia*, have inhibited the HepG2 immortalized cancer cell lines. These fruits were found to be rich in phenolic compounds (Omena et al. 2012). Similarly, wild berries were also shown to inhibit the proliferation of different tumor cells under *in vitro* assays. *Rubus rosifolius*, rich in compounds such as hyptatic acid B and 4-epi-nigaichigoside F1, inhibited colon tumor cells; *Vaccinium angustifolium*, with proanthocyanidins, inhibited prostate and liver tumor cells; *Rubus ellipticus* in cervical tumor cells; and three species of *Fragaria* (*F. virginiana*, *F. chiloensis*, *F. xananassa*) in lung epithelial tumor cells (Li et al. 2016). Furthermore, antitumor activities have been reported with the extracts of WEFs, namely *Rubus caesium*, *Viburnum lantana*, *Crataegus monogyna*, *Crataegus tanacetifolia*, *Canthium parviflorum*, *Carissa carandus*, *Syzgium cumini*, and *Opuntia ficus-indica* (Turker et al. 2012, Pratap et al. 2023).

Several WEF extracts were tested for the inhibition of the A549 cell line (human lung adenocarcinoma cell line). A study on six WEFs of Mizoram, India, including *Averrhoa carambola*, *Eleocarpus tectorius*, *Embelia subcoriacea*, *Garcennia lanceifolia*, *Elaeagnus caudata*, and *Cissus obovata* showed significant cytotoxicity (IC_{50} 18.2–203.2 μg/mL) against the A549 cell lines, and quantification of bioactive compounds through HPLC showed the abundance of gallic acid, quercetin, rutin, hesperidin, and naringin (Hazarika et al. 2012). Similarly, methanolic extracts of seven WEFs were evaluated against the A549 cell line through an MTT assay. *Tamarindus indica* (IC_{50} of 199.77 ± 0.20 μg/mL) and *Citrus grandis* (IC_{50} of 176.63 ± 10.66 μg/mL) showed higher inhibition activity compared to other fruits (Ngurthankhumi et al. 2024).

9.4.3.5 Neuroprotective Potential

Polyphenolic phytochemicals of wild fruits with antioxidant properties, such as anthocyanins, tannins, and steroidal saponins, are particularly beneficial in neuroprotection. They reduce oxidative stress and fight infection and disease more effectively by supporting the immune system from oxidative damage and enhancing its capabilities (Achhami et al. 2025). Wild fruits, such as *Cleistocalyx nervosum*, *Rubus brigantinus*, *Rubus vagabundus*, *Prunus spinosa*, *Aegle marmelos*, *Arbutus unedo*, *Momordica cymbalaria*, *Rosa canina*, *Ziziphus mauritiana*, and *Rosa micrantha*, have been reported to contain neuroprotective potential (Pratap et al. 2023). Inhibition of acetylcholinesterase activity due to the presence of chlorogenic acid in several wild fruits, such as *Sorbus torminalis*, *Genipa americana*, *Spondia tuberosa*, and *Spondia purpurea*, was reported (Li et al. 2016, Santacruz et al. 2012).

In the same way, acetylcholinesterase inhibition has been reported in the polyphenol-rich extract of wild blueberry (*Vaccinium angustifolium*) (Papandreou et al. 2009). As the acetylcholinesterase enzyme metabolizes the neurotransmitter acetylcholine, it enhances the probability of Alzheimer's disease progression (Patel et al. 2018). Thus, the inhibition of acetylcholinesterase activity is particularly beneficial in the treatment of Alzheimer's disease (Papandreou et al. 2009, Suwardi and Navia 2023).

9.4.3.6 Other Applications

Other protective therapeutic applications of WEFs are mostly related to the protection of the digestive tract, liver, and DNA (Li et al. 2016). These bioactivities are also greatly attributed to the presence of diverse secondary metabolites (Semwal et al. 2023). Some prominent studies to explore the additional therapeutic activities are listed in Table 4.

Table 4. Bioactivities of different Wild Edible Fruits (WEFs).

Scientific Name	Bioactivity	Test performed	Major phytoconstituents	References
Aristotelia chilensis (Molina) Stuntz	Gastroprotective	Gastroprotective activity in rat's stomach mucosa	Phenolics, aglycone	(Cespedes et al. 2010)
Canarium subulatum Guillaumin	Antidiabetic	Alpha-amylase and alpha-glucosidase inhibition	Phenolics, flavonoids	(Liyanaarachchie et al. 2021)
Elaeagnus latifolia L.	DNA Protective	DNA protection assay	Catechin, purpurin, catechin, tannic acid, rutin, and reserpine	(Panja et al. 2014)
Prunus bracteopadus Koehne	Hepato protective	*In vivo* hepato-ameliorating activity test	Gallic acid, methyl gallate, purpurin, reserpine, tannic acid, rutin, catechin, and ascorbic acid	(Chaudhuri et al. 2015)
Rubus idaeus L.	Diuretic	*In vivo* diuretic test	-	(Zhang et al. 2011)

9.5 Safety Considerations and Future Directions

In the future, detailed exploration, including isolation and identification of bioactive compounds present in wild fruits, their mechanism of action, and toxicity testing in targeted therapy, is necessary to evaluate efficacy and human safety (Li et al. 2016). For instance, hepatic and renal complications were reported in the fruits of *Antidesma bunius*, due to the presence of toxic alkaloids (Chowtivannakul et al. 2016, Islam et al. 2018), and hence, in-depth studies are needed to identify the constituents responsible for toxicity (Muñoz et al. 2021).

On the other hand, antinutrients are another factor to consider in WEFs, as these compounds might interfere with nutrient absorption and sometimes cause toxicity (Rout and Basak 2014). Phytic acid, saponin, oxalate, and several alkaloids are the antinutrients recorded in WEFs that reduce the bioavailability of micronutrients and other essential chemicals (Ngurthankhumi et al. 2024). Other antinutrients and toxic compounds include oxalic acid, pyrrolizidine alkaloids, cucurbitacins, anthracene derivatives, monoterpenes, phenylpropanoids, and prenylflavonoids (González-Zamorano et al. 2025).

Considering both nutraceutical values as well as toxicity-related factors of WEFs, their sustainable utilization along with conservation is needed (Suwardi and Navia 2023). This can be established through a comprehensive approach, including germplasm preservation, genetic improvement, and integrating the WEFs with the socio-economic practices of indigenous people. Thus, domestication, increased production, and conservation might be the potential future strategies for the WEFs (Melaku and Ebrahim 2021).

9.6 Conclusion

Wild edible fruits are widely utilized as both nutrients and therapeutic agents due to the abundance and diversity of macronutrients, micronutrients, and bioactive compounds. These fruits have been traditionally explored by the local people of different geographical regions to support their food and nutritional requirements and as a source of income. Wild edible fruits not only help to diversify the local food systems and provide a healthy diet, but due to their rich nutritional content and diverse therapeutic properties, they can significantly contribute to addressing the growing burden of malnutrition in many developing countries. Notable presence of vitamins like β-carotene, ascorbic acid, niacin, folate, riboflavin, and tocopherol, as well as minerals such as Na, K, Ca, Mg, Fe, P, and Zn makes them suitable for functional foods. Bioactive phyto-constituents, such as polyphenols and flavonoids, were shown to hold therapeutic potential, like antioxidant, antimicrobial, anti-inflammatory, anti-tumor, neuroprotective, among many other health benefits. Still, the comprehensive evaluation of bioactive compounds, considering their mechanism of action, efficacy, and safety in targeted therapy, remains to be explored. Thus, recognizing the optimal nutritive and bioactive potential will help to integrate sustainable utilization practices of wild edible fruits. Incorporation of WEFs in the national nutritional, agricultural, and forest policies, plans, and programs and focused research in the area of novel nutraceutical and therapeutic product development could contribute to livelihood improvement and enhance nutritional security.

References

Abbasi, A. M., Khan, M. A., Khan, N. and Shah, M. H. (2013). Ethnobotanical survey of medicinally important wild edible fruits species used by tribal communities of Lesser Himalayas-Pakistan. Journal of Ethnopharmacology 148(2): 528–536.

Abedin, M. S., Nuruddin, M. M., Ahmed, K. U. and Hossain, A. (2012). Nutritive compositions of locally available jackfruit seeds (*Artocarpus heterophyllus*) in Bangladesh.

Achaglinkame, M. A., Aderibigbe, R. O., Hensel, O., Sturm, B. and Korese, J. K. (2019). Nutritional characteristics of four underutilized edible wild fruits of dietary interest in Ghana. Foods 8(3): 104.

Achhami, H., Pachhai, B. B., Chaudhary, S., Manandhar, P. and Bhatt, L. R. (2025). Physicochemical, nutritional, antinutritional and antioxidant properties of juice and wines from *Rhododendron arboreum* Sm. Petals. Applied Food Research 100929. https://doi.org/10.1016/j.afres.2025.100929.

Adepoju, O. T. (2009). Proximate composition and micronutrient potentials of three locally available wild fruits in Nigeria. African Journal of Agricultural Research 4(9): 887–892.

Aderinola, O., Omoyajowo, K., Babalola, O., Ogidan, J. and Amiolemen, S. (2018). Comparative nutritional analysis of Phoenix dactylifera and Phoenix reclinata seeds. Journal of Research and Review in Science (JRRS) 4(1): 10. https://doi.org/10.36108/jrrslasu/7102/40(0150).

Aganga, A. A. and Mosase, K. W. (2001). Tannin content, nutritive value and dry matter digestibility of *Lonchocarpus capassa*, *Zizyphus mucronata*, *Sclerocarya birrea*, *Kirkia acuminata* and *Rhus lancea* seeds. Animal Feed Science and Technology 91(1): 107–113. https://doi.org/10.1016/S0377-8401(01)00235-8.

Ajayi, I. A. and Adesanwo, O. (2009). Comparative study of the mineral element and fatty acid composition of Dacryodes edulis pulp and seed. World Journal of Agricultural Sciences 5(3): 279–283.

Alae-Carew, C., Nicoleau, S., Bird, F. A., Hawkins, P., Tuomisto, H. L., Haines, A. et al. (2020). The impact of environmental changes on the yield and nutritional quality of fruits, nuts and seeds: A systematic review. Environmental Research Letters: ERL 15(2): 023002.

Alpsoy, L. and Yalvac, M. E. (2011). Key roles of vitamins A, C, and E in aflatoxin B1-induced oxidative stress. Vitamins & Hormones 86: 287–305.

Anand, S. P., Deborah, S. and Velmurugan, G. (2017). Antimicrobial activity, nutritional profile and phytochemical screening of wild edible fruit of Catunaregam spinosa (Thunb.) triveng. Pharma Innovation 6(10): 106–109.

Anhwange, Asen, B., Ikyenge, B. A., Nyiatagher, D. T. and Ageh, J. T. (2010). Chemical analysis of Citrullus lanatus (Thunb.), Cucumeropsis mannii (Naud.) and Telfairia occidentalis (Hook F.) seeds oils. The Journal of Applied Sciences Research.

Anwar, R., Borbi, M. and Rakha, A. (2024). Significance and the use of legumes in developing weaning foods with a balanced nutrition—a review. Legume Science 6(3): e249. https://doi.org/10.1002/leg3.249.

Asfaw, A., Lulekal, E., Bekele, T., Debella, A., Tessema, S., Meresa, A. et al. (2023). Ethnobotanical study of wild edible plants and implications for food security. Trees, Forests and People 14: 100453. https://doi.org/10.1016/j.tfp.2023.100453.

Ayessou, N. C., Ndiaye, C., Cissé, M., Gueye, M., Sakho, M. and Dornier, M. (2014). Nutrient composition and nutritional potential of wild fruit *Dialium guineense*. Journal of Food Composition and Analysis 34(2): 186–191. https://doi.org/10.1016/j.jfca.2014.01.002.

Bahukhandi, A., Barola, A. and Sekar, K. C. (2020). Antioxidant activity and polyphenolics of Fragaria nubicola: A Wild Edible Fruit Species of Himalaya. Proceedings of the National Academy of Sciences, India Section B: Biological Sciences 90(4): 761–767. https://doi.org/10.1007/s40011-019-01142-5.

Bajramova, A. and Spégel, P. (2022). A comparative study of the fatty acid profile of common fruits and fruits claimed to confer health benefits. Journal of Food Composition and Analysis 112: 104657.

Bala, A. (2021). Assessment of iron bioaccessibility in Karonda (Carissa spp.) and its products [PhD Thesis, Punjab Agricultural University, Ludhiana]. https://krishikosh.egranth.ac.in/server/api/core/bitstreams/b693e797-0738-4d72-a5d6-ff8d79ef6cc4/content.

Banwo, K., Olojede, A. O., Adesulu-Dahunsi, A. T., Verma, D. K., Thakur, M., Tripathy, S. et al. (2021). Functional importance of bioactive compounds of foods with Potential Health Benefits: A review on recent trends. Food Bioscience 43: 101320.

Bayang, J. P., Laya, A., Kolla, M. C. and Koubala, B. B. (2021). Variation of physical properties, nutritional value and bioactive nutrients in dry and fresh wild edible fruits of twenty-three species from Far North region of Cameroon. Journal of Agriculture and Food Research 4: 100146.

Benhura, C. M. A. N., Muchuweti, M. A. N. and Nyagura, M. (2012). Proximate analysis of Parinari curatellifolia fruit pulp of fruit from parts of Harare and a rural area in Zimbabwe. Pakistan Journal of Nutrition 11(7): 541–544.

Bernhoft, A. (2010). A brief review on bioactive compounds in plants. Bioactive Compounds in Plants-Benefits and Risks for Man and Animals 50: 11–17.

Bhandari, R., Achhami, H., Pachhai, B. B., Chaudhary, S., Pant, S. P. and Bhatt, L. R. (2025). Physicochemical, proximate composition, and antioxidant activity of fresh and fermented juices of selected wild edible fruits from the Gulmi District, Western Nepal. LWT 118307.

Bhatt, I. D., Rawat, S., Badhani, A. and Rawal, R. S. (2017). Nutraceutical potential of selected wild edible fruits of the Indian Himalayan region. Food Chemistry 215: 84–91.

Bhatt, L. R. (2021). Wild Edible Fruits of Nepal. Nepal Academy of Science and Technology, Khumaltar, Lalitpur, Nepal, pp 252.

Bhusal, S., Pant, D. R., Joshi, G. P., Adhikari, M., Raut, J. K., Pandey, M. R. et al. (2020). Antioxidant activity and nutraceutical potential of selected Nepalese wild edible fruits. Scientific World 13(13): 8–13.

Bhuyan, D. J., Alsherbiny, M. A., Perera, S., Low, M., Basu, A., Devi, O. A. et al. (2019). The odyssey of bioactive compounds in avocado (Persea americana) and their health benefits. Antioxidants 8(10): 426.

Biswas, S. C., Dutta, P., Barua, P., Misra, T. K. and Das, S. (2021). Ethnobotanical study of underutilized wild edible fruits and their antibacterial activity against human pathogens. Journal of Pure & Applied Microbiology 15(4): 1976–1986. doi: 10.22207/JPAM.15.4.19.

Borelli, T., Güzelsoy, N. A., Hunter, D., Tan, A., Karabak, S., Uçurum, H. Ö. et al. (2022). Assessment of the nutritional value of selected wild food plants in Türkiye and their promotion for improved nutrition. Sustainability 14(17): 11015.

Bowen-Forbes, C. S., Zhang, Y. and Nair, M. G. (2010). Anthocyanin content, antioxidant, anti-inflammatory and anticancer properties of blackberry and raspberry fruits. Journal of Food Composition and Analysis 23(6): 554–560.

Brooks, R., Goldson-Barnaby, Andrea and Bailey, D. (2020). Nutritional and medicinal properties of Phyllanthus Acidus L. (Jimbilin). International Journal of Fruit Science 20(sup3): S1706–S1710. https://doi.org/10.1080/15538362.2020.1830013.

Calder, P. C. (2015). Functional roles of fatty acids and their effects on human health. Journal of Parenteral and Enteral Nutrition 39(1S). https://doi.org/10.1177/0148607115595980.

Cámara, M., Fernández-Ruiz, V. and Ruiz-Rodríguez, B. M. (2016). Wild edible plants as sources of carotenoids, fibre, phenolics and other non-nutrient bioactive compounds. pp. 187–205. *In*: Sánchez-Mata, M. D. C. and Tardío, J. (eds.). Mediterranean Wild Edible Plants. Springer New York. https://doi.org/10.1007/978-1-4939-3329-7_9.

Carlos Ruiz-Ruiz, J., Ramón-Sierra, J., Arias-Argaez, C., Magaña-Ortiz, D. and Ortiz-Vázquez, E. (2017). Antibacterial activity of proteins extracted from the pulp of wild edible fruit of *Bromelia pinguin* L. International Journal of Food Properties 20(1): 220–230. https://doi.org/10.1080/10942912.2016.1154572.

Cechinel-Filho, V. (2012). Plant Bioactives and Drug Discovery: Principles, Practice, and Perspectives. John Wiley & Sons.

Cespedes, C. L., Alarcon, J., Avila, J. G. and El-Hafidi, M. (2010). Anti-inflammatory, antioedema and gastroprotective activities of Aristotelia chilensis extracts, part 2. Boletín Latinoamericano y Del Caribe de Plantas Medicinales y Aromáticas 9(6): 432–439.

Chaka, B. and Osano, A. M. (2019). Analysis of selected nutrient levels at different growth stages of Dovyalis caffra (Kei-apple) fruits.

Chaudhary, S., Achhami, H., Rai, E., Shyaula, M., Ghimire, M., Adhikari, A. et al. (2025). Diversity and spatiotemporal trends of wild edible plants in Nepal: a review on wild edible plants contribution to food security. Flora 152853.

Chakravarty, S., Bhutia, K. D., Suresh, C. P., Shukla, G. and Pala, N. A. (2016). A review on diversity, conservation and nutrition of wild edible fruits. Journal of Applied and Natural Science 8(4): 2346.

Chalise, J. P., Acharya, K., Gurung, N., Bhusal, R. P., Gurung, R., Skalko-Basnet, N. et al. (2010). Antioxidant activity and polyphenol content in edible wild fruits from Nepal. International Journal of Food Sciences and Nutrition 61(4): 425–432. https://doi.org/10.3109/09637481003591590.

Chandler, S. (1995). The nutritional value of bananas. pp. 468–480. *In*: Gowen, S. (ed.), Bananas and Plantains. Springer Netherlands. https://doi.org/10.1007/978-94-011-0737-2_16.

Chaudhuri, D., Ghate, N. B., Panja, S., Das, A. and Mandal, N. (2015). Wild edible fruit of Prunus nepalensis Ser.(Steud), a potential source of antioxidants, ameliorates iron overload-induced hepatotoxicity and liver fibrosis in mice. PloS One 10(12): e0144280.

Choudhary, M. and Grover, K. (2019). Palm (Elaeis guineensis Jacq.) Oil. pp. 789–802. *In*: Ramadan, M. F. (ed.). Fruit Oils: Chemistry and Functionality. Springer International Publishing. https://doi.org/10.1007/978-3-030-12473-1_42.

Chowtivannakul, P., Srichaikul, B. and Talubmook, C. (2016). Hypoglycemic and hypolipidemic effects of seed extract from Antidesma bunius (L.) Spreng in streptozotocin-induced diabetic rats. Pakistan Journal of Biological Sciences: PJBS 19(5): 211–218.

Compaoré, M. and Kiendrebeogo, M. (2018). Free radical scavenging capacity and anti-biofilm potentiality of six wild edible fruits from Burkina Faso. Int. J. Curr. Microbiol. App. Sci. 7(9): 2085–2093.

Cosmulescu, S., Trandafir, I. and Nour, V. (2017). Phenolic acids and flavonoids profiles of extracts from edible wild fruits and their antioxidant properties. International Journal of Food Properties 20(12): 3124–3134. https://doi.org/10.1080/10942912.2016.1274906.

Cowan, A. K. (2017). Occurrence, metabolism, transport and function of seven-carbon sugars. Phytochemistry Reviews 16(1): 137–157. https://doi.org/10.1007/s11101-016-9463-3.

Cristina, R. and Jordao, A. A. (2024). Wild edible fruits: sources of bioactive compounds and food sovereignty in brazil–a short-review. Current Trends in Food Science 1(1).

Cuevas-Rodríguez, E. O., Dia, V. P., Yousef, G. G., García-Saucedo, P. A., López-Medina, J., Paredes-López, O. et al. (2010). Inhibition of pro-inflammatory responses and antioxidant capacity of Mexican Blackberry (Rubus spp.) extracts. Journal of Agricultural and Food Chemistry 58(17): 9542–9548. https://doi.org/10.1021/jf102590p.

Dawadi, P., Shrestha, R., Mishra, S., Bista, S., RAUT, J. K., Joshi, T. P. et al. (2022). Nutritional value and antioxidant properties of Viburnum mullaha Buch.-Ham. Ex D. Don fruit from central Nepal. Turkish Journal of Agriculture and Forestry 46(5): 781–789.

Dawadi, P., Siddiqui, M. A., Belbase, S., Syangtan, G., Kronenberg, B., Rana, K. et al. (2023). Characterizing nutritional, antioxidant and antimicrobial values of Diploknema butyracea (Roxburgh) H. J. Lam from the Chepang Community, Makwanpur, Nepal. Nepal Journal of Biotechnology 11(2): Article 2. https://doi.org/10.54796/njb.v11i2.292.

de Almeida Caetano, R., Santos, É., Poian, R., Carvalho, A., Silva, R. and Medeiros, P. (2023). Wild food plants with the potential to improve food and nutrition security may be threatened by timber extraction: A systematic review of the Brazilian context. Ethnobiology and Conservation 12. https://doi.org/10.15451/ec2023-07-12.15-1-35.

Deshmukh, B. S. and Waghmode, A. (2011). Role of wild edible fruits as a food resource: Traditional knowledge. International Journal of Pharmacy & Life Sciences 2(7).

Dreher, M. L. (2018). Whole fruits and fruit fiber emerging health effects. Nutrients 10(12): 1833.

Duguma, H. T. (2020). Wild edible plant nutritional contribution and consumer perception in Ethiopia. International Journal of Food Science 2020: 1–16. https://doi.org/10.1155/2020/2958623.

Dulal, K., Chaudhary, S., Uprety, Y., Shrestha, N., Shakya, S. and Munankarmi, N. (2022). Ethnomedicinal plants used by the local people of Changunarayan Municipality, Bhaktapur, Nepal. Ethnobotany Research and Applications 23: 1–27.

Efimtseva, E. A. and Chelpanova, T. I. (2020). Apples as a source of soluble and insoluble dietary fibers: effect of dietary fibers on appetite. Human Physiology 46(2): 224–234. https://doi.org/10.1134/S036211972002005X.

Egea, I., Sánchez-Bel, P., Romojaro, F. and Pretel, M. T. (2010). Six edible wild fruits as potential antioxidant additives or nutritional supplements. Plant Foods for Human Nutrition 65(2): 121–129. https://doi.org/10.1007/s11130-010-0159-3.

El-Sohaimy, S. A. and Hafez, E. E. (2010). Biochemical and nutritional characterizations of date palm fruits (Phoenix dactylifera L.). J. Appl. Sci. Res. 6(6): 1060–1067.

FAO. (2019). The state of food and agriculture: Moving forward sustainably. Food and Agriculture Organization of the United Nations. https://goodfoodsrilanka.lk/wp-content/uploads/2023/01/The-state-of-FOOD-AND-AGRICULTURE.pdf.

Fernandes, A., Mateus, N. and de Freitas, V. (2023). Polyphenol-dietary fiber conjugates from fruits and vegetables: nature and biological fate in a food and nutrition perspective. Foods 12(5): Article 5. https://doi.org/10.3390/foods12051052.

Ferreira, I. C., Morales, P. and Barros, L. (2017). Wild Plants, Mushrooms and Nuts: Functional Food Properties and Applications. John Wiley & Sons.

Gairola, S., Singh, K., Kumar, P., Kumar, B. and Sharma, Y. P. (2021). The wild edible plants of Paddar valley, Jammu division, Jammu and Kashmir, India. Ethnobotany Research and Applications 22: 1–21.

Garekae, H. and Shackleton, C. M. (2020). Foraging wild food in urban spaces: The contribution of wild foods to urban dietary diversity in South Africa. Sustainability 12(2): 678.

González-Zamorano, L., Cámara, R. M., Morales, P. and Cámara, M. (2025). Harnessing edible wild fruits: sustainability and health aspects. Nutrients 17(3): 412. https://doi.org/10.3390/nu17030412.

Goosen, N. J., Oosthuizen, D., Stander, M. A., Dabai, A. I., Pedavoah, M.-M. and Usman, G. O. (2018). Phenolics, organic acids and minerals in the fruit juice of the indigenous African sourplum (*Ximenia caffra*, Olacaceae). South African Journal of Botany 119: 11–16. https://doi.org/10.1016/j.sajb.2018.08.008.

Guo, Z., Jia, X., Zheng, Z., Lu, X., Zheng, Y., Zheng, B. et al. (2018). Chemical composition and nutritional function of olive (Olea europaea L.): A review. Phytochemistry Reviews 17(5): 1091–1110. https://doi.org/10.1007/s11101-017-9526-0.

Gurashi, N., Kordofanip, M., Abdelgadir, K. and Salih, A. (2016). Variation in chemical composition of Baobab (Adansonia digitata L) fruits pulp in relation to fruit shape types and locations. International Journal of Applied Science and Engineering Research 2: 106–119.

Gwatidzo, L., Chowe, L., Musekiwa, C. and Mukaratirwa-Muchanyereyi, N. (2018). *In vitro* anti-inflammatory activity of Vangueria infausta: An edible wild fruit from Zimbabwe. African Journal of Pharmacy and Pharmacology 12(13): 168–175.

Hazarika, T. K., Lalramchuana and Nautiyal, B. P. (2012). Studies on wild edible fruits of Mizoram, India used as ethno-medicine. Genetic Resources and Crop Evolution 59(8): 1767–1776. https://doi.org/10.1007/s10722-012-9799-5.

Hegazy, A. K., Al-Rowaily, S. L., Faisal, M., Alatar, A. A., El-Bana, M. I. and Assaeed, A. M. (2013). Nutritive value and antioxidant activity of some edible wild fruits in the Middle East. Journal of Medicinal Plant Research 7(15): 938–946.

Hewlings, S. (2020). Coconuts and health: Different chain lengths of saturated fats require different consideration. Journal of Cardiovascular Development and Disease 7(4): 59.

Hui, C. Y. (2020). Effects of fiber-degrading enzymatic treatments on structural and functional properties of pink guava (*Psidium guajava* linn.) Wastes [PhD Thesis, UTAR]. http://eprints.utar.edu.my/id/eprint/5247.

Ishaku, G. A., Ardo, B. P., Abubakar, H. and Peingurta, F. A. (2016). Nutritional Composition of Tamarindus indica fruit pulp. J. Chem. Sci. 6(8): 695–699.

Islam, S., Ahammed, M. S., Sukorno, F. I., Koly, S. F., Biswas, M. M. and Hossain, S. (2018). A review on phytochemical and pharmacological potentials of Antidesma bunius. J. Anal. Pharm. Res. 7(5): 602–604.

Islary, A., Sarmah, J. and Basumatary, S. (2018). Anti-bacterial property and anti-nutritional contents of five wild fruits of Assam, India. Journal of Applied Pharmaceutical Science 8(12): 170–175.

Jaakkola, M., Korpelainen, V., Hoppula, K. and Virtanen, V. (2012). Chemical composition of ripe fruits of *Rubus chamaemorus* L. grown in different habitats. Journal of the Science of Food and Agriculture 92(6): 1324–1330. https://doi.org/10.1002/jsfa.4705.

Jabłońska-Ryś, E., Zalewska-Korona, M. and Kalbarczyk, J. (2009). Antioxidant capacity, ascorbic acid and phenolics content in wild edible fruits. J. Fruit Ornam. Plant Res. 17(2): 115–120.

Jain, C., Khatana, S. and Vijayvergia, R. (2019). Bioactivity of secondary metabolites of various plants: A review. Int. J. Pharm. Sci. Res. 10(2): 494–504.

Jiru, N. A., Fekadu Gemede, Habtamu and Keyata, E. O. (2023). Nutritional composition and antioxidant properties of selected underutilized wild edible fruits in East Wollega Zone, Western Ethiopia. International Journal of Fruit Science 23(1): 34–45. https://doi.org/10.1080/15538362.2023.2166649.

Joseph, K. S., Bolla, S., Joshi, K., Bhat, M., Naik, K., Patil, S. et al. (2017). Determination of chemical composition and nutritive value with fatty acid compositions of African Mangosteen (Garcinia Livingstonei). Erwerbs-Obstbau 59(3): 195–202. https://doi.org/10.1007/s10341-016-0311-9.

Ju, Y., Zhuo, J., Liu, B. and Long, C. (2013). Eating from the wild: Diversity of wild edible plants used by Tibetans in Shangri-la region, Yunnan, China. Journal of Ethnobiology and Ethnomedicine 9(1): 28. https://doi.org/10.1186/1746-4269-9-28.

Jurgiel-Małecka, G., Gibczyńska, M., Siwek, H. and Buchwał, A. (2017). Comparison of fruits chemical composition of selected cultivars wild strawberry (Fragaria vesca L.). Eur. J. Hortic. Sci. 82(4): 204–210.

Kalt, W., Cassidy, A., Howard, L. R., Krikorian, R., Stull, A. J., Tremblay, F. et al. (2020). Recent research on the health benefits of blueberries and their anthocyanins. Advances in Nutrition 11(2): 224–236.

Kapoor, B., Kapoor, D., Gautam, S., Singh, R. and Bhardwaj, S. (2021). Dietary polyunsaturated fatty acids (PUFAs): Uses and potential health benefits. Current Nutrition Reports 10(3): 232–242. https://doi.org/10.1007/s13668-021-00363-3.

Karasawa, M. M. G. and Mohan, C. (2018). Fruits as prospective reserves of bioactive compounds: a review. Natural Products and Bioprospecting 8(5): 335–346. https://doi.org/10.1007/s13659-018-0186-6.

Karki, N., Achhami, H., Pachhai, B. B., Bhattarai, S., Shahi, D. K., Bhatt, L. R. et al. (2024). Evaluating citrus juice: A comparative study of physicochemical, nutraceutical, antioxidant, and antimicrobial properties of citrus juices from Nepal. Heliyon 10(23). https://doi.org/10.1016/j.heliyon.2024.e40773.

Kashyap, K., Hait, M., Roymahapatra, G. and Vaishnav, M. M. (2022). Proximate and elemental analysis of Careya arborea Roxb plant's root. ES Food & Agroforestry 7(8): 41–47.

Kaurinovic, B. and Vastag, D. (2019). Flavonoids and phenolic acids as potential natural antioxidants. Antioxidants 2(1): 1–14.

Khan, M. P. Z., Ahmad, M., Zafar, M., Sultana, S., Ali, M. I. and Sun, H. (2015). Ethnomedicinal uses of edible wild fruits (EWFs) in Swat Valley, Northern Pakistan. Journal of Ethnopharmacology 173: 191–203.

Khomdram, S. and Guruaribam, S. (2010). Determination of antioxidant activity and vitamin C of some wild fruits of Manipur. Bioscan, 5.

Knox, Y. M., Suzutani, T., Yosida, I. and Azuma, M. (2003). Anti-influenza virus activity of crude extract of *Ribes nigrum* L. Phytotherapy Research 17(2): 120–122. https://doi.org/10.1002/ptr.1053.

Komor, E. (2000). Source physiology and assimilate transport: The interaction of sucrose metabolism, starch storage and phloem export in source leaves and the effects on sugar status in phloem. Functional Plant Biology 27(6): 497–505.

Koubala, B. B., Bayang, J. P., Wangso, H., Kolla, M. C. and Laya, A. (2021). Variation of Phenolics (Bound and Free), minerals, and antioxidant activity of twenty-eight wild edible fruits of twenty-three species from Far North Region of Cameroon. BioMed Research International 2021(1): 4154381. https://doi.org/10.1155/2021/4154381.

Kris-Etherton, P. M. (1999). Monounsaturated fatty acids and risk of cardiovascular disease. Circulation 100(11): 1253–1258. https://doi.org/10.1161/01.CIR.100.11.1253.

Lamien-Meda, A., Lamien, C. E., Compaoré, M. M., Meda, R. N., Kiendrebeogo, M., Zeba, B. et al. (2008). Polyphenol content and antioxidant activity of fourteen wild edible fruits from Burkina Faso. Molecules 13(3): 581–594.

Lara-Arevalo, J. (2023). Nutrient-Rich Foods in Western African Food Supply: Applying Nutrient Profiling Models to the FAO Food Composition Table for Western Africa (WAFCT 2019). University of Washington.

Lau, F. C., Bagchi, M., Zafra-Stone, S. and Bagchi, D. (2009). The benefits of antioxidant-rich fruits on skin health. pp. 217–232. In Nutritional Cosmetics. Elsevier.

Lebaka, V. R., Wee, Y.-J., Ye, W. and Korivi, M. (2021). Nutritional composition and bioactive compounds in three different parts of mango fruit. International Journal of Environmental Research and Public Health 18(2): 741.

Levey, D. J., Bissell, H. A. and O'Keefe, S. F. (2000). Conversion of nitrogen to protein and amino acids in wild fruits. Journal of Chemical Ecology 26(7): 1749–1763. https://doi.org/10.1023/A:1005503316406.

Li, Y., Zhang, J.-J., Xu, D.-P., Zhou, T., Zhou, Y., Li, S. et al. (2016). Bioactivities and health benefits of wild fruits. International Journal of Molecular Sciences 17(8): 1258.

Liyanaarachchie, C. P. T., Gunatilake, M., Jayasinghe, L. and Bandara, B. M. R. (2021). *In vitro* antidiabetic and *in vivo* hypoglycaemic activities and toxicity of *Canarium zeylanicum* Bark extracts. Journal of Biologically Active Products from Nature 11(3): 242–253. https://doi.org/10.1080/22311866.2021.1916595.

Lunn, J. and Buttriss, J. L. (2007). Carbohydrates and dietary fibre. Nutrition Bulletin 32(1): 21–64. https://doi.org/10.1111/j.1467-3010.2007.00616.x.

Magaia, T., Uamusse, A., Sjöholm, I. and Skog, K. (2013). Proximate analysis of five wild fruits of Mozambique. The Scientific World Journal 2013(1): 601435. https://doi.org/10.1155/2013/601435.

Mahapatra, A. K., Mishra, S., Basak, U. C. and Panda, P. C. (2012). Nutrient analysis of some selected wild edible fruits of deciduous forests of India: An explorative study towards non conventional bio-nutrition. Advance Journal of Food Science and Technology 4(1): 15–21.

Maikhuri, R. K., Parshwan, Dalbeer S., Kewlani, Pushpa, Negi, Vikram S., Rawat, Sandeep and Rawat, L. S. (2021). Nutritional composition of seed kernel and oil of Wild Edible Plant Species from Western Himalaya, India. International Journal of Fruit Science 21(1): 609–618. https://doi.org/10.1080/15538362.2021.1907009.

Maliehe, S. T. (2015, January 1). An evaluation of nutraceutical components of syzygium cordatum fruits for the treatment of gastrointestinal tract infections. SciSpace - Paper; University of Zululand.

Mateus-Reguengo, L., Barbosa-Pereira, L., Rembangouet, W., Bertolino, M., Giordano, M., Rojo-Poveda, O. et al. (2020). Food applications of *Irvingia gabonensis* (Aubry-Lecomte ex. O'Rorke) Baill., the 'bush mango': A review. Critical Reviews in Food Science and Nutrition 60(14): 2446–2459. https://doi.org/10.1080/10408398.2019.1646704.

Mbhele, Z., Zharare, G. E., Zimudzi, C., Mchunu, C. N. and Ntuli, N. R. (2025). Variation in nutritional composition of Strychnos spinosa Lam. Morphotypes in KwaZulu-Natal, South Africa. Genetic Resources and Crop Evolution 72(1): 359–373. https://doi.org/10.1007/s10722-024-01982-9.

Mehmood, M. H., Anila, N., Begum, S., Syed, S. A., Siddiqui, B. S. and Gilani, A.-H. (2014). Pharmacological basis for the medicinal use of Carissa carandas in constipation and diarrhea. Journal of Ethnopharmacology 153(2): 359–367.

Melaku, A. and Ebrahim, M. A. (2021). Critical review on wild-edible fruit species in Ethiopia. International Journal of Forestry Research 2021: 1–12. https://doi.org/10.1155/2021/8538188.

Molino, S., De Lellis, L. F., Morone, M. V., Cordara, M., Larsen, D. S., Piccinocchi, R. et al. (2025). Improving Irritable Bowel Syndrome (IBS) symptoms and quality of life with quebracho and chestnut tannin-based supplementation: a single-centre, randomised, double-blind, placebo-controlled clinical trial. Nutrients 17(3): 552.

Moodley, R., Koorbanally, N. and Jonnalagadda, S. B. (2012). Elemental composition and fatty acid profile of the edible fruits of Amatungula (Carissa macrocarpa) and impact of soil quality on chemical characteristics. Analytica Chimica Acta 730: 33–41.

Morales, P., Ferreira, I. C. F. R., Carvalho, A. M., Fernández-Ruiz, V., Sánchez-Mata, M. C., Cámara, M. et al. (2013). Wild edible fruits as a potential source of phytochemicals with capacity to inhibit lipid peroxidation. European Journal of Lipid Science and Technology 115(2): 176–185. https://doi.org/10.1002/ejlt.201200162.

Morris, A. L. and Mohiuddin, S. S. (2020). Biochemistry, nutrients. https://europepmc.org/books/nbk554545.

Mothapo, M. J., Mafeo, T. P. and Mamphiswana, N. D. (2014). Physico-Chemical Properties and Selected Nutritional Components of Wild Medlar (*Vangueria infausta*) Fruit Harvested at Two Harvesting Times.

Motti, R. (2022). Wild edible plants: a challenge for future diet and health. Plants 11(3): Article 3. https://doi.org/10.3390/plants11030344.

Muhammad, N., Luo, Z., Yang, M., Liu, Z. and Liu, M. (2022). The nutritional, medicinal, and drought-resistance properties of Ziziphus Mill. Make it an important candidate for alleviating food insecurity in arid regions—A case of Pakistan. Horticulturae 8(10): 867.

Mukhtar, E. (2016). The diversity of wild edible fruit plants and traditional knowledge in West Aceh region, Indonesia. Environment 21: 28.

Munankarmi, N. N., Chaudhary, S., Neupane, S., Shyaula, S. L., Zhang, W. and Gauchan, D. P. (2025). Ethnomedicinal uses of plant resources in Bethanchowk Rural Municipality of Kavrepalanchowk District, Central Nepal. Ethnobotany Research and Applications 30: 1–47.

Muñoz, M. N. M., Alvarado, U. G., Reyes, J. I. L. and Watanabe, K. (2021). Acute oral toxicity assessment of ethanolic extracts of Antidesma bunius (L.) Spreng fruits in mice. Toxicology Reports 8: 1289–1299.

Murillo, E., Britton, G. B. and Durant, A. A. (2012). Antioxidant activity and polyphenol content in cultivated and wild edible fruits grown in Panama. Journal of Pharmacy and Bioallied Sciences 4(4): 313–317.

Muthai, K. U., Karori, M. S., Muchugi, A., Indieka, A. S., Dembele, C., Mng'omba, S. et al. (2017). Nutritional variation in baobab (Adansonia digitata L.) fruit pulp and seeds based on Africa geographical regions. Food Science & Nutrition 5(6): 1116–1129. https://doi.org/10.1002/fsn3.502.

Mwamatope, B., Chikowe, I., Tembo, D. T., Kamanula, J. F., Masumbu, F. F. F. and Kumwenda, F. D. (2023). Phytochemical composition and antioxidant activity of edible wild fruits from Malawi. BioMed Research International 2023(1): 2621434. https://doi.org/10.1155/2023/2621434.

Napoli, J. L. (2016). Functions of intracellular retinoid binding-proteins. pp. 21–76. *In*: Asson-Batres, M. A. and Rochette-Egly, C. (eds.). The Biochemistry of Retinoid Signaling II: The Physiology of Vitamin A - Uptake, Transport, Metabolism and Signaling. Springer Netherlands. https://doi.org/10.1007/978-94-024-0945-1_2.

Nawaz, M., Warriach, A., Muhammad, U., Akram, M., Ashfaq, F. and Raza, A. (2015). Proximate Composition of Apricot (*Prunus armeniaca*) Fruit and Kernel. https://doi.org/10.5829/idosi.aejaes.2015.15.10.12810.

Nayak, J. and Basak, U. C. (2015). Analysis of some nutritional properties in eight wild edible fruits of Odisha, India. Int. J. Curr. Sci. 14: 55–62.

Ncube, N. S., Afolayan, A. J. and Okoh, A. I. (2008). Assessment techniques of antimicrobial properties of natural compounds of plant origin: Current methods and future trends. African Journal of Biotechnology 7(12). https://www.ajol.info/index.php/ajb/article/view/58804.

Newman, D. J., Cragg, G. M. and Snader, K. M. (2000). The influence of natural products upon drug discovery. Natural Product Reports 17(3): 215–234.

Ngemakwe, P. N., Remize, F., Thaoge, M. L. and Sivakumar, D. (2017). Phytochemical and nutritional properties of underutilised fruits in the southern African region. South African Journal of Botany 113: 137–149.

Ngurthankhumi, R., Hazarika, T. K., Lalruatsangi, E., Lalhmachhuani, H. and Debbarma, P. (2024). Anticancer screening of few wild edible fruits in Mizoram, Northeast India using MTT Assay. Indian Journal of Agricultural Research 1: 6.

Nicolétis, É., Caron, P., El Solh, M., Cole, M., Fresco, L. O., Godoy-Faúndez, A. et al. (2019). Agroecological and other innovative approaches for sustainable agriculture and food systems that enhance food security and nutrition. A report by the High Level Panel of Experts on Food Security and Nutrition of the Committee on World Food Security. https://agritrop.cirad.fr/604473/1/604473.pdf.

Nikolaeva-Glomb, L., Mukova, L., Nikolova, N., Badjakov, I., Dincheva, I., Kondakova, V. et al. (2014). *In vitro* antiviral activity of a series of wild berry fruit extracts against representatives of *Picorna-, Orthomyxo*—and *Paramyxoviridae*. Natural Product Communications 9(1): 1934578X1400900116. https://doi.org/10.1177/1934578X1400900116.

Ogunlaja, O. O., Moodley, R., Baijnath, H. and Jonnalagadda, S. B. (2020). Elemental distribution and health risk assessment of the edible fruits of two ficus species, Ficus sycomorus L. and Ficus burtt-davyi Hutch. Biological Trace Element Research 198(1): 303–314. https://doi.org/10.1007/s12011-020-02048-4.

Ojha Khatri, S., Chaudhary, S., Shrestha, N. and Munankarmi, N. N. (2021). Ethnomedicinal study and phytochemical screening of selected plants in Jhule, Dolakha District, Nepal. Vegetos 34(4): 834–846. https://doi.org/10.1007/s42535-021-00266-2.

Omena, C. M. B., Valentim, I. B., Guedes, G. da S., Rabelo, L. A., Mano, C. M., Bechara, E. J. H. et al. (2012). Antioxidant, anti-acetylcholinesterase and cytotoxic activities of ethanol extracts of peel, pulp and seeds of exotic Brazilian fruits: Antioxidant, anti-acetylcholinesterase and cytotoxic activities in fruits. Food Research International 49(1): 334–344.

Padulosi, S., Thompson, J. and Rudebjer, P. G. (2013). Fighting poverty, hunger and malnutrition with neglected and underutilized species: Needs, challenges and the way forward. https://cgspace.cgiar.org/items/30d45d24-eeac-49d5-93cb-3d3188143dbb.

Pandey, Y., Upadhyay, S., Bhatt, S. S., Sharma, L., Manivannan, S. and Chanbisana, C. (2018). Nutritional compositions of Baccaurea sapida and Eleaocarpus sikkimnesis of Sikkim Himalaya. Int. J. Curr. Microbiol. App. Sci, 7(2): 2101–2106.

Panja, S., Chaudhuri, D., Ghate, N. B., Le Minh, H. and Mandal, N. (2014). *In vitro* assessment of phytochemicals, antioxidant and DNA protective potential of wild edible fruit of Elaeagnus latifolia Linn. Fruits 69(4): 303–314.

Panta, S., Parajulee, D., Subedi, G. and Giri, B. (2021). Ethnobotanical study of wild edible plants in Pyuthan, Nepal. International Journal of Environment, Agriculture and Biotechnology 6(4): 079–087. DOI: 10.22161/ijeab.64.10.

Papandreou, M. A., Dimakopoulou, A., Linardaki, Z. I., Cordopatis, P., Klimis-Zacas, D., Margarity, M. et al. (2009). Effect of a polyphenol-rich wild blueberry extract on cognitive performance of mice, brain antioxidant markers and acetylcholinesterase activity. Behavioural Brain Research 198(2): 352–358.

Patel, S. S., Raghuwanshi, R., Masood, M., Acharya, A. and Jain, S. K. (2018). Medicinal plants with acetylcholinesterase inhibitory activity. Reviews in the Neurosciences 29(5): 491–529. https://doi.org/10.1515/revneuro-2017-0054.

Patil, R. P., Pai, S. R., Pawar, N. V., Shimpale, V. B., Patil, R. M. and Nimbalkar, M. S. (2012). Chemical Characterization, mineral analysis, and antioxidant potential of two underutilized berries (*Carissa carandus* and *Eleagnus conferta*) from the Western Ghats of India. Critical Reviews in Food Science and Nutrition 52(4): 312–320. https://doi.org/10.1080/10408398.2010.500227.

Patra, A. K. (2012). An overview of antimicrobial properties of different classes of phytochemicals. pp. 1–32. *In*: Patra, A. K. (ed.). Dietary Phytochemicals and Microbes. Springer Netherlands. https://doi.org/10.1007/978-94-007-3926-0_1.

Perera, S., Silva, A. B. G., Amarathunga, Y., De Silva, S., Jayatissa, R., Gamage, A. et al. (2022). Nutritional composition and antioxidant activity of selected underutilized fruits grown in Sri Lanka. Agronomy 12(5): Article 5. https://doi.org/10.3390/agronomy12051073.

Prakash, D., Upadhyay, G., Gupta, C., Pushpangadan, P. and Singh, K. K. (2012). Antioxidant and free radical scavenging activities of some promising wild edible fruits. International Food Research Journal 19(3).

Pratap, G. K., Shantaram, M., Raghavendra, H. L., Kumar, S., Harshini, V. K. and Poornima, D. V. (2023). A comprehensive analysis of wild edible fruits from the Rangayyanadurga four-horned antelope wildlife sanctuary. Biomedicine 43(02): 545–553.

Punchay, K., Inta, A., Tiansawat, P., Balslev, H. and Wangpakapattanawong, P. (2020). Traditional knowledge of wild food plants of Thai Karen and Lawa (Thailand). Genetic Resources and Crop Evolution 67(5): 1277–1299. https://doi.org/10.1007/s10722-020-00910-x.

Punjabi, K., Chitalia, V. K., Mukadam, T., Sharma, J., Maniar, J., Vasave, S. et al. (2017). Evaluation of the antimicrobial activity of gooseberry and wild apple fruit juices of the northeast region of India. Journal of Medicinal Herbs and Ethnomedicine 3: 1–7.

Rai, E. (2024). Evaluation of Nutrient Content, Antioxidant Potential and Antimicrobial Activity of Selected Wild Edible Fruits [PhD Thesis]. https://elibrary.tucl.edu.np/bitstreams/86564e52-54e1-4210-a44a-4349f3451023/download.

Rana, Y. S., Tiwari, O. P., Krishan, R. and Sharma, C. M. (2018). Determination of nutritional potential of five important wild edible fruits traditionally used in Western Himalaya. Int. J. Life Sci. 6(1): 79–86.

Ranasinghe, R. A. S. N., Maduwanthi, S. D. T. and Marapana, R. A. U. J. (2019). Nutritional and health benefits of Jackfruit (*Artocarpus heterophyllus* Lam.): A review. International Journal of Food Science 2019: 1–12. https://doi.org/10.1155/2019/4327183.

Rawat, S., Jugran, A., Giri, L., Bhatt, I. D. and Rawal, R. S. (2011). Assessment of antioxidant properties in fruits of *Myrica esculenta*: A popular wild edible species in Indian Himalayan Region. Evidence-Based Complementary and Alternative Medicine 2011(1): 512787. https://doi.org/10.1093/ecam/neq055.

Riaz, M., Khalid, R., Afzal, M., Anjum, F., Fatima, H., Zia, S. et al. (2023). Phytobioactive compounds as therapeutic agents for human diseases: A review. Food Science & Nutrition 11(6): 2500–2529. https://doi.org/10.1002/fsn3.3308.

Roberts, D. P. and Mattoo, A. K. (2018). Sustainable agriculture—Enhancing environmental benefits, food nutritional quality and building crop resilience to abiotic and biotic stresses. Agriculture 8(1): 8.

Rogez, H., Buxant, R., Mignolet, E., Souza, J. N. S., Silva, E. M. and Larondelle, Y. (2004). Chemical composition of the pulp of three typical Amazonian fruits: Araça-boi (Eugenia stipitata), bacuri (Platonia insignis) and cupuaçu (Theobroma grandiflorum). European Food Research and Technology 218(4): 380–384. https://doi.org/10.1007/s00217-003-0853-6.

Rout, P. and Basak, U. C. (2014). Evaluation of anti-nutritional factors in sixteen wild edible fruits of Odisha, India. Int. J. Curr. Sci. 13: 34–42.

Rymbai, H., Verma, V. K., Talang, H., Assumi, S. R., Devi, M. B., Vanlalruati, Sangma, R. H. C. et al. (2023). Biochemical and antioxidant activity of wild edible fruits of the eastern Himalaya, India. Frontiers in Nutrition 10: 1039965.

Sadia, H., Ahmad, M., Belgacem, A. O., Zafar, M., Baig, M. B., Sultana, S. et al. (2022). The potential of edible wild fruits as alternative option to ensure food security in a changing climate: a case study from Pakistan. pp. 213–250. *In*: Behnassi, M., Gupta, H., Barjees Baig, M. and Noorka, I. R. (eds.). The Food Security, Biodiversity, and Climate Nexus. Springer International Publishing. https://doi.org/10.1007/978-3-031-12586-7_11.

Sahoo, G., Swamy, S. L., Rout, S., Wani, A. M. and Mishra, A. (2021). Exploitation of wild leafy vegetables and under-utilized fruits: Consequences for food and nutritional security. Annals of the Romanian Society for Cell Biology 25(6): 5656–5668.

Saklani, S. and Chandra, S. (2012). *In vitro* antimicrobial activity, nutritional profile and phytochemical screening of wild edible fruit of Garhwal Himalaya (Ficus auriculata). Int. J. Pharm. Sci. Rev. Res. 12(2): 61–64.

Sánchez-Mata, M. de C., Matallana-González, M. C. and Morales, P. (2016). The contribution of wild plants to dietary intakes of micronutrients (I): Vitamins. Mediterranean Wild Edible Plants: Ethnobotany and Food Composition Tables 111–139.

Sangsopha, J., Moongngarm, A., Pornchaiyaphum, S., Bandorn, A., Khaejonlarp, R. and Pluemjia, B. (2023). Chemical compositions, phytochemicals, and biological activity of dried edible wild fruits. Pakistan Journal of Agricultural Sciences 60(3): 299–306.

Santacruz, L., Carriazo, J. G., Almanza, O. and Osorio, C. (2012). Anthocyanin composition of Wild Colombian fruits and antioxidant capacity measurement by electron paramagnetic resonance spectroscopy. Journal of Agricultural and Food Chemistry 60(6): 1397–1404. https://doi.org/10.1021/jf2042533.

Seal, T. (2011). Antioxidant activity of some wild edible fruits of Meghalaya state in India. Advances in Biological Research 5(3): 155–160.

Seal, T. and Chaudhuri, K. (2017). High Performance Liquid Chromatography Method for the Estimation of Water Soluble Vitamin in Five Wild Edible Fruits Consumed by the Tribal People of North-Eastern Region in India.

Semwal, P., Painuli, S., Jamloki, A., Rauf, A., Rahman, Md. M., Olatunde, A. et al. (2023). Himalayan wild fruits as a strong source of nutraceuticals, therapeutics, food and nutrition security. Food Reviews International 39(9): 6500–6536. https://doi.org/10.1080/87559129.2022.2121407.

Shin, T., Fujikawa, K., Moe, A. Z. and Uchiyama, H. (2018). Traditional knowledge of wild edible plants with special emphasis on medicinal uses in Southern Shan State, Myanmar. Journal of Ethnobiology and Ethnomedicine 14(1): 48. https://doi.org/10.1186/s13002-018-0248-1.

Shrestha, R., Dawadi, P., Bhusal, S. and Bhatt, L. R. (2021). Nutritional value and antioxidant properties of Diospyros malabarica (Desr.) Kostel., fruit from midhills of western Nepal. Nepal Journal of Science and Technology 20(1): 113–125.

Sibiya, N. P., Kayitesi, E. and Moteetee, A. N. (2021). Proximate analyses and amino acid composition of selected wild indigenous fruits of Southern Africa. Plants 10(4): 721.

Silva, M. L., Rita, K., Bernardo, M. A., Mesquita, M. F. de, Pintão, A. M. and Moncada, M. (2023). Adansonia digitata L. (Baobab) bioactive compounds, biological activities, and the potential effect on glycemia: A narrative review. Nutrients 15(9): 2170.

Singh, A. K., Elango, D., Raigne, J., Van der Laan, L., Rairdin, A., Soregaon, C. et al. (2025). Plant-based protein crops and their improvement: Current status and future perspectives. Crop Science 65(1): e21389.

Singh, J., Rajasekaran, A., Negi, A. K., Pala, N. A., Panwar, V. P., Bussmann, R. W. et al. (2023). Potential of wild edible fruits for nutrition in indigenous communities of Northwest Himalaya, India. Ethnobotany Research and Applications 25: 1–15.

Singh, P., Kesharwani, R. K. and Keservani, R. K. (2017). 6 - Protein, carbohydrates, and fats: energy metabolism. pp. 103–115. *In*: Bagchi, D. (ed.). Sustained Energy for Enhanced Human Functions and Activity. Academic Press. https://doi.org/10.1016/B978-0-12-805413-0.00006-5.

Singh, U., Devaraj, S. and Jialal, I. (2005). Vitamin E, oxidative stress, and inflammation. Annual Review of Nutrition 25(1): 151–174. https://doi.org/10.1146/annurev.nutr.24.012003.132446.

Siyum, Z. H. and Meresa, T. A. (2021). Physicochemical properties and nutritional values of *Carissa Spinarum* L."AGAM" fruit. International Journal of Fruit Science 21(1): 826–834. https://doi.org/10.1080/15538362.2021.1936348.

Song, Y., Wei, X.-Q., Li, M.-Y., Duan, X.-W., Sun, Y.-M., Yang, R.-L. et al. (2018). Nutritional composition and antioxidant properties of the fruits of a Chinese Wild Passiflora foetida. Molecules 23(2): Article 2. https://doi.org/10.3390/molecules23020459.

Souilem, F., Dias, M. I., Barros, L., Calhelha, R. C., Alves, M. J., Harzallah-Skhiri, F. et al. (2019). Amantagula Fruit (Carissa macrocarpa (Eckl.) A.DC.): Nutritional and phytochemical characterization. Plant Foods for Human Nutrition 74(1): 76–82. https://doi.org/10.1007/s11130-018-0703-0.

Stadlmayr, B., Charrondière, U. R., Eisenwagen, S., Jamnadass, R. and Kehlenbeck, K. (2013). Nutrient composition of selected indigenous fruits from sub-Saharan Africa. Journal of the Science of Food and Agriculture 93(11): 2627–2636. https://doi.org/10.1002/jsfa.6196.

Stephen, J. and Radhakrishnan, M. (2022). Avocado (*Persea americana* Mill.) fruit: Nutritional value, handling and processing techniques, and health benefits. Journal of Food Processing and Preservation 46(12): e17207. https://doi.org/10.1111/jfpp.17207.

Sundriyal, M. and Sundriyal, R. C. (2001). Wild edible plants of the Sikkim Himalaya: Nutritive values of selected species. Economic Botany 55(3): 377–390. https://doi.org/10.1007/BF02866561.

Surukite, O., Kafeelah, Y., Olusegun, F. and Damola, O. (2013). Qualitative studies on proximate analysis and characterization of oil from Persea americana (Avocado Pear). Journal of Natural Sciences Research 3(2): 68–74.

Suwardi, A. B. and Navia, Z. I. (2023). Sustainable use and management of wild edible fruit plants: a case study in the Ulu Masen Protected Forest, West Aceh, Indonesia. Journal of Sustainable Forestry 42(8): 811–830. https://doi.org/10.1080/10549811.2022.2123355.

Tshikalange, T. E., Modishane, D. C. and Tabit, F. T. (2017). Antimicrobial, antioxidant, and cytotoxicity properties of selected wild edible fruits of traditional medicinal plants. Journal of Herbs, Spices & Medicinal Plants 23(1): 68–76. https://doi.org/10.1080/10496475.2016.1261387.

Turker, A. U., Yildirim, A. B. and Karakas, F. P. (2012). Antibacterial and antitumor activities of some wild fruits grown in Turkey. Biotechnology & Biotechnological Equipment 26(1): 2765–2772. https://doi.org/10.5504/BBEQ.2011.0156.

Tyler, G. and Olsson, T. (2001). Plant uptake of major and minor mineral elements as influenced by soil acidity and liming. Plant and Soil 230(2): 307–321. https://doi.org/10.1023/A:1010314400976.

Uzoekwe, N. M. and Mohammed, J. J. (2015). Phytochemical, proximate and mineral contents of leaves and back of Ficus capensis. Journal of Applied Sciences and Environmental Management 19(4): 633–637.

Valvi, S. R. and Rathod, V. S. (2011). Mineral composition of some wild edible fruits from Kolhapur district. International Journal of Applied Biology and Pharmaceutical Technology 2(1): 392–396.

Vemmos, S. N., Petri, E. and Stournaras, V. (2013). Seasonal changes in photosynthetic activity and carbohydrate content in leaves and fruit of three fig cultivars (*Ficus carica* L.). Scientia Horticulturae 160: 198–207. https://doi.org/10.1016/j.scienta.2013.05.036.

Vincent, L., Soorianathasundaram, K., Dinesh, M. R., Shivashankara, K. S. and C, V. (2022). Change in physiological parameters and partitioning of sugar upon Papaya ringspot virus infection in *Carica papaya* and its wild relatives. South African Journal of Botany 151: 466–474. https://doi.org/10.1016/j.sajb.2022.10.033.

Vitalini, S., Iriti, M., Puricelli, C., Ciuchi, D., Segale, A. and Fico, G. (2013). Traditional knowledge on medicinal and food plants used in Val San Giacomo (Sondrio, Italy)—An alpine ethnobotanical study. Journal of Ethnopharmacology 145(2): 517–529.

Wilson, A.-L. and Downs, C. T. (2012). Fruit nutritional composition and non-nutritive traits of indigenous South African tree species. South African Journal of Botany 78: 30–36.

Wu, G. (2009). Amino acids: Metabolism, functions, and nutrition. Amino Acids 37(1): 1–17. https://doi.org/10.1007/s00726-009-0269-0.

Wyszkowska, J., Boros-Lajszner, E. and Kucharski, J. (2024). The impact of soil contamination with lead on the biomass of maize intended for energy purposes, and the biochemical and physicochemical properties of the soil. Energies 17(5): 1156.

Yang, B. and Kallio, H. P. (2001). Fatty acid composition of lipids in Sea Buckthorn (*Hippophaë rhamnoides* L.) Berries of different origins. Journal of Agricultural and Food Chemistry 49(4): 1939–1947. https://doi.org/10.1021/jf001059s.

Zhang, Y., Zhang, Z., Yang, Y., Zu, X., Guan, D. I. and Guan, Y. (2011). Diuretic activity of Rubus idaeus L (Rosaceae) in rats. Tropical Journal of Pharmaceutical Research 10(3). https://www.ajol.info/index.php/tjpr/article/view/67933.

Zhitkovich, A. (2020). Nuclear and cytoplasmic functions of vitamin C. Chemical Research in Toxicology 33(10): 2515–2526. https://doi.org/10.1021/acs.chemrestox.0c00348.

Zielińska, A. and Nowak, I. (2017). Abundance of active ingredients in sea-buckthorn oil. Lipids in Health and Disease 16(1): 95. https://doi.org/10.1186/s12944-017-0469-7.

Chapter 10

Nutritional and Therapeutic Potential of Brazilian Cerrado Fruits

Challenges and Future Perspectives

Darlisson Slag Neri Silva,[1] *Hudson de Carvalho Silva,*[1] *Tiago Linus Silva Coelho,*[1] *Ronaldo Cunha Coelho*[2] and *Edivan Carvalho Vieira*[1]

10.1 Introduction

Brazil is the fifth-largest country in the world, with approximately 8.5 million km^2. Due to its large territorial extension, Brazil covers six different biomes, each with its climatic singularities: Amazon, Cerrado, Atlantic Forest, Caatinga, Pampa, and Pantanal (Batlle-Bayer et al. 2010, Flach et al. 2021). The Brazilian Cerrado is one of the richest and most diverse biomes in the world, covering approximately 2 million square kilometers and harboring an extraordinary variety of plant species. Among its vast biodiversity, native fruits such as buriti (*Mauritia flexuosa*), pequi (*Caryocar brasiliense*), murici (*Byrsonima crassifolia*), guabiraba (*Campomanesia xanthocarpa*), tucum (*Astrocaryum vulgare*), and tuturubá (*Pouteria macrophylla*) stand out for their nutritional and therapeutic potential (Souza et al. 2012, Carvalho et al. 2020).

[1] Interdisciplinary Chemistry Laboratory, Department of Chemistry, Federal University of Piauí – UFPI, 64049-550, Teresina, PI, Brazil.
[2] Department of Teacher Training, Federal Institute of Piauí, 64000-040 Teresina, Piauí, Brazil.
* Corresponding author: darlisson.ufpi@hotmail.com

DOI: 10.1201/9781003668169-12

Despite their high nutritional value, many of these fruits remain underutilized and largely unknown to the general population due to a lack of awareness, difficulties in commercialization, and limited scientific dissemination (Neri-Numa et al. 2018, Silva and Rial 2025). Cerrado fruits are valuable sources of bioactive compounds such as carotenoids, phenolic compounds, fibers, and essential minerals (Alexandre et al. 2025). Carotenoids, for instance, are potent antioxidants that contribute to immune system enhancement, cardiovascular health, and reduced risk of chronic diseases (Milanez et al. 2018, Saini et al. 2022).

Similarly, phenolic compounds are well known for their anti-inflammatory and antimicrobial properties, playing a significant role in disease prevention and health promotion (Fiedor and Burda 2014, Borges et al. 2022a). In addition to their bioactive composition, these fruits provide essential micronutrients such as calcium, iron, potassium, and magnesium, which are crucial for human nutrition and metabolic regulation (Ribeiro Neto et al. 2020, Silva and Rial 2025).

Beyond their nutritional attributes, Cerrado fruits exhibit significant therapeutic potential. Several studies have demonstrated their antioxidant, anti-inflammatory, and antimicrobial properties, making them promising candidates for nutraceutical and pharmaceutical applications (Raphaelli et al. 2021, Machado et al. 2025). For example, guabiraba has been associated with cardiovascular benefits, including cholesterol reduction and oxidative stress mitigation, while buriti is recognized for its high β-carotene content and skin-protective effects (Otero et al. 2017, Silva et al. 2021).

However, despite their evident potential, the utilization of these fruits faces multiple challenges. Limited agricultural investment, logistical difficulties, and the absence of large-scale processing technologies hinder their commercialization and broader incorporation into the global market (Souza and Viana 2018, Borges et al. 2022a). Additionally, knowledge gaps regarding optimal cultivation practices and post-harvest management further restrict their economic viability (Oliveira et al. 2017, Monteiro et al. 2025).

The purpose of this chapter is to provide a comprehensive review of the nutritional and therapeutic value of unique fruits from the Brazilian Cerrado. Emphasis will be placed on their bioactive composition, health benefits, existing challenges, and future prospects for their application in nutrition and medicine.

10.2 Nutritional Composition of Cerrado Fruits

The Brazilian Cerrado (Figure 1) harbors an impressive variety of native fruits with high nutritional potential. These fruits contain a rich composition of macronutrients (carbohydrates, proteins, and lipids) and micronutrients (vitamins and minerals), in addition to a wide range of bioactive compounds such as phenolic compounds, carotenoids, and dietary fibers.

Cerrado fruits exhibit a remarkable balance of essential nutrients that contribute to a healthy diet. For instance, jatobá-do-cerrado (*Hymenaea stigonocarpa*) has a high carbohydrate content (34.1%), making it a potential energy source, while also being rich in dietary fiber (44.3%) (Martins et al. 2015). In contrast, buriti and pequi are distinguished by their lipid content, particularly monounsaturated and

Figure 1. Map of Brazil showing the location of Brazilian biomes.

polyunsaturated fatty acids, which contribute to cardiovascular health (Melo et al. 2014, Caldeira et al. 2021).

Several Cerrado fruits are excellent sources of vitamin C, with gabiroba (guabiraba) containing up to 234 mg/100 g, exceeding common citrus fruits. Mangaba (*Hancornia speciosa*) is also rich in vitamin C (165.8 mg/100 g) and folates (98.3 μg/100 g), essential for DNA synthesis and prenatal health (Schiassi et al. 2018). Jatobá-do-cerrado is particularly abundant in calcium (850 mg/100 g) and magnesium (3173 mg/100 g), making it an important dietary source of these minerals (Bailão et al. 2015, Arruda et al. 2022, Borges et al. 2022b).

Bioactive compounds are organic substances found mainly in plants and fruits, characterized by their remarkable antioxidant properties, which result in a wide range of contributions to human health. Many of these benefits range from combating the development of heart disease to reducing the risk of certain types of cancer, as well as mitigating oxidative stress and protecting DNA from damage (Carvalho et al. 2025). The bioactive compounds present in Cerrado fruits (Figure 2) also contribute significantly to their health benefits. These include:

- *Phenolic compounds*: Found in high concentrations in gabiroba, cagaita (*Eugenia dysenterica*), and mangaba, these compounds exhibit strong antioxidant, anti-inflammatory, and antimicrobial properties. Phenolic compounds (Figure 3)

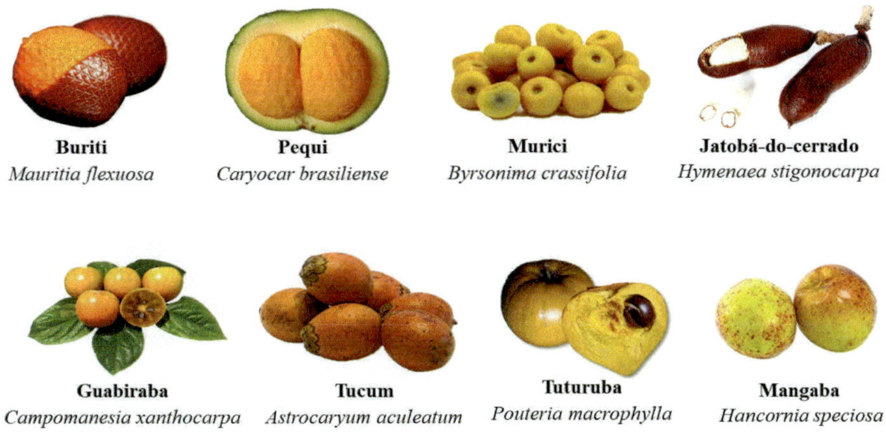

Figure 2. Some fruits from the Brazilian Cerrado.

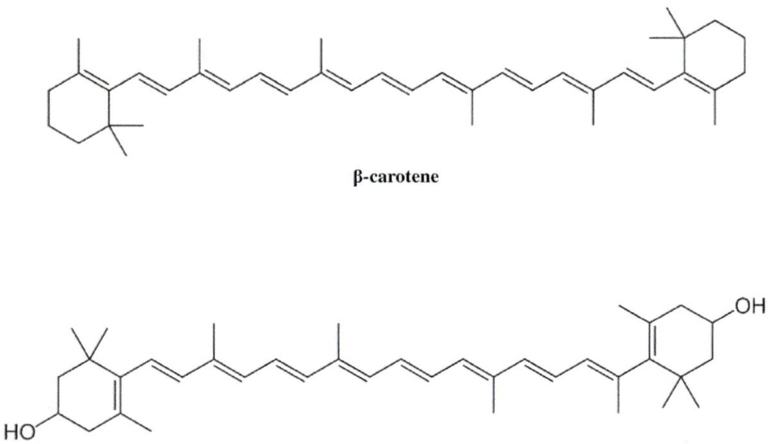

Figure 3. Structural representation of β-carotene (hydrocarbon carotenoid) and zeaxanthin (oxygenated carotenoid).

are part of an important specific class of radical neutralizers. These species can eliminate free radicals produced during normal cellular metabolism, preventing autoxidation, in addition to serving as signaling molecules (Chen et al. 2019). Recent findings reinforce the multifunctional role of phenolic compounds in fruit-based matrices. For example, high total phenolic content (43.5 ± 2.12 and 23 ± 3.54 mg GAE/g, respectively) has been associated with significant antioxidant activity, evaluated by DPPH and ABTS assays. The fruits also demonstrated antimicrobial effects, indicating stronger antibacterial potential than *B. aethiopum*. Interestingly, the latter showed enhanced performance when combined with standard antimicrobial agents. These findings highlight the synergistic potential of natural fruit extracts with pharmaceuticals and support their future application as food preservatives or adjunctive therapeutic agents

(Dzigbor et al. 2025). The ability of these antioxidant compounds to react with reactive oxygen and nitrogen species gives them beneficial properties for human health, such as anti-inflammatory, cardioprotective, neuroprotective, anticancer, and renal-protective activities, in addition to acting in the inhibition of the growth process of some microorganisms. In addition, these compounds play a crucial role in the sensory and nutritional quality of a wide range of plants, contributing to the color, aroma, and astringency of the fruit or vegetable, as well as enhancing oxidative stability, as mentioned above (Li et al. 2020, Martín-Gómez et al. 2020). Their ability to eliminate free radicals produced during normal cellular metabolism is described through three main mechanisms: transfer of H atoms to peroxyl radicals, transfer of electrons coupled to protons, and sequential transfer of electrons with loss of protons (Losada-Barreiro and Bravo-Díaz 2017). During food processing, the concentration and activity of phenolic compounds may undergo significant transformations, as observed in lactic acid fermentation of fruit and vegetable (F&V) matrices. Studies have identified four main pathways affecting free phenolic levels: adsorption to plant cell walls, microbial adsorption, microbial biotransformation, and reduced stability of covalent bonds between bound phenolics and the cell wall matrix. While adsorption processes tend to reduce the content of free phenolics, microbial biotransformation enhances their release and potentially their bioactivity. The interaction of these pathways creates a dynamic equilibrium described as a "tug-of-war," which varies according to the fermentation stage and the physicochemical composition of the plant matrix. These findings illustrate the complex and non-linear behavior of phenolic compounds during food processing and reinforce their functional relevance in fermented fruit-based products (Ganjeh et al. 2025).

- *Carotenoids*: Particularly abundant in buriti and pequi, which are among the richest natural sources of provitamin A, essential for vision and immune function. Responsible for the red, orange, and yellow colors in fruits and vegetables, carotenoids (Figure 4) are known to have a great influence on human health largely due to their structural characteristics of conjugated double bonds (Mapelli-Brahm et al. 2020). The classification of carotenoids is governed by their chemical structures. There are two groups of carotenoids: xanthophylls and carotenes. The first group has as its main structural characteristic the presence of oxygenated groups, including hydroxyls, carboxyls, and epoxy groups (Mariutti and Mercadante 2018). Xanthophylls are the oxidized versions of carotenoids and are present in many forms of life, such as bacteria, fungi, and algae, constituting the most common pigments in nature (Aziz et al. 2020). For example, in orange-pigmented fruits, like many found in the Cerrado, the xanthophylls violaxanthin, β-cryptoxanthin, antheraxanthin, and their isomers are the major carotenoids (Zacarías-García et al. 2021). Buriti stands out among Cerrado fruits for its carotenoid-rich pulp, and recent optimization studies have reinforced this potential (Silva et al. 2022). Using response surface methodology, it was found that total carotenoid content in the pulp reached up to 3.92 mg/kg, with extraction efficiency influenced by ethanol concentration, temperature, and pulp percentage. Optimal conditions involved 20% pulp and

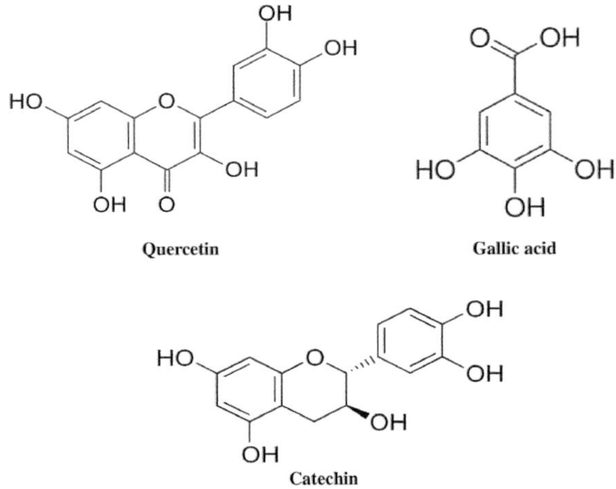

Figure 4. Structural representation of quercetin, gallic acid, and catechin-, three phenolic compounds with distinct antioxidant properties.

temperatures around 70°C, favoring carotenoid recovery. While the pulp also exhibited notable levels of phenolics and flavonoids, the carotenoid fraction is especially relevant due to its antioxidant activity and applicability in functional food formulations (Nogueira et al. 2025).

- *Dietary fibers*: Present in jatobá-do-cerrado and mangaba, dietary fibers have great potential to be used in functional foods and nutraceuticals. They are classified into soluble (e.g., pectin) and insoluble (e.g., cellulose) fibers and present functional and health-promoting properties (Haris et al. 2023), crucial for digestive health and the regulation of blood sugar levels. Recent studies have highlighted the fiber-rich semi-defatted cake derived from oil extraction as a promising byproduct for functional applications. This residual biomass, previously underutilized, contains significant levels of dietary fiber along with proteins and carbohydrates, making it suitable for the development of enriched food formulations. The incorporation of this fibrous material into value-added products exemplifies a sustainable strategy that reduces waste, adds economic value to the buriti production chain, and aligns with bioeconomy principles by promoting the full use of native fruit resources (Gonçalves et al. 2025).

10.2.1 Comparative Analysis of Nutritional Composition

The fruits of the Brazilian Cerrado are rich sources of bioactive compounds with significant therapeutic properties, particularly due to their high content of phenolic compounds, carotenoids, fatty acids, and dietary fibers (Berto et al. 2015). These phytochemicals contribute to their antioxidant, anti-inflammatory, antimicrobial, and pharmacological activities, positioning Cerrado fruits as potential candidates for nutraceutical and functional food applications (Gonçalves et al. 2010).

The nutritional profile of Cerrado fruits highlights their diversity and potential for dietary and functional applications. These fruits provide a rich source of carbohydrates, lipids, proteins, vitamins, and essential minerals, each offering unique contributions to human nutrition. Among the fruits, jatobá-do-cerrado stands out as the richest source of carbohydrates (34.1%), making it a potential energy-dense fruit. This characteristic suggests its suitability for sports nutrition and high-energy dietary formulations, similar to carbohydrate-rich fruits like bananas (Udomkun et al. 2019).

Lipids, also known as fats, are one of the primary nutrients in food and perform several vital functions in the body, such as the production of hormones and the absorption of fat-soluble vitamins (Cavalcante et al. 2017). Pequi and buriti exhibit the highest lipid concentrations (57.0% and 50.3%, respectively), with a high proportion of monounsaturated and polyunsaturated fatty acids, which are known for their cardioprotective and anti-inflammatory properties (Alves et al. 2014). These fruits can be considered excellent sources of natural lipids, similar to avocados and olives, commonly used in heart-healthy diets (Aguiar and Souza 2017, Cândido and Silva 2017, Rajan and Bhat 2019). The presence of essential fatty acids, such as oleic acid, in pequi further reinforces its potential benefits for cardiovascular health (Serra et al. 2019).

Although Cerrado fruits are not traditionally known for their protein content, buriti (7.7%) and jatobá-do-cerrado (5.6%) provide moderate amounts, which may contribute to daily protein intake, particularly in plant-based diets. While they do not match legumes in protein concentration, their amino acid composition and digestibility warrant further investigation. The vitamin C content among these fruits varies widely, with gabiroba (234 mg/100 g) and mangaba (165.8 mg/100 g) exceeding the vitamin C content of common citrus fruits. This positions them as potent natural antioxidants, capable of enhancing immune function, reducing oxidative stress, and preventing chronic diseases (Curimbaba et al. 2020). Buriti (45.5 mg/100 g) and pequi (2.3 mg/100 g) contain lower amounts but still contribute to overall antioxidant intake.

Regarding mineral composition, jatobá-do-cerrado is an exceptional source of calcium (850 mg/100 g) and magnesium (3173 mg/100 g), surpassing many commonly consumed nuts and seeds. Magnesium is the fourth most abundant cation in the human body and the second most present electrolyte within cells, making it extremely important for the activity of cellular enzymes, especially those that act in glycolysis and the stimulation of ATPases (Ajith 2020). Nascimento et al. (2020) reported high values of magnesium for araticum (*Annona crassiflora*) (56.7 ± 2.4 mg/100 g) and cagaita (66.0 ± 6.1 mg/100 g); on the other hand, caja (*Spondias mombin*), showed comparatively lower concentrations (45.5 ± 2.1 mg/100 g), highlighting the variability in magnesium levels among Cerrado fruits.

These minerals are crucial for bone health, muscle function, and enzymatic reactions, making jatobá a promising functional ingredient in dietary supplements. Meanwhile, buriti (56 mg/100 g of calcium, 24.5 mg/100 g of magnesium) and pequi (31 mg/100 g of calcium, 1447 mg/100 g of magnesium) also contribute to mineral intake, particularly in plant-based and vegetarian diets (Grillo et al. 2020, Shokunbi et al. 2023).

Examining the calcium concentration in other fruits of the Cerrado, murici, buriti, and tuturubá present higher concentrations compared to yellow mombin

(*Spondias mombin*) (40.3 ± 1.5 mg/100 g), marolo (*Annona crassiflora*) (39.3 ± 3.2 mg/100 g), cagaita (22.5 ± 0.3 mg/100 g), and mangaba (31.0 ± 2.4 mg/100 g) (Schiassi et al. 2018). According to Nascimento et al. (2020), buriti had a total calcium concentration of 36.9 mg/100 g, and murici showed 78.0 mg/100 g (Silva et al. 2008).

When compared to other popular fruits, such as avocado (8 mg/100 g), pineapple (22 mg/100 g), açaí (35 mg/100 g), banana (8 mg/100 g), and papaya (25 mg/100 g), the Cerrado fruits are good sources of total calcium (TACO 2011). Jatobá-do-cerrado (44.3%), buriti (38.7–48.9%), and mangaba (11.6%) are notable for their high fiber content, which plays a crucial role in gut health, glycemic control, and cholesterol metabolism. The soluble and insoluble fibers in these fruits may aid in digestive health, promoting beneficial gut microbiota and potentially preventing metabolic disorders such as diabetes and obesity. Iron is essential for transporting oxygen throughout the body, supplying oxygen to cells. It is an important component of hemoglobin and myoglobin, which store oxygen in blood and muscle (Merwe et al. 2019).

Some fruits found in the Brazilian Cerrado are interesting protein sources. The pequi collected in the states of Tocantins, Goiás, and Mato Grosso presented concentrations in the range of 3.73 to 4.09 g/100 g (Alves et al. 2014). Silva et al. (2008) reported protein contents that vary significantly among the analyzed Cerrado fruits. Chichá (*Sterculia striata*) exhibited the highest protein content, with 19.58 g/100 g, standing out as an excellent protein source. Murici also showed relevant content, with 0.72 g/100 g, followed by cagaita, which presented 0.82 g/100 g. Gabiroba recorded a protein content of 0.50 g/100 g, while mangaba and puçá showed slightly higher values, with 1.20 g/100 g and 1.02 g/100 g, respectively. These data reinforce the importance of these fruits not only as sources of bioactive compounds but also as contributors to protein intake, especially in regional diets or in strategies for the nutritional utilization of native fruits. These data emphasize the nutritional diversity of Cerrado fruits, demonstrating their potential for incorporation into functional foods and nutraceuticals.

Beyond the commonly consumed portions of Cerrado fruits, growing attention has been directed toward agro-industrial byproducts such as the pulp of baru (*Dipteryx alata* Vogel). Despite being frequently discarded during processing, this pulp has demonstrated notable nutritional value and bioactive potential. Although the seed is commercially appreciated, representing around 5% of the fruit, the pulp contains high levels of sugars (41.86%), dietary fiber (29.12%), and essential minerals. It also presents considerable concentrations of trigonelline (139.10 mg/100 g), tannins (429.16 mg TAE/100 g), vitamin C (109.57 mg/100 g), and phenolic compounds like trans-cinnamic acid (5.91 mg/100 g), chlorogenic acid, and gallic acid (1.07 mg/100 g). The volatile fraction is dominated by sesquiterpenes such as α-copaene and bicyclogermacrene (42.75%), followed by germacrene D (11.69%), aromadendrene (9.05%), α-cubebene (6.84%), β-elemene (5.90%), and ledene (5.82%), many of which are widely used in the essential oil industry. These findings reinforce the potential of baru pulp as a sustainable, low-cost source of functional ingredients, promoting waste valorization and economic opportunities in the Cerrado biome (Monteiro et al. 2025).

10.3 Therapeutic Potential of Cerrado Fruits

One of the most remarkable health benefits of these fruits is their antioxidant and anti-inflammatory potential, which plays a crucial role in preventing and mitigating the effects of oxidative stress-related diseases. Phenolic compounds such as flavonoids, tannins, and phenolic acids are abundant in species like cagaita (*Eugenia dysenterica*), guabiraba (*Campomanesia xanthocarpa*), and mangaba (*Hancornia speciosa*), exhibiting strong radical-scavenging activity and the ability to modulate inflammatory pathways (Schiassi et al. 2018). In particular, quercetin and kaempferol, present in cagaita, have been shown to inhibit key pro-inflammatory mediators such as tumor necrosis factor-alpha (TNF-α) and nuclear factor-kappa B (NF-κB), both of which are associated with chronic inflammatory diseases. Similarly, the high ellagic acid and tannin content in gabiroba contributes to reducing lipid peroxidation and modulating oxidative stress-induced inflammatory responses (Stafussa et al. 2021, Nogueira et al. 2022).

Recent advances in metabolomic profiling have provided valuable insights into the phenolic composition of Cerrado fruits, particularly guabiroba. In a recent study employing LC-ESI-QToF-MS/MS techniques, a total of 62 phenolic compounds were tentatively identified in different parts of the fruit, including peel, pulp, and seeds. Among them, 44 were flavonoids, such as catechin, epicatechin, and quercitrin, highlighting the dominance of this class in the guabiroba matrix. Quantitative analysis revealed high concentrations of ellagic acid (172 to 6,276 μg/g) and gallic acid (23 to 7,955 μg/g), especially in bound phenolic fractions. Notably, the seeds exhibited the highest levels of bioactive compounds, reinforcing their antioxidant potential. These findings demonstrate the need to explore all anatomical parts of the fruit, including often-discarded fractions, and validate their bioactivity for functional and nutraceutical applications (Machado et al. 2025).

The genus Eugenia, with more than a hundred species, includes cagaita, which has long been utilized in traditional medicine by local communities of Brazil to treat ailments such as diarrhea, diabetes, and jaundice. The bioactive compounds present in its fruit and leaves exhibit significant pharmacological potential (Santana et al. 2022). Notably, the phenolic compounds found in cagaita have demonstrated strong antioxidant activity *in vitro*, effectively neutralizing free radicals and reducing oxidative stress. Additionally, these bioactives play a crucial role in modulating carbohydrate metabolism by inhibiting key enzymatic pathways, thereby mitigating postprandial glucose spikes and reducing the risk of obesity and type 2 diabetes. Such properties highlight the potential of *E. dysenterica* as a functional food ingredient for metabolic health management, reinforcing its relevance in both traditional medicine and modern therapeutic applications (Hwang et al. 2016, Araujo et al. 2021a).

Buriti has been traditionally used for its medicinal properties in several regions, especially in the Amazon. Costa et al. (2021) investigated the anti-COVID-19 activity of some components of buriti oil through computer simulations. Figure 5 illustrates the molecular interaction between 13-cis-β-carotene, a major carotenoid in buriti oil, and the main protease of SARS-CoV (2GTB). The compound binds to the enzyme through hydrophobic interactions with key amino acids, suggesting potential antiviral activity by inhibiting viral replication.

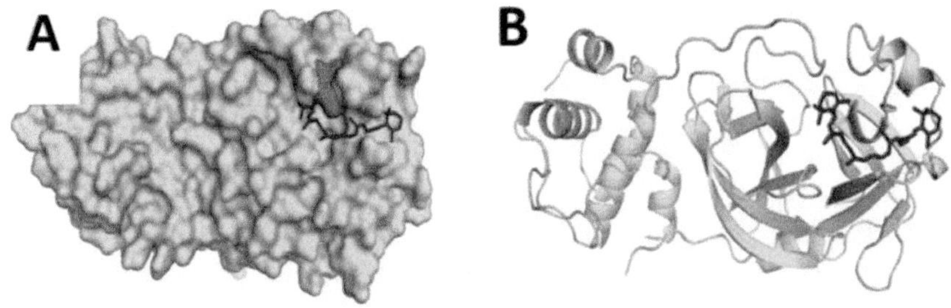

Figure 5. Molecular interaction between 13-cis-β-carotene, a major carotenoid in buriti oil, and the main protease of SARS-CoV (2GTB). The hydrophobic binding with key amino acid residues suggests a potential inhibitory effect on viral replication. Adapted from Costa et al. (2021).

The results suggest that certain compounds, such as α-carotene, interact favorably with SARS-CoV's main peptidase. These theoretical findings can guide experimental studies for developing new drugs that can inhibit the action of the COVID-19 virus. In addition to its potential in COVID-19 research, the ability of buriti oil to inhibit the main proteolytic enzymes present in the venom of snakes of the genus *Bothrops* is a significant advance in research into treatments for snake poisoning in Brazil. Furthermore, this oil has demonstrated the potential to assist in the cellular repair of injuries caused by the venom, making it a possible promising therapeutic option (Rodrigues et al. 2023).

The carotenoids present in Cerrado fruits, particularly in buriti and pequi, further enhance their anti-inflammatory and protective effects. These fruits contain high levels of provitamin A (β-carotene) and lutein, which have been linked to improved immune function and neuroprotection (Ferruzzi et al. 2020). Studies demonstrate that pequi oil has hepatoprotective and antidiabetic effects. This means that it can help reduce liver damage caused by toxic agents, such as carbon tetrachloride, and possesses properties that can be beneficial for diabetes control by regulating blood glucose levels (Torres et al. 2016, Caldeira et al. 2021, Silva and Rial 2025). Caldeira et al. (2021) investigated the antidiabetic potential of pequi fruit peels. The results indicated that an extract and fractions of pequi peels demonstrated the ability to inhibit the release of cytokines associated with insulin resistance. Chemical analysis identified compounds such as corilagin and geranine in the pequi peels. These findings highlight the potential of pequi peels in the development of functional foods for managing type 2 diabetes.

The oils extracted from Cerrado fruits and oilseeds have gained increasing scientific interest due to their therapeutic potential and industrial versatility. Recent studies have highlighted the therapeutic and technological relevance of oils extracted from Cerrado fruits and oilseeds such as pequi, buriti, cagaita, and mangaba. These oils exhibit high levels of oleic acid, a monounsaturated fatty acid with anti-inflammatory and cardiovascular benefits, as well as phytosterols, tocopherols, and carotenoids (Donado-Pestana et al. 2018). Pequi oil stands out for its antioxidant and anti-inflammatory properties, while buriti oil, rich in beta-carotene and oleic acid, has demonstrated potential for skin and hair care (Arruda et al. 2022). Cagaita oil contains high concentrations of linoleic acid, contributing to antimicrobial action, and mangaba oil shows antihypertensive potential due to its unsaturated lipid profile. Despite these promising characteristics, challenges remain in scaling up extraction and ensuring sustainable industrial use. The growing demand for natural, functional, and eco-friendly products positions Cerrado oils as valuable bioresources, not only for nutrition and health but also for use in cosmetics, pharmaceuticals, and sustainable development strategies in local communities (Silva and Rial 2025).

In addition to their known antioxidant and anti-inflammatory properties, monomeric phenols such as catechins, chlorogenic acid, gallic acid, quercetin, resveratrol, and rosmarinic acid—found in many Cerrado fruits—have demonstrated significant anti-obesity effects. These compounds inhibit adipogenesis, increase lipolysis, suppress appetite, and reduce inflammation, contributing to weight management and metabolic health. Innovative extraction methods such as microwave-assisted extraction (MAE), ultrasound-assisted extraction (UAE), and pressurized liquid extraction (PLE) enhance the recovery of these bioactives, optimizing their application in nutraceuticals and functional foods. Exploring the therapeutic potential of these compounds may offer a promising path for developing safe, plant-based anti-obesity interventions (Ganjeh et al. 2025).

The complex chemical composition of pequi has aroused great interest in the scientific community due to its potential for developing new drugs (Baptista et al. 2018, Cornelio-Santiago et al. 2021). A promising example is lipid-core polymeric nanocapsules loaded with pequi oil for dermal wound healing. These nanocapsules enable gradual release of the oil, providing sustained action at the injury site and promoting the regeneration of affected tissues. This advancement represents a promising approach in the field of regenerative medicine and has the potential to offer significant benefits for dermal wound healing (Pires et al. 2020).

The molecular dynamics of marolo (*Annona crassiflora* Mart.) throughout its development have been elucidated through integrative proteomic and metabolomic analyses. The ripe fruit exhibits a rich profile of bioactive compounds, including carotenoids, vitamin C, alkaloids, phenolic compounds, and volatile esters. Enzymes involved in cell wall degradation and oxidative balance were detected alongside those linked to the biosynthesis of these metabolites. Notably, the antioxidant system remained active across all developmental stages, suggesting a stable protective mechanism. This molecular-level understanding reinforces marolo's potential as a source of health-promoting compounds and supports future biotechnological applications aimed at enhancing its functional properties and commercial value (Alexandre et al. 2025).

The tucum fruit is an excellent source of nutrients and bioactive compounds, which confer a series of benefits to human and animal health. The fruit also contains antioxidant compounds, such as carotenoids and flavonoids, which help prevent cell damage and reduce the risk of chronic diseases (Bhatt and Patel 2020, Soares et al. 2020, Machado et al. 2022, Menezes et al. 2022). In this way, the polyphenols present—bioactive plant compounds known for their antioxidant and anti-inflammatory properties—are associated with a series of health benefits, such as reducing the risk of chronic diseases, including cancer and cardiovascular and neurodegenerative diseases (Araujo et al. 2021b, Machado et al. 2022).

Tucum-do-cerrado fruit has presented antitumor effects against the Ehrlich mammary tumor model in mice through the modulation of oxidative stress and inflammation by its polysaccharides glucomannan and glucuronoarabinoxylan (Oliveira et al. 2024).

As with buriti, it was reported that the fruits of *Astrocaryum vulgare* have inhibitory properties against snake venoms, which the authors associate with their phospholipase A2 inhibitory activity (Soares et al. 2005). Tucumã oil has exhibited notable antibacterial activity, with a minimum inhibitory concentration (MIC) of 800 μg/mL against a range of pathogens, including *Acinetobacter baumannii*, *Enterococcus faecalis*, and inhibitory effects on *Enterococcus cecorum* and *Clostridium perfringens* (Rossato et al. 2019).

Several studies have highlighted the importance of *Campomanesia xanthocarpa*, demonstrating that extracts from guabirobeira, whether purified or in their natural state, derived from the fruit, leaf, or seed, have various proven therapeutic activities. These activities include anti-inflammatory properties (Silva et al. 2016), antiprotozoal effects (Brandelli et al. 2013), anti-diarrheal properties (Moreira et al. 2011), antimicrobial, antiplatelet, antithrombotic effects (Otero et al. 2017), fibrinolytic, and hypotensive effects (Sant'Anna et al. 2017). However, to date, no studies have consolidated data regarding the nutraceutical and pharmaceutical potential of guabirobeira (Raphaelli et al. 2021).

Klafke et al. (2010) reported that the use of capsules containing crushed dried leaves of *C. xanthocarpa*, at doses of 250 and 500 mg/kg, demonstrated a reduction in total cholesterol levels and low-density lipoprotein (LDL) cholesterol, as well as a decrease in plasma oxidative stress in patients with hypercholesterolemia. Bioactive compounds such as saponins, tannins, terpenes, and flavonoids were also identified. These results highlight the therapeutic potential of guabirobeira and emphasize the importance of its use in promoting cardiovascular health (Raphaelli et al. 2021). The development of corn starch and gelatin-based films with guabiraba pulp is being explored for applications in food packaging (Malherbi et al. 2019, Rodrigues et al. 2020).

Murici is a native fruit, and its harvesting usually takes place in areas of natural occurrence. So far, studies on the genetic diversity of murici have been limited, with research primarily conducted in Mexico and Brazil (Martínez et al. 2013, Garza et al. 2021). In local culture, murici has been widely used as a home remedy for various health conditions. Traditionally, it is employed in treating gastrointestinal inflammations, gynecological issues, and skin infections. In folk medicine, it is common to prepare teas, juices, or poultices using the fruit. In these cases, murici is mashed and applied directly to the affected skin or consumed as infusions (Souza et al. 2017).

Some believe that murici has anti-inflammatory, antiseptic, and wound-healing properties, which could make it helpful in treating conditions such as gastritis, ulcers, vaginal discharge, and skin infections (Martínez-Vázquez et al. 1999, Medina-Torres et al. 2004, Pérez Gutiérrez and Muñiz Ramirez 2013, Gutiérrez-Molina et al. 2021). However, it is important to note that the efficacy of these medicinal uses has yet to be scientifically proven, and it is always recommended to seek proper medical guidance (Irías-Mata et al. 2018).

The tuturubá, also known as abiu-cutite or cutite, is a fruit native to South America, found mainly in countries such as Brazil, Peru, Colombia, and Venezuela (León and Williams 2009, Santos et al. 2015, Cruz 2019). It is scientifically known as *Pouteria macrophylla* and belongs to the Sapotaceae family, which comprises approximately 50 genera and 1,000 species widely distributed across tropical and subtropical regions of the globe (Lorenzi 2002, Ricker et al. 2016, Serrano et al. 2021).

The Sapotaceae family is home to several species recognized for their medicinal properties, both in the scientific literature and in popular knowledge (Gordon et al. 2011, Silva et al. 2012). These plants have been studied due to their potential therapeutic effects. Among the effects of these plants, antibacterial, antifungal, antiviral, antitumor, anthelmintic, analgesic, and anti-inflammatory activities stand out (Montenegro et al. 2006, Santos et al. 2015).

Among the species of the genus *Pouteria*, which belongs to the Sapotaceae family, some have proven biological activities (Boleti et al. 2007). A notable example is *Pouteria caimito*, which has demonstrated antioxidant and photoprotective activity against UVA and UVB rays. This property makes *Pouteria caimito* promising for skin protection against damage caused by solar radiation. *Pouteria ramiflora*, on the other hand, has anti-inflammatory and antioxidant activity and also has photoprotective effects against UVA and UVB rays (Santos et al. 2014).

Tuturubá also showed antimicrobial activity and toxicity against brine shrimp larvae. Finally, *Pouteria torta* demonstrated antimicrobial and antifungal activity. These studies indicate that species of the genus *Pouteria* have promising potential in their biological activities (Boleti et al. 2007, Santos et al. 2014). A class of common substances found in the Sapotaceae family is the triterpenes, which are related to several of the mentioned biological activities (Montenegro et al. 2006). Triterpenes are chemical compounds with significant medicinal properties (Rodrigues et al. 2017). Furthermore, the presence of alkaloids, flavonoids, benzenoids, phenylpropanoids, and naphthoquinones has also been reported in several species of Sapotaceae (Christe et al. 2021, Baky et al. 2022).

Additionally, jatobá-do-cerrado has been reported to have gastroprotective properties, as its catechin- and gallic acid-rich extracts accelerate the healing of gastric ulcers by reducing oxidative stress and modulating inflammatory responses in the digestive tract (Martins et al. 2015). Beyond their antioxidant and anti-inflammatory roles, many Cerrado fruits demonstrate remarkable antimicrobial properties, making them valuable for both medicinal and food preservation applications (Koolen et al. 2013). The phenolic-rich extracts of cagaita have demonstrated antibacterial activity against foodborne pathogens such as *Staphylococcus aureus* and *Listeria monocytogenes* (Baptista et al. 2018). Moreover, chalcones and flavanones found in gabiroba have exhibited inhibitory effects against *Mycobacterium tuberculosis*, suggesting their potential as complementary therapies for infectious diseases (Cardozo et al. 2018). Additionally, lobeira (*Solanum lycocarpum*) extracts have shown potent antifungal properties against *Candida* species, reinforcing their potential for natural antifungal formulations (Bailão et al. 2015).

The pharmacological applications of Cerrado fruits extend to the prevention and management of metabolic disorders, cardiovascular diseases, and neurodegenerative conditions. Cagaita juice, for instance, has been linked to an acute hypoglycemic effect, reducing postprandial glucose levels and showing potential in type 2 diabetes management (Hirschmann 1995, Cardozo et al. 2018). Moreover, cagaita oil is rich in linoleic acid, with antimicrobial properties and roles in blood pressure regulation, and also presents alpha-linolenic acid, responsible for anti-inflammatory and cardioprotective potential (Bailão et al. 2015, Lawrence 2021).

The phenolic extracts from species such as *Hymenaea stigonocarpa* (jatobá-do-cerrado) and *Hancornia speciosa* (mangaba) have been shown to improve lipid metabolism, reduce LDL oxidation, and lower systemic inflammation, contributing to cardiovascular health. Furthermore, the biodiversity of the Cerrado fosters a unique phytochemical profile in its fruits, making them valuable resources for the food and pharmaceutical industries. Enhancing the utilization of these bioactive-rich fruits through innovative processing and product development could unlock new opportunities for functional foods and nutraceutical applications (Rodrigues et al. 2023).

In addition to the already reported Cerrado species, recent studies have highlighted the bioactivity of other underutilized tropical fruits, such as *Borassus flabellifer* and *Borassus aethiopum*. These fruits demonstrated considerable total phenolic content (TPC), 43.5 ± 2.12 and 23 ± 3.54 mg GAE/g dried extract, respectively, alongside notable antioxidant activities, with *Borassus flabellifer* achieving $28.76 \pm 0.56\%$ and $57.37 \pm 2.85\%$ in DPPH and ABTS assays, respectively. Beyond their antioxidant profile, these fruits also exhibited antimicrobial properties, with minimum inhibitory concentrations (MIC) ranging from 0.10 to 12.50 mg/mL against foodborne pathogens. Notably, *B. aethiopum*, despite presenting lower standalone activity, showed synergistic effects when combined with conventional antimicrobial drugs, thus reinforcing its potential use as a natural food preservative or therapeutic agent (Monteiro et al. 2025). These findings underscore the relevance of tropical fruits as alternative sources of functional compounds for health promotion and industrial applications.

The high carotenoid content in buriti and pequi is also associated with neuroprotective effects. Indeed, the oil obtained from the buriti fruit has placed the species, in recent years, in the category of important raw materials for the development of drugs, with healing, anthelmintic, and antimicrobial properties, used in the treatment of various diseases. Buriti oil is highly valued and explored by the cosmetics market, being involved in the production of creams, soaps, and shampoos (Mansur et al. 2020).

Pequi oil, in addition to having a large amount of carotenoids and vitamin C, is rich in oleic acid, which is a monounsaturated fatty acid well known for its anti-inflammatory potential and for promoting cardiovascular health due to the reduction of LDL cholesterol levels (Carneiro et al. 2023). Its cardiovascular benefits have been confirmed in animal and human models, as well as its antimicrobial and antifungal properties against several strains (Coutinho et al. 2020).

Similar to pequi oil, mangaba oil has also demonstrated benefits in preventing cardiovascular diseases due to its high concentration of phytosterols, which reduce cholesterol absorption, and oleic acid content. Besides, mangaba oil is a good source of bioactive compounds with antioxidant properties, such as α-tocopherol and β-carotene, which indicates that the oil can neutralize oxidative stress (Lawrence 2021, Nattagh-Eshtivani et al. 2022).

Given their broad spectrum of therapeutic properties, Cerrado fruits present promising opportunities for nutraceutical and functional food applications. Their bioactive compounds, when incorporated into supplements, beverages, or pharmaceutical formulations, could offer natural alternatives for disease prevention and health promotion. However, despite their potential, these fruits remain underutilized, and further clinical studies and technological innovations are needed to enhance their bioavailability, stability, and commercial viability (Martins et al. 2015). Expanding research on the mechanisms of action, synergistic effects, and long-term benefits of Cerrado fruit-derived compounds will be essential for fully harnessing their therapeutic value and integrating them into modern dietary and medical practices.

10.4 Challenges in the Utilization of Cerrado Fruits

Despite their remarkable nutritional and therapeutic potential, fruits from the Brazilian Cerrado face significant barriers to commercialization and large-scale utilization. These challenges stem from logistical limitations, difficulties in the extraction and preservation of bioactive compounds, and low consumer awareness, all of which hinder their integration into global food and pharmaceutical markets (Schiassi et al. 2018).

One of the primary logistical and commercial challenges is the remote location of Cerrado fruit-producing areas, which often lack the necessary infrastructure for efficient harvesting, transportation, and storage. Many native fruit species are wild-harvested rather than cultivated on an industrial scale, making their supply inconsistent and highly seasonal. For instance, buriti, pequi, and murici are traditionally collected from natural ecosystems, leading to fluctuations in availability and price (Cândido et al. 2009, Calvano et al. 2019). Additionally, the lack of structured distribution channels means that these fruits are often sold in local markets at low prices or wasted due to their high perishability. The absence of adequate refrigeration and storage technologies exacerbates this issue, leading to post-harvest losses that can reach up to 40% in some regions (Siviero et al. 2021).

In addition to the precarious infrastructure in local communities, another critical aspect of the use and development of the Cerrado fruit production chain concerns the lack of investment in training for local producers, limiting the commercialization and generation of sustainable value for Cerrado fruit products (Lima et al. 2020). Traditional processing techniques, such as thermal pasteurization, often lead to losses in bioactive content, reducing their nutraceutical efficacy. In contrast, advanced methods such as ultrasound-assisted extraction (UAE) and supercritical fluid extraction (SFE) have shown promising results in maximizing the retention of bioactive compounds, but their high costs and limited availability restrict their widespread adoption. Additionally, the standardization of bioactive content in Cerrado fruit-based products remains a challenge, limiting their application in functional foods and pharmaceuticals (Silva et al. 2022).

Beyond logistical and processing constraints, consumer awareness and market acceptance present further obstacles to the large-scale utilization of Cerrado fruits. Many of these species, including tucum, guabiraba, and tuturubá, are virtually unknown to the general public, both in Brazil and internationally (Raphaelli et al. 2021). The lack of marketing campaigns, branding strategies, and regulatory frameworks for these fruits has contributed to their underrepresentation in mainstream food markets. Additionally, some Cerrado fruits have intense flavors or unconventional textures, which may require product reformulation or consumer education efforts to enhance their acceptance (Machado et al. 2022).

10.5 Future Prospects for the Utilization of Cerrado Fruits

The valorization of Cerrado fruits represents an opportunity for economic, nutritional, and pharmaceutical advancements, given their rich bioactive composition and potential health benefits (Cardozo et al. 2018). However, their large-scale utilization requires strategic investments, technological development, and supportive public policies to integrate them effectively into global markets. The following perspectives highlight potential pathways to expand the use of Cerrado fruits in the food and pharmaceutical industries, the necessity for research and innovation, and the role of public policies in promoting sustainability and economic growth (Reis et al. 2021, Borges et al. 2022a).

The incorporation of Cerrado fruits into functional foods, dietary supplements, and pharmaceutical formulations has gained traction due to their high content of phenolic compounds, carotenoids, and dietary fibers, which are associated with antioxidant, anti-inflammatory, and metabolic regulatory effects. However, scalability remains a challenge, primarily due to irregular supply chains, lack of standardized extraction techniques, and limited processing infrastructure (Melo et al. 2021).

One of the most promising approaches is the development of value-added products, such as freeze-dried fruit powders, nutraceutical capsules, and bioactive extracts for medicinal use. These innovations could enhance market stability and increase consumer accessibility (Silva et al. 2018, Raphaelli et al. 2021, Kultys and Kurek 2022). Additionally, the pharmaceutical industry could leverage bioactive-rich extracts from fruits like buriti and pequi, which have been linked to neuroprotective and cardioprotective properties (Vargas-Murga et al. 2016).

In the food industry, functional beverages, protein bars, and fermented products enriched with Cerrado fruit extracts could serve as a gateway for global market penetration. Emerging biotechnological techniques, such as microencapsulation and nanotechnology, can improve bioavailability and stability of the bioactive compounds, facilitating their incorporation into longer shelf-life products (Market Reports 2021, Saini et al. 2022).

Beyond pharmacological validation, agricultural research must focus on improving cultivation techniques to ensure higher productivity and sustainability. Advances in agroforestry systems (AFS), which integrate Cerrado fruit-bearing trees with traditional crops, could increase yield stability while preserving biodiversity (Martinelli et al. 2019, Melo et al. 2021). Furthermore, post-harvest technologies such as controlled-atmosphere storage and cold-chain logistics need further development to minimize post-harvest losses and extend distribution networks.

To further enhance commercialization, public-private partnerships should support marketing campaigns emphasizing the health benefits and cultural significance of Cerrado fruits. Educational programs targeting both consumers and industry stakeholders can bridge the knowledge gap and facilitate consumer acceptance in domestic and export markets.

10.6 Conclusion

The biodiversity of the Brazilian Cerrado presents an invaluable nutritional and therapeutic reservoir, with native fruits offering a rich composition of essential nutrients and bioactive compounds. The high content of phenolic compounds, carotenoids, dietary fibers, and essential minerals confers antioxidant, anti-inflammatory, antimicrobial, and metabolic-regulating properties, positioning these fruits as potential functional foods and nutraceuticals. The unique biochemical composition of species such as buriti, pequi, jatobá-do-cerrado, and gabiroba highlights their potential contributions to human health, particularly in the prevention of chronic diseases.

Despite their scientific and economic relevance, Cerrado fruits remain underutilized and undervalued in both national and international markets. The lack of investment in processing infrastructure, logistical barriers, and limited consumer awareness restricts their widespread adoption. Future research should focus on comprehensive clinical trials, validating the therapeutic efficacy of bioactive compounds and optimizing processing methods to enhance stability and bioavailability. Cerrado fruits have the potential to play a pivotal role in global food security and health promotion. Expanding their utilization requires a collective effort from the scientific community, industry stakeholders, and policymakers to bridge the gap between traditional knowledge and modern food systems.

References

Aguiar, J. P. L. and Souza, F. das C. do. (2017). Dehydration and spraying of buriti pulp (*Mauritia flexuosa* L.): Shelf-life evaluation. Revista Brasileira de Fruticultura 39(1): 1–7. https://doi.org/10.1590/0100-29452017.

Ajith, T. A. (2020). Possible therapeutic effect of magnesium in ocular diseases. Journal of Basic and Clinical Physiology and Pharmacology 31(2): 1–6. https://doi.org/10.1515/jbcpp-2019-0107.

Alexandre, C., Vasconcelos, S., Borges, R., Guilherme, L., Gustavo, G., Machado, L. et al. (2025). Proteome and metabolome of Annona crassiflora Mart . fruit and their interaction during development. Scientia Horticulturae 339(November 2024). https://doi.org/10.1016/j.scienta.2024.113809.

Alves, A. M., Fernades, D. C., Sousa, A. G. de O., Naves, R. V. and Naves, M. M. V. (2014). Physical and nutritional characteristics of pequi fruits from Tocantins, Goiás e Minas Gerais. Brazilian Journal of Food Technology 17(3): 198–203.

Araujo, R. L. de, Tomás-Barberán, F. A., Santos, R. F. dos, Martinez-Blazquez, J. A. and Genovese, M. I. (2021a). Postprandial glucose-lowering effect of cagaita (Eugenia dysenterica DC) fruit juice in dysglycemic subjects with metabolic syndrome: An exploratory study. Food Research International 142(September 2020). https://doi.org/10.1016/j.foodres.2021.110209.

Araujo, N. M. P., Arruda, H. S., Marques, D. R. P., de Oliveira, W. Q., Pereira, G. A. and Pastore, G. M. (2021b). Functional and nutritional properties of selected Amazon fruits: A review. Food Research International 147(February). https://doi.org/10.1016/j.foodres.2021.110520.

Arruda, H. S., Araújo, M. V. L. and Marostica Junior, M. R. (2022). Underexploited Brazilian Cerrado fruits as sources of phenolic compounds for diseases management: A review. Food Chemistry: Molecular Sciences 5(November). https://doi.org/10.1016/j.fochms.2022.100148.

Aziz, E., Batool, R., Akhtar, W., Rehman, S., Shahzad, T., Malik, A. et al. (2020). Xanthophyll: Health benefits and therapeutic insights. Life Sciences 240(November 2019). https://doi.org/10.1016/j.lfs.2019.117104.

Bailão, E. F. L. C., Devilla, I. A., da Conceição, E. C. and Borges, L. L. (2015). Bioactive compounds found in Brazilian cerrado fruits. International Journal of Molecular Sciences 16(10): 23760–23783. https://doi.org/10.3390/ijms161023760.

Baky, M. H., Elsaid, M. B. and Farag, M. A. (2022). Phytochemical and biological diversity of triterpenoid saponins from family Sapotaceae: A comprehensive review. Phytochemistry 202(August): 113345. https://doi.org/10.1016/j.phytochem.2022.113345.

Baptista, A., Gonçalves, R. V., Bressan, J. and do Carmo Gouveia Pelúzio, M. (2018). Antioxidant and antimicrobial activities of crude extracts and fractions of cashew (Anacardium occidentale L.), cajui (Anacardium microcarpum), and pequi (Caryocar brasiliense C.): A systematic review. Oxidative Medicine and Cellular Longevity 13. https://doi.org/10.1155/2018/3753562.

Batlle-Bayer, L., Batjes, N. H. and Bindraban, P. S. (2010). Changes in organic carbon stocks upon land use conversion in the Brazilian Cerrado: A review. Agriculture, Ecosystems and Environment 137(1–2): 47–58. https://doi.org/10.1016/j.agee.2010.02.003.

Berto, A., da Silva, A. F., Visentainer, J. V., Matsushita, M. and de Souza, N. E. (2015). Proximate compositions, mineral contents and fatty acid compositions of native Amazonian fruits. Food Research International 77: 441–449. https://doi.org/10.1016/j.foodres.2015.08.018.

Bhatt, T. and Patel, K. (2020). Carotenoids: Potent to prevent diseases review. Natural Products and Bioprospecting 10(3): 109–117. https://doi.org/10.1007/s13659-020-00244-2.

Boleti, A. P. D. A., Freire, M. D. G. M., Coelho, M. B., Da Silva, W., Baldasso, P. A., Gomes, V. M. et al. (2007). Insecticidal and antifungal activity of a protein from Pouteria torta seeds with lectin-like properties. Journal of Agricultural and Food Chemistry 55(7): 2653–2658. https://doi.org/10.1021/jf0636317.

Borges, M. S., Barbosa, R. S., Rambo, M. K. D., Rambo, M. C. D. and Scapin, E. (2022a). Evaluation of residual biomass produced in Cerrado Tocantinense as potential raw biomass for biorefinery. Biomass Conversion and Biorefinery 12(8): 3055–3066. https://doi.org/10.1007/s13399-020-00892-x.

Borges, P. S. B., Edelenbos, M., Larsen, E., Hernandes, T., Nunes, E. E., de Barros Vilas Boas, E. V. et al. (2022b). The bioactive constituents and antioxidant activities of ten selected Brazilian Cerrado fruits. Food Chemistry: X 14(February). https://doi.org/10.1016/j.fochx.2022.100268.

Brandelli, C. L. C., Vieira, P. D. B., Macedo, A. J. and Tasca, T. (2013). Remarkable anti-trichomonas vaginalis activity of plants traditionally used by the mbyá-guarani indigenous group in Brazil. BioMed Research International 2013. https://doi.org/10.1155/2013/826370.

Caldeira, A. S. P., Mbiakop, U. C., Pádua, R. M., van de Venter, M., Matsabisa, M. G., Campana, P. R. V. et al. (2021). Bioguided chemical characterization of pequi (Caryocar brasiliense) fruit peels towards an anti-diabetic activity. Food Chemistry 345(November 2020): 1–10. https://doi.org/10.1016/j.foodchem.2020.128734.

Calvano, M. P. C. A., Brumatti, R. C., Garcia, M. V., Barros, J. C. and Andreotti, R. (2019). Economic efficiency of Rhipicephalus microplus control and effect on beef cattle performance in the Brazilian Cerrado. Experimental and Applied Acarology 79(3–4): 459–471. https://doi.org/10.1007/s10493-019-00446-5.

Cândido, P. de A., Malafaia, G. da C. and Rezende, M. L. (2009). A exploração do pequi na região norte de Minas Gerais : abordagem por meio do Sistema Agroalimentar Localizado. Interfaces Em Desenvolvimento, Agricultura e Sociedade 5(2): 118–138.

Cândido, T. L. N. and Silva, M. R. (2017). Comparison of the physicochemical profiles of buriti from the Brazilian Cerrado and the Amazon region. Food Science and Technology (Brazil) 37: 78–82. https://doi.org/10.1590/1678-457X.32516.

Cardozo, C. M. L., Inada, A. C., Marcelino, G., Figueiredo, P. S., Arakaki, D. G., Aiko Hiane, P. et al. (2018). Therapeutic potential of Brazilian cerrado campomanesia species on metabolic dysfunctions. Molecules 23(9): 1–17. https://doi.org/10.3390/molecules23092336.

Carneiro, C. R., Alhaji, A. M., da Silva, C. A. S., de Sousa, R. de C. S., Monteiro, S. and Coimbra, J. S. dos R. (2023). Potential challenges of the extraction of carotenoids and fatty acids from Pequi (Caryocar brasiliense) oil. Foods 12(9): 1–21. https://doi.org/10.3390/foods12091907.

Carvalho, H. J. M., Barcia, M. T. and Schmiele, M. (2025). Underexploited fruits from the Brazilian Cerrado: Biodiversity, phenolic composition and biological activities. Food Bioscience 66(December 2024). https://doi.org/10.1016/j.fbio.2025.106269.

Carvalho, L. M. de S., Lemos, M. C. M., Sanches, E. A., da Silva, L. S., de Araújo Bezerra, J., Aguiar, J. P. L. et al. (2020). Improvement of the bioaccessibility of bioactive compounds from Amazon fruits treated using high energy ultrasound. Ultrasonics Sonochemistry 67(April): 105148. https://doi.org/10.1016/j.ultsonch.2020.105148.

Cavalcante, R. B. M., Araújo, M. A. D. M., Rocha, M. D. M. and Moreira-Araújo, R. S. D. R. (2017). Effect of thermal processing on chemical compositions, bioactive compounds, and antioxidant activities of cowpea cultivars. Revista Caatinga 30(4): 1050–1058. https://doi.org/10.1590/1983-21252017v30n426rc.

Chen, G. L., Fan, M. X., Wu, J. L., Li, N. and Guo, M. Q. (2019). Antioxidant and anti-inflammatory properties of flavonoids from lotus plumule. Food Chemistry 277(April 2018): 706–712. https://doi.org/10.1016/j.foodchem.2018.11.040.

Christe, C., Boluda, C. G., Koubínová, D., Gautier, L. and Naciri, Y. (2021). New genetic markers for Sapotaceae phylogenomics: More than 600 nuclear genes applicable from family to population levels. Molecular Phylogenetics and Evolution 160. https://doi.org/10.1016/j.ympev.2021.107123.

Cornelio-Santiago, H. P., Bodini, R. B. and Oliveira, A. L. de. (2021). Potential of oilseeds native to amazon and Brazilian cerrado biomes: benefits, chemical and functional properties, and extraction methods. JAOCS, Journal of the American Oil Chemists' Society 98(1): 3–20. https://doi.org/10.1002/aocs.12452.

Costa, A. N., de Sá, É. R. A., Bezerra, R. D. S., Souza, J. L. and Lima, F. das C. A. (2021). Constituents of buriti oil (Mauritia flexuosa L.) like inhibitors of the SARS-Coronavirus main peptidase: an investigation by docking and molecular dynamics. Journal of Biomolecular Structure and Dynamics 39(13): 4610–4617. https://doi.org/10.1080/07391102.2020.1778538.

Coutinho, D. de S., Pires, J., Gomes, H., Pohlmann, A. R., Guterres, S. S., E Silva, P. M. R. et al. (2020). Pequi (Caryocar brasiliense cambess)-loaded nanoemulsion, orally delivered, modulates inflammation in lps-induced acute lung injury in mice. Pharmaceutics 12(11): 1–17. https://doi.org/10.3390/pharmaceutics12111075.

Cruz, E. D. (2019). Germinação de sementes de espécies amazônicas: tuturubá [Pouteria venosa (Mart.) Baehni]. In Embrapa Comunicado Técnico (p. 6).

Curimbaba, T. F. S., Almeida-Junior, L. D., Chagas, A. S., Quaglio, A. E. V., Herculano, A. M. and Di Stasi, L. C. (2020). Prebiotic, antioxidant and anti-inflammatory properties of edible Amazon fruits. Food Bioscience 36(October 2018): 100599. https://doi.org/10.1016/j.fbio.2020.100599.

da Silva, B. A., Gordon, A., Jungfer, E., Marx, F. and Maia, J. G. S. (2012). Antioxidant capacity and phenolics of Pouteria macrophylla, an under-utilized fruit from Brazilian Amazon. European Food Research and Technology 234(5): 761–768. https://doi.org/10.1007/s00217-012-1684-0.

da Silva, E. K. A. and Rial, R. C. (2025). Fatty acid profile, nutritional and therapeutic properties of vegetable oils from the Brazilian Cerrado. Journal of Food Composition and Analysis 137(PA): 106819. https://doi.org/10.1016/j.jfca.2024.106819.

da Silva, É. R. S., Salmazzo, G. R., da Silva Arrigo, J., Oliveira, R. J., Kassuya, C. A. L. and Cardoso, C. A. L. (2016). Anti-inflammatory evaluation and toxicological analysis of Campomanesia xanthocarpa Berg. Inflammation 39(4): 1462–1468. https://doi.org/10.1007/s10753-016-0378-3.

Donado-Pestana, C. M., dos Santos-Donado, P. R., Daza, L. D., Belchior, T., Festuccia, W. T. and Genovese, M. I. (2018). Cagaita fruit (Eugenia dysenterica DC.) and obesity: Role of polyphenols on already established obesity. Food Research International 103: 40–47. https://doi.org/10.1016/j.foodres.2017.10.011.

Dzigbor, A., Neglo, D., Dzah, C. S. and Sraha, R. (2025). Total phenolic content, phytochemical screening, antioxidant and antimicrobial activities of Borassus flabellifer and Borassus aethiopum fruits. Food Chemistry Advances 7(November 2024): 100937. https://doi.org/10.1016/j.focha.2025.100937.

Ferruzzi, M. G., Kruger, J., Mohamedshah, Z., Debelo, H. and Taylor, J. R. N. (2020). Insights from *in vitro* exploration of factors influencing iron, zinc and provitamin A carotenoid bioaccessibility and intestinal absorption from cereals. Journal of Cereal Science 96: 103126. https://doi.org/10.1016/j.jcs.2020.103126.

Fiedor, J. and Burda, K. (2014). Potential role of carotenoids as antioxidants in human health and disease. Nutrients 6(2): 466–488. https://doi.org/10.3390/nu6020466.

Flach, R., Abrahão, G., Bryant, B., Scarabello, M., Soterroni, A. C., Ramos, F. M. et al. (2021). Conserving the Cerrado and Amazon biomes of Brazil protects the soy economy from damaging warming. World Development 146: 105582. https://doi.org/10.1016/j.worlddev.2021.105582.

Ganjeh, A. M., Pinto, C. A., Casal, S., Habanova, M., Holovicova, M. and Saraiva, J. A. (2025). Novel methods for extraction of monomeric phenols from plants and fruits with potential anti-obesity activity. Journal of Food Composition and Analysis 141(November 2024): 107383. https://doi.org/10.1016/j.jfca.2025.107383.

Garza, C. S. A.-D. la, Balois-Morales, R., Berumen-Varela, G., León-Fernández, A. E., Bautista-Rosales, P. U., López-Guzmán, G. G. et al. (2021). Physicochemical characterization and dietary fiber of 15 Nance (Byrsonima crassifolia L.) fruits selections from Nayarit. Scientia Horticulturae 289(February). https://doi.org/10.1016/j.scienta.2021.110460.

Gonçalves, A. E. D. S. S., Lajolo, F. M. and Genovese, M. I. (2010). Chemical composition and antioxidant/antidiabetic potential of brazilian native fruits and commercial frozen pulps. Journal of Agricultural and Food Chemistry 58(8): 4666–4674. https://doi.org/10.1021/jf903875u.

Gonçalves, M. L. M. B. B., Gonçalves, D., Sampaio Neto, O. Z., Meirelles, A. J. A., Batista, E. A. C., Sampaio, K. A. et al. (2025). Valorization of Amazonian buriti (Mauritia flexuosa) semi-defatted residual cake as a circular economy strategy: A techno-economical analysis. Industrial Crops and Products 226(June 2024): 120629. https://doi.org/10.1016/j.indcrop.2025.120629.

Gordon, A., Jungfer, E., Da Silva, B. A., Maia, J. G. S. and Marx, F. (2011). Phenolic constituents and antioxidant capacity of four underutilized fruits from the amazon region. Journal of Agricultural and Food Chemistry 59(14): 7688–7699. https://doi.org/10.1021/jf201039r.

Grillo, A. de C., Guedes, I. M. da S., Nicolai, J. C. and Fernandez, W. S. (2020). Importância E Atuação Dos Sais Minerais No Organismo. Revista Científica Eletrônica De Enfermagem Da Faef 5(3): 248–253. http://faef.revista.inf.br/imagens_arquivos/arquivos_destaque/BY50V66CJgicZcz_2020-7-7-8-45-37.pdf.

Gutiérrez-Molina, J., Corona-Rangel, M. L., Ventura-Aguilar, R. I., Barrera-Necha, L. L., Bautista-Baños, S. and Correa-Pacheco, Z. N. (2021). Chitosan and Byrsonima crassifolia-based nanostructured coatings: Characterization and effect on tomato preservation during refrigerated storage. Food Bioscience 42(June). https://doi.org/10.1016/j.fbio.2021.101212.

Haris, S., Alam, M., Galiwango, E., Mohamed, M. M., Kamal-Eldin, A. and Al-Marzouqi, A. H. (2023). Characterization analysis of date fruit pomace: An underutilized waste bioresource rich in dietary fiber and phenolic antioxidants. Waste Management 163(March): 34–42. https://doi.org/10.1016/j.wasman.2023.03.027.

Hirschmann, G. S. (1995). Flavonoids from Calycorectes, Campomanesia, Eugenia and Hexachlamys species. Fitoterapia 66: 373–374.

Hwang, K. A., Hwang, Y. J. and Song, J. (2016). Antioxidant activities and oxidative stress inhibitory effects of ethanol extracts from Cornus officinalis on raw 264.7 cells. BMC Complementary and Alternative Medicine 16(1): 1–9. https://doi.org/10.1186/s12906-016-1172-3.

Irías-Mata, A., Jiménez, V. M., Steingass, C. B., Schweiggert, R. M., Carle, R. and Esquivel, P. (2018). Carotenoids and xanthophyll esters of yellow and red nance fruits (Byrsonima crassifolia (L.) Kunth) from Costa Rica. Food Research International 111(2017): 708–714. https://doi.org/10.1016/j.foodres.2018.05.063.

Klafke, J. Z., da Silva, M. A., Panigas, T. F., Belli, K. C., de Oliveira, M. F., Barichello, M. M. et al. (2010). Effects of Campomanesia xanthocarpa on biochemical, hematological and oxidative stress parameters in hypercholesterolemic patients. Journal of Ethnopharmacology 127(2): 299–305. https://doi.org/10.1016/j.jep.2009.11.004.

Koolen, H. H. F., da Silva, F. M. A., Gozzo, F. C., de Souza, A. Q. L. and de Souza, A. D. L. (2013). Antioxidant, antimicrobial activities and characterization of phenolic compounds from buriti (Mauritia flexuosa L. f.) by UPLC-ESI-MS/MS. Food Research International 51(2): 467–473. https://doi.org/10.1016/j.foodres.2013.01.039.

Kultys, E. and Kurek, M. A. (2022). Green extraction of carotenoids from fruit and vegetable byproducts: a review. Molecules 27(2). https://doi.org/10.3390/molecules27020518.

Lawrence, G. D. (2021). Perspective: The saturated fat-unsaturated oil dilemma: Relations of dietary fatty acids and serum cholesterol, atherosclerosis, inflammation, cancer, and all-cause mortality. Advances in Nutrition 12(3): 647–656. https://doi.org/10.1093/advances/nmab013.

León, H. and Williams, J. (2009). Anatomía del xilema secundario de 14 especies del género pouteria Aubl. (Sapotaceae) en Venezuela. Acta Botanica Venezuelica 32(2): 433–451.

Li, J., Huang, S. Y., Deng, Q., Li, G., Su, G., Liu, J. et al. (2020). Extraction and characterization of phenolic compounds with antioxidant and antimicrobial activities from pickled radish. Food and Chemical Toxicology 136(December 2019): 111050. https://doi.org/10.1016/j.fct.2019.111050.

Lima, R. da S., Nunes, I. L. and Block, J. M. (2020). Ultrasound-assisted extraction for the recovery of carotenoids from guava's pulp and waste powders. Plant Foods for Human Nutrition 75(1): 63–69. https://doi.org/10.1007/s11130-019-00784-0.

Lorenzi, H. (2002). Árvores brasileiras: manual de identificação e cultivo de plantas arbóreas nativas do Brasil (3rd ed.). Instituto Plantarum.

Losada-Barreiro, S. and Bravo-Díaz, C. (2017). Free radicals and polyphenols: The redox chemistry of neurodegenerative diseases. European Journal of Medicinal Chemistry 133: 379–402. https://doi.org/10.1016/j.ejmech.2017.03.061.

Machado, A. P. da F., Nascimento, R. de P. do, Alves, M. da R., Reguengo, L. M. and Marostica Junior, M. R. (2022). Brazilian tucumã-do-Amazonas (Astrocaryum aculeatum) and tucumã-do-Pará (Astrocaryum vulgare) fruits: bioactive composition, health benefits, and technological potential. Food Research International 151(October 2021): 110902. https://doi.org/10.1016/j.foodres.2021.110902.

Machado, P. G., Londero, D. S., Alves, E. E., Farias, C. A. A., Pudenzi, M. A., Barcia, M. T. et al. (2025). Combining data from two mass spectrometric analyzers coupled to liquid chromatography (LC-ESI-QToF-MS/MS and LC-ESI-QqQ-MS/MS) to characterize the free and bound phenolic compounds from peel, pulp, and seeds of guabiroba fruits. Microchemical Journal 209(January): 112757. https://doi.org/10.1016/j.microc.2025.112757.

Malherbi, N. M., Schmitz, A. C., Grando, R. C., Bilck, A. P., Yamashita, F., Tormen, L. et al. (2019). Corn starch and gelatin-based films added with guabiroba pulp for application in food packaging. Food Packaging and Shelf Life 19(November 2018): 140–146. https://doi.org/10.1016/j.fpsl.2018.12.008.

Mansur, M. C. P. P. R., Campos, C., Vermelho, A. B., Nobrega, J., da Cunha Boldrini, L., Balottin, L. et al. (2020). Photoprotective nanoemulsions containing microbial carotenoids and buriti oil: Efficacy and safety study. Arabian Journal of Chemistry 13(8): 6741–6752. https://doi.org/10.1016/j.arabjc.2020.06.028.

Mariutti, L. R. B. and Mercadante, A. Z. (2018). Carotenoid esters analysis and occurrence: What do we know so far? Archives of Biochemistry and Biophysics 648(March): 36–43. https://doi.org/10.1016/j.abb.2018.04.005.

Market Reports. (2021). Carotenoids Market by Type (Astaxanthin, Beta-carotene, Lutein, Lycopene, Canthaxanthin, and Zeaxanthin), Application (Feed, Food & Beverages, Dietary Supplements, Cosmetics, and Pharmaceuticals), Source, Formulation, and Regeion - Global Forecast to 2026. MarketsandMarkets. https://www.marketsandmarkets.com/Market-Reports/carotenoid-market-158421566.html.

Martín-Gómez, J., Varo, M. Á., Mérida, J. and Serratosa, M. P. (2020). Influence of drying processes on anthocyanin profiles, total phenolic compounds and antioxidant activities of blueberry (Vaccinium corymbosum). Lwt 120(May 2019): 108931. https://doi.org/10.1016/j.lwt.2019.108931.

Martinelli, C., Maria, M., Parron, M., Vogel, E. and Favarini, C. (2019). Environmental performance of agroforestry systems in the Cerrado. World Development 122: 339–348. https://doi.org/10.1016/j.worlddev.2019.06.003.

Martínez-Vázquez, M., González-Esquinca, A. R., Cazares Luna, L., Moreno Gutiérrez, M. N. and García-Argáez, A. N. (1999). Antimicrobial activity of Byrsonima crassifolia (L.) H.B.K. Journal of Ethnopharmacology 66(1): 79–82. https://doi.org/10.1016/S0378-8741(98)00155-X.

Martínez, M. E., Lesher, J. M. G., Castañón, G. N., De La Cruz, E. L. and Zapata, C. H. (2013). Genetic variability of nanche in tabasco, mexico, determined with rapds. Phyton-International Journal of Experimental Botany 82: 209–214. https://doi.org/10.32604/phyton.2013.82.209.

Martins, J. L. R., Rodrigues, O. R. L., de Sousa, F. B., Fajemiroye, J. O., Galdino, P. M., Florentino, I. F. et al. (2015). Medicinal species with gastroprotective activity found in the Brazilian Cerrado. Fundamental and Clinical Pharmacology 29(3): 238–251. https://doi.org/10.1111/fcp.12113.

Medina-Torres, R., Salazar-García, S. and Gómez-Aguilar, J. R. (2004). Fruit quality indices in eight nance [Byrsonima crassifolia (L.) H.B.K.] selections. HortScience 39(5): 1070–1073. https://doi.org/10.21273/hortsci.39.5.1070.

Melo, C. De, Aparecido, L., Santos, C., Souza, G. De, Martins, T. D. O. and Castro, T. (2021). Bees in agroforestry systems in the Cerrado. Journal of Agricultural Research 0(0): 1–5. https://doi.org/10.1080/00218839.2021.1907977.

Melo, R. D., Fernandes, D. C., Alves, A. M. and Naves, M. M. V. (2014). Carotenoids are related to the colour and lipid content of the pequi (Caryocar brasiliense Camb.) pulp from the Brazilian Savanna. Food Science and Technology 34(3): 507–512. https://doi.org/10.1590/1678-457x.6369.

Menezes, E. G. O., Barbosa, J. R., Pires, F. C. S., Ferreira, M. C. R., de Souza e Silva, A. P., Siqueira, L. M. M. et al. (2022). Development of a new scale-up equation to obtain Tucumã-of-Pará (Astrocaryum vulgare Mart.) oil rich in carotenoids using supercritical CO_2 as solvent. Journal of Supercritical Fluids 181(October 2021): 105481. https://doi.org/10.1016/j.supflu.2021.105481.

Merwe, R. van der, Kruger, J., Ferruzzi, M. G., Duodu, K. G. and Taylor, J. R. N. (2019). Improving iron and zinc bioaccessibility through food-to-food fortification of pearl millet with tropical plant foodstuffs (moringa leaf powder, roselle calyces and baobab fruit pulp). Journal of Food Science and Technology 56(4): 2244–2256. https://doi.org/10.1007/s13197-019-03711-y.

Milanez, J. T., Neves, L. C., Colombo, R. C., Shahab, M. and Roberto, S. R. (2018). Bioactive compounds and antioxidant activity of buriti fruits, during the postharvest, harvested at different ripening stages. Scientia Horticulturae 227(December 2016): 10–21. https://doi.org/10.1016/j.scienta.2017.08.045.

Monteiro, G. de M., Carvalho, E. E. N., Lago, R. C. do, Silva, L. G. M. da, Souza, L. R. de, Costa, C. A. R. da et al. (2025). Compositional analysis of baru (Dipteryx alata Vogel) pulp highlighting its industrial potential. Food Research International 201(November 2024). https://doi.org/10.1016/j.foodres.2024.115548.

Montenegro, L. H. M., Oliveira, P. E. S., Conserva, L. M., Rocha, E. M. M., Brito, A. C., Araújo, R. M. et al. (2006). Terpenóides e avaliação do potencial antimalárico, larvicida, anti-radicalar e anticolinesterásico de Pouteria venosa (Sapotaceae). Revista Brasileira de Farmacognosia 16: 611–617. https://doi.org/10.1590/s0102-695x2006000500005.

Moreira, T. M. S., Salvagnini, L. E., Santos, E., Silva, V. Y. A., Moreira, R. R. D., Salgado, H. R. N. et al. (2011). Antidiarrheal activity of campomanesia xanthocarpa fruit. Journal of Medicinal Food 14(5): 528–531. https://doi.org/10.1089/jmf.2009.0278.

Nascimento, A. L. A. A., Brandi, I. V., Durães, C. A. F., Lima, J. P. De, Soares, S. B. and Mesquita, B. M. A. D. C. (2020). Chemical characterization and antioxidant potential of native fruits of the Cerrado of northern Minas Gerais. Brazilian Journal of Food Technology 23: 1–9. https://doi.org/10.1590/1981-6723.29619.

Nattagh-Eshtivani, E., Barghchi, H., Pahlavani, N., Barati, M., Amiri, Y., Fadel, A. et al. (2022). Biological and pharmacological effects and nutritional impact of phytosterols: A comprehensive review. Phytotherapy Research 36(1): 299–322. https://doi.org/10.1002/ptr.7312.

Neri-Numa, I. A., Soriano Sancho, R. A., Pereira, A. P. A. and Pastore, G. M. (2018). Small Brazilian wild fruits: Nutrients, bioactive compounds, health-promotion properties and commercial interest. Food Research International 103: 345–360. https://doi.org/10.1016/j.foodres.2017.10.053.

Nogueira, G. M., Cardoso, F. A. R., Perdoncini, M. R. F. G., da Silva, M. V. and Rigobello, E. S. (2025). Evaluation of the extraction of antioxidant compounds from buriti pulp (Mauritia flexuosa L.) by response surface methodology. Heliyon 11(2). https://doi.org/10.1016/j.heliyon.2025.e41926.

Nogueira, L. A., Figueiredo, Y. G., Ramos, A. L. C. C., Correia, V. T. da V., Nunes, B. V., Ribeiro, L. V. et al. (2022). The presence of flavonoids in some products and fruits of the Genus Eugenia: an integrative review. Frontiers in Food Science and Technology 2(May): 1–12. https://doi.org/10.3389/frfst.2022.899492.

Oliveira, C. S., Gonçalves, L. E. N., Coutinho, M. P., Peixoto, N. and Gatto, A. (2017). Social and Environmental Aspects of Pequi Marketing in Goiás. Floresta e Ambiente 24: 1–6. https://doi.org/10.1590/2179-8087.058213.

Oliveira, K. M., Abboud, K. Y., Radulski, D. R., Faria, B. C., Galindo, C. M., Pereira, G. S. et al. (2024). Polysaccharides extracted from tucum-do-cerrado fruits (Bactris setosa Mart) have antineoplastic effects in mice while preserving hepatic gluconeogenesis. International Journal of Biological Macromolecules 278(July). https://doi.org/10.1016/j.ijbiomac.2024.134590.

Otero, J. S., Hirsch, G. E., Klafke, J. Z., Porto, F. G., de Almeida, A. S., Nascimento, S. et al. (2017). Inhibitory effect of Campomanesia xanthocarpa in platelet aggregation: Comparison and synergism with acetylsalicylic acid. Thrombosis Research 154: 42–49. https://doi.org/10.1016/j.thromres.2017.03.020.

Pérez Gutiérrez, R. M. and Muñiz Ramirez, A. (2013). Hexane extract of the seeds of Byrsonima crassifolia accelerates wound healing in streptozotocin-induced diabetic rats. Chinese Journal of Integrative Medicine 1–7. https://doi.org/10.1007/s11655-013-1556-x.

Pires, J., Cargnin, S. T., Costa, S. A., Sinhorin, V. D. G., Damazo, A. S., Sinhorin, A. P. et al. (2020). Healing of dermal wounds property of Caryocar brasiliense oil loaded polymeric lipid-core nanocapsules: formulation and *in vivo* evaluation. European Journal of Pharmaceutical Sciences 150(March). https://doi.org/10.1016/j.ejps.2020.105356.

Rajan, N. S. and Bhat, R. (2019). Bioactive compounds in underutilized fruits and nuts. In Reference Series in Phytochemistry. https://doi.org/10.1007/978-3-030-06120-3_36-1.

Raphaelli, C. de O., Pereira, E. dos S., Camargo, T. M., Ribeiro, J. A., Pereira, M. C., Vinholes, J. et al. (2021). Biological activity and chemical composition of fruits, seeds and leaves of guabirobeira (Campomanesia xanthocarpa O. Berg – Myrtaceae): A review. Food Bioscience 40(July 2020). https://doi.org/10.1016/j.fbio.2021.100899.

Reis, J. C. dos, Rodrigues, G. S., Barros, I. de, Ribeiro Rodrigues, R. de A., Garrett, R. D., Valentim, J. F. et al. (2021). Integrated crop-livestock systems: A sustainable land-use alternative for food production in the Brazilian Cerrado and Amazon. Journal of Cleaner Production 283: 124580. https://doi.org/10.1016/j.jclepro.2020.124580.

Ribeiro Neto, J. A., Pimenta Tarôco, B. R., Batista dos Santos, H., Thomé, R. G., Wolfram, E. and Maciel de A Ribeiro, R. I. (2020). Using the plants of Brazilian Cerrado for wound healing: From traditional use to scientific approach. Journal of Ethnopharmacology 260(January). https://doi.org/10.1016/j.jep.2020.112547.

Ricker, M., Valencia-Avalos, S., Hernández, H. M., Gómez-Hinostrosa, C., Martínez-Salas, E. M., Alvarado-Cárdenas, L. O. et al. (2016). Especies arbóreas y arborescentes de México: Apocynaceae, Cactaceae, Ebenaceae, Fagaceae y Sapotaceae. Revista Mexicana de Biodiversidad 87(4): 1189–1202. https://doi.org/10.1016/j.rmb.2016.07.018.

Rodrigues, H. C. S., Silveira, M. P., Helm, C. V., de Matos Jorge, L. M. and Jorge, R. M. M. (2020). Gluten free edible film based on rice flour reinforced by guabiroba (Campomanesia xanthocarpa) pulp. Journal of Applied Polymer Science 137(41): 1–11. https://doi.org/10.1002/app.49254.

Rodrigues, P. M., Dutra Gomes, J. V., Jamal, C. M., Cunha Neto, Á., Santos, M. L., Fagg, C. W. et al. (2017). Triterpenes from Pouteria ramiflora (Mart.) Radlk. Leaves (Sapotaceae). Food and Chemical Toxicology 109: 1063–1068. https://doi.org/10.1016/j.fct.2017.05.026.

Rodrigues, P. S. de M., Cirqueira Martins, H., Falcão, M. S., Trevisan, M., Portaro, F. C. V., da Silva, L. G. et al. (2023). Effects of Mauritia flexuosa L. f. buriti oil on symptoms induced by Bothrops moojeni snake envenomation. Journal of Ethnopharmacology 313(March). https://doi.org/10.1016/j.jep.2023.116612.

Rossato, A., Silveira, L. da S., Lopes, L. Q. S., Filho, W. P. D. S., Schaffer, L. F., Santos, R. C. V. et al. (2019). Evaluation *in vitro* of antimicrobial activity of tucumã oil (Astrocaryum Vulgare). Archives in Biosciences & Health 1(1): 99–112. https://doi.org/10.18593/abh.19701.

Saini, R. K., Prasad, P., Lokesh, V., Shang, X., Shin, J., Keum, Y. S. et al. (2022). Carotenoids: Dietary sources, extraction, encapsulation, bioavailability, and health benefits—a review of recent advancements. Antioxidants 11(4). https://doi.org/10.3390/antiox11040795.

Sant'anna, L. S., Merlugo, L., Ehle, C. S., Limberger, J., Fernandes, M. B., Santos, M. C. et al. (2017). Chemical composition and hypotensive effect of Campomanesia xanthocarpa. Evidence-Based Complementary and Alternative Medicine 2017. https://doi.org/10.1155/2017/1591762.

Santana, L. F., Sasso, S., Aquino, D. F. S., de Cássia Freitas, K., de Cássia Avellaneda Guimarães, R., Pott, A. et al. (2022). Nutraceutic potential of bioactive compounds of Eugenia dysenterica DC in metabolic alterations. Molecules 27(8): 1–19. https://doi.org/10.3390/molecules27082477.

Santos, R. F. dos E. P., Silva Silva, I. S. de M., Hendges, E. A., da Silva, A. L. L., Barbosa, A. M., Santos, K. S. et al. (2014). Evaluation of antimicrobial potential and cytotoxic of Pouteria venosa species. BMC Proceedings 8(S4): 5–6. https://doi.org/10.1186/1753-6561-8-s4-p6.

Santos, R. F. E. P., Silva, I. S. M., Veríssimo, R. C. S. S., Lúcio, I. M. L., Campesatto, E. A., Conserva, L. M. et al. (2015). Estudo do potencial antimicrobiano e citotóxico da espécie Pouteria venosa (Sapotaceae). Revista Brasileira de Plantas Medicinais 17(3): 367–373. https://doi.org/10.1590/1983-084X/13_078.

Schiassi, M. C. E. V., Souza, V. R. de, Lago, A. M. T., Campos, L. G. and Queiroz, F. (2018). Fruits from the Brazilian Cerrado region: Physico-chemical characterization, bioactive compounds, antioxidant activities, and sensory evaluation. Food Chemistry 245(October 2017): 305–311. https://doi.org/10.1016/j.foodchem.2017.10.104.

Serra, J. L., Rodrigues, A. M. da C., de Freitas, R. A., Meirelles, A. J. de A., Darnet, S. H. and Silva, L. H. M. da. (2019). Alternative sources of oils and fats from Amazonian plants: Fatty acids, methyl tocols, total carotenoids and chemical composition. Food Research International 116: 12–19. https://doi.org/10.1016/j.foodres.2018.12.028.

Serrano, J., Richardson, J. E., Milne, R. I., Mondragon, G. A., Hawkins, J. A., Bartish, I. V. et al. (2021). Andean orogeny and the diversification of lowland neotropical rain forest trees: A case study in Sapotaceae. Global and Planetary Change 201(April). https://doi.org/10.1016/j.gloplacha.2021.103481.

Shokunbi, O. S., Adepoju, O. T., Ramaite, I. D. I., Shokunbi, O. S., Mojapelo, P. E. L. and Akinyele, I. O. (2023). Potassium, sodium, calcium and magnesium levels of commonly consumed foods and estimates of dietary intakes of selected Nigerian adults. Heliyon 9(3): e13729. https://doi.org/10.1016/j.heliyon.2023.e13729.

Silva, D. S. N., Silva, M. de S., Coelho, T. L. S., Dantas, C., Lopes Júnior, C. A., Caldas, N. M. et al. (2022). Combining high intensity ultrasound and experimental design to improve carotenoid extraction efficiency from Buriti (Mauritia flexuosa). Ultrasonics Sonochemistry 88(June): 106076. https://doi.org/10.1016/j.ultsonch.2022.106076.

Silva, E. K. A. da and Rial, R. C. (2025). Fatty acid profile, nutritional and therapeutic properties of vegetable oils from the Brazilian Cerrado. Journal of Food Composition and Analysis 137(PA): 106819. https://doi.org/10.1016/j.jfca.2024.106819.

Silva, L. L. da, Almeida, R. de, Silva, F. T. e and Verícimo, M. A. (2021). Review on the therapeutic activities of the genus Trichilia. Research, Society and Development 10(5): e29610514916. https://doi.org/10.33448/rsd-v10i5.14916.

Silva, M. R., Lacerda, D. B. C. L., Santos, G. G. and Martins, D. M. D. O. (2008). Chemical characterization of native species of fruits from savanna ecosystem. Ciencia Rural 38(6): 1790–1793. https://doi.org/10.1590/s0103-84782008000600051.

Silva, R. S., de L. Santos, C., Mar, J. M., Kluczkovski, A. M., de A. Figueiredo, J., Borges, S. V. et al. (2018). Physicochemical properties of tucumã (Astrocaryum aculeatum) powders with different carbohydrate biopolymers. Lwt 94(April): 79–86. https://doi.org/10.1016/j.lwt.2018.04.047.

Siviero, M. A., Rushel, A. R., Yared, J. A. G., Aguiar, O. J. R., Pereira, P. C. G., Vieira, S. B. et al. (2021). Harvesting criteria application as a forests: a case study from brazilian for management of degraded tropical technical and financial alternative amazon. pp. 117–136. In Silvicultura e Manejo Florestal: Técnicas de Utilização e Conservação da Natureza (Vol. 2). https://doi.org/10.37885/210404392.

Soares, A. M., Ticli, F. K., Marcussi, S., Lourenco, M. V., Januario, A. H., Sampaio, S. V. et al. (2005). Medicinal plants with inhibitory properties against snake venoms. Current Medicinal Chemistry 12(22): 2625–2641. https://doi.org/10.2174/092986705774370655.

Soares, C. M. D. S., Morais, R. A., dos Santos, A. L., Sousa, H. M. S., Teles, J. S., Silva, R. R. da et al. (2020). Quantificação De Compostos Fenólicos Totais Em Frutos Não Convencionais Cultivados No Cerrado. DESAFIOS - Revista Interdisciplinar Da Universidade Federal Do Tocantins 7(Especial): 65–71. https://doi.org/10.20873/uftsupl2020-8607.

Souza, R. O. de, Alves, G. de A. D., Forte, A. L. S. A., Marquele-Oliveira, F., da Silva, D. F., Rogez, H. et al. (2017). Byrsonima crassifolia extract and fraction prevent UVB-induced oxidative stress in keratinocytes culture and increase antioxidant activity on skin. Industrial Crops and Products 108(June): 485–494. https://doi.org/10.1016/j.indcrop.2017.07.015.

Souza, V. R. De, Pereira, P. A. P., Queiroz, F., Borges, S. V. and De Deus Souza Carneiro, J. (2012). Determination of bioactive compounds, antioxidant activity and chemical composition of Cerrado Brazilian fruits. Food Chemistry 134(1): 381–386. https://doi.org/10.1016/j.foodchem.2012.02.191.

Souza, N. F. da S. and Viana, D. dos S. F. (2018). ASPECTOS ECOLÓGICOS E POTENCIAL ECONÔMICO DO BURITI (Mauritia flexuosa). Agrarian Academy 5(9): 535. https://doi.org/10.18677/Agrarian_Academy_2018a52.

Stafussa, A. P., Maciel, G. M., Bortolini, D. G., Maroldi, W. V., Ribeiro, V. R., Fachi, M. M. et al. (2021). Bioactivity and bioaccessibility of phenolic compounds from Brazilian fruit purees. Future Foods 4(July): 100066. https://doi.org/10.1016/j.fufo.2021.100066.

TACO. (2011). NÚCLEO DE ESTUDOS E PESQUISAS EM ALIMENTAÇÃO – NEPA - Tabela Brasileira de Composição de Alimentos (NEPA-UNICAMP (ed.); 4th ed.).

Torres, L. R. d. O., Santana, F. C. d., Torres-Leal, F. L., Melo, I. L. P. d., Yoshime, L. T., Matos-Neto, E. M. et al. (2016). Pequi (Caryocar brasiliense Camb.) almond oil attenuates carbon tetrachloride-induced acute hepatic injury in rats: Antioxidant and anti-inflammatory effects. Food and Chemical Toxicology 97: 205–216. https://doi.org/10.1016/j.fct.2016.09.009.

Udomkun, P., Tirawattanawanich, C., Ilukor, J., Sridonpai, P., Njukwe, E., Nimbona, P. et al. (2019). Promoting the use of locally produced crops in making cereal-legume-based composite flours: An assessment of nutrient, antinutrient, mineral molar ratios, and aflatoxin content. Food Chemistry 286(October 2018): 651–658. https://doi.org/10.1016/j.foodchem.2019.02.055.

Vargas-Murga, L., de Rosso, V. V., Mercadante, A. Z. and Olmedilla-Alonso, B. (2016). Fruits and vegetables in the Brazilian Household Budget Survey (2008–2009): Carotenoid content and assessment of individual carotenoid intake. Journal of Food Composition and Analysis 50: 88–96. https://doi.org/10.1016/j.jfca.2016.05.012.

Zacarías-García, J., Lux, P. E., Carle, R., Schweiggert, R. M., Steingass, C. B., Zacarías, L. et al. (2021). Characterization of the Pale Yellow Petal/Xanthophyll Esterase gene family in citrus as candidates for carotenoid esterification in fruits. Food Chemistry 342(October 2020): 128322. https://doi.org/10.1016/j.foodchem.2020.128322.

Section III

Regulations and Future Functional Foods

Chapter 11

Regulatory Framework for Bioactive Compounds in Fruits

João Paulo Rodrigues da Silva,[1,*] *Thayson Rodrigues Lopes,*[1] *Pelrry da Silva Costa,*[1] *Leonne Lopes Brito,*[1] *Gabriel e Silva Sales*[1] and *Valdiléia Teixeira Uchôa*[2]

11.1 Introduction

Fruit production is a major component of global agriculture and plays a vital role in supporting human nutrition and cognitive development. The numerous benefits of fruit for human health are associated with its chemical composition, which contains various bioactive compounds such as polyphenols, carotenoids, alkaloids, lipids, polysaccharides, vitamins, and mineral salts (Galanakis 2019). Polyphenols, which include flavonoids, tannins, and phenolic compounds, play an important role in neutralizing reactive species by donating electrons to free radicals and activating antioxidant biomolecules (Karimi et al. 2021). These actions contribute to reducing inflammation, helping to prevent some diseases such as diabetes, obesity, Parkinson's, Alzheimer's, and others (Araujo et al. 2021). The bioactive compounds present in fruit have gained prominence not only because of their health benefits, but also because of the growing demand for functional and natural products on the market (Maqsood et al. 2019).

Brazil stands out for being one of the largest producers and exporters of fruit on the planet, with a climate and soil conducive to various species (Paz et al. 2015).

[1] Graduate Program in Chemistry, Chemistry Department, Federal University of Piauí - UFPI, P.O. Box 64049-550, Teresina, PI, Brazil.

[2] Graduate Program in Chemistry PPGQ-GERATEC, Universidade Estadual do Piauí-UESPI Joao Cabral N.223, P.O. Box 381, CEP-64002-150, Teresina, PI, Brazil.

* Corresponding author: joaopaulor.dasilva@hotmail.com

DOI: 10.1201/9781003668169-14

However, the regulation of these products is essential to ensure consumer safety and the effectiveness of health claims, creating an environment conducive to innovation in the sector. This chapter discusses the main regulatory aspects involving bioactive compounds in fruit, focusing on legislation and guidelines.

This chapter aims to present and discuss the main regulatory frameworks involving bioactive compounds present in fruit, although most do not have specific regulations for bioactive compounds, but for their carriers—in this case, fruit—considering different legal and geographical contexts. The regulations in force in countries such as Brazil, the United States and the members of the European Union, Canada, India, and China are analyzed, as well as the guidelines established by the OECD, with a focus on requirements related to labeling, consumer safety, and functional and health claims. By comparing these regulatory frameworks, the aim is to understand how they influence scientific development, technological innovation, and the marketing of fruit with potential health benefits on a global scale.

11.2 The Importance of Regulatory Supervision in the Safety of Bioactive Compounds in Fruit

The growing interest in fruits rich in bioactive compounds, both in scientific circles and in the production sector, is directly related to the recognition of their possible positive effects on human health. However, for these benefits to be effectively delivered to the consumer safely, effective regulatory mechanisms are essential (Shashirekha et al. 2015).

Despite their natural origin, many of these compounds can be toxic when consumed in high quantities or by groups with greater physiological sensitivity (Nath et al. 2022). In addition, there is a risk of adverse effects resulting from interaction with drugs or other components of the diet, which requires rigorous toxicological and clinical assessments before releasing these ingredients in food products (Vilas-Boas et al. 2021, Duda-Chodak and Tarko 2023).

Another relevant challenge concerns the variability of bioactive content in fruit. Factors such as species, variety, agricultural practices, climatic conditions, processing, and storage methods can have a significant impact on the chemical composition of the food. This variability makes it difficult to standardize products and compromises the consistency of functional claims, making it even more necessary to adopt clear criteria for quality control (Vidinamo et al. 2021).

In this context, regulatory frameworks must guarantee that products with health appeals, such as fresh fruit with functional properties, juices, pulps, and extracts, are subject to standardization and validation processes, ensuring constant and safe levels of bioactive compounds. The role of regulatory agencies, therefore, is central to promoting transparency in labeling, protecting consumer health, and encouraging responsible practices on the part of the industry (Granato et al. 2023).

In this way, regulatory supervision plays a fundamental role not only in preventing risks to public health, but also in the scientific and commercial valorization of functional foods derived from fruit, creating a more reliable environment for innovation and consumption.

11.3 Regulation Around the World

The regulation of bioactive compounds in fruit, and food in general, shows significant variations between different countries and economic blocs, reflecting not only the legislative particularities of each nation, but also public health priorities, available scientific advances, and the degree of maturity of the food sector. In general, international regulatory frameworks aim to ensure consumer safety, standardize functional claims, and avoid misleading marketing practices, especially when it comes to substances with a potential physiological impact (Martins et al. 2023).

Although there is a growing consensus on the need for scientific substantiation to support the use of bioactive compounds and their respective claims, the criteria required, the regulatory categories adopted (such as functional foods, supplements, or nutraceuticals), and the approval procedures vary widely (Cansanção et al. 2025, Fernandes et al. 2022).

On the other hand, developing countries face additional challenges related to the lack of specific standards, the scarcity of local studies, and the need to harmonize their legislation with international standards, especially in contexts of export and global competitiveness (Cansanção et al. 2025).

11.3.1 United States

In the United States of America (USA), the regulation of bioactive compounds in foods, including fruits with functional properties, is mainly conducted by the Food and Drug Administration (FDA). The US regulatory approach clearly distinguishes between conventional foods, food supplements, and drugs, and the classification of a product determines the stringency of the applicable legal requirements (Komala et al. 2023). Bioactive compounds that act to promote health or prevent disease can be incorporated into food supplements, as long as they are considered safe based on their GRAS (Generally Recognized as Safe) status or are approved as food ingredients through formal toxicological evaluation processes (Son and Martirosyan 2024, Zovi et al. 2025).

The regulation of bioactive compounds in foods, including fruits and their derivatives, is mainly governed by the Dietary Supplement Health and Education Act (DSHEA), enacted in 1994 (Wallace 2015). This legislation established the legal basis for the marketing of dietary supplements, allowing the inclusion of bioactive substances such as phytosterols, flavonoids, and natural antioxidants present in fruit, provided that no therapeutic indications are claimed, which would otherwise place the product within the regulatory definition of a medicinal product.

Fresh fruit, processed fruit, or fruit used as a base for functional products may contain claims of functional or health properties on their labels. However, these claims must be notified to the Food and Drug Administration (FDA) and accompanied by the standard warning stating that the product has not been evaluated by the agency for the prevention or treatment of disease (Tafese Awulachew 2024). Specific claims, such as the association between the consumption of a certain fruit with bioactive compounds and the reduction or cure of diseases, require robust scientific support, usually based on controlled clinical studies, and are subject to pre-approval by the FDA when applied to conventional foods (Wallace 2015).

Another fundamental aspect of US legislation is the focus on clear and precise labeling. All the bioactive compounds present in the product must be duly identified on the label, and the claims must not mislead the consumer (Martirosyan and Ekblad 2022). The FDA takes a reactive approach to enforcement, acting mainly on complaints or cases that pose a risk to public health. When necessary, the agency can issue warnings, demand the withdrawal of products from the market, or apply legal sanctions (Fortin 2009).

Despite offering greater regulatory flexibility to the industry, the US system imposes on the manufacturer the responsibility of scientifically proving the efficacy and safety of the compounds added to the products, which makes investment in quality scientific research essential (Taylor 2024). This model therefore represents an opportunity, but also a challenge for the export and marketing of fruit and functional foods in the United States, requiring strict compliance with the criteria established by the FDA.

11.3.2 European Union

In the European Union, the bioactive compounds present in fruit and fruit derivatives are regulated by Regulation (EC) No. 1924/2006, which establishes guidelines on nutrition and health claims associated with food. This legal framework ensures that claims made on labels, packaging, or promotional materials are based on rigorous scientific evidence to avoid misleading practices and guarantee consumer protection (European Union 2006).

Fruits rich in phytochemicals, such as polyphenols, anthocyanins, flavonoids, and phytosterols, can have their functional properties highlighted in products marketed in the European Union, provided they meet the requirements set by the European Food Safety Authority (EFSA) (Ververis et al. 2020). EFSA is responsible for scientifically assessing the efficacy and safety of bioactive compounds used in food, based on data from clinical and toxicological studies (Vettorazzi et al. 2020). These studies must consistently demonstrate that the claimed benefits are real and achievable by consuming the amounts declared on the labels.

In addition, Directive 2001/18/EC deals with the deliberate release of genetically modified organisms (GMOs), including transgenic fruit developed to have higher levels of functional compounds (Paraskevopoulos and Federici 2021). The regulation requires an in-depth analysis of the safety of these products, considering not only toxicity but also possible interactions between bioactive compounds and drugs, an aspect of growing relevance in the scientific literature (Schuler et al. 2019).

Although the European Union has a robust regulatory system, it still faces challenges such as the need to harmonize interpretations between member states, the scarcity of scientific data for some natural substances present in fruit, and the rapid pace of technological innovation in the functional food sector. Even so, the European regulatory environment is considered favorable to scientific research and the development of new products, especially in the context of fruit, offering opportunities for the introduction of foods with recognized health benefits and in compliance with current legislation.

11.3.3 Regulations in Brazil

The legal basis for food control in Brazil was established by Decree-Law No. 986 of October 21, 1969, which instituted basic rules on food to guarantee standards of identity, quality, and safety for human consumption. This initial regulatory framework outlined the principles that guide, to this day, the formulation of public policies, inspection, and regulation of food products, including those containing bioactive compounds (Brasil 1969). Since then, the country has expanded and improved its legislation to keep up with scientific, technological, and commercial advances in the food and nutrition sector.

Brazilian regulation is characterized by the integrated action of two sectors of the Public Administration: the agricultural sector, represented by the Ministry of Agriculture, Livestock and Supply (MAPA), and the health sector, under the responsibility of the National Health Surveillance Agency (Anvisa). The competence of these bodies is defined by a complex set of legal and infra-legal rules, ensuring that supervision and control cover all stages of the production chain, from cultivation to marketing.

MAPA is responsible for the stages related to primary production, plant classification, and inspection of beverages, while the National Health Surveillance Agency (ANVISA) acts more directly in the sanitary regulation of industrialized foods, labeling, use of additives, new ingredients, and functional claims (BRASIL 1969, ANVISA 1999, MAPA 2018). In the case of bioactive compounds present in fruit, inspection can involve both spheres (agriculture and health) depending on the type of product and its degree of processing, and competence is often shared between agencies.

Regarding the visual quality and minimum standard for consumption of fruit and vegetables, MAPA Normative Instruction 69/2018 establishes mandatory criteria for classifying these products. According to this regulation, vegetables must be whole, clean, firm, physiologically developed, and free of visible pests, with no foreign odors, deep damage, signs of rot, dehydration, wilting, or freezing allowed. This standardization aims to make it easier to identify products that are fit for consumption, both for consumers and for agricultural inspectors, who receive specific training to identify items outside the established standards (MAPA 2018).

In the context of foods of plant origin, especially fruit containing bioactive compounds, Brazilian regulations are clear on the need for scientific backing for any functional or health property claims. Anvisa requires these claims to be supported by robust scientific evidence, which may include clinical trials, toxicological studies, and epidemiological data, depending on the nature of the compound and the claimed effect (ANVISA 1999).

Fruits or derivatives with characteristics such as the presence of bioactive compounds, including antioxidants, phytosterols, or other phytochemicals, can fall into specific regulatory categories such as foods with functional and health claims, novel foods, or novel ingredients, according to the criteria established by ANVISA

(ANVISA 1999). Each of these categories requires scientific documentation, toxicological safety data, and, in some cases, clinical trials, as well as compliance with their own technical and procedural requirements. The labeling of these products is also subject to strict rules and must contain clear information on the bioactive compound present, its effective quantity required to obtain the claimed effect, as well as mandatory warnings whenever necessary to ensure consumer safety (ANVISA 1999).

Thus, compliance with the requirements established by MAPA and ANVISA is an important prerequisite for ensuring the safety, standardization, and functional potential of the plant raw materials used in the food, supplement, and natural product industries, and consequently of their bioactive products.

11.3.4 Regulations of Bioactive Compounds in Canada

In Canada, products with bioactive compounds can be regulated under different categories according to their characteristics and functional claims. One of the main legal frameworks is Natural Health Products (NHPs), which are overseen by Health Canada through the Natural Health Products and Non-Prescription Drugs Directorate (NNHPD) (Health Canada 2020a).

Since the Natural Health Products Regulations came into force in January 2004, more than 43,000 products have been legally licensed for sale in the country (Health Canada 2013). NHPs include a wide variety of substances: vitamins, minerals, herbal remedies, traditional medicines, homeopathics, probiotics, amino acids, and essential fatty acids (Spagnuolo 2020).

To obtain a marketing license, Natural Health Products (NHPs) must go through a rigorous pre-market approval process. This process requires a full description of the medicinal and non-medicinal ingredients present in the formulation, details of the manufacturing process, presentation of the proposed claims, and submission of scientific evidence proving the safety, efficacy, and quality of the health product (Health Canada 2020b).

Health Canada accepts a variety of sources as the basis for this assessment, including clinical trials, scientific publications, pharmacopoeias, and traditional knowledge. The type and strength of evidence required vary according to the potential risk and the nature of the health claim. Products that claim, for example, to improve cardiovascular function or offer antioxidant protection based on fruit extracts need to scientifically demonstrate such effects (Health Canada 2020c).

In addition to initial approval, NHP manufacturers must have a local operating license and follow strict requirements for good manufacturing practices. They are also required to carry out pharmacovigilance, monitoring and reporting adverse reactions, especially those considered serious, to Health Canada (Health Canada 2020a).

This Canadian regulatory model demonstrates a balance between encouraging innovation in the functional foods and natural supplements sector and rigorous protection of public health, serving as an example of good regulatory practices for the trade in products with bioactive compounds.

11.3.5 Regulations of Bioactive Compounds in India

In India, the regulation of functional foods, nutraceuticals, dietary supplements, and related products such as bioactives in fruits is conducted by the Food Safety and Standards Authority of India (Dhara 2021, Intrasook et al. 2024). Since the promulgation of the Food Safety and Standards (Health Supplements, Nutraceuticals, Food for Special Dietary Use, Food for Special Medical Purpose, Functional Food and Novel Food) Regulations in 2016, the FSSAI has established detailed guidelines for the composition, labeling, safety, and marketing of these products (FSSAI 2016).

Products with bioactive fruit compounds, such as extracts or pulps, are regulated by the FSSAI under strict rules regarding purity, labeling, and the prohibition of therapeutic claims. Functional claims are only allowed with scientific proof assessed by a technical panel (FSSAI 2016, Kumari et al. 2021).

The implementation of the Food Safety and Standards Act, 2006 brought several structural benefits to food regulation in India, mainly by consolidating eight previous laws into a single piece of legislation, which represented a significant advance toward regulatory harmonization and legal clarity on conventional, functional, and novel foods (Government of India 2006). This unification provided alignment with international standards, greater uniformity in areas such as innovative foods, a scientific basis for establishing safety standards, and contributed to the reduction of irregular practices and corruption in the sector (Keservani et al. 2024).

Despite the initial challenges, such as the need for effective supervision, technical qualifications, and adaptation of the industry to stricter standards, the benefits of the new regulations in India have outweighed these limitations (Dhara 2021). Unlike the United States, which has the DSHEA, India is still consolidating its regulatory system for supplements and nutraceuticals, demonstrating a commitment to food safety and consumer protection (Keservani et al. 2024).

11.3.6 Regulation of Bioactive Compounds in China

In China, the bioactive compounds present in fruits do not have specific regulations but are evaluated according to their intended function. The regulation of these compounds in China is structured through the Food Safety Law of the People's Republic of China (FSL), revised in 2015. This legislation consolidated previously dispersed rules and began to centrally supervise food safety and involves various regulatory bodies (Buijs et al. 2023).

The main regulatory bodies that structure the legal framework are the National Health Commission (NHC), the State Administration for Market Regulation (SAMR), and the China National Center for Food Safety Risk Assessment (CFSA) (Xi 2017). The NHC is responsible for establishing the list of authorized functional ingredients, as well as approving new food ingredients and novel food resources (equivalent to the European Union's "novel food"), in addition to publishing technical and health regulations (Buijs et al. 2023).

SAMR oversees labeling, registration, marketing, and health claims, as well as being responsible for registering food supplements and functional foods. The

CFSA is responsible for carrying out toxicological, nutritional, and functional safety assessments (Lin 2017).

In March 2016, the China Food and Drug Administration (CFDA) issued the *Administrative Measures for the Registration and Deposit of Health Food*, which came into effect on July 1, 2016. The CFDA also released a comprehensive interpretation of this measure. This measure was developed under the relevant provisions of Articles 74, 75, 76, 77, 78, and 82 of China's Food Safety Law (Belik 2019). The Law aims to ensure food safety and protect the health and safety of the public. It consists of 154 articles, divided into 10 chapters.

This means that the registration, filing, and supervision of health foods in China must comply with these measures. Therefore, there is a direct impact on the regulation of bioactive compounds in China, since bioactive functional foods, in addition to their basic nutritional function, must offer additional health benefits due to the presence of bioactive compounds (Buijs et al. 2023).

It is also important to remember that these measures also specify the information required on health food registration certificates and their annexes. In addition, the measures reinforce the penalties for infringements related to the registration and filing of health food registrations.

Other important standards that impact the regulation of bioactive compounds present in food in China include: GB 16740-2014 – National standard for food supplements; GB 2760-2014 – National list of food additives; and the "*Catalogue of Functional Ingredients*" (updated by the NHC), which lists approved ingredients with functional claims (Buijs et al. 2023).

11.4 Relationship between Different Regulations

Although this chapter focuses on the regulatory aspects of bioactive compounds in fruit, it is important to note that the main legal frameworks in force, both in Brazil and in international spheres such as the United States, the European Union, and the OECD, do not establish specific guidelines aimed exclusively at fruit, but rather general guidelines applicable to all foods or ingredients with functional properties.

Anvisa in Brazil considers bioactive compounds in the context of claims of functional and health properties, without specifying them biochemically, through Resolution RDC No. 54/2012, which establishes the requirements for labeling and proof of efficacy. However, this regulation does not define the food origin of bioactive compounds, applying the same criteria to substances extracted from fruit, cereals, legumes, or other plant sources (BRASIL 2012).

The US system, on the other hand, although also based on safety standards, transfers greater responsibility to producers and companies, allowing more flexibility in the formulation and marketing of products containing bioactive compounds. The regulations favor market freedom and transparency in labeling, which facilitates innovation and access to new ingredients, including fruit derivatives, provided they do not involve therapeutic claims, which could classify them differently.

In contrast, the European Union and Canada adopt one of the most rigorous and technically demanding models. Due to their strict health standards, bioactive

Table 1. Comparative summary of the main regulatory aspects applicable to bioactive compounds in food, with an emphasis on fruit, in the regulatory frameworks of the European Union, the United States, Brazil, Canada, India, and China.

Aspect	United States	European Union	Canada	Brazil	India	China
Main regulatory body	FDA (Food and Drug Administration)	EFSA (European Food Safety Authority)	NNHPD (Natural and Non-prescription Health Products Directorate)	ANVISA (industrialized products) and MAPA (fresh products)	FSSAI (Food Safety and Standards Authority of India)	NHC (National Health Commission), SAMR, CFSA (China National Center for Food Safety Assessment)
Nature of regulation	Flexible, based on the manufacturer's responsibility	Scientific, preventive and harmonized	Flexible, with proof required on request	Technical, preventive and risk-based	Technical and preventive regulatory reform	Technical, preventive and based on national standards (GB standards)
Requirement for allegations	Moderate - requires notification and disclaimer (DSHEA)	High mandatory prior scientific assessment by EFSA	High scientific proof and registration	High - requires robust scientific evidence	Moderate - data-based registration and internal review	High - mandatory technical assessment and prior authorization
Focus on fruit	Yes, especially in the form of supplements	Yes - emphasis on fruit bioactive compounds	Yes - fruit is used in natural products and nutraceuticals	Yes - both fresh and in pulps and derivatives	Yes - regulated as part of functional foods	Yes - focus on fresh fruit and its bioactive compounds
Barriers to innovation	Low environment is more open to innovation	High strict efficacy and safety criteria	High - requirements for robust evidence and lengthy approval	High bureaucratic processes and time-consuming	High - requirements for clinical and toxicological data	High - rigorous technical review and strict scientific criteria
Consumer protection	Moderate - based on the manufacturer's self-control	Very high - focus on risk assessment and transparency	Very high - strict protocols and scientific review	High - focus on safety and labeling	Moderate - system still being consolidated	High - strong state control and focus on food safety

Source: Research data.

compounds must go through a formal scientific evaluation process, based on efficacy and the absence of risks. This model sets a significant technical hurdle for the industry, which encourages higher quality standards and strengthens credibility with European consumers, who are generally more demanding regarding the quality and origin of the food they consume.

In Asia, India stands out as an emerging country that is adapting to international quality rules and becoming stricter in terms of fruit import and production processes, increasingly becoming one of the world's leading fruit consumption powers. China, as one of the largest importers and consumers of fruit, has very strict criteria for buying and storing fruit and consequently for their bioactive compounds. Many exporting countries, such as Brazil, have adapted to Chinese standards for exporting fruit, which has strengthened trade relations between the two countries.

The differences between the regulatory frameworks require commercial bodies, producers, and researchers to be attentive to the specificities of each market, especially when it comes to inserting fruits and their derivatives with functional claims into international contexts. This comparison highlights not only the technical and regulatory challenges faced in the fruit production chain, but also the opportunities for strategic positioning and valorization of native species and their bioactive compounds, provided that the safety, quality, and compliance requirements defined in each system are respected. Table 1 summarizes the main regulatory aspects of bioactive compounds in food, with an emphasis on fruit, considering the regulatory frameworks of the European Union, the United States, Brazil, Canada, India, and China.

The Te comparative analysis reveals that, even though their common objective is to protect public health and consumer confidence, each body prioritizes different elements. Brazil and the European Union adopt more restrictive stances, focusing on the prior evaluation of claims, while the United States gives greater autonomy to the industry, as long as it complies with transparent labeling. The OECD, for its part, does not regulate bioactive compounds but contributes to the international alignment of visual quality standards, which is crucial for the global competitiveness of fruit.

This regulatory diversity poses challenges but also offers strategic opportunities for Brazilian producers and researchers, who can adapt their products according to the target markets, always considering the scientific, legal, and commercial criteria required.

11.5 Conclusions

The regulation of bioactive compounds in fruit is a dynamic and constantly evolving field. Understanding the regulatory aspects is essential for producers, researchers, and consumers. As science advances and new evidence emerges, it will be essential to adopt regulatory policies and practices that guarantee the safety and efficacy of bioactive compounds, thus promoting public health and innovation in the sector.

The role of fruit in the human diet goes far beyond simply providing vitamins and minerals. Today, it is understood that they are extremely rich sources of bioactive compounds with the potential to promote health and prevent various diseases. These compounds, such as polyphenols, carotenoids, and flavonoids, have attracted increasing interest from science and the food industry due to their antioxidant, anti-inflammatory, and protective effects against chronic diseases such as diabetes, obesity, and neurodegenerative disorders.

In this scenario, ensuring that these foods reach the consumer safely and with quality becomes essential. In Brazil, the work of Anvisa and MAPA has been fundamental in structuring and supervising the production, processing, and marketing of fruit and, consequently, bioactive products derived from it. Brazilian legislation has evolved to keep up with scientific innovations and meet market demands, especially with regard to scientific proof of functional claims. This control is important to protect the consumer and, at the same time, to encourage innovation in the sector, as the country is one of the largest exporters in the world.

When it comes to exports, the challenges are even greater. As one of the world's leading fruit producers, Brazil must meet the quality and certification standards required by different countries. Joining the OECD program, for example, has helped standardize quality criteria and facilitated access for Brazilian fruit to international markets. This not only strengthens the country's image as a reliable producer but also encourages the constant search for improvements at all stages of the production chain.

In the global context, however, there are still major differences between each country's legislation. In many places, especially developing countries, there is a lack of specific standards for bioactive compounds and limited resources to carry out robust scientific research to support regulations. This disparity can hinder international trade and limit the use of the functional potential of these foods.

In view of this, investing in science, regulatory infrastructure, and technical education is essential if Brazil and other countries are to not only guarantee the safety of the food they offer, but also make the most of the benefits these compounds can provide. The combination of scientific knowledge, efficient regulation, and a commitment to quality is what will allow the sector to continue to grow with responsibility and a focus on public health.

This chapter highlights that while there are no direct and specific regulations for bioactive compounds in fruits, various national laws and frameworks exist to ensure the quality and safety of fruits and their derivatives that carry these compounds. Each country applies its own regulatory approach, generally aligned with internationally accepted standards. More than ever, fruit and fruit products play a strategic role in promoting health, innovating in the food industry, and strengthening the economy. By consolidating a solid regulatory base and investing in research that proves the functional benefits of bioactive compounds, Brazil is taking important steps toward a future in which health and development go hand in hand.

References

ANVISA. (1999, May 3). Resolution RDC nº 18, of April 30, 1999: Approves the technical regulation on foods with claims of functional and/or health properties. Federal Official Gazette, section 1, Brasília, DF.

Araújo, F. F. de, Farias, D. D. P. and Neri-Numa, I. A. (2021). Polyphenols and their applications: An approach in food chemistry and innovation potential. Food Chemistry 338: 127535. https://doi.org/10.1016/j.foodchem.2020.127535.

Aryee, A. N. A., Ofori, K. F., Otchere, E., Dare, K. O. e and Akanbi, T. O. (2024). Food-derived bioactive components and health claims: The role of regulatory agencies. pp. 498–529. In Plant Food Phytochemicals and Bioactive Compounds in Nutrition and Health. CRC Press.

Belik, Walter. (2019). Food and nutrition security in China: historical aspects and current challenges. pp. 37–55. *In*: Marcos Costa Lima and Eduardo Matos Oliveira (eds.). Foreignization of Land and Food and Nutrition Security: Brazil and China in Perspective. Recife: FASA.

Buijs, J., van der Meulen, B. M. and Sun, J. (eds.). (2023). Chinese Food Legislation: A Legal-Systematic Analysis (Vol. 15). BRILLIANT.

Brazil (1969, October 24). Decree-Law No. 986, of October 21, 1969: Institutes basic rules on food. Diário Oficial da União, section 1, Brasília, DF.

Brazil. Ministry of Agriculture, Livestock and Supply. (2018, December 26). Normative Instruction No. 69, of December 21, 2018: Establishes the standards for plant classification of fruits, vegetables and greens. Federal Official Gazette, section 1, Brasília, DF.

Brazil. Ministry of Agriculture, Livestock and Supply (2018, November 6). Normative Instruction No. 69 of 06/11/2018. https://www.defesa.agricultura.sp.gov.br/legislacoes/instrucao-normativa-69-de-06-11-2018,1195.html.

Cansanção, A. S. da C., França, G. F. C. O. and Júnior, W. de L. M. (2025). A systemic approach to food security, focusing on the human right to adequate food. Aracê 7: 1105–1127. https://doi.org/10.56238/arev7n1-067.

Dhara, D., Biswas, S., Das, S. and Biswas, O. (2021). Status of food safety and food security in India from the perspective of FSSAI. Indian Journal of Animal Health 60(2): 167–173.

Duda-Chodak, A. and Tarko, T. (2023). Possible side effects of polyphenols and their interactions with drugs. Moléculas 28(6): 2536.

Faolex. (2025). Database | Food and Agriculture Organization of the United Nations. Available at: https://faolex.fao.org/docs/pdf/chn154228.pdf. Accessed on: June 8, 2025.

Fernandes Martins, K., Teixeira, D. and de Oliveira Corrêa, R. (2022). Gains in sustainability using voluntary sustainability standards: A systematic review. Cleaner Logistics and Supply Chain 3: 100084. https://doi.org/10.1016/j.clscn.2022.100084.

Fortin, N. D. (2009). Food Regulation: Law, Science, Policy, and Practice. John Wiley & Sons.

FSSAI - Food Safety and Standards Authority of India. (2016). Food Safety and Standards (Health Supplements, Nutraceuticals, Food for Special Dietary Use, Food for Special Medical Purpose, Functional Food and Novel Food) Regulations, 2016. Retrieved from: https://www.fao.org/faolex/results/details/fr/c/LEX-FAOC168163/.

Galanakis, C. M. (ed.). (2019). The Role of Alternative and Innovative Food Ingredients and Products in Consumer Wellness. Academic Press.

Government of India. (2006). The Food Safety and Standards Act, 2006 (No. 34 of 2006). Ministry of Law and Justice. Retrieved from: https://www.indiacode.nic.in/handle/123456789/2027?view_type=browse#:~:text=An%20Act%20to%20consolidate%20the,and%20wholesome%20food%20for%20human.

Granato, D., Zabetakis, I. and Koidis, A. (2023). Sustainability, nutrition, and scientific advances of functional foods under the new EU and global legislation initiatives. Journal of Functional Foods 109: 105793.

Health Canada. (2013). Guidance document: The use of food health claims. https://www.canada.ca/en/health-canada/services/food-nutrition.html.

Health Canada (2020a, December 1). Drugs and health products. https://www.hc-sc.gc.ca/dhp-mps/prodnatur/legislation/acts-lois/prodnatur/index-eng.php.

Health Canada (2020b, December 1). Information kit - Regulation for Natural Health Products. https://www.canada.ca/en/health-canada/services/drugs-health-products/natural-non-prescription/regulation/information-kit.html.

Health Canada (2020c, December 1). Natural Health Products - Over 43,000 products authorized for sale and growing. https://www.canada.ca/en/health-canada/services/drugs-health-products/natural-non-prescription/regulation/43-000-products-authorized-sale-growing.html.

Hu, L., Zheng, Y., Woods, T. A., Kusunose, Y., Buck, S. and Saad, S. (2023). The market for private food safety certifications: Conceptual framework, review, and future research directions. Applied Economic Perspectives and Policy 45: 197–220. https://doi.org/10.1002/aepp.13226.

Intrasook, J., Tsusaka, T. W. and Anal, A. K. (2024). Current food safety trends and regulations and policies for functional foods and beverages containing botanicals. Journal of Food and Drug Analysis 32(2): 112.

Karimi, A., Kazemi, M., Samani, S. A. and Simal-Gandara, J. (2021). Bioactive compounds from by-products of eggplant: Functional properties, potential applications and advances in valorization methods. Trends in Food Science & Technology 112: 518–531. https://doi.org/10.1016/j.tifs.2021.04.027.

Keservani, R. K., Sharma, A. K., Ahmad, F. and Baig, M. E. (2014). Nutraceutical and functional food regulations in India. pp. 353–365. *In*: Bagchi, D. (ed.). Nutraceutical and Functional Food Regulations in the United States and Around the World (2nd ed.). Academic Press. https://doi.org/10.1016/B978-0-12-405870-5.00019-0.

Komala, M. G., Ong, S. G., Qadri, M. U., Elshafie, L. M., Pollock, C. A. and Saad, S. (2023). Investigating the regulatory process, safety, efficacy and product transparency for nutraceuticals in the USA, Europe and Australia. Foods 12(2): 427. https://doi.org/10.3390/foods12020427.

Lin, Sy (2017). State capitalism and the governance of Chinese food security. Japanese Journal of Political Science 18(1): 106–138. doi:10.1017/S1468109916000335.

Maqsood, S., Adiamo, O., Ahmad, M. and Mudgil, P. (2020). Bioactive compounds from date fruit and seed as potential nutraceutical and functional food ingredients. Food Chemistry 308: 125522. https://doi.org/10.1016/j.foodchem.2019.125522.

Martins, M. M. V., Denny, D. M. T., Cechin, A., Bispo, S. Q. A., Nonnenberg, M. J. B. and Carneiro, F. L. (2023). Voluntary sustainability standards and their importance in international trade. Bulletin of International Economics and Policy 137–162. https://doi.org/10.38116/bepi34art6.

Martirosyan, D. and Ekblad, M. (2022). Functional foods classification system: Exemplifying through analysis of bioactive compounds. Functional Foods Science 2(6): 247–263.

Nath, H., Samtiya, M. and Dhewa, T. (2022). Beneficial attributes and adverse effects of major plant-based foods anti-nutrients on health: A review. Human Nutrition & Metabolism 28: 200147.

OECD. (2021). International Standards for Fruit and Vegetables. OECD Publishing. https://www.oecd.org/agriculture/fruit-vegetables.

Paraskevopoulos, K. and Federici, S. (2021). Overview of EFSA and European national authorities' scientific opinions on the risk assessment of plants developed through new genomic techniques. EFSA Journal 19(6): e06314. https://doi.org/10.2903/j.efsa.2021.6314.

European Parliament & Council of the European Union. (2006). Regulation (EC) No. 1924/2006 of December 20, 2006. Official Journal of the European Union. https://eur-lex.europa.eu/legal-content/PT/TXT/PDF/?uri=CELEX:32006R1924.

Paz, M., Gúllon, P., Barrozo, M. F., Carvalho, A. P., Domingues, V. F., Gomes, A. M. et al. (2015). Brazilian fruit pulps as functional foods and additives: Evaluation of bioactive compounds. Food Chemistry 172: 462–468. https://doi.org/10.1016/j.foodchem.2014.09.102.

Schuler, L., Zust, D., Vybiral, D. and Hau, P. (2019). GM food regulations in the EU. In Reference Module in Food Science. Elsevier. https://doi.org/10.1016/B978-0-08-100596-5.22609-7.

Shashirekha, M. N., Mallikarjuna, S. E. and Rajarathnam, S. (2015). Status of bioactive compounds in foods, with focus on fruits and vegetables. Critical Reviews in Food Science and Nutrition 55(10): 1324–1339.

Son, J. and Martirosyan, D. (2024). Salient features for GRAS status affirmation. Functional Food Science 4(8): 299–308. https://doi.org/10.31989/ffs.v4i8.1417.

Spagnuolo, P. A. (ed.). (2020). Nutraceuticals and human health: the food-to-supplement paradigm. Royal Society of Chemistry.

Tafese Awulachew, M. (2024). Functional foods: Functional ingredients, sources and classification, health claims, food intolerance, and allergy. https://doi.org/10.5772/intechopen.114157.

Taylor, M. (2024). Whither the Regulator: Food and Drug Law, The Natural Health Product Regulations and The Erosion of Safety, Efficacy and Quality [Doctoral dissertation, Dalhousie University, Schulich School of Law].

Ververis, E., Ackerl, R., Azzollini, D., Colombo, P. A., de Sesmaisons, A., Dumas, C. et al. (2020). Novel foods in the European Union: Scientific requirements and challenges of the risk assessment process by the European Food Safety Authority. Food Research International 137: 109515. https://doi.org/10.1016/j.foodres.2020.109515.

Vettorazzi, A., Cerain, A., Sanz-Serrano, J., Gil, A. and Azqueta, A. (2020). European regulatory framework and safety assessment of food-related bioactive compounds. Nutrients 12(3): 613. https://doi.org/10.3390/nu12030613.

Vidinamo, F., Fawzia, S. and Karim, M. A. (2021). Effect of drying methods and storage with agro-ecological conditions on phytochemicals and antioxidant activity of fruits: a review. Critical Reviews in Food Science and Nutrition 62(2): 353–361.

Vilas-Boas, A. A., Pintado, M. and Oliveira, A. L. (2021). Natural bioactive compounds from food waste: Toxicity and safety concerns. Foods 10(7): 1564.

Wallace, T. C. (2015). Twenty years of the Dietary Supplement Health and Education Act-How should dietary supplements be regulated? The Journal of Nutrition 145(8): 1683–1686. https://doi.org/10.3945/jn.115.211102.

Xi, Jinping. (2017). The Governance of China II. Beijing: Foreign Languages Press Co. Ltd.

Zovi, A., Vitiello, A., Sabbatucci, M., Musazzi, U. M., Sagratini, G., Cifani, C. et al. (2025). Food supplements marketed worldwide: a comparative analysis between the European and the U.S. Regulatory Frameworks. J. Diet Suppl. 22: 25–40. https://doi.org/10.1080/19390211.2024.2389397.

Chapter 12

Future Directions in Bioactive Compounds and Functional Foods

Julienne Cavalcante, Francisco Thiago Bandeira Silva, Sara de Sousa Campelo, Tayllon Rocha de Sousa and *Anderson Nogueira Mendes**

12.1 Introduction

In 2009, at the 6th International Conference on "Functional Foods for Chronic Diseases," the Functional Foods Center (FFC) proposed a definition for functional foods (FFs), which was approved by the audience (Williamson et al. 2009). After several years, the definition was refined, and the current nomenclature is: "Natural or processed foods containing biologically active compounds that, in specifically defined, effective, and non-toxic quantities, provide a clinically proven and documented health benefit, using specific biomarkers to promote optimal health, thereby reducing the risk of chronic/viral diseases and enabling symptom control" (Gur et al. 2018, Martirosyan et al. 2022a). However, it is widely agreed that the definition, production, classification, and regulation of FFs remain ambiguous, and major manufacturers exhibit variations in their approaches (de Boer and Bast 2015, Jones et al. 2008, Kapustian et al. 2023). In 28 jurisdictions, the majority of economically developed countries, as identified by Boer and Bast (2015), have regulations on nutrition, health-related claims, and programs that oversee design and application (de Boer and Bast 2015).

The predominant concepts and arguments in the scientific literature state that FFs must contain specific components that have a positive impact on human life

Laboratory of Science in Innovation and Technology - LACITEC, Department of Biophysics and Physiology, Federal University of Piauí, 64049-550, Teresina, Piauí, Brazil.

* Corresponding author: anderson.mendes@ufpi.edu.br

DOI: 10.1201/9781003668169-15

routines, with effects validated by biomarkers. Additionally, the research emphasizes the importance of determining the quantity, efficacy, and potential toxicity of active components, as well as establishing the central role of foods in preventive medicine, particularly considering that effective and conventional medications have high costs and adverse effects (Gur et al. 2018, Martirosyan et al. 2021). FFs offer promises in this regard, as they are composed of a variety of biologically active markers or bioactive compounds (BCs) with pharmaceutical and biomedical potential (Domínguez Díaz et al. 2020, Kapustian et al. 2023, Teodoro 2019).

Vlaicu et al. (2023) highlight processing methods and operations used to obtain functional foods, such as fortification or enrichment, fermentation, nanotechnology, 3D printing, spray drying, and enzymatic hydrolysis. Additionally, the selection and design of FFs should aim to enhance the bioavailability of BCs, thus ensuring the FF's potential. The industry must also examine variables related to stability, chemical composition, functional properties, sensory appeal, price, as well as health claims, which are important issues because such claims assert that a food product or nutrient provides a specific health benefit (e.g., "helps lower cholesterol") and can stimulate innovation and competition among companies (Domínguez Díaz et al. 2020, Sanz Merino 2022).

Since FFs is a relatively new field with some divergence in its definition and regulation, gaps in health claims remain, leading consumers to rely on the information provided on labels specified by corporations. To address this issue and guide the development of reliable and sustainable products, the FFC proposed a 16-step protocol for the development of functional foods (Martirosyan et al. 2022b).

Bioactive compounds are molecules such as polysaccharides, phytochemicals (including lutein and carotenoids), DHA, phenolic compounds, tannins, peptides, lipids, enzymes, terpenoids, vitamins, and minerals (Teodoro 2019) and, according to Martirosyan et al. (2022a), are constituents of foods that provide health benefits by controlling, preventing, or treating diseases such as obesity, insulin resistance, hypertension, chronic inflammation, dyslipidemia, oxidative stress, and diabetes mellitus (Martirosyan et al. 2021). These compounds can be derived from animals or found in plants. BCs can also be categorized by function, biological effect, or biochemical pathways, assisting clinicians, pharmacists, and toxicologists (Kurek et al. 2022, Nigam et al. 2019).

Due to their multifunctional nature and diversity of properties, bioactive compounds represent a significant area in scientific research. There are reports of phenolic compounds in cruciferous vegetables that demonstrate potential in the prevention of gastrointestinal cancers and other related types (Mitra et al. 2022). Proteins, immunoglobulins, growth factors, lysozymes, lactoferrin, and lactoperoxidase in bovine colostrum play a crucial role in autoimmune disorders or respiratory infections after physical training (Poonia and Shiva 2022). Vitamin C, found in citrus fruits, has antioxidant properties and acts as a cofactor (Leem and Martirosyan 2019).

Bioactive compounds derived from probiotics have antioxidant, antibacterial, and antiviral activities (Guo et al. 2022) and can act in the activation of different neurotransmitters, resulting in benefits for mental health. The probiotic yeast *Pichia kudriavzevii*, for example, has the ability to modulate inflammation and increase epithelial defense (Maione et al. 2025). The unsaturated free fatty acids present in breast milk promote intestinal health in infants (Waller et al. 2025). The ethanolic extract of propolis improves oxidative stress and inhibits hepatic inflammation, regulating insulin resistance (Liu et al. 2025).

In the field of personalized nutrition (PN), the correlation between bioactive compounds and individual responses is explored to optimize health outcomes, involving nutritional counseling, products, or services to formulate nutritional approaches focused on the prevention, management, and treatment of diseases (Chen and Chen 2022). This area of study recognizes that genetic variations, metabolic and cultural differences—such as food intolerances, allergies, and preferences, lifestyle factors, and other individual characteristics—exert a significant influence on how individuals process and respond to dietary components. The creation of a PN approach requires phenotypic data, body composition measurements, physical activity levels, clinical indicators, biochemical markers assessing nutritional status, and genomic information, obtained through the integration of sensor technologies such as continuous glucose monitoring, transcriptomics, proteomics, and metabolomics. Additionally, advancements in bioinformatics are crucial for integrating and interpreting the vast amounts of data generated by these technologies, enabling the investigation of complex interactions between genes and nutrients (Chen and Chen 2022).

Biomarkers are essential for planning PN as they provide objective insights into individual responses to dietary interventions. According to Lopez (2020), the anti-inflammatory effects of polyphenols, modulated by epigenetic changes, can be tracked using biomarkers, revealing their mechanisms of action, for example, the impact of daily resveratrol intake from grape extract on the regulation of the inflammatory response (Ramos-Lopez et al. 2021). Other studies are also exploring the grouping of individuals with comparable metabolic profiles, called "metabotyping," to identify subgroups that respond differentially to dietary interventions using (poly) phenols and to examine associations with cardiometabolic risk factors (Narduzzi et al. 2022).

Although promising, the use of bioactive compounds requires careful dosing before being incorporated into food products to avoid harmful side effects (Martirosyan and Sanchez 2022). Additionally, they often exhibit instability when applied as food ingredients and during storage due to low solubility, low bioavailability, or chemical instability (Martirosyan and Sanchez 2022, Wen et al. 2017). Several studies have investigated solutions to this issue, including the selection of an appropriate food vehicle, which can be done through encapsulation, such as using polymeric materials, to protect and release the compounds at the right time, along with the use of technologies like nanoencapsulation, nanoprecipitation, lyophilization, spray drying, inclusion complexation, supercritical fluid, and emulsification (Wang and Wu 2021, Wen et al. 2017).

Although medicines or foods cannot offer a cure for all diseases, bioactive compounds, together with functional foods, can provide symptomatic relief and improve the quality of life for those living with such conditions (Gur et al. 2018).

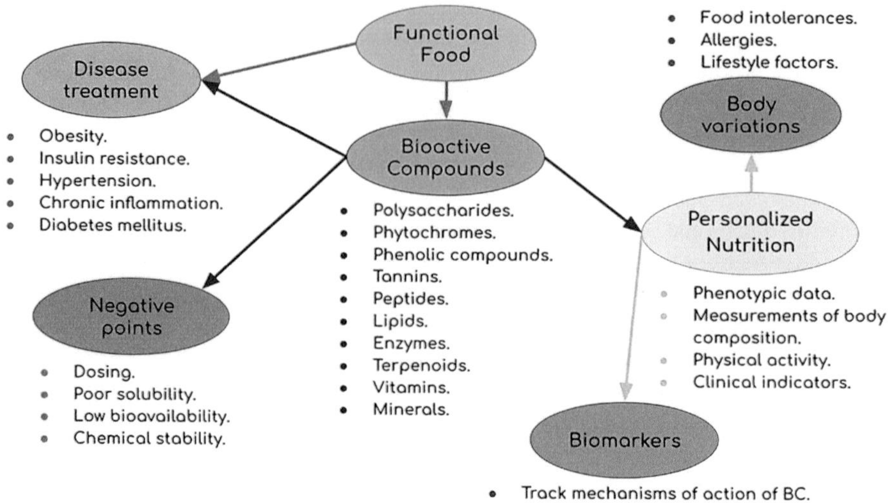

Figure 1. The benefits and harms of bioactive compounds present in functional foods.

In light of these considerations, the subsequent sections will explore the benefits, existing types of bioactive compounds and functional foods, and emerging research trends associated with this field. Figure 1 summarizes the benefits and drawbacks of bioactive compounds present in functional foods.

12.2 Benefits and Actions of Bioactive Compounds and Functional Foods

The bioactive compounds present in functional foods have revolutionized the perception of nutrition, offering benefits that go far beyond the supply of energy and essential nutrients, with an increasingly recognized role in the prevention and management of chronic non-communicable diseases (Tashiro and Imoto 2012). These substances, which encompass a wide range of molecules such as polyphenols, carotenoids, omega-3 fatty acids, dietary fiber, prebiotics, and probiotics, are at the center of research that seeks to unravel their potential for human health and their innovative application in the food industry (Mittal et al. 2024). Natural sources such as berries (blueberries, rich in anthocyanins; raspberries, with ellagitannins), colorful vegetables (carrots, source of beta-carotene; pumpkin, with alpha-carotene), seeds (flaxseed, rich in alpha-linolenic acid; chia, with soluble fiber), oilseeds (walnuts, containing omega-3), turmeric (antioxidant curcumin), amaranth (fibers and squalene), and basil (phenolic acids) exemplify the botanical richness that supports the development of functional foods, pointing to a future where biodiversity will be even more explored (Dixit et al. 2023, Mittal et al. 2024, Peng et al. 2020, Vignesh et al. 2024).

The antioxidant action of bioactive compounds is one of their most celebrated pillars, combating oxidative stress that is at the root of pathologies such as cardiovascular diseases, type 2 diabetes, and various types of cancer (Mittal et al.

2024). Flavonoids found in foods such as green tea (catechins), cocoa (procyanidins), and açaí (anthocyanins) demonstrate anti-inflammatory and cardioprotective effects, increasing the functionality of HDL cholesterol, reducing lipid peroxidation, and attenuating vascular inflammation (Luna-Castillo et al. 2021). Polyphenols from extra virgin olive oil (hydroxytyrosol) and carotenoids from tomatoes (lycopene) and spinach (lutein) regulate the lipid profile, reducing oxidized LDL and promoting arterial health (da C. Pinaffi-Langley et al. 2024, Luna-Castillo et al. 2021, Stankovic et al. 2025). Soluble fiber from oats (beta-glucans) and saponins from legumes such as lentils and chickpeas modulate blood glucose, reduce insulin resistance, and strengthen metabolism, offering a natural barrier against type 2 diabetes and metabolic syndrome (Noce et al. 2021).

Furthermore, spirulina, with its phycocyanins and polysaccharides, exhibits anticancer, neuroprotective, and immunomodulatory properties, inhibiting the proliferation of tumor cells and protecting neurons from oxidative stress (Bortolini et al. 2022), while omega-3 fatty acids from sources such as salmon (DHA), sardines (EPA), and microalgae (docosahexaenoic acid) attenuate neuroinflammation, being allies in the prevention of neurodegenerative diseases such as Alzheimer's and Parkinson's (Elbandy 2023). Sunlight, although beneficial, can pose a threat to the skin, our natural protective barrier, due to the phototoxicity caused by overexposure to ultraviolet radiation (UVR), which results in burns, accelerates photoaging, and favors skin cancer. In this context, carotenoids such as those present in carrots and tomatoes, associated with polyphenols and flavonoids, act as natural photoprotectors, exhibiting antioxidant, antimutagenic, and anti-inflammatory properties that alleviate skin damage, with recent studies exploring their application in nanoemulsions and liposomes to increase bioavailability and stability (Flieger et al. 2024).

Gut health, supported by the trillions of microorganisms that make up the human microbiome, is another area profoundly impacted by these compounds (Al-Habsi et al. 2024). Prebiotics such as inulin (found in artichokes, chicory, and green bananas), oligofructose (from onions), and galactooligosaccharides (from peas and beans) feed beneficial bacteria such as *Bifidobacterium*, resulting in the production of short-chain fatty acids (SCFA) such as butyrate, which reinforces the integrity of the intestinal barrier and reduces systemic inflammation (Al-Habsi et al. 2024, Anand et al. 2025, González-Velázquez et al. 2024).

Probiotics found in commercial products such as Yakult (*Lactobacillus casei* Shirota), Activia (*Bifidobacterium lactis* DN-173 010), and artisanal kefir (*Lactobacillus kefiranofaciens*) balance the microbiota, alleviate symptoms of irritable bowel syndrome, strengthen immunity, and reduce intestinal permeability (Al-Habsi et al. 2024). Recent research shows that supplements such as Beneo Orafti® Inulin and specific strains such as *Lactobacillus rhamnosus* GG improve mental health, decreasing anxiety and depression by increasing serotonin production via the gut–brain axis (Alonso-Allende et al. 2024, Bistas and Tabet 2023, Van Harsselaar et al. 2024). In women, probiotics such as *Lactobacillus reuteri* RC-14 prevent urogenital infections, while prebiotics such as galactooligosaccharides promote calcium absorption, reducing the risk of osteoporosis (de Vrese 2009). Postbiotics, defined as preparations of inactivated microorganisms or their metabolites (Salminen et al. 2021), emerge as promising alternatives, with studies indicating their role in the regulation of inflammatory cytokines and in the control of obesity and type 2 diabetes through metabolites such as lactic acid and propionate (Li et al. 2021).

In the clinical sphere, the consumption of probiotic yogurts (e.g., Danone Bio) and prebiotic supplements (e.g., Fibersol-2) is associated with a reduction of up to 32% in the risk of progression of chronic kidney disease in women and elderly people over 55 years of age, according to data from NHANES (2011–2020), possibly by reducing uremic toxins derived from the microbiome (Liu et al. 2022). Plant compounds such as quercetin (from apples and onions), catechins (from black tea), and polysaccharides from citrus peels inhibit xanthine oxidase and promote the excretion of uric acid, offering natural and safe options for hypouricemia with fewer side effects than allopurinol (Jiang et al. 2020).

In the food industry, agro-industrial waste such as grape skins (resveratrol), coffee grounds (chlorogenic acid), sugarcane bagasse (insoluble fibers), and mango skins (polyphenols) are transformed into natural antioxidants and antimicrobials, being incorporated into cereal bars, functional beverages, and supplements, with the potential to reduce food waste by up to 25% in production chains (Perez-Gregorio and Simal-Gandara 2017). Sea buckthorn (*Hippophae rhamnoides*), with its almost 200 bioactive compounds, including omega-7 and superoxide dismutase, stands out for its antioxidant, antidiabetic, and anti-obesity effects, being an ideal raw material for low-cost functional oils and juices (Wang et al. 2022).

Growing awareness of the relationship between diet and health is driving changes in global dietary patterns, with systematic reviews (2011–2021) indicating that foods such as flaxseed (lignans), kefir (probiotics), green banana (resistant starch), and fermented products (e.g., kombucha) improve intestinal motility, prevent constipation, and strengthen the microbiota (Silva et al. 2024). Diets rich in fruits (e.g., pomegranate, with punicalagins), vegetables (e.g., broccoli, with sulforaphane), and fiber (e.g., amaranth and quinoa) significantly reduce the risk of chronic diseases such as hypertension and colorectal cancer, promoting metabolic homeostasis and immune response (Del Cornò et al. 2021, Jiang et al. 2021, Luvián-Morales et al. 2022, Stephen et al. 2023, Vernia et al. 2021). Hydrocolloids extracted from algae (e.g., carrageenan), seeds (e.g., guar gum), and fruits (e.g., apple pectin) not only improve the texture and stability of foods, but also increase the bioavailability of bioactives, reduce LDL cholesterol, and modulate glycemia, opening doors for formulations of protein bars and functional gels (Manzoor et al. 2020).

The discovery of new bioactive compounds broadens the horizons of functional nutrition. Betalains from beetroot (antioxidants), fucoidans from brown algae (immunomodulators), chitosan from crustaceans (hypocholesterolemic), and exopolysaccharides from lactic bacteria (prebiotics) represent innovative sources that can be incorporated into beverages, capsules, and fortified foods (Manzoor et al. 2020, Michalak 2022). Biotechnology plays a transformative role in this scenario: *in vitro* cultures of *Withania somnifera*, a plant widely explored in Ayurvedic medicine for its immunomodulatory, anticancer, antistress, and neuroprotective properties, produce withanolides on a large scale for supplements, meeting the growing demand for high-quality phytoconstituents in the face of low seed viability and the insufficiency of traditional cultivation (Namdeo and Ingawale 2021).

Meanwhile, plants such as camelina modified via CRISPR/Cas9 synthesize omega-3 (DHA and EPA) at levels comparable to fish, reducing dependence on marine stocks (Khan et al. 2017). Bioreactors optimize the production of probiotics and postbiotics, such as butyric acid from *Clostridium butyricum*, for applications in infant formulas and therapeutic diets (Al-Habsi et al. 2024). Sustainability gains strength with the valorization of by-products such as *Spartina anglica*, which improves the cecal microbiota in birds with 6% supplementation in the diet (Zhou et al. 2025), and rice husks, rich in fiber and silica, used in functional breads, contributing to a circular economy that can reduce carbon emissions by up to 15% in industrial processes (Singh et al. 2020).

Personalized nutrition emerges as a game changer, integrating bioactive compounds into individual profiles. Nutrigenomics reveals how omega-3 benefits carriers of *FADS1* gene variants, increasing the conversion of ALA to EPA, while probiotics such as *Bifidobacterium longum* adjust diets for lactose-intolerant individuals based on genetic and microbial analyses (Kolodziejczyk et al. 2019). On the horizon, technologies such as 3D food printing can incorporate tailored bioactives, such as neuroprotective peptides, into personalized snacks, while microbiome-based therapies such as synbiotic-optimized fecal transplants promise to treat conditions such as obesity and inflammatory bowel disease (Kolodziejczyk et al. 2019). This convergence of biotechnology, sustainability, and personalization points to a future where functional foods not only prevent disease but also optimize

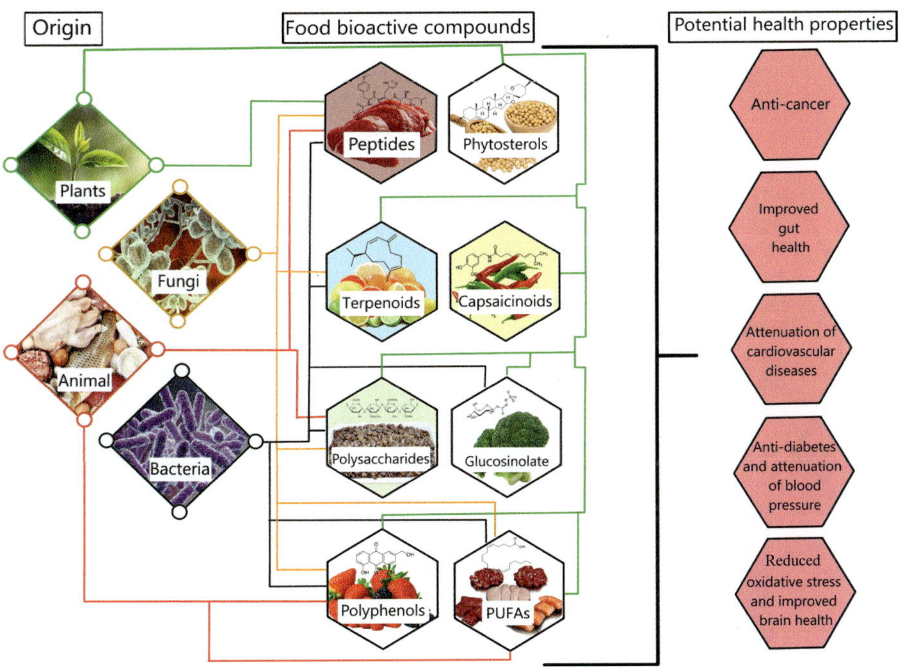

Figure 2. Main sources, bioactive compounds and their health benefits.

health in an individualized and environmentally responsible way. The main sources of bioactive compounds and their health benefits are highlighted in Figure 2.

The diversity of plants and microbes in nature has numerous applications for improving human well-being, requiring the use of a wide range of extraction methods (Pai et al. 2022). In this context, Figure 3 presents a flowchart with the steps for obtaining bioactive compounds from natural sources. Optimization in obtaining these bioactive compounds is directly linked to the chosen processing method,

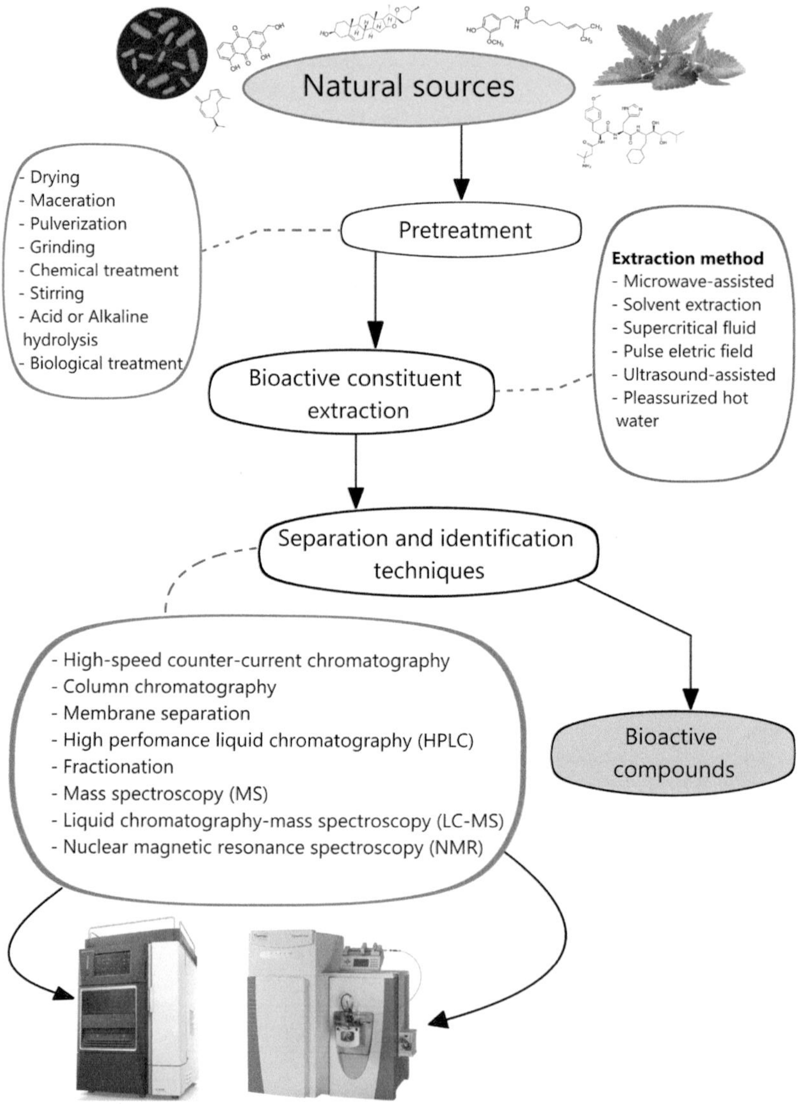

Figure 3. Flowchart of the process of extraction, separation, and identification of bioactive compounds.

leading to more sustainable production and preserving the abundance and richness of the species used in research (Anibogwu et al. 2021).

In the scientific field, we have the so-called advanced extraction techniques, which have made it possible to achieve greater selectivity, yield, speed, and less use of solvents during the process, especially compared to conventional processes that are often used in the pre-treatment phase to obtain bioactive compounds today (Bhadange et al. 2024). In general, supercritical fluid extraction and pressurized liquid extraction are ubiquitous high-pressure techniques for separating bioactive compounds from natural sources (Anibogwu et al. 2021). In fact, these techniques have considerably high extraction efficiencies, having great importance in this field today.

Pressurized liquid extraction can be summarized as the use of solvents at high temperatures—however, relatively lower than their respective critical points—in order to keep them in the liquid state. This process exploits mass transfer properties at high temperatures and pressures (Zakaria et al. 2022). On the other hand, supercritical fluid extraction is a technique that uses supercritical fluids, such as carbon dioxide, to extract bioactive components. This process offers advantages such as minimal residue, adjustable solvent power, and low toxicity (Bhadange et al. 2024, Nguyen et al. 2024).

12.3 Challenges and Opportunities for Innovation in the Food Industry

12.3.1 Food Security in Strategic Policies

As a cross-cutting and multidisciplinary approach, the concept of State security includes, among others, energy, cyber, and food security. Countries that are major producers or importers of food consider food security to be one of the main pillars of their national security. Within this context, the implementation of strategic food security policies plays a decisive role in maintaining sovereignty, power stability, and expanding a country's economic and geopolitical influence (Bertolini et al. 2023).

The modern concept of security, considered multidimensional, addresses aspects that are not only intrinsic to the security interests of a modern State, but also re-discusses new ways of thinking about national security, suggesting strategies that encompass cross-cutting issues, with preventive approaches to the escalation of conflicts (Annan 2000). In this context, national security is not necessarily directly related to the ideas of territorial protection typical of military thought, but rather includes seven specific categories: (i) economic security; (ii) food security; (iii) health security; (iv) environmental security; (v) personal security; (vi) public security; and (vii) political security (Ballesteros Martín 2016).

It is observed that the food security policies adopted by countries depend on their respective political models, their role in ensuring global food security, and their concept of self-sufficiency. Due to its leading role in the context of global food security, Brazil needs a food security policy inserted in a State macro-strategy aimed at influencing other countries economically and politically in its favor (Bertolini et al. 2023).

As an example of a producing country, we can highlight the Chinese government, which seeks to control not only domestic food production and quality but also imports and exports through internal regulation. In 2009, China's Food Safety Law replaced the Food Hygiene Law and established the Food Safety Committee of the State Council, together with the China Food and Drug Administration (Lin 2017). In 2013, the State Council was restructured, and the State Council Food Safety Committee and the China Food and Drug Administration were merged and upgraded into a new ministerial-level agency of the China Food and Drug Administration, with the authority and resources to regulate trade activities and manage food market safety (Lin 2017). Thus, the State, through its control over domestic regulation, acts selectively on the type of food that will be imported into China, in the choice of the exporting country, and even in the choice of the producer that will export to the Chinese market.

The importance of a food security policy for the globalization of markets, food imports, and, therefore, global food security has become an issue linked to geopolitical relations. After the First World War, food shortages and concerns about dependence on external supplies motivated European governments to adopt emergency actions, which included, in addition to imposing regulations on food distribution, encouraging the recovery of agriculture and sectoral intervention by the State (Zobbe 2001).

12.3.2 Regulatory Barriers

The food industry contributes substantially to global economic and social development. However, several complications are present, such as food safety, traceability, durability, and food quality, thus requiring the application of barriers of different natures to ensure human health safety (Dong and Jiang 2022). The main barriers existing in the food industry are categorized into six main groups: technological, economic, regulatory, environmental, organizational, and market (Gupta et al. 2025).

In this context, regulatory barriers refer to standards, laws, and requirements established by governments and regulatory entities to ensure food safety, quality, and compliance. Despite the benefits, such obstacles can prevent or limit the introduction of new products into the market, both nationally and internationally (Dong and Jiang 2022, Gupta et al. 2025). In addition, strict regulations vary according to the country, further complicating food safety practices (Kanwal et al. 2025).

Regulatory barriers represent a real and solid challenge in the production and integration of new food technologies. Krzywonos et al. (2025) highlight in their study the urgent need for clear regulations and standardized guidelines to facilitate the successful adoption of olive leaf extract (OLE) or other recycled foods in the food industry, leading to the development of more sustainable food systems. Such regulations should be even more thorough with regard to nanotechnologies, given their safety for the consumer (Payal et al. 2025).

Finally, the industry is currently at the intersection of several developing trends and challenges that have the potential to reshape its future trajectory (Apeh and Nwulu 2025). To this end, reviewing and establishing new standardized safety protocols and regulatory guidelines is essential to ensure that the use of nanomaterials, nano coatings, waste, by-products, and even certain meat-derived peptides does not compromise human health while maximizing their advantages (Payal et al. 2025).

12.4 Challenges and Opportunities for Innovation in the Food Industry

The food industry is at a critical juncture, challenged by issues such as climate change, population growth, nutritional transition, and sustainability demands. Figure 4 illustrates the main opportunities for innovation in the sector, highlighting areas such as health, conservation, ingredient functionality, biotechnology, process engineering, and valorization of traditional foods. This interconnection highlights the need for interdisciplinary solutions to meet the needs of a changing market that demands healthier, safer, more functional, and sustainably produced foods (Azra et al. 2021, Garnett 2013).

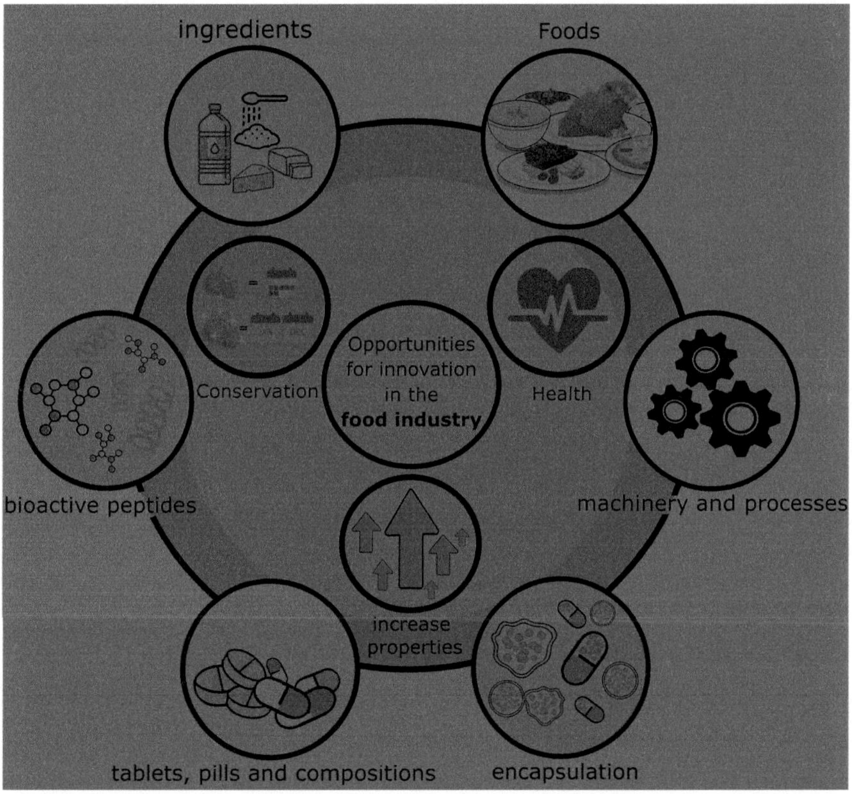

Figure 4. Main areas of opportunity for innovation in the food industry, including conservation, health, functional properties, use of peptides and bioactives, application of microorganisms, modernization of machinery, and valorization of traditional ingredients.

The growing demand for functional and personalized foods has driven the development of bioactive ingredients, such as peptides with antioxidant, antimicrobial, and anti-inflammatory properties, as well as probiotics and prebiotics that modulate the intestinal microbiota. These compounds are increasingly being applied as nutritional strategies to prevent chronic non-communicable diseases, such as type 2 diabetes and cardiovascular diseases (Hendijani and Akbari 2018, Naseri et al. 2022, Olas 2020, Oniszczuk et al. 2021). In parallel, innovations in food preservation, such as encapsulation techniques, smart packaging, and the use of nanotechnology, aim to maintain product stability without resorting to artificial additives, extending shelf life and reducing losses throughout the production chain.

From a technological point of view, the advancement of machinery and processes based on artificial intelligence, 3D food printing, and precision fermentations represents a revolution in food production systems, allowing greater traceability, automation, and customization of products (Canatan et al. 2025, Dancausa Millán and Millán Vázquez de la Torre 2024, Jeong and Kim 2025, Jiang et al. 2025, Waseem et al. 2024). The challenges faced by the food industry also create a fertile environment for science-driven innovation, with the potential to promote advances in public health, food security, and global sustainability.

Figure 5 illustrates the application of nanotechnology as a strategic tool for the controlled delivery of bioactive compounds derived from food. Several compounds present in food matrices—such as polysaccharides, phytochemicals, polyunsaturated fatty acids (DHA), phenolic compounds, tannins, peptides, lipids, enzymes, terpenoids, vitamins, and minerals—have recognized functional potential but often face limitations regarding their stability, solubility, or bioavailability. In this context, encapsulation techniques, especially in nanoparticulate structures such as nanoliposomes, nanoemulsions, polymeric nanoparticles, and nanotubes, are used to protect these compounds during processing, storage, and transport in the body.

After encapsulation, bioactive compounds can be administered by different routes, with controlled release by mechanisms such as diffusion, deformation, or absorption in different systems of the human body, including skin, bloodstream, and gastrointestinal tract. This approach, known as drug delivery, allows encapsulated compounds to be released efficiently at the site of action, optimizing their therapeutic or functional effects. In addition to increasing the efficacy of bioactives, these technologies promote greater safety and precision in delivery, and are promising not only for the pharmaceutical sector but also for functional foods and nutraceuticals. The use of nanoparticles thus represents an innovative frontier in the development of foods with properties aimed at promoting health (Chavda et al. 2022, Petrovic and Barbinta-Patrascu 2023).

Technological innovations in the food industry have promoted significant advances in the development of functional, safe, and sustainable products. The analysis of recent patent applications shows a strong trend toward the incorporation of bioactive compounds, natural ingredients, and processing methods that increase the nutritional value of foods Table 1.

Technologies such as the microencapsulation of anthocyanins from eggplant peel and bioactive peptides from whey demonstrate potential for the creation of

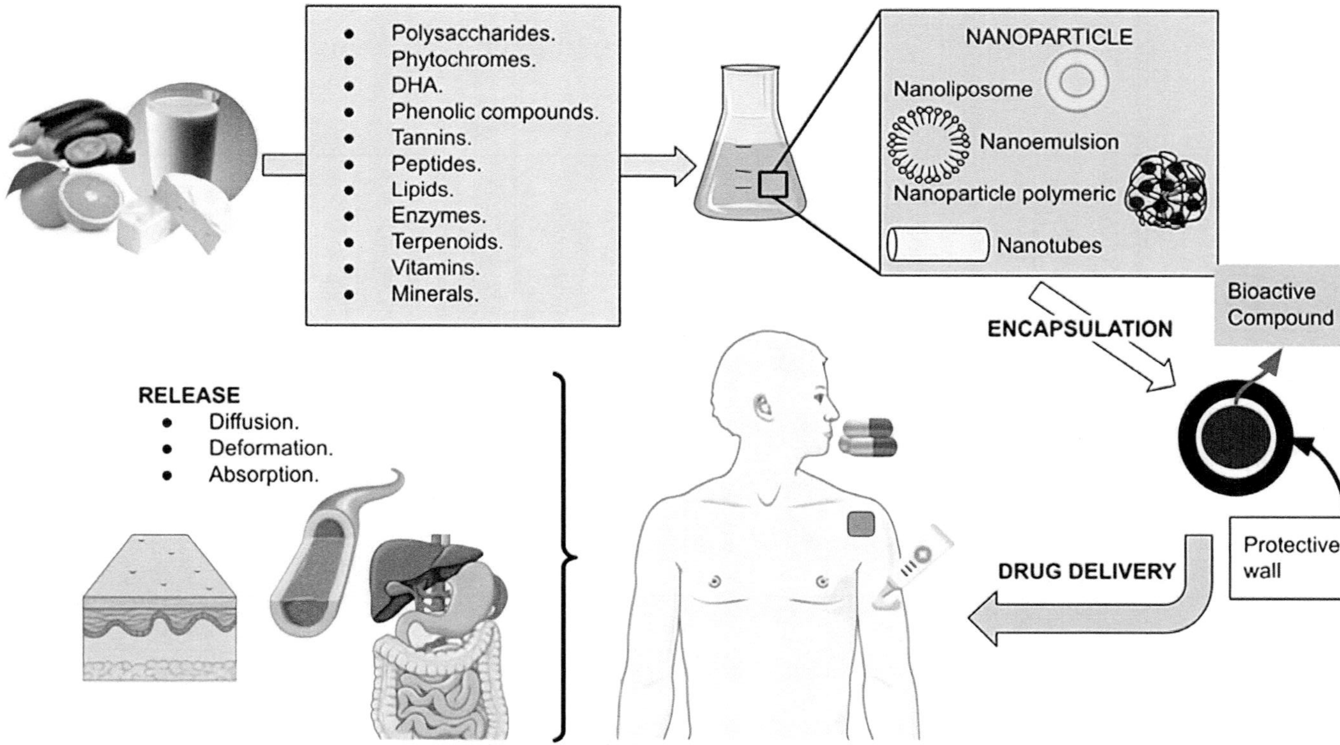

Figure 5. Application of nanotechnology in the encapsulation and controlled release of bioactive compounds derived from food.

Table 1. Patent registrations with the possibility of commercial products for the food industry.

Patent Technology	Description	Reference
Highly functional microencapsulated ingredients based on anthocyanin extracts from eggplant peel and bioactive whey peptides, for use in the food industry.	Process for obtaining microencapsulated ingredients by freeze-drying, based on anthocyanin extracts from eggplant peel and whey protein hydrolysates, incorporated into foods or nutraceuticals that enable the production of highly functional and healthy foods.	(Condurache et al. 2018)
Method for improving the light stability of phycocyanin.	The phycocyanin, subjected to heating and ultrasonic treatment, results in the utilization rate of phycocyanin in the food industry being improved and the broad application scenario of natural bioactive color being expanded.	(Shengfeng et al. 2023)
High-nutritional, high-fiber vegetable protein cheese.	The high-protein, high-fiber vegetable cheese bar is composed only of vegetable oil, natural oil gel, natural hydrogel and water, the components are green and safe, no chemical additives are added, and the health requirements of consumers for cheese bar snacks are met.	(Shuang et al. 2024)
Snail-derived umami peptide with a salinity-enhancing effect.	Two new snail-derived umami peptides PIYEGY and TVPIYEG are separated from a snail meat hydrolysate. The snail-derived umami peptide is expected to be used as a new umami enhancer in the food industry.	(Jianyin and Kaiqi 2024)
Compound plant peptide tablet with functions of dispelling alcohol effects and protecting the liver.	The compound plant peptide tablet candy contains natural extracts and purified bioactive components, and has the effective functions of dispelling alcohol effects and protecting the liver; the components used are all natural derivatives and can be eaten safely.	(Jiandong et al. 2021)
The system and process for converting winemaking derivatives into bioactive nutrient-rich products and/or food ingredients.	The system and process are used to convert winemaking derivatives into nutrient-rich bioactive products and/or food ingredients containing antioxidants and other bioactive molecules from the pomace. They can be used as health supplements and in the cosmetic industry.	(Clancy et al. 2022)
Dry form of a biologically active food additive based on cysts of a crustacean of the genus artemia.	A more nutrient-rich and absorbable (bioavailable) biologically active additive, based on cysts (eggs) of a crustacean of the genus Artemia, for use in food.	(Alexandrovna et al. 2014)
Bioactive drink is an aqueous solution based on drinking or mineral water, amino acids L-arginine and/or taurine.	The drink contains amino acids in subminiature doses. The health level improves, and it is economically viable to produce bioactive drinks due to the low concentrations of the active substances.	(Trofimova et al. 2008)

Table 1 contd. ...

...Table 1 contd.

Patent Technology	Description	Reference
Long-acting tablet to keep food fresh.	The tablet keeps food fresh and releases active substances stably and quickly. It has a high capacity to decompose harmful substances such as bacteria and viruses, with high stability and environmental friendliness.	(Yuling 2016)
Method of producing iodized bread.	It consists of the preliminary iodination of food chitosan, which is one of the components of the dough, ensuring a high degree of constant iodine stability, overall improving the beneficial properties for the consumer of the functional product.	(Mamtsev et al. 2011)
Magnetic nanoparticle for separation method of lysozyme from egg white.	This patent describes a method for preparing magnetic ferrous-ferric oxide nanoparticles modified with carboxymethyl chitosan by a chemical coprecipitation process. These magnetic particles can be used as an adsorption support for the separation and/or purification of lysozyme from egg white, with applications in the food industry.	(Yanjun et al. 2013)

highly stable and highly functional ingredients, facilitating their incorporation into foods or nutraceuticals (Condurache et al. 2018). These microencapsulated systems can preserve sensitive compounds, such as antioxidants, allowing their controlled and targeted release. Through physicochemical treatments, such as heating and ultrasound, it is possible to significantly improve the light stability of phycocyanin, expanding its use in food and beverages (Shengfeng et al. 2023). In parallel, the development of umami peptides derived from snail meat represents an innovative approach to intensify the salty flavor in foods with lower sodium content, promoting reformulation strategies of industrialized products with a focus on public health (Jianyin and Kaiqi 2024).

The growing demand for foods with nutritional and functional appeal drives the development of products based on plant-based and natural ingredients. Technologies such as plant-based cheese rich in proteins and fibers, developed without chemical additives (Shuang et al. 2024), and plant peptide tablets with hepatoprotective effects and the ability to reduce the effects of alcohol stand out (Jiandong et al. 2021). Bioactive beverages containing amino acids such as L-arginine and taurine in minimum concentrations (Trofimova et al. 2008) also demonstrate industrial viability, low production cost, and functional appeal, responding to the growing demand for beverages with therapeutic and preventive properties. It is observed that the advancement of technologies integrates sustainability and innovation, such as the process of converting winemaking by-products into bioactive ingredients rich in antioxidants (Clancy et al. 2022) and the production of food additives based on Artemia crustacean cysts, with high bioavailability (Alexandrovna et al. 2014). Other innovations include chitosan iodination methods for the production of functional breads with high iodine content (Mamtsev et al. 2011), prolonged-release tablets for food preservation (Yuling 2016), and magnetic nanoparticles for separating lysozyme from egg white (Yanjun et al. 2013). Together, these technologies demonstrate how applied research can transform food systems, promoting health, sustainability, and industrial efficiency.

12.5 Conclusion and Future Perspectives

Bioactive compounds and functional foods have emerged as promising tools in the prevention and management of non-communicable chronic diseases, promoting a nutritional approach that goes beyond the basic supply of essential nutrients. The wide diversity of natural sources, including plants, algae, seeds, agro-industrial by-products, and microorganisms, combined with advancements in technologies such as encapsulation, 3D food printing, precision fermentation, and nanotechnology, has driven innovation and personalization within the food industry.

Concurrently, personalized nutrition, grounded in genomic, phenotypic, and microbiome data, offers the potential to maximize the effects of bioactive compounds in an individualized manner, contributing to more effective and preventive nutritional medicine. However, the safe and effective application of these compounds still requires careful attention to their bioavailability, chemical stability, and dose standardization—challenges that can be addressed through the use of controlled-release systems and bioprocessing strategies.

Future directions for the sector point to a convergence of health, sustainability, and technology, fostering a scenario in which functional foods not only enhance quality of life but also serve as catalysts for transformation within global food systems. Upcoming advances are expected to focus on clearer and more harmonized regulations, the reinforcement of food security policies, and the expansion of access to clean and efficient technologies.

In summary, the future of functional food science will depend on interdisciplinary collaboration among researchers, industry stakeholders, policymakers, and consumers, with a shared goal of building a healthier, more resilient, and equitable food system. This trajectory highlights a shift toward evidence-based nutrition, with a strong emphasis on public health promotion, sustainable innovation, and the valorization of biodiversity.

References

Alexandrovna, N., Dobrynina, A., Ivankov Vadim, I. and Bogatyrenk, N. (2014). Dry form of a biologically active food additive based on cysts of a crustacean of the genus artemia. WO2016039656A1.

Al-Habsi, N., Al-Khalili, M., Haque, S. A., Elias, M., Olqi, N. A. I. and Al Uraimi, T. (2024). Health benefits of prebiotics, probiotics, synbiotics, and postbiotics. Nutrients 16(22). https://doi.org/10.3390/nu16223955.

Alonso-Allende, J., Milagro, F. I. and Aranaz, P. (2024). Health effects and mechanisms of inulin action in human metabolism. Nutrients 16(17). https://doi.org/10.3390/nu16172935.

Anand, A., Manjula, S. N., Fuloria, N. K., Sharma, H. and Mruthunjaya, K. (2025). Inulin as a prebiotic and its effect on gut microbiota. pp. 113–135. *In*: Akram, W., Mishra, N. and Haider, T. (eds.). Inulin for Pharmaceutical Applications: A Versatile Biopolymer. Springer Nature Singapore. https://doi.org/10.1007/978-981-97-9056-2_6.

Anibogwu, R., Jesus, K. De, Pradhan, S., Pashikanti, S., Mateen, S. and Sharma, K. (2021). Extraction, isolation and characterization of bioactive compounds from Artemisia and their biological significance: a review. Molecules 26(22). https://doi.org/10.3390/molecules26226995.

Annan, K. A. (2000). We the Peoples : The Role of the United Nations in the 21st Century. United Nations, Dept. of Public Information.

Apeh, O. O. and Nwulu, N. I. (2025). Improving traceability and sustainability in the agri-food industry through blockchain technology: A bibliometric approach, benefits and challenges. Energy Nexus 17: 100388. https://doi.org/https://doi.org/10.1016/j.nexus.2025.100388.

Azra, M. N., Okomoda, V. T., Tabatabaei, M., Hassan, M. and Ikhwanuddin, M. (2021). The contributions of shellfish aquaculture to global food security: assessing its characteristics from a future food perspective. Frontiers in Marine Science 8. https://doi.org/10.3389/fmars.2021.654897.

Ballesteros Martín, M. A. (2016). En busca de una estrategia de seguridad nacional. Ministerio de Defensa.

Bhadange, Y. A., Carpenter, J. and Saharan, V. K. (2024). A comprehensive review on advanced extraction techniques for retrieving bioactive components from natural sources. ACS Omega 9(29): 31274–31297. https://doi.org/10.1021/acsomega.4c02718.

Bistas, K. G. and Tabet, J. P. (2023). The benefits of prebiotics and probiotics on mental health. Cureus 15(8): e43217. https://doi.org/10.7759/cureus.43217.

Bortolini, D. G., Maciel, G. M., Fernandes, I. de A. A., Pedro, A. C., Rubio, F. T. V., Branco, I. G. et al. (2022). Functional properties of bioactive compounds from Spirulina spp.: Current status and future trends. Food Chemistry: Molecular Sciences 5: 100134. https://doi.org/https://doi.org/10.1016/j.fochms.2022.100134.

Canatan, M., Alkhulaifi, N., Watson, N. and Boz, Z. (2025). Artificial intelligence in food manufacturing: a review of current work and future opportunities. Food Engineering Reviews. https://doi.org/10.1007/s12393-024-09395-1.

Chavda, V. P., Patel, A. B., Mistry, K. J., Suthar, S. F., Wu, Z. X., Chen, Z. S. et al. (2022). Nano-drug delivery systems entrapping natural bioactive compounds for cancer: recent progress and future challenges. In Frontiers in Oncology (Vol. 12). Frontiers Media S.A. https://doi.org/10.3389/fonc.2022.867655.

Chen, R. and Chen, G. (2022). Personalized nutrition for people with diabetes and at risk of diabetes has begun. Journal of Future Foods 2(3): 193–202. https://doi.org/https://doi.org/10.1016/j.jfutfo.2022.06.001.

Clancy, S., Strachan, G. and Broddy, W. (2022). System, process, use and food product or ingredient derived from fruit wine derivatives. CA3171955A1.

Condurache, N. N., Bahrim, G. E., Râpeanu, G. and Stănciuc, N. (2018). Highly functional microencapsulated ingredients based on anthocyanin extracts from eggplant peel and bioactive peptides from whey, to be used in food industry. RO134083A2.

Bertolini, A. C., de Assis Reis Neto, S. and de Lima Andrade, J. R. (2023). SEGURANÇA ALIMENTAR NAS POLÍTICAS ESTRATÉGICAS DA CHINA E DOS ESTADOS UNIDOS: APORTES PARA O BRASIL. Revista Tempo Do Mundo 31: 319–343. https://doi.org/10.38116/rtm31art11.

da C. Pinaffi-Langley, A. C., Tarantini, S., Hord, N. G. and Yabluchanskiy, A. (2024). Polyphenol-derived microbiota metabolites and cardiovascular health: a concise review of human studies. Antioxidants 13(12). https://doi.org/10.3390/antiox13121552.

Dancausa Millán, M. G. and Millán Vázquez de la Torre, M. G. (2024). 3D food printing: Technological advances, personalization and future challenges in the food industry. International Journal of Gastronomy and Food Science 37: 100963. https://doi.org/https://doi.org/10.1016/j.ijgfs.2024.100963.

de Boer, A. and Bast, A. (2015). International legislation on nutrition and health claims. Food Policy 55: 61–70. https://doi.org/https://doi.org/10.1016/j.foodpol.2015.06.002.

de Vrese, Michael. (2009). Health benefits of probiotics and prebiotics in women. Menopause International 15(1): 35–40. https://doi.org/10.1258/mi.2009.009008.

Del Cornò, M., Varì, R., Scazzocchio, B., Varano, B., Masella, R. and Conti, L. (2021). Dietary fatty acids at the crossroad between obesity and colorectal cancer: fine regulators of adipose tissue homeostasis and immune response. Cells 10(7). https://doi.org/10.3390/cells10071738.

Dixit, V., Joseph Kamal, S. W., Bajrang Chole, P., Dayal, D., Chaubey, K. K., Pal, A. K. et al. (2023). Functional foods: exploring the health benefits of bioactive compounds from plant and animal sources. Journal of Food Quality 2023(1): 5546753. https://doi.org/https://doi.org/10.1155/2023/5546753.

Domínguez Díaz, L., Fernández-Ruiz, V. and Cámara, M. (2020). An international regulatory review of food health-related claims in functional food products labeling. Journal of Functional Foods 68: 103896. https://doi.org/https://doi.org/10.1016/j.jff.2020.103896.

Dong, X. and Jiang, B. (2022). The market effectiveness of regulatory certification for sustainable food supply: A conjoint analysis approach. Sustainable Production and Consumption 34: 300–309. https://doi.org/https://doi.org/10.1016/j.spc.2022.09.020.

Elbandy, M. (2023). Anti-inflammatory effects of marine bioactive compounds and their potential as functional food ingredients in the prevention and treatment of neuroinflammatory disorders. Molecules 28(1). https://doi.org/10.3390/molecules28010002.

Flieger, J., Raszewska-Famielec, M., Radzikowska-Büchner, E. and Flieger, W. (2024). Skin protection by carotenoid pigments. International Journal of Molecular Sciences 25(3). https://doi.org/10.3390/ijms25031431.

Garnett, T. (2013). Food sustainability: problems, perspectives and solutions. Proceedings of the Nutrition Society 72(1): 29–39. https://doi.org/DOI: 10.1017/S0029665112002947.

González-Velázquez, G., Aguirre-Garrido, J. F., Oros-Pantoja, R., Salinas-Velarde, I. D., Contreras, I., Estrada, J. A. et al. (2024). Supplementation with inulin reverses cognitive flexibility alterations and modulates the gut microbiota in high-fat-fed mice. Frontiers in Behavioral Neuroscience 18. https://www.frontiersin.org/journals/behavioral-neuroscience/articles/10.3389/fnbeh.2024.1445154.

Guo, J., Qi, M., Chen, H., Zhou, C., Ruan, R., Yan, X. et al. (2022). Macroalgae-derived multifunctional bioactive substances: the potential applications for food and pharmaceuticals. Foods 11(21). https://doi.org/10.3390/foods11213455.

Gupta, N., Garg, P. and Ahuja, N. (2025). An integrated pythagorean fuzzy delphi-AHP-CoCoSo approach for exploring barriers and mitigation strategies for sustainable supply chain in the food industry. Supply Chain Analytics 10: 100105. https://doi.org/https://doi.org/10.1016/j.sca.2025.100105.

Gur, J., Mawuntu, M. and Martirosyan, D. (2018). FFC's advancement of functional food definition. Functional Foods in Health and Disease 8(7): 385–397. https://doi.org/10.31989/ffhd.v8i7.531.

Hendijani, F. and Akbari, V. (2018). Probiotic supplementation for management of cardiovascular risk factors in adults with type II diabetes: A systematic review and meta-analysis. Clinical Nutrition 37(2): 532–541. https://doi.org/https://doi.org/10.1016/j.clnu.2017.02.015.

Jeong, J. and Kim, S. (2025). Application of machine learning for quantitative analysis of industrial fermentation using image processing. Food Science and Biotechnology 34(2): 373–381. https://doi.org/10.1007/s10068-024-01744-4.

Jiandong, F., Weiqiang, L., Shuimei, C. and Xiumei, C. (2021). Composite plant peptide tablet candy with functions of dispelling effects of alcohol and protecting liver and preparation method thereof. CN115812829A.

Jiang, L.-L., Gong, X., Ji, M.-Y., Wang, C.-C., Wang, J.-H. and Li, M.-H. (2020). Bioactive compounds from plant-based functional foods: a promising choice for the prevention and management of hyperuricemia. Foods 9(8). https://doi.org/10.3390/foods9080973.

Jiang, Q., Bao, Y., Ma, T., Tsuchikawa, S., Inagaki, T., Wang, H. et al. (2025). Intelligent monitoring of post-processing characteristics in 3D-printed food products: A focus on fermentation process of starch-gluten mixture using NIR and multivariate analysis. Journal of Food Engineering 388: 112357. https://doi.org/https://doi.org/10.1016/j.jfoodeng.2024.112357.

Jiang, S., Liu, H. and Li, C. (2021). Dietary regulation of oxidative stress in chronic metabolic diseases. Foods 10(8). https://doi.org/10.3390/foods10081854.

Jianyin, M. and Kaiqi, M. (2024). Snail-derived umami peptide with saltiness-enhancing effect, preparation method thereof, and application thereof in freshness-enhancing and salt-reducing foods. CN118146307A.

Jones, P. J. H., Asp, N.-G. and Silva, P. (2008). Evidence for health claims on foods: how much is enough? introduction and general remarks12. The Journal of Nutrition 138(6): 1189S–1191S. https://doi.org/https://doi.org/10.1093/jn/138.6.1189S.

Kanwal, N., Zhang, M., Zeb, M., Hussain, M. and Wang, D. (2025). Revolutionizing food safety in the airline industry: AI-powered smart solutions. Trends in Food Science & Technology 159: 104970. https://doi.org/https://doi.org/10.1016/j.tifs.2025.104970.

Kapustian, A., Cherno, N., Naumenko, P. K., Osolina, P. S. and Science, A. P. Tech. (2023). Regulation of functional foods in ukraine and the world. prospects for the use of postbiotics as functional ingredients. Food Science and Technology. https://api.semanticscholar.org/CorpusID:264048354.

Khan, W. A., Chun-Mei, H., Khan, N., Iqbal, A., Lyu, S.-W. and Shah, F. (2017). Bioengineered plants can be a useful source of omega-3 fatty acids. BioMed Research International 2017(1): 7348919. https://doi.org/https://doi.org/10.1155/2017/7348919.

Kolodziejczyk, A. A., Zheng, D. and Elinav, E. (2019). Diet–microbiota interactions and personalized nutrition. Nature Reviews Microbiology 17(12): 742–753. https://doi.org/10.1038/s41579-019-0256-8.

Kurek, M., Benaida-Debbache, N., Elez Garofulić, I., Galić, K., Avallone, S., Voilley, A. et al. (2022). Antioxidants and bioactive compounds in food: critical review of issues and prospects. Antioxidants 11(4). https://doi.org/10.3390/antiox11040742.

Krzywonos, M., Difonzo, G. and Pasqualone, A. (2025). Challenges and technological requirements in agri-food waste upcycling: The case study of olive leaf extract. Future Foods 11: 100547. https://doi.org/https://doi.org/10.1016/j.fufo.2025.100547.

Leem, C. and Martirosyan, D. M. (2019). The bioactive compounds of probiotic foods/supplements and their application in managing mental disorders. pp. 206–220. *In*: Bioactive Compounds in Health and Disease (Vol. 2, Issue 10). Functional Food Institute. https://doi.org/10.31989/bchd.v2i10.431.

Li, H.-Y., Zhou, D.-D., Gan, R.-Y., Huang, S.-Y., Zhao, C.-N., Shang, A. et al. (2021). Effects and mechanisms of probiotics, prebiotics, synbiotics, and postbiotics on metabolic diseases targeting gut microbiota: a narrative review. Nutrients 13(9). https://doi.org/10.3390/nu13093211.

Lin, S. Y. (2017). State capitalism and chinese food security governance. Japanese Journal of Political Science 18(1): 106–138. https://doi.org/DOI: 10.1017/S1468109916000335.

Liu, J., Tian, S., Wang, J., Sang, Q., Su, W., Wu, J. et al. (2025). Hepatoprotective effects of propolis ethanol extract on D-galactose-induced aging mice: A metabolomics and transcriptomics analysis focusing on insulin resistance pathways. Journal of Functional Foods 125: 106668. https://doi.org/https://doi.org/10.1016/j.jff.2025.106668.

Liu, X., Gao, W., Yang, J., Mao, G., Lu, H. and Xing, W. (2022). Association between probiotic, prebiotic, and yogurt consumption and chronic kidney disease: The NHANES 2010–2020. Frontiers in Nutrition 9. https://doi.org/10.3389/fnut.2022.1058238.

Luna-Castillo, K. P., Lin, S., Muñoz-Valle, J. F., Vizmanos, B., López-Quintero, A. and Márquez-Sandoval, F. (2021). Functional food and bioactive compounds on the modulation of the functionality of HDL-C: A narrative review. Nutrients 13(4). https://doi.org/10.3390/nu13041165.

Luvián-Morales, J., Francisco Omar, V.-C., Laura, F.-C., Lucely, C.-P. and Castro-Eguiluz, D. (2022). Functional foods modulating inflammation and metabolism in chronic diseases: a systematic review. Critical Reviews in Food Science and Nutrition 62(16): 4371–4392. https://doi.org/10.1080/10408398.2021.1875189.

Maione, A., Imparato, M., Buonanno, A., Galdiero, M., de Alteriis, E., Guida, M. et al. (2025). Protective and immunomodulatory effects of the novel probiotic yeast Pichia kudriavzevii isolated from a home-made kefir during infection in human colon epithelial cells: An exploratory study. Journal of Functional Foods 125: 106666. https://doi.org/https://doi.org/10.1016/j.jff.2025.106666.

Mamtsev, A. N., Ponomarev, E. E., Kozlov, V. N., Ponomareva, L. F. and Sokolnikov, M. V. (2011). Iodinated bread production method. RU2474123C1.

Manzoor, M., Singh, J., Bandral, J. D., Gani, A. and Shams, R. (2020). Food hydrocolloids: Functional, nutraceutical and novel applications for delivery of bioactive compounds. International Journal of Biological Macromolecules 165: 554–567. https://doi.org/https://doi.org/10.1016/j.ijbiomac.2020.09.182.

Martirosyan, D., Kanya, H. and Nadalet, C. (2021). Can functional foods reduce the risk of disease? Advancement of functional food definition and steps to create functional food products. Functional Foods in Health and Disease 11(5): 213–221. https://doi.org/10.31989/ffhd.v11i5.788.

Martirosyan, D., Lampert, T. and Ekblad, M. (2022a). Classification and regulation of functional food proposed by the Functional Food Center. Functional Food Science 2(2): 25–46. https://doi.org/10.31989/ffs.v2i2.890.

Martirosyan, D., Lampert, T. and Lee, M. (2022b). A comprehensive review on the role of food bioactive compounds in functional food science. pp. 64–78. In Functional Food Science (Vol. 2, Issue 3). Functional Food Institute. https://doi.org/10.31989/ffs.v2i3.906.

Martirosyan, D. and Sanchez, S. S. (2022). Quantum Theory of Functional Food Science: Establishment of dosage of bioactive compounds in functional food products. Functional Food Science 2(3): 79–93. https://doi.org/10.31989/ffs.v2i3.915.

Michalak, M. (2022). Plant-derived antioxidants: significance in skin health and the ageing process. International Journal of Molecular Sciences 23(2). https://doi.org/10.3390/ijms23020585.

Mitra, S., Emran, T. Bin, Chandran, D., Zidan, B. M. R. M., Das, R., Mamada, S. S. et al. (2022). Cruciferous vegetables as a treasure of functional foods bioactive compounds: Targeting p53 family in gastrointestinal tract and associated cancers. Frontiers in Nutrition 9. https://www.frontiersin.org/journals/nutrition/articles/10.3389/fnut.2022.951935.

Mittal, R. K., Mishra, R., Sharma, V. and Purohit, P. (2024). Bioactive exploration in functional foods: unlocking nature's treasures. Current Pharmaceutical Biotechnology 25(11): 1419–1435. https://doi.org/https://doi.org/10.2174/0113892010282580231120041659.

Namdeo, A. G. and Ingawale, D. K. (2021). Ashwagandha: Advances in plant biotechnological approaches for propagation and production of bioactive compounds. Journal of Ethnopharmacology 271: 113709. https://doi.org/https://doi.org/10.1016/j.jep.2020.113709.

Narduzzi, L., Agulló, V., Favari, C., Tosi, N., Mignogna, C., Crozier, A. et al. (2022). (Poly)phenolic compounds and gut microbiome: new opportunities for personalized nutrition. Microbiome Research Reports 1(3). https://doi.org/10.20517/mrr.2022.06.

Naseri, K., Saadati, S., Yari, Z., Asbaghi, O., Hezaveh, Z. S., Mafi, D. et al. (2022). Beneficial effects of probiotic and synbiotic supplementation on some cardiovascular risk factors among individuals with prediabetes and type 2 diabetes mellitus: A grade-assessed systematic review, meta-analysis, and meta-regression of randomized clinical trials. Pharmacological Research 182: 106288. https://doi.org/https://doi.org/10.1016/j.phrs.2022.106288.

Nguyen, T. L., Ora, A., Häkkinen, S. T., Ritala, A., Räisänen, R., Kallioinen-Mänttäri, M. et al. (2024). Innovative extraction technologies of bioactive compounds from plant by-products for textile colorants and antimicrobial agents. Biomass Conversion and Biorefinery 14(20): 24973–25002. https://doi.org/10.1007/s13399-023-04726-4.

Nigam, M., Atanassova, M., Mishra, A. P., Pezzani, R., Devkota, H. P., Plygun, S. et al. (2019). Bioactive compounds and health benefits of Artemisia species. In Natural Product Communications (Vol. 14, Issue 7). SAGE Publications Inc. https://doi.org/10.1177/1934578X19850354.

Noce, A., Di Lauro, M., Di Daniele, F., Pietroboni Zaitseva, A., Marrone, G., Borboni, P. et al. (2021). Natural bioactive compounds useful in clinical management of metabolic syndrome. Nutrients 13(2). https://doi.org/10.3390/nu13020630.

Olas, B. (2020). Probiotics, prebiotics and synbiotics—a promising strategy in prevention and treatment of cardiovascular diseases? International Journal of Molecular Sciences 21(24). https://doi.org/10.3390/ijms21249737.

Oniszczuk, A., Oniszczuk, T., Gancarz, M. and Szymańska, J. (2021). Role of gut microbiota, probiotics and prebiotics in the cardiovascular diseases. Molecules 26(4). https://doi.org/10.3390/molecules26041172.

Pai, S., Hebbar, A. and Selvaraj, S. (2022). A critical look at challenges and future scopes of bioactive compounds and their incorporations in the food, energy, and pharmaceutical sector. Environmental Science and Pollution Research 29(24): 35518–35541. https://doi.org/10.1007/s11356-022-19423-4.

Payal, A., Sandeep, G. D. S., Bammidi, M., Narayandas, A., Syed, I. and Rao, M. V. (2025). Recent advances in plant protein-based edible coatings for shelf-life enhancement of perishable and high nutritive value foods – A review. Food Packaging and Shelf Life 48: 101465. https://doi.org/https://doi.org/10.1016/j.fpsl.2025.101465.

Peng, M., Tabashsum, Z., Anderson, M., Truong, A., Houser, A. K., Padilla, J. et al. (2020). Effectiveness of probiotics, prebiotics, and prebiotic-like components in common functional foods.

Comprehensive Reviews in Food Science and Food Safety 19(4): 1908–1933. https://doi.org/https://doi.org/10.1111/1541-4337.12565.

Perez-Gregorio, R. and imal-Gandara, J. (2017). A critical review of bioactive food components, and of their functional mechanisms, biological effects and health outcomes. Current Pharmaceutical Design 23(19): 2731–2741. https://doi.org/10.2174/1381612823666170317122913.

Petrovic, S. M. and Barbinta-Patrascu, M.-E. (2023). Organic and biogenic nanocarriers as bio-friendly systems for bioactive compounds' delivery: state-of-the art and challenges. Materials 16(24). https://doi.org/10.3390/ma16247550.

Poonia, A. and Shiva. (2022). Bioactive compounds, nutritional profile and health benefits of colostrum: a review. Food Production, Processing and Nutrition 4(1): 26. https://doi.org/10.1186/s43014-022-00104-1.

Ramos-Lopez, O., Milagro, F. I., Riezu-Boj, J. I. and Martinez, J. A. (2021). Epigenetic signatures underlying inflammation: an interplay of nutrition, physical activity, metabolic diseases, and environmental factors for personalized nutrition. Inflammation Research 70(1): 29–49. https://doi.org/10.1007/s00011-020-01425-y.

Salminen, S., Collado, M. C., Endo, A., Hill, C., Lebeer, S., Quigley, E. M. M. et al. (2021). The International Scientific Association of Probiotics and Prebiotics (ISAPP) consensus statement on the definition and scope of postbiotics. Nature Reviews Gastroenterology & Hepatology 18(9): 649–667. https://doi.org/10.1038/s41575-021-00440-6.

Sanz Merino, N. (2022). Policy styles and epistemic policies in the regulation of health claims. a comparison of Europe, the United States, and Japan. Social Epistemology 36(4): 449–465. https://doi.org/10.1080/02691728.2021.2004467.

Shengfeng, P., Jun, L., Wei, L., Lei, Z., Yuman, W. and Yu, L. (2023). Method for improving light stability of phycocyanin. CN117024576A.

Shuang, Z., Minghao, Z., Xue, L., Qingkui, Z. and Bailin, Z. (2024). A high-nutrition, high-fiber plant protein cheese capable of replacing cheese, and a preparation method and application thereof. CN118436000A.

Silva, L. E. C., Santos, T. M. dos, Ferriani, L. O., Silva, D. A. and Sena, G. G. S. (2024). Efeitos de alimentos/compostos funcionais na saúde: uma revisão integrativa. Brazilian Journal of Health Review 7(4): e71222. https://doi.org/10.34119/bjhrv7n4-066.

Singh, B., Singh, J. P., Kaur, A. and Singh, N. (2020). Phenolic composition, antioxidant potential and health benefits of citrus peel. Food Research International 132: 109114. https://doi.org/https://doi.org/10.1016/j.foodres.2020.109114.

Stankovic, S., Mutavdzin Krneta, S., Djuric, D., Milosevic, V. and Milenkovic, D. (2025). Plant polyphenols as heart's best friends: from health properties, to cellular effects, to molecular mechanisms of action. International Journal of Molecular Sciences 26(3). https://doi.org/10.3390/ijms26030915.

Stephen, J., Manoharan, D. and Radhakrishnan, M. (2023). Immune boosting functional components of natural foods and its health benefits. Food Production, Processing and Nutrition 5(1): 61. https://doi.org/10.1186/s43014-023-00178-5.

Tashiro, E. and Imoto, M. (2012). Target identification of bioactive compounds. Bioorganic & Medicinal Chemistry 20(6): 1910–1921. https://doi.org/https://doi.org/10.1016/j.bmc.2011.10.081.

Teodoro, A. J. (2019). Bioactive compounds of food: their role in the prevention and treatment of diseases. Oxidative Medicine and Cellular Longevity 2019(1): 3765986. https://doi.org/https://doi.org/10.1155/2019/3765986.

Trofimova, S. V., Havinson, V. K., Ivko, O. M., Trofimov, A. V. and Kudryavtseva, T. A. (2008). Bioactive beverage. RU2377936C1.

Van Harsselaar, J., Mödinger, Y., Dharsono, T., Menzel, D., Theis, S. and Schön, C. (2024). Prebiotic effect of oligofructose after 2 weeks supplementation with a low dose: A randomized, double-blind, placebo-controlled, cross-over study. Journal of Functional Foods 119: 106356. https://doi.org/https://doi.org/10.1016/j.jff.2024.106356.

Vernia, F., Longo, S., Stefanelli, G., Viscido, A. and Latella, G. (2021). Dietary factors modulating colorectal carcinogenesis. Nutrients 13(1). https://doi.org/10.3390/nu13010143.

Vignesh, A., Amal, T. C., Sarvalingam, A. and Vasanth, K. (2024). A review on the influence of nutraceuticals and functional foods on health. Food Chemistry Advances 5: 100749. https://doi.org/https://doi.org/10.1016/j.focha.2024.100749.

Vlaicu, P. A., Untea, A. E., Varzaru, I., Saracila, M. and Oancea, A. G. (2023). Designing nutrition for health—incorporating dietary by-products into poultry feeds to create functional foods with insights into health benefits, risks, bioactive compounds, food component functionality and safety regulations. Foods 12(21). https://doi.org/10.3390/foods12214001.

Waller, M. E., Gutierrez, A., Ticer, T. D., Glover, J. S., Baatz, J. E., Wagner, C. L. et al. (2025). Profiling the response of individual gut microbes to free fatty acids (FFAs) found in human milk. Journal of Functional Foods 125: 106664. https://doi.org/https://doi.org/10.1016/j.jff.2025.106664.

Wang, K., Zhenzhen, X. and Liao, X. (2022). Bioactive compounds, health benefits and functional food products of sea buckthorn: a review. Critical Reviews in Food Science and Nutrition 62(24): 6761–6782. https://doi.org/10.1080/10408398.2021.1905605.

Wang, P. and Wu, Y. (2021). A review on colloidal delivery vehicles using carvacrol as a model bioactive compound. Food Hydrocolloids 120: 106922. https://doi.org/https://doi.org/10.1016/j.foodhyd.2021.106922.

Waseem, M., Tahir, A. U. and Majeed, Y. (2024). Printing the future of food: The physics perspective on 3D food printing. Food Physics 1: 100003. https://doi.org/https://doi.org/10.1016/j.foodp.2023.100003.

Wen, P., Zong, M.-H., Linhardt, R. J., Feng, K. and Wu, H. (2017). Electrospinning: A novel nano-encapsulation approach for bioactive compounds. Trends in Food Science & Technology 70: 56–68. https://doi.org/https://doi.org/10.1016/j.tifs.2017.10.009.

Williamson, G., Sies, H., Heber, D., Keen, C. L., Macdonald, I. A., Actis-Gorreta, L. et al. (2009). Functional foods for health promotion: state-of-the-science on dietary flavonoids Extended abstracts from the 12th Annual Conference on Functional Foods for Health Promotion, April 2009. Nutrition Reviews 67(12): 736–743. https://doi.org/10.1111/j.1753-4887.2009.00257.x.

Yanjun, Y., Yujie, S., Jun, S. and Shengqi, R. (2013). Preparation of novel magnetic nano-particle and method for separation of lysozyme from egg white. CN103394338A.

Yuling, W. (2016). Long-acting food antistaling tablet. CN106805073A.

Zakaria, S. M., Mustapa Kamal, S. M., Harun, M. R., Omar, R. and Siajam, S. I. (2022). Extraction of phenolic compounds from Chlorella sp. microalgae using pressurized hot water: kinetics study. Biomass Conversion and Biorefinery 12(6): 2081–2089. https://doi.org/10.1007/s13399-020-01027-y.

Zhou, X., Wang, L., Dai, J., Jia, H., Shi, K., Zhao, J. et al. (2025). Dietary fiber-rich Spartina anglica improves intestinal health and antioxidant capacity of Zhedong White Geese. Antioxidants 14(1). https://doi.org/10.3390/antiox14010087.

Zobbe, H. (2001). Give to AgEcon Search The Economic and Historical Foundation of the Common Agricultural Policy in Europe. The Economic and Historical Foundation of the Common Agricultural Policy in Europe. http://ageconsearch.umn.edu.

Chapter 13

The Future of Fruits as Next-Generation Functional Foods

Lina Clara Gayoso e Almendra Ibiapina Moreno, Luana de Moura Monteiro, Paulo de Oliveira Gomes Filho, Samara Daniele Carlos da Silva, Joao Gabriel Albuquerque de Moura Marques and *Anderson Nogueira Mendes**

13.1 Introduction

Over the past few years, there has been a growing concern about healthy eating. People have been looking for foods that have a better nutritional profile and that contain specific compounds to prevent or treat health problems, improving both the quality of life and life expectancy of the population (Manoj et al. 2023). In this context, "functional foods" and "nutraceuticals" emerged, defined as foods that offer additional health benefits beyond their basic nutritional value, containing bioactive compounds that can improve health and reduce the risk of developing chronic diseases (Frumuzachi et al. 2025).

Fruits, in addition to containing high levels of macronutrients, show a high content of bioactive compounds, including polyphenols (especially flavonoids), carotenoids, fibers, prebiotics, and sulfur compounds (Rodríguez-Cortina and Hernández-Carrión 2025). These compounds exhibit remarkable activities, such as antioxidant, anti-inflammatory, antibacterial, anticancer, anti-obesity,

Laboratory of Science in Innovation and Technology - LACITEC, Department of Biophysics and Physiology, Federal University of Piauí, 64049-550, Teresina, Piauí, Brazil.
* Corresponding author: e-mail: anderson.mendes@ufpi.edu.br

DOI: 10.1201/9781003668169-16

antihypertensive, antidiabetic, and neuroprotective effects, promoting several health benefits (Sakulrang et al. 2025, Teibo et al. 2021).

Furthermore, substances present in fruits such as anthocyanins—belonging to the flavonoid class—are widely used as natural colorants in the food industry, offering a sustainable alternative to synthetic additives in food products and nutritional supplements (Mohammadi et al. 2024). This approach supports food producers in offering healthier diets, a movement known as "clean label".

It is important to notice that in order to obtain the benefits of functional foods, they need to be consumed regularly. Therefore, in addition to fresh fruit, they can be ingested after being subjected to a drying process—once dried, the fruits are not perishable, which adds high value to the products for use in various food situations that respond to the demand and needs of consumers—or in preparations such as cereals or drinks such as juices and shakes (de Oliveira et al. 2025).

13.2 Health Benefits of Bioactive Compounds Occurring in Fruits

Polyphenols are a group of plant secondary metabolites composed of more than 8000 known structures, with diverse conformation and chemical functions, widely distributed in fruits (Wang et al. 2024). They comprise a variety of molecules that contain a classic phenolic ring design (i.e., several hydroxyl groups in aromatic rings), whose structure can vary from simple molecules to complex polymers (Câmara et al. 2020). Polyphenols can be divided into five groups: flavonoids, phenolic acids, stilbenes, tannins, and lignans (Wojtunik-Kulesza et al. 2020), which are subdivided into several subgroups (Figure 1).

Polyphenols play crucial roles in plant defense systems, showing antibacterial, antifungal, and antiviral activity, in addition to being responsible for the color and flavor of many plant tissues (Luca et al. 2020). These compounds, especially flavonoids and phenolic acids, generally accumulate in the vacuoles of fruits such as strawberries, blackberries, raspberries, blueberries, grapes, oranges, watermelons, and açaí.

Carotenoids are a group of tetraterpenoid pigments with over 700 characterized structures, ranging from yellow to red, widely distributed in plants, algae, fungi, and bacteria (Sue et al. 2025). Many flowers, fruits, and roots owe their vibrant orange, yellow, and red colors to carotenoids. Carotenoids can be classified into two groups: xanthophylls and carotenes. Thus, carotenoids composed only of carbon and hydrogen are called carotenes, and carotenoids that contain oxygen are known as xanthophylls (Qui et al. 2024). Examples of carotenoids include alpha-carotene, beta-carotene, lycopene, lutein, zeaxanthin, and astaxanthin, which are found in fruits such as papaya, mango, pineapple, red guava, tangerine, and watermelon (Zhang et al. 2024).

These compounds play an essential role in photosynthesis and photoprotection of plants (Hashimoto et al. 2016), provide precursors for the biosynthesis of the phytohormones abscisic acid (ABA) and strigolactones (Al-Babili and Bouwmeester 2015), and their derivatives act as signaling molecules to mediate plant development and their responses to environmental stimuli (Hou et al. 2016).

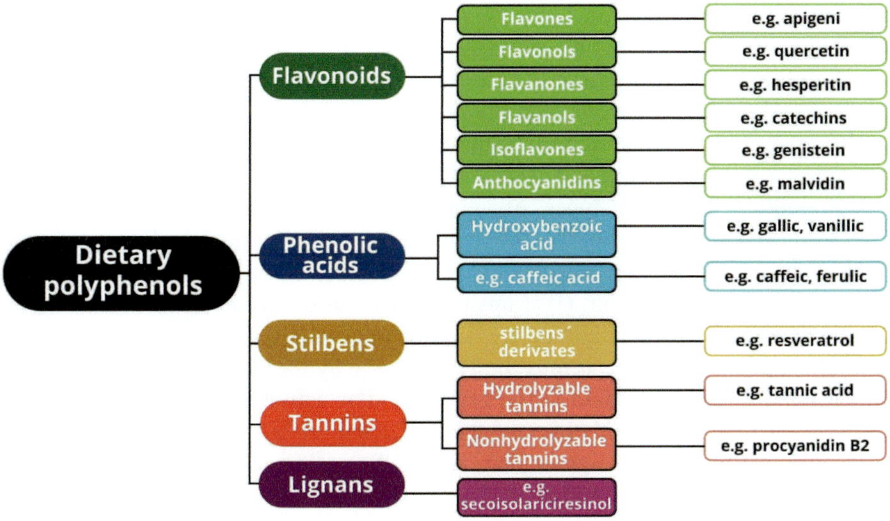

Figure 1. Representative examples of dietary polyphenols.

In addition, the impact of carotenoids goes beyond pigmentation and plant growth regulation, as these compounds also constitute crucial nutrients for human health (Liu et al. 2025). They provide dietary sources of provitamin A and act as antioxidants to reduce the onset of some chronic diseases, such as cardiovascular disease, cancers, and age-related eye diseases (Sun et al. 2018).

Continuing with the theme of bioactive components in fruits, we should also mention processing residues, such as peels, seeds, and pomace, which are rich in dietary fiber, constituting a potential source of fermentable carbohydrates and compounds with biological activity (de Oliveira et al. 2024). The Food and Drug Administration (FDA) has defined dietary fiber as non-digestible carbohydrates with three or more monomeric units, along with lignin, that are intrinsic and intact in plants (Turner and Lupton 2021). This definition also includes isolated or synthetic carbohydrates with three or more monomeric units that the FDA has determined provide beneficial physiological effects.

Several methods have been used to classify dietary fiber types, including their role in plants, polysaccharide type, solubility under gastrointestinal conditions, site of digestion, and digestion products. Currently, the most accepted approach differentiates fibers based on their solubility in a buffer at a defined pH and their *in vitro* fermentability using an enzymatic solution that mimics human digestive enzymes, a method that classifies dietary fibers into soluble and insoluble fibers (Samami et al. 2025).

Thus, insoluble fibers do not dissolve in water and remain largely intact after passing through the digestive system. They effectively retain water, promoting bowel movements and preventing constipation. Their fibrous texture requires more chewing, and they are generally less fermentable. Examples of these types of fiber include cellulose and hemicellulose (Dhingra et al. 2012). In contrast, soluble fibers, including pectins, gums, and mucilages, dissolve in water and form a gel-like substance in the digestive tract, contributing minimally to fecal bulk but are actively fermented by gut bacteria (Lu et al. 2023).

Soluble dietary fibers have the ability to bind to toxic molecules, reduce blood glucose and cholesterol levels, and improve the intestinal microbiota profile, acting in the prevention of diabetes and cardiovascular diseases, while insoluble fibers act by improving intestinal transit (Bhatt and Gupta 2024).

Prebiotics are other types of fermented ingredients that can modify the composition and/or activity of the gastrointestinal microbiota, thus conferring health benefits (Gibson et al. 2017). A compound is classified as a prebiotic if it meets the following criteria: (i) it must be stable in the acidic environment of the stomach, resistant to degradation by digestive enzymes, and not absorbed in the gastrointestinal system; (ii) it must be fermentable by the intestinal microbiota; and (iii) it must selectively promote the growth and/or activity of beneficial intestinal bacteria, providing a health improvement (Gibson et al. 2010). The best-known prebiotics include fructooligosaccharides (FOS), galactooligosaccharides (GOS), and transgalactooligosaccharides (TOS).

Volatile sulfur compounds, also known as thiocompounds, are organic structures that have one or more sulfur atoms linked to the carbon chain, giving the juicy, fresh, and authentic aroma to many varieties of fruits, such as passion fruit, yellow monbin, jabuticaba, and pitanga (Cannon and Ho 2018). Among the best-known thiocompounds are thiols or thioalcohols and thioethers, but there are also thioesters, thioketones, and thiophenols (Dordevic et al. 2023). From a pharmacological point of view, these compounds have anti-inflammatory, neutrophil immunomodulatory, antihypertensive, antioxidant, and hypocholesterolemic activities, in addition to promoting protection against carcinogenesis and mutagenesis (Farhadi et al. 2020).

Dietary polyphenols play an important role in promoting long-term health, as these compounds have been shown to reduce the risk of chronic and degenerative diseases (Nowak et al. 2022). Polyphenols exhibit antioxidant, anti-inflammatory, antidiabetic, anticancer, cardioprotective, and gastroprotective properties, in addition to exhibiting important antimicrobial effects against human pathogenic bacteria (Köpsel et al. 2025).

Polyphenols are the most studied natural antioxidants, and their anti-aging properties, in addition to their ability to combat age-related diseases, such as cardiovascular disease, cancer, neurodegenerative diseases, type 2 diabetes mellitus, and osteoporosis, have been extensively investigated (Xu et al. 2023). Studies have shown that they promote the regulation of oxidative damage, inflammation, autophagy, and anti-aging effects of the intestinal microbiota (Wu et al. 2025). Research has shown that the antioxidant capacity of polyphenols may be correlated with their phenolic hydroxyl groups. For example, kaempferol-3,7,4'-trimethyl ether has one phenolic hydroxyl group with antioxidant capacity equivalent to that of Trolox. Kaempferol-3,4'-dimethyl ether has two phenolic hydroxyl groups, and its antioxidant capacity is twice that of Trolox. Subsequently, researchers found a similar trend in kaempferol-7-neohesperidoside and kaempferol (Cao et al. 1997).

Tumor necrosis factor alpha (TNF-α), interleukins 1β (IL-1β), 2 (IL-2), 6 (IL-6), 8 (IL-8), and 12 (IL-12), and inflammatory mediators such as NO, inducible NO synthase (iNOS), and cyclooxygenase 2 (COX-2) are common targets of polyphenols, such as curcumin, quercetin, and genistein. These enzymes are also considered potential targets for polyphenols in the treatment of diseases such as cancer, cardiovascular diseases, and neurodegenerative diseases (Yahfoufi et al. 2018). In this sense, research has shown that quercetin, catechin, and epigallocatechin gallate (EGCG) exert anticancer effects through classical inflammatory signaling pathways, such as nuclear factor-kappa B (NF-κB), mitogen-activated protein kinases (MAPK), and arachidonic acid metabolism. Polyphenols have also been shown to be effective agents in the treatment and prevention of cardiovascular and neurodegenerative diseases by playing an important role in inhibiting the pro-inflammatory signaling pathways described above.

Autophagy promotes the elimination of aberrant aggregates and the maintenance of protein quality control, which helps combat aging and diseases. Polyphenol consumption activates autophagy through the adenosine 5-monophosphate-activated protein kinase–mammalian target of rapamycin (AMPK-mTOR) signaling pathways. This activation contributes to the preservation of health and the delay of aging in several animal models. For example, resveratrol promoted increased autophagy through inhibition of the phosphatidylinositol 3-kinases–mammalian target of rapamycin (PI3K/AKT/mTOR) pathway in rats (Wan et al. 2016), while kaempferol induced the same effect in human endothelial cells (Che et al. 2017), and gallic acid increased autophagy in both CCD-18Co cells and rats (Kim et al. 2017). Some studies have shown that polyphenols may exert anti-aging effects by modulating the gut microbiota. The health benefits of most polyphenols are partly derived from the interplay between remodeling the gut microbiota to a healthy composition and modifying the bioavailability of microbial metabolites in the host (Wu et al. 2025). As previously mentioned, carotenoids are natural bioactive compounds that exert their activity mainly through their antioxidant properties and the regulation of inflammatory response mechanisms (Oduro-Obeng et al. 2024). These properties allow the elimination of lipid peroxyl radicals and singlet oxygen species, which contribute to the etiology of several chronic noncommunicable diseases (Kopec and Failla 2018). During digestion, carotenoids released from various food matrices can eliminate and prevent the generation of reactive oxygen species (ROS) in the gastrointestinal tract. Thus, the intake of foods rich in carotenoids is associated with a reduced risk of cardiovascular diseases, some types of cancer, and respiratory diseases.

Studies have shown an inverse relationship between blood concentrations of lycopene and cancer risk in humans (Xu et al. 2016). Thus, lycopene supplementation has been considered potentially useful as an adjuvant therapy against cancer due to its anticancer, anti-inflammatory, antioxidant, antiangiogenesis properties, in addition to the effect of inhibiting cell proliferation, pro-apoptotic activity, and immune modulation (Jurado-Fasoli et al. 2023). Lycopene also showed positive effects on cardiovascular risk factors (Cheng et al. 2017).

On the other hand, research has shown that lung function is related to blood levels of lycopene, α-carotene and β-carotene (Islam et al. 2022). In this sense, the level of carotenoids found in the blood was considered positively correlated with respiratory function indexes called forced expiratory volume in one second (FEV1) and forced vital capacity (FVC) in population samples.

Additionally, carotenoids, which are characterized as UV-protective pigments, have been shown to promote defenses against photooxidative damage and protect proteins, lipids, and DNA from oxidative stress induced by UV irradiation (Contreras et al. 2025).

Recent research suggests that dietary fiber, which is a non-digestible carbohydrate that serves as a valuable source of nutrients for the gut microbiota, may play a crucial role in modulating immune responses and influencing cancer progression (Samami et al. 2025).

Soluble fibers found in apples and citrus fruits promote fermentation that produces short-chain fatty acids (SCFAs) such as acetate, propionate, and butyrate, which play important roles in regulating metabolism, the immune system, and host cell growth (Yusuf et al. 2022). Although SCFA levels in the bloodstream are relatively low, they act as chemical messengers and regulate several physiological functions in the body (O'Riordan et al. 2022).

Soluble fibers also bind to bile acids in the small intestine, preventing their absorption and leading to reduced blood cholesterol levels (Ramos et al. 2022), and inhibit sugar absorption (Giuntini et al. 2022).

Most sulfur-containing compounds exhibit very strong affinities for biological targets, such as enzymes and receptors, since the presence of sulfur in heterocycles installs unique characteristics, such as increased lipophilicity, amplified metabolic stability, ability to form a variety of chemical bonds and interactions, including hydrogen bonds, and high affinity for metal ions, making sulfur compounds significant modulators of different biochemical pathways (Hamad 2025).

In this line, studies have shown that sulfur compounds, such as diallyl disulfide, allicin, diallyl trisulfide, ajoenes, vinyldithiins, and S-allylcysteine, have therapeutic properties, as they act as antibacterial, antiviral, fungicide, anti-cardiovascular disease, antidiabetes, antihypertensive, anti-atherosclerosis, anti-rheumatism and anti-hyperlipidemia agents, in addition to having anticancer activities (Jikah and Edo 2023).

One of the main ways in which sulfur-containing heterocyclic compounds produce anticancer effects is by inhibiting cancer cell survival and proliferation by targeting enzymes that cancer cells employ for their own survival (Pathania et al. 2019). For example, many sulfur-containing compounds serve as topoisomerase inhibitors, which causes DNA replication and repair enzymes to be deactivated, resulting in DNA damage in cancer cells. Consequently, apoptosis is triggered. Sulfur-containing heterocyclic compounds also inhibit kinases, signaling proteins critical for growth, preventing cancer cells from proliferating and surviving (Ward and DeNicola 2019). Another mechanism of action for sulfur-containing heterocyclic compounds in fighting cancer includes inducing oxidative stress within cancer cells (Carradori et al. 2022).

13.3 Future Scope in Research and Market Potential

13.3.1 Advances in Functional Food Research

Nanotechnology plays a fundamental role in food science, particularly in the nanoencapsulation of these bioactive compounds, as an emerging technology implemented in the food production chain, facilitating the delivery and regulation of functional ingredients at the desired location, protecting them from unfavorable environmental factors, increasing their stability, bioavailability, and therapeutic potential. Almost all of these active components are very sensitive compounds and, without protective treatments, lose their functional properties. Encapsulation becomes an effective tool for the administration of therapeutic agents and sensitive compounds (Klojdová et al. 2023).

A growing number of studies are providing a broader analysis of this technique for encapsulating bioactive compounds, emphasizing the characteristics, food applications, and implications for human health (Rezagholizade-shirvan et al. 2024). Encapsulation technologies, which administer a variety of sensing agents, have been used for functional food preparations, including capsules, particles, emulsions, liposomes, and lipid nanoparticles, among others. Extrusion technology offers considerable operational flexibility, often employed for volatile and thermosensitive bioactive chemicals. The active ingredients are combined with a hydrocolloid solution and forced through a nozzle to form a gelling solution, resulting in the formation of microcapsules (Emon et al. 2025).

Freeze-drying consists of freezing a solution by reducing the surrounding pressure, allowing the frozen solvent to sublime directly from the solid to the gaseous state. It is a technique that maintains the structural integrity and activity of sensitive compounds, making it suitable for encapsulating substances sensitive to high temperatures, such as enzymes (Sereti 2025, Rezagholizade-shirvan et al. 2024). The coacervation method is used to create microcapsules or nanoparticles, whose approach consists of the separation of liquid–liquid phases, in which one or more hydrocolloids present in a polymeric solution form a protective layer around an active ingredient suspended in this solution (Rezagholizade-shirvan et al. 2024). Liposomal microcapsules are artificial vehicles with phospholipid bilayers, similar to natural membranes, with an amphiphilic nature, which trap numerous bioactive compounds. They release encapsulated compounds at specific targets, allowing a controlled delivery system. Their liposome formation involves several steps designed to optimize encapsulation efficiency and the stability of encapsulated compounds. They also present positive results in terms of stability during storage (Sereti 2025). Products that can be encapsulated contribute to the treatment and prevention of specific diseases, and thus this material can be used for both therapies and personalized diets (Rezagholizade-shirvan et al. 2024).

The future outlook for encapsulation in the food industry is vast and promising. With the growing demand for healthier and longer-lasting food products, encapsulation technology plays a crucial role in improving food quality, increasing the stability of bioactive compounds, and creating functional foods. As previously mentioned, encapsulation strategies aim to improve the stability, solubility, bioavailability, sensory properties, retention of bioactive properties, and shelf life of encapsulated compounds (Klojdová et al. 2023).

In the integration of intelligent digital technologies in industrial production, where automation and digitalization of entities and processes are fundamental, tools such as artificial intelligence (AI) have become increasingly essential, offering innovative solutions in numerous domains. Nutrition, in this approach, is no exception to the fields influenced by this integration. In the nutritional field, conceptual algorithms, machine learning, and machine learning techniques can be applied for the purposes of analysis, interpretation, and decision-making, informed by various information related to nutritional data, dietary patterns, and other health factors. This facilitates technological advances that reshape the profile of dietary interventions. These AI techniques are immensely promising in the sense of revolutionizing the way nutritional results are interpreted (Theodore Armand et al. 2024).

ChatGPT, for example, according to Chatelan (2023), can inform users about how to eat as healthily as possible, in addition to functions for prescribing diets, planning recipes, and making shopping lists, only by considering, for example, the description that users give about what they have in their pantry or refrigerator. Those individuals who use ChatGPT should be aware that it does not reference the sources of information when providing answers, so it is not possible to know whether the sources are evidence-based or not. And if asked for references, it will replicate what is most mentioned on the web. So this feature can be useful for a quick second opinion and for clinical decision-making. In the case of personalized nutrition approaches, fluctuations in factors within the same individual over time, used to design diets, are not known, and the initial algorithms may need to be updated with the use of these tools (Detopoulou et al. 2023).

Artificial intelligence can also enhance clinical judgment of outcomes within the professional/patient relationship or direct-to-consumer health advice, but these aspects raise ethical and legal issues. These considerations are vast and complicated. Recognizing that artificial intelligence will play a greater role in all of our lives is paramount, and as such we should be encouraged to understand its benefits and limitations, leading to broader levels of debate and use (Chatelan 2023, Bond et al. 2023).

Furthermore, there is limited data on the long-term effectiveness of AI-assisted nutritional interventions. AI-assisted personalized nutrition needs further justification, while the regulatory framework requires constant updating to keep up with scientific advances and address ethical concerns (Detopoulou et al. 2023).

The lack of interpretability and explainability in AI models is a limitation highlighted in several studies. Understanding how they work is increasingly challenging as these models become more complex. This limitation is of concern both in clinical and healthcare settings, where one of the key points for developing the professional/patient relationship is to gain trust and facilitate informed decision-making (Theodore Armand et al. 2024).

Nutrition research has offered innovative tools with promising potential to improve both individual and public health outcomes.

13.3.2 The Growth of the Functional Foods Market

There is a growing trend in developed economies towards healthier eating habits, which means that consumers are acquiring new needs, specifically for high-quality products with numerous functions and nutritional properties that bring benefits to their health. The concern with maintaining health, together with the studies that involve it, has led the population to seek healthier habits and thus these functional foods have gained greater attention, given their long-term performance (Safraid et al. 2024).

Functional foods emerge from this search and are of great importance for competition between food companies, especially in economies where food consumption is already stagnant. Thus, companies focus their growth plans on the food market, with differentiated services and adding quality to products. Several studies indicate that consuming foods enriched with functional ingredients reduces the risk of chronic diseases and improves health as a whole. Understanding the factors that lead to consumption is essential for the successful launch of a product and, consequently, for the development of marketing strategies in the emerging functional food market. This is due to the nature of this type of market (Sgroi et al. 2024).

Functional foods have experienced rapid growth, accounting for US$ 280.7 billion in 2021, but are estimated to reach values equivalent to US$ 547.2 billion by 2030 - Chart 1 (Sereti et al. 2025) (Figure 2). Asia holds the largest market for functional foods, with the Japanese spending around 150 euros per person per year on this niche. The USA and Canada come in second, collectively reaching 25% of the global level. Europe has a wide distribution of functional foods, with the main markets including Germany (14%), France (13%), and Italy (11%). Dairy and dairy-based products hold the largest market share (38%), followed by bakery and cereals (22.7%). Europe‘s share of this market is around 2.8 million euros globally, with a projected growth rate of 14% in the coming years. Understanding the benefits of functional foods for a healthy diet demonstrates how promising this market is, reinforcing the need for measures to encourage consumer confidence to be increasingly adopted (Sgroi et al. 2024, Safraid et al. 2024).

As consumers seek alternative healthy foods, there is a growing demand for plant-based functional foods that offer additional health benefits beyond basic nutrition. These foods demonstrate physiological benefits due to the presence of functional ingredients or bioactive compounds, such as proteins, amino acids, polyphenols, vitamins, and other antioxidant compounds that are not present in conventional animal-based food products. The plant-based functional food market is growing exponentially and is expected to reach US$ 137.1 billion by 2026 and over US$ 275 billion by the year 2030. This significant increase is due to a number of factors, including increased awareness of the link between diet and health and increased availability of functional food ingredients (Anyiam et al. 2025).

Within this trend of seeking sustainable alternatives, focusing on health, the environment, and food safety, the plant-based market has emerged, seeking proteins that have consumers aware of the impact of their food choices, in addition to also considering animal welfare and their own health. In vegan diets, alternative possibilities have been included such as legumes, grains, nuts, seeds, mycoproteins,

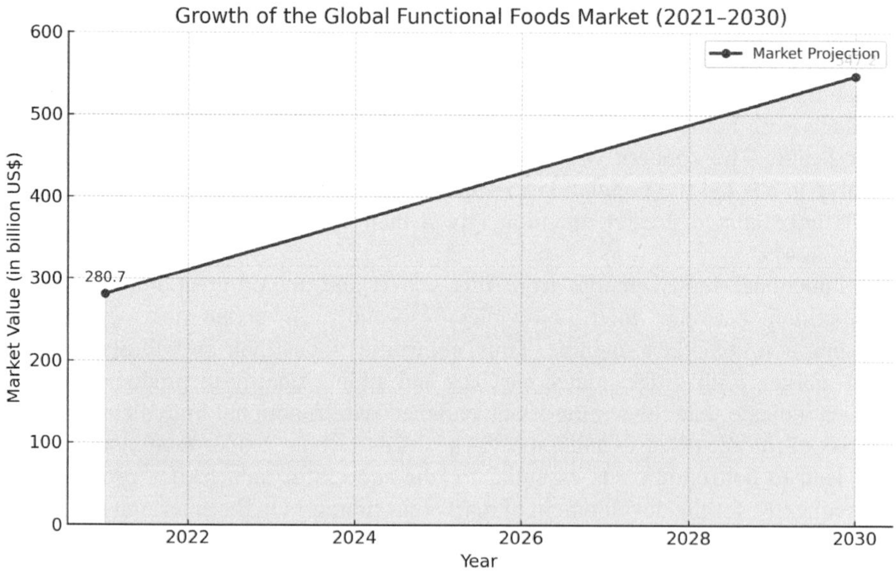

Figure 2. Projected growth of the global functional food market (2021–2030).

and algae, where the latter has emerged as very promising among functional foods. The benefits go beyond reducing the environmental impact, as they reduce the use of water, land, and energy when compared to the use of animal protein, in addition to reducing gas emissions and decreasing the possibility of spreading diseases from animals to humans (Ulhas et al. 2023, Saeed et al. 2025).

The Food and Drug Administration does not have an established regulation for functional foods; therefore, their commercialization is under the existing regulatory options. The development of regulations for the documentation of the safety of ingredients and dietary supplements is an imminent concern. The Nutrition Labeling and Education Act (NLEA) developed in the 1990s introduced food labels, which justify the relationship between foods or food components with diseases or health conditions, and all these references to health benefits must undergo FDA approval before marketing (Essa et al. 2023).

In 1997, the FDA Modernization Act (FDAMA) was implemented. This act streamlined the FDA's pre-approval process by implementing "authoritative statements" on food labels as health claims. These statements must be published by U.S. government agencies responsible for public health, such as the National Institutes of Health (NIH). Manufacturers intending to use the authoritative statement must notify the FDA 120 days prior to marketing. The most impactful change in regulation was instituted by the Dietary Supplement Health and Education Act of 1994 (DSHEA), which considered food additives such as vitamins, for example, as capable of improving the health of the population. They introduced an agreement that includes reporting the safety and efficacy of these compounds in the living system. If the compound claims to treat any disease involving nutrient deficiency, the prevalence of the disease must be investigated in the United States (Essa et al. 2023).

13.4 Role of Bioactive Compounds in Sustainable Diets

13.4.1 Fruits and Food Sustainability

Vegetables and fruits, when consumed, provide essential nutrients and consequently reduce the onset of numerous chronic diseases. In Europe, neither adults nor children can consume sufficient quantities of these foods to meet the World Health Organization's recommendations of at least 400 g of vegetables and fruits daily. This is a real gap between what is recommended and what is consumed, and is mainly present in industrialized countries. Thus, it becomes a concern that leads to many public health interventions and initiatives (Helgegren et al. 2025).

Ultra-processed foods consist of refined ingredients, low nutritional quality, low levels of essential micronutrients, unsaturated fatty acids and dietary fiber, in addition to having a high amount of synthetic compounds and being industrially processed. Their consumption is closely related to the origin of chronic non-communicable diseases, cardiovascular diseases, type 2 diabetes, cancer and several other diseases. These conditions, combined with dietary and lifestyle patterns, have been a leading global cause of mortality and have played a critical role in their pathogenesis (Babalola et al. 2025).

Ultra-processed foods (UPFs) provide a much higher energy content, accounting for a significant proportion of total energy intake in several countries. In New Zealand and Canada, UPFs contribute to 60% of total calorie intake; Norway and Belgium, about 59%. In the United States, a figure of 58% of calorie intake, and in the United Kingdom, the percentage is 57%. Sweden follows with 55% of calorie intake, Spain a little lower at 52%, while Chile and Brazil are less dependent on these foods, representing 30% of their calorie intake (Agjei et al. 2025).

There is no policy to reduce the consumption of ultra-processed foods. Many sectors in the food and beverage industry are opposed to placing a barrier and often lobby against such measures, citing potential economic damage and job losses. Low- and middle-income countries rely heavily on these sectors for their economic development and employment, and are therefore reluctant to impose stricter regulations. In addition, inadequate infrastructure and governance in these low- and middle-income countries make it difficult to impose policies that implement this increased regulatory capacity with regard to ultra-processed foods (Babalola et al. 2025). These foods are also more accessible, with a cost advantage compared to other foods.

Globally, the food system is a major contributor to the environmental problems caused by humanity. One of the key strategies for keeping the global food system within planetary boundaries is to reduce food waste (Malefors 2025). Waste levels are at an all-time high, with 14% of the world's food lost after it is harvested and before it reaches stores. Approximately 3.1 billion individuals face difficulties in accessing an adequate diet, with more than 690 million people suffering from chronic hunger (FAO 2020, Casson et al. 2024).

Initiatives emerge in this scenario to recover food that is still healthy and safe, through donations, preventing it from being wasted: collection networks, community and charity programs, shelters, soup kitchens, among others (Casson et al. 2024).

These attitudes prevent environmental impact and are related to the practice of avoiding the emission of carbon and other gases that cause the greenhouse effect, introducing an important aspect that is the assessment of environmental balance, combining the concept of carbon reduction with all impact categories (Casson et al. 2024). It is important, when considering environmental sustainability, to reduce food waste as a whole and reduce carbon emissions. The approach to carbon emissions is essential because it provides the necessary information to guide public policies so that specific actions can be initiated. Even if the carbon emissions results are negligible with regard to other issues and aspects, efforts should still be directed to where there is greater sustainable potential (Malefors et al. 2025).

Sustainable food value chains are built on three pillars: economic sustainability in profitable regions, social sustainability in regions that are favorable to society, and environmental sustainability in natural habitats (Maalouf et al. 2024). Campaigns that raise awareness and educate about food waste should focus on the environmental and nutritional consequences that fall under high-impact items. All of these efforts will contribute to environmental sustainability, improving nutritional outcomes, and contributing to reducing food waste (Malefors et al. 2025).

13.4.2 Sustainable Production and Genetic Improvement

In the contemporary environmental panorama, human carelessness has played a fundamental role in the consequences of its development model and its excessive consumption (Kazazian 2005). Proportionally, the more humans consume, the more packaging is produced and, subsequently, more packaging is discarded into the environment (Moreira 2015). Recently, the problems of inadequate disposal of solid waste have intensified excessively, thanks to new technologies for producing products and packaging (Cortez 2011), presenting a real challenge to sustainable production in the global industry.

Sustainable development is defined as covering the needs of contemporary society, without harming or affecting the quality of life of future generations, within the scope of satisfying their main needs (Farias 2024). The main properties of sustainability are the efficient management of natural resources, in addition to environmental protection, as well as of all forms of life, whether human or not, providing their protection for current and future generations. The main characteristics of sustainability involve social aspects, which seek balance and equality; economic aspects, which seek to transform comparative advantages into competitive ones; and environmental aspects, which seek to maintain and conserve the planet's natural resources (Mello 2007).

To promote plant health, the health of the soil in which they will grow must first be ensured. In this context, the use of organic fertilizers to increase soil fertility can be a great resource, as they are materials of natural origin and are derived from plants, minerals, and animals. These fertilizers become a sustainable alternative to chemical fertilizers, since they provide essential nutrients in a progressive manner, ensuring soil moisture, in addition to enabling the health of the entire surrounding ecosystem. In the context of plant cultivation, organic fertilizer is complete and functional, as it has a large abundance of nutrients and is produced through composting, a process that is based on the decomposition of various natural components, and uses various abundant and low-cost resources such as dry leaves, plant residues, fruit and vegetable remains, as well as animal manure (Farias 2024).

Within agroecological production, genetic improvement seeks to develop a variety of plants that resemble the concepts of agroecology, developing sustainability, the production of health-focused foods, and environmental resilience, where the balanced incorporation of these practices in the agroecological context becomes crucial to generating agricultural systems that are productive but also environmentally responsible. In addition to genetic improvement, biotechnology has also made progress in terms of application in agricultural production sectors, with improvements in the production process being observed over the years, with advances being noted mainly in the production of soybeans, wheat, rice, corn, and several other commodities. However, contemporary food systems face obstacles caused by excessive dependence on staple crops, creating an ecologically unsustainable scenario and also generating climate vulnerability (Marques 2008).

When it comes to sustainable rural development, genetic improvement cannot be excluded from this process, since it is a fundamental technique for the modernization of agriculture in general. The use of this technique allows the selection of interesting and desirable properties of a given species, making it possible to produce diversities with great resistance to both abiotic and biotic stresses, in addition to increasing the quality and volume of productivity. For the program to be successful, it is important to have a large range of genetic variability, as well as favorable production variables, such as high seed production, a high ratio of female flowers to male flowers, reduced size, resilience to pests and diseases, standardization, early maturation, resilience to drought, and improvement of the chemical and physical characteristics of the oil. Genetic modifications have enormous potential, as they promote and enable the introduction of desirable genes from species, genera, families, and genes that are of non-plant origin (Marques 2008).

Population growth and climate factors are always unpredictable, and this is why international agricultural systems are shaken. To provide food to meet the needs of the global population, which is continually growing in the face of climate change, it is necessary to apply improvements in genetic improvement techniques, aiming to strengthen the resilience of agricultural crops amidst various climate changes. Advances in genetic improvement provide not only targeted plant properties, but also an increase in overall productivity and greater adaptive capacity, thus reducing the threat of food instability. In the current scenario, it is necessary to provide genetic variability and not compromise it, considering the role of cultivars adjusted to the most recent climate conditions, being able to provide nutritious food in a sustainable manner (Ferreira et al. 2023).

A fundamental solution to the problem of soil degradation was the development of regenerative agriculture, since this problem originates in soil erosion, which makes it unfeasible to support crops, as well as the loss of organic matter and biodiversity, due to the difficulty of propagating species. The problem of erosion was drastically intensified by the misuse of intensive agriculture, which is very aggressive to soil health, since it uses pesticides, synthetic fertilizers, and continuous harvest cycles, without even allowing the soil to recover its nutrients (Blum et al. 2019). All of these facts contributed to the increased speed of deterioration of soils subjected to this system, which consequently reduced the productivity of crops and worsened the scenario of global food insecurity (Palacios-López 2025).

A characteristic of the regenerative agriculture model is its focus on restoring or improving soil health, using techniques that generate biodiversity and capture carbon in its cycle in the soils of plantations, enabling the restoration of plantations and the agricultural production system. To achieve this rehabilitation, several strategies are used, such as crop rotation, which allows the restoration of nutrients present in the planting region, as well as cover crops (Khangura et al. 2023). The regenerative method, if applied correctly and appropriately, has the power to improve production yields, increase the resilience of plantations with regard to climate variations, and ensure the production of healthier foods (Vega 2024).

Analyzing the aspects of agriculture around the world, the science of agroecology emerges to revolutionize the sustainability of damaged planting areas, using conscious and ecologically correct means (Gliessman 2021), and this science is an important and vital point to restore the integrity of the environment, previously taken by inadequate means. There are three possible classifications for the production system based on agroecology: integral, sustainable, and endogenous, where all share the characteristic of active human participation within the process (Souza 2024). The adoption of a model that understands and respects the fact that natural resources are finite ends up guaranteeing the quality of life for both future and contemporary generations, presenting itself as an alternative for sustainability and social equality, since ecological agriculture has the potential for transformations (Guzmán 2013, Gliessman 2015, Titonell 2019).

It is important to mention, within the agroecological aspects, some properties such as crop diversification, which consists of changing what is produced on a land at a certain period of time, promoting improvement in biodiversity and mitigating the appearance of pests, in addition to soil management, which uses crop rotations, addition of organic matter to the soil, and methods for water retention, all focused on eliminating soil erosion. It is also important to mention the conservation and economy of water resources, seeking to solve problems of access to water and safeguarding the availability and good use of this resource (Titonell 2019). Another important point of these agroecological aspects is the planting of trees, shrubs, and other crops among the plantations, with the purpose of providing an ecosystem that strengthens and self-regulates, optimizing production and the environment (Gliessman 2000, Guzmán 2013). The last aspect is described as the reduction of external inputs, consisting of preferably using organic inputs that reduce the need to use chemical products such as pesticides or unnatural fertilizers that only contribute to the weakening of the soil and its nutrients (Souza 2021, 2022, 2023, 2024).

The climate is a factor that greatly influences the quality of agriculture, since it alters access to water, the availability of nutrients in the soil, the absence or not of pests, the incidence of light, and all of these factors can affect the efficiency of food production (Badawy et al. 2022). Seeking to solve these imbroglios and variables, biotechnology and genomics have progressed enormously, opening up new possibilities, such as genetic improvement. Genetic engineering, marker-assisted selection, and gene editing using the CRISPR/Cas9 system are technologies that enable the use of specific properties more efficiently and with a high degree of precision (Ayanoğlu et al. 2020). Technological advances in this area have led to significant advances in food production, in addition to solving problems caused by climate change (Silva 2024), introducing foods with transgenic technology, for example. In recent years, the beginning of processes that use this crop have been intensely changing agriculture, mainly through the use of biotechnologies to improve food, causing some concern among some scholars who fear consequences for the health and physical integrity of people who consume such foods (Zhang and Zhu 2024).

One of the resources that biotechnology can use is precision gene editing, which is directly related to zinc finger nuclease technologies, also known as (ZFNs), transcription activator-like effector nucleases (TALENs), as well as clustered regularly interspaced short palindromic repeats (CRISPR), which, if they are type 9 proteins, are called (CRISPR-Cas9) (Zhang 2023). The concept of precision gene editing is something that exists in nature, with mutations occurring spontaneously over the years due to the most diverse factors imaginable, where mutations that adapt best to the environment end up prevailing in nature.

With the development of this technique, new possibilities have been unlocked, which could revolutionize the global food issue, since gene manipulation, especially CRISPR-Cas9, has provided an improvement in the quality of vegetables and fruits. Thanks to the ingenuity and simplicity of the CRISPR-Cas9 process, it is possible to alter RNA sequences, enabling the transfer of genes to new locations, simplifying the process, since there is no need to manufacture proteins, which also makes this method more economically accessible when compared to the others mentioned (Zhang 2023).

The CRISPR-Cas9 technique has a wide range of uses and possibilities, enabling its application in both cells and species, and can activate, eliminate, inhibit, or cause specific mutations in a chosen gene. With these alternatives, the genes modified in a given fruit crop can improve its protection against diseases (Zhang 2023), its resistance to climate and its variations, and its strengthening against pests, where combined, these properties undeniably improve the productivity and overall performance of the crop. Therefore, both gene editing and biotechnology work together to develop food security for the world's population, in addition to safeguarding the nutritional quality of the fruits that are produced.

13.4.3 Functional Fruits and Global Food Security

Throughout human history, diseases have been a hindrance and have caused irreparable losses. Given this reality, the study of functional substances can be an efficient response to this problem, since they can help prevent diseases. These functional foods have the ability to provide essential substances for the functioning of the human body, such as nutrients or moderation of LDL cholesterol, for example, thus minimizing the chances of a person having diseases such as cancer, as well as allowing control of blood pressure and glucose (Haser 2020).

Functional foods only have this ability because they have certain substances in their composition, such as polyphenols, carotenoids, fibers, prebiotics, sulfur compounds, as well as fatty acids. These substances can be found in whole grains, vegetable oils, legumes, vegetables, and especially in fruits (Vidal et al. 2012). When it comes to vegetables and fruits, an important fruit to mention is kiwi, since it is a functional fruit, as it has a large amount of flavonoids in its composition, which gives it antioxidant, anti-inflammatory, and antihypertensive potential, acting as a fundamental regulator of human health (Hettihewa et al. 2018). Another fruit worth mentioning is the strawberry, due to its large presence of micronutrients such as magnesium, manganese, iron, and phosphorus, for example. Strawberries also have the ability to act in the regulation of glucose present in the blood, in the protection against obesity, as well as in the maintenance of the anti-inflammatory system (Giampieri et al. 2015).

Talking about hunger implies understanding that it manifests itself as a social phenomenon and not just as a physiological response of the organism to eat, which violates human rights and profoundly affects survival, due to malnutrition (Ribeiro Junior 2021). In this sense, the importance of food security has been discussed for decades, especially with the outbreak of the First World War, when the need to protect the food supply for the combatant and civilian population became imminent for the governments involved, since this scenario led to the call-up of men who worked in agricultural production to fight, generating a reduction and increase in prices, in addition to the naval blockades that were implemented due to the war scenario that interfered with the delivery of food imports (Hungria 2024). Therefore, combating hunger and promoting food security is a humanitarian challenge that requires global efforts.

Food and nutritional security has been affected by climate change with regard to the production and nutritional potential of food, leading to an increase in prices and an adjustment in consumption by the population, making social inequalities even more evident (Wenczenovicz et al. 2024). In the meantime, it is understood that mitigation actions are essential for these environmental problems that have caused droughts, floods, heat waves, the spread of pests and agricultural diseases, as well as the impact on water availability, reducing agricultural productivity and consequently reducing the food supply. It is also worth noting that the poor distribution of income has also widened the economic gap for this access, especially for the poorest/vulnerable, due to increasingly expensive food, leaving great challenges in this regard, since the food supply has become more unstable and the global population has grown. Thus, adopting sustainable management, prevention, and production practices can guarantee the nutritional value of food and the maintenance of life (Wenczenovicz et al. 2024).

Food waste around the world is one of the main challenges facing global agriculture. In other words, reducing this loss generates serious social and economic impacts, especially in the fruit and vegetable sector. It is important to highlight that this situation worsens the issue of food insecurity, which affects approximately 2.33 billion people worldwide, according to the 2023 UN Report (See Food Brasil 2023), causing impairment in the development and functioning of the body due to deficiencies in nutrients such as iron (Fe), zinc (Zn), iodine (I), and vitamin A (Duarte 2021, Vergutz 2016). In this sense, efficient and sustainable biofortification techniques have been developed to increase the bioavailable levels of vitamins and functional nutrients in agricultural products (Silva 2024). Furthermore, there has been genetic improvement (genetic biofortification) in increasing the levels of carotenoids, iron, and zinc (White and Broadley 2009), which promotes the nutritional quality of staple foods in different parts of plants such as roots, leaves, and grains, in addition to reducing the costs of post-harvest supplementation, especially in developing countries, where there is the potential to provide continuous benefits (Graham et al. 2007).

This genetic improvement has contributed to the fight against so-called "hidden hunger," that is, it has reduced the lack of vitamins and minerals, not to mention that it has increased resistance to droughts, pests, and even the issue of food production in poor soils, especially in those regions most affected by climate change and with scarce resources, such as some African countries such as Chad and Somalia, and Southeast Asian countries such as Syria.

Around the world, there is a consensus on the concern about the problem of hunger, since food and public health should be basic and indisputable rights for any human being. In Brazil, for example, there is the Unified Health System, also known as SUS, which is responsible for managing and ensuring the well-being of the population with regard to public health. According to the law, this system must promote the implementation of food and nutrition programs, aiming to improve the immunological and nutritional quality and well-being of its citizens. Worldwide, the function of politics is to provide favorable food conditions; otherwise, the health of the population cannot be preserved (Bortolini 2025). To ensure this purpose, a viable solution is the use of functional foods, specifically functional fruits, since they provide nutrients and, in general, substances that the human body needs to survive and maintain its organism, thus providing food safety and quality.

13.5 Future Perspectives and Challenges

13.5.1 The Impact of Nutritional Personalization on the Consumption of Functional Fruits

Personalization of nutrition, currently referred to as precision nutrition, seeks to adapt an individual's eating plan, taking into account their specific goals, needs, and expectations. To date, definitions of personalized nutrition have focused primarily on genetic variations and/or the gut microbiome, often adding a limited set of additional information, such as anthropometric measurements or food intake (Linseisen et al. 2025). Precision nutrition has advanced in this regard by including more complete phenotype data and incorporating discoveries from omics technologies, such as metabolomics, epigenetics, and proteomics.

The growing appreciation for health and well-being has driven interest in personalized nutrition, an approach that adapts dietary recommendations to each person's individual characteristics, considering genetic, metabolic, and behavioral factors. In this context, functional fruits, rich in bioactive compounds such as polyphenols and carotenoids, have gained prominence due to their potential health benefits, including antioxidant, anti-inflammatory, and cardioprotective properties (Liu 2013). The integration of biomedical, psychobehavioral, and environmental data is essential for the development of effective personalized nutrition strategies (Linseisen et al. 2025). In this context, there is successive evidence that a healthy eating strategy with a greater number of plant-based foods plays an important role in the prevention of chronic diseases such as heart disease, stroke, cancer, diabetes, and several other diseases that are caused by age-related functional decline (Liu 2013).

The use of advanced software has been increasingly employed to incorporate genetic, behavioral, and clinical information, with the aim of developing individualized eating plans, thus achieving greater adherence and effectiveness in the diets developed. Food4Me, a European project, showed that actions based on personalized nutrition with the support of digital platforms became more effective in encouraging changes in eating habits and improving health indicators, highlighting the importance of large-scale integrated data analysis (Celis-Morales et al. 2017). These advances highlight the great potential of integrating artificial intelligence in providing more precise and individualized diets.

In the context of functional foods, fruits in particular are natural sources of biologically active compounds such as polyphenols and anthocyanins, which play a fundamental role in modulating physiological processes, oxidative stress, inflammation, and fat cell metabolism. Peels and seeds, which are usually discarded parts of fruits, are sources from which these compounds can be extracted, strengthening their sustainable potential and their purpose in the development of functional foods (Kanait et al. 2022).

Some fruits such as *Annona crassiflora*, known as araticum, *Vaccinium myrtillus*, known as blueberry, and *Punica granatum*, popularly known as pomegranate, have gained great popularity in the international literature due to their specific benefits. Araticum, a typical fruit of the Brazilian cerrado, has been examined for its comprehensive composition of phenolic groups and anti-inflammatory and antioxidant activities, demonstrating its potential for applications not only in food but also in pharmacology (Arruda and Pastore 2019). Blueberry, widely studied in Western communities, has shown improvements in vascular function in individuals at cardiological risk, with its effects mediated by specific flavonoids (Rodriguez-Mateos et al. 2013). The constant ingestion of pomegranate juice has been linked to a decrease in the thickness of the walls of the carotid arteries, blood pressure, and oxidation of low-density lipoprotein, which is a lipoprotein that transports cholesterol and triglycerides in the blood, in patients with carotid stenosis, highlighting its role in the development of cardiovascular health (Aviram et al. 2004).

13.5.2 Barriers to the Adoption of Functional Foods in Everyday Life

Despite the growing recognition of the health benefits of functional foods, their widespread incorporation into the population's daily diet still faces several obstacles. These limitations can be classified into five broad categories: regulatory, economic, sensory, informational, and educational barriers, which interact with each other and compromise the effective adoption of these foods in everyday life.

One of the main limitations highlighted in the literature refers to the rigidity and complexity of the regulatory frameworks that control health claims on functional food labels. As emphasized by Domínguez et al. (2020), the European regulatory system is highly rigorous, requiring robust scientific evidence for any functional claim. This represents an obstacle for companies, especially small and medium-sized ones, which face high costs to conduct the required clinical studies. This situation discourages the development of new functional products and limits consumer access to innovative and potentially beneficial options.

From the consumer's perspective, sensory factors and food preferences also stand out as important barriers. Studies indicate that many consumers are not willing to compromise on the taste, texture, and appearance of foods in exchange for functional benefits (Mogendi et al. 2016). Thus, even when functional foods are available, their acceptance still depends on a balance between functionality and sensory pleasure.

Furthermore, the lack of knowledge and understanding about the benefits of functional foods represents a significant obstacle. Although compounds such as polyphenols, flavonoids, and carotenoids are widely present in fruits and vegetables, many consumers are unaware of their physiological effects. This is related to the lack of effective food education campaigns that address the functionality of foods in an accessible and practical way (Enríquez et al. 2023).

Another relevant barrier involves the way in which health claims are communicated on labels. Even when a functional product contains beneficial compounds, technical or ambiguous language used in its labeling can make it difficult for the general public to understand. Domínguez et al. (2020) suggest that clearer, standardized claims based on nutritional education would help consumers make informed and conscious decisions.

Finally, socioeconomic aspects also influence the consumption of functional foods. The generally higher cost of these products makes them less accessible to a significant portion of the population. To overcome this challenge, authors such as Santos et al. (2022) advocate the valorization of local ingredients and agro-industrial by-products as a strategy to produce more accessible and sustainable functional foods.

Together, these factors show that the full adoption of functional foods in everyday life depends on integrated actions involving more accessible regulation, investment in food education, technological innovations that improve the sensory experience of products, and public policies that guarantee equitable access to them.

13.5.3 The Role of Nutritional Education in Promoting the Consumption of Fruits Rich in Bioactive Compounds

In this scenario, nutritional education plays a crucial role in promoting the consumption of fruits rich in bioactive compounds. Educational programs that provide personalized guidance, taking into account the socioeconomic and cultural context of individuals, have proven effective in improving diet quality (Briazu et al. 2024). In addition, initiatives that aim to increase familiarity with and acceptance of plant-based foods can facilitate the transition to healthier and more sustainable diets (Abe-Inge et al. 2024).

According to Enríquez et al. (2023), one of the biggest obstacles to the regular consumption of functional fruits and vegetables is consumers' lack of knowledge about the bioactive compounds present in these foods and their mechanisms of action in the body. Even when these foods are accessible, their adoption in daily life depends on the appreciation of their effects in disease prevention, which is only possible through continuous educational strategies integrated into the daily lives of communities.

In this context, nutritional education must go beyond the transmission of basic information about nutrients. It needs to address the physiological role of bioactive compounds, the importance of regular and varied fruit consumption, and the full use of foods, including peels, seeds, and by-products, which often contain high levels of these compounds. Educational initiatives should also work on the cultural redefinition of food and encourage communities to take a leading role in choosing healthy habits (Enríquez et al. 2023).

Santos et al. (2022) reinforce this understanding when discussing the use of byproducts from fruit processing—such as peels, seeds, and residual pulp—in functional food formulations, such as flours, snacks, and beverages. However, the acceptance of these products depends on the degree of consumer knowledge about their benefits. Therefore, food education campaigns are essential to promote the understanding and acceptance of these functional products, contributing not only to public health but also to environmental sustainability by reducing waste.

In addition, the inclusion of content on bioactive compounds in school curricula, community programs, and social media can generate lasting impacts, especially among young people and children. School environments, for example, are conducive to the formation of healthy habits and regular contact with functional fruits in a playful and contextualized way.

In short, food education is a strategic pillar to promote the conscious and daily consumption of functional fruits. It acts as a mediator between scientific knowledge and dietary practices, being fundamental for promoting health, strengthening food sovereignty, and building more sustainable and resilient food systems.

13.5.4 Future Prospects

Several scientific evidences demonstrate that regular fruit consumption is associated with a reduced risk of cardiovascular disease, cancer, type 2 diabetes, and other inflammatory conditions (Liu 2013). This functional activity is attributed to the synergistic action between phytochemicals and nutrients present in whole foods. Fruits such as pitaya stand out for their antioxidant, anti-inflammatory, hepatoprotective, healing, and antimicrobial properties, as shown by Chatterjee et al. (2024), thanks to compounds such as betalains, flavonoids, and terpenoids.

The growing trend towards food personalization has driven the use of emerging technologies to create customized functional foods. The study by Briazu et al. (2024) demonstrated that personalized dietary advice based on sociodemographic, cognitive, and sensory variables is more effective in promoting fruit and vegetable consumption. In addition, technologies such as 3D and 4D printing are being applied to create fruit-based products with shapes, textures, and nutrient release adapted to different consumer profiles (Chauhan et al. 2025, Tomasevic et al. 2021). For Liu et al. (2017), 3D printing offers advantages such as shape customization, simplification of the production chain, and potential for integrating digital and sensory nutrition, representing a concrete trend for the future of functional foods. These "smart" foods respond to environmental stimuli such as temperature and pH, optimizing their functionality.

Sustainability is also becoming an essential criterion in the formulation of functional foods. The use of by-products from fruit processing, such as functional flours rich in fiber and phenolic compounds, represents a circular economy strategy with a positive impact on health and the environment (Santos et al. 2022, Romero-Martínez et al. 2025).

The fermentation process has gained prominence as an alternative to increase the nutritional, sensory, and functional value of fruits. According to Yuan et al. (2024), fermentation with lactic acid bacteria and yeast improves stability, extends shelf life, and intensifies the content of polyphenols, organic acids, and bioactive peptides, contributing to antioxidant, anti-inflammatory, and intestinal microbiota modulation properties. Another important trend is fruit-based probiotic functional beverages. Manoj et al. (2023) highlight that these beverages are effective vehicles for delivering beneficial microorganisms to the body, especially for vegan or lactose-intolerant consumers. Furthermore, they combine attractive sensory properties with positive physiological effects, such as immunomodulation and microbiota balance.

Species such as *Vasconcellea stipulata*, *V. pubescens* and *V. pentagona*, studied by Aguirre-Rodríguez et al. (2024), demonstrate remarkable functional potential, with compounds such as esters, tocopherols, and isothiocyanates, which have antioxidant, antiparasitic, and anticancer activities. These Andean fruits are promising candidates for the development of new functional foods, especially in Latin America.

Fruits play a central role in the evolution of functional foods. Their future as a health promotion tool depends on the integration of technological advances, sustainability, nutritional personalization, and effective regulation. Investments in research and food education will be crucial to consolidate fruits as protagonists of functional, healthy, and inclusive food systems.

13.6 Conclusion

The need for healthy eating has led to the search for better nutritional profiles, among which functional foods stand out, which, in addition to nutrients, contain bioactive compounds that are valuable to human health. Fruits are foods that contain these beneficial micronutrients to the population, with their antioxidant, anti-inflammatory, antibacterial properties, among others, but they need to be consumed regularly to promote well-being and health.

The bioactives present in them combine the most diverse activities. Polyphenols, carotenoids, prebiotics, dietary fibers, and sulfur compounds have numerous uses, ranging from their role in defense systems, through their importance in photosynthesis, to providing food for the intestinal microbiota. The role of these bioactives in promoting long-term health, with their antioxidant role and suppression of inflammatory responses, leads to their association with a longer and increasingly healthier life.

Studies have focused on advancing the most modern methods of administration and use of these compounds, as can be seen in nanotechnology. The growth of the functional market and the demands for increasingly present technologies are real. The growing financial estimate in the coming years depends on maximum possible use, with rigorous and efficient monitoring. Therefore, the environmental impact and control of this market must permeate everything from the understanding of the food system and its waste to sustainable food. Everything converges with the guidance of the value chains that guide this system and therefore directly influence the world's food supply.

The personalization of food drives the evolution of technologies, sustainability, and market control, making them a global trend. Fruits play an important role in the evolution of functional foods, being fundamental in promoting health and in the long-term repercussions on the economy and health of the world's population.

References

Anyiam, P. N., Phongthai, S., Grossmann, L., Jung, Y. H., Sai-Ut, S., Onsaard, E. et al. (2025) Potential plant proteins for functional food ingredients: Composition, utilization and its challenges. NFS Journal 38: 100216.

Abe-Inge, V., Aidoo, R., Moncada De La Fuente, M. and Kwofie, E. M. (2024). Plant-based dietary shift: current trends, barriers, and carriers. Trends in Food Science & Technology 143: 104292.

Aguirre-Rodríguez, A., Duarte-Casar, R., Rojas-Le-Fort, M. and Romero-Benavides, J. C. (2024). Food uses, functional activities, and bioactive compounds of three Ecuadorian Vasconcellea fruits: bibliometric analysis and review. Journal of Agriculture and Food Research 17: 101244.

Al-Babili, S. and Bouwmeester, H. J. (2015). Strigolactones, a novel carotenoid-derived plant hormone. Annual Review of Plant Biology 66: 161–186.

Anyiam, P. N., Phongthai, S., Grossmann, L., Jung, Y. H., Sai-Ut, S., Onsaard, E. et al. (2025). Potential plant proteins for functional food ingredients: Composition, utilization and its challenges. NFS Journal 38: 100216.

Arruda, H. S. and Pastore, G. M. (2019). Araticum (*Annona crassiflora* Mart.) as a source of nutrients and bioactive compounds for food and non-food purposes: a comprehensive review. Food Research International (Ottawa, Ont.) 123: 450–480.

Aviram, M., Rosenblat, M., Gaitini, D., Nitecki, S., Hoffman, A., Dornfeld, L. et al. (2004). Pomegranate juice consumption for 3 years by patients with carotid artery stenosis reduces common carotid intima-media thickness, blood pressure and LDL oxidation. Clinical Nutrition (Edinburgh, Scotland) 23(3): 423–433.

Ayanoğlu, F. B., Elçin, A. E. and Elçin, Y. M. (2020, fevereiro). Functional foods: their role in disease prevention and health promotion. Institute of Food Technologists. Expert Panel on Food Safety and Nutrition.

Babalola, O. O., Akinnusi, E., Ottu, P. O., Bridget, K., Oyubu, G., Ajiboye, S. A. et al. (2025). The impact of ultra-processed foods on cardiovascular diseases and cancer: Epidemiological and mechanistic insights. Aspects of Molecular Medicine 5: 100072.

Badawy, A., Elmahdi, A., El-Hafez, S. A. A. and Kineber, A. F. (2022, janeiro 30). Water profitability analysis to improve food security and climate resilience: a case study in the Egyptian Nile Delta. Multidisciplinary Digital Publishing Institute 10(2): 17.

Bhatt, S. and Gupta, M. (2024). Explicating the effect of extraction methods on the techno-functional, structural, and *in vitro* prebiotic potential of soluble dietary fibers from mango and pomegranate peel. Sustainable Food Technology 2: 1506–1516.

Blum, W. E. H., Brevik, E. C. and Zehetner, F. (2019). The importance of soil biodiversity in agricultural ecosystems. Soil Systems 3(1): 11–25.

Bond, A., Mccay, K. and Lal, S. (2023). Artificial intelligence & clinical nutrition: What the future might have in store. Clinical Nutrition ESPEN 57: 542–549.

Bortolini, G. A., Basso, C. and Jaime, P. C. (2025). Recomendações para o fortalecimento da implementação da Política Nacional de Alimentação e Nutrição. Universidade de São Paulo, São Paulo, SP, Brasil.

Briazu, R. A., Bell, L., Dodd, G. F., Blackburn, S., Massri, C., Chang, B. et al. (2024). The effectiveness of personalised food choice advice tailored to an individual's socio-demographic, cognitive characteristics, and sensory preferences. Appetite 201: 107600.

Cannon, R. J. and Ho, C. (2018). Volatile sulfur compounds in tropical fruits. Journal of Food and Drug Analysis 26: 445–468.

Câmara, J. S., Albuquerque, B. R., Aguiar, J., Corrêa, R. C. G., Gonçalves, J. L., Granato, D. et al. (2020). Food bioactive compounds and emerging techniques for their extraction: polyphenols as a case study. Foods 10: 10.3390.

Cao, G., Sofic, E. and Prior, R. L. (1997). Antioxidant and prooxidant behavior of flavonoids: structure-activity relationships. Free Radical Biology and Medicine 22: 749–760.

Carradori, S., Guglielmi, P., Luisi, G. and Secci, D. (2022). Nitrogen-and sulfur-containing Heterocycles as dual anti-oxidant and anti-cancer agents. Handbook of Oxidative Stress in Cancer: Mechanistic Aspects 1: 1–18.

Celis-Morales, C., Livingstone, K. M., Marsaux, C. F., Macready, A. L., Fallaize, R., O'Donovan, C. B. et al. (2017). Effect of personalized nutrition on health-related behaviour change: evidence from the Food4Me European randomized controlled trial. International Journal of Epidemiology 46(2): 578–588.

Chatterjee, D., Behera, D. and Singh, N. (2024). Therapeutic potential of various functional components present within dragon fruit. Hybrid Advances 6: 100185.

Chauhan, P., Kaushal, M., Vaidya, D., Gupta, A., Kashyap, S. and Suthar, P. (2024). Personalized nutrition to intelligent foods with 3D and 4D printing. Trends in Food Science & Technology 120: 1–14.

Chauhan, P., Kaushal, M., Vaidya, D., Gupta, A., Kashyap, S. and Suthar, P. (2025). Personalized nutrition to intelligent foods with 3D and 4D printing. Applied Food Research 5: 100863.

Che, J., Liang, B., Zhang, Y., Wang, Y., Tang, J. and Shi, G. (2017). Kaempferol alleviates ox-LDL-induced apoptosis by up-regulation of autophagy via inhibiting PI3K/Akt/mTOR pathway in human endothelial cells. Cardiovascular Pathology 31: 57–62.

Chen, J., Song, Y. and Zhang, L. (2013). Effect of lycopene supplementation on oxidative stress: an exploratory systematic review and meta-analysis of randomized controlled trials. Journal of Medicinal Food 16: 361–374.

Contreras, B. V. P., Vermelho, A. B., Casanova, L., Lage, C. A. S., Vilela, C. L. S., Cardoso, V. S. et al. (2025). Enhanced UV-B photoprotection activity of carotenoids from the novel *Arthrobacter* sp. strain LAPM80 isolated from King George Island, Antarctica. Heliyon 11: e41400.

Cortez, A. T. (2011). Embalagens: o que fazer com elas? Revista Geográfica de América Central, Número Especial: EGAL. Costa Rica.

da Silva, C. M., de Azevedo, G. A., Frade, L. F. S., Barreto, A. V. R., Soares, J. V., da Silva, M. P. et al. (2024). Genetic improvement of plants for adaptability to climate change. Revista de Gestão e Secretariado, São José dos Pinhais, Paraná.

de Oliveira, A. C., Cruz, J. M. A., Mar, J. M., Corrêa, R. F., Rodrigues, J. V. S., Mendes, T. A. O. et al. (2025). Abiu-do-Mato (Pouteria caimito, Sapotaceae): A promising Amazonian fruit with rich chemical composition, antioxidant potential, and cytotoxic safety. Food Research International 205: 115984.

de Oliveira, J. N., de Albuquerque, T. M. R., Lima, M. S., de Andrade, E. R. S., Monteiro, M., do Nascimento, Y. M. et al. (2024). Evaluating the potential prebiotic effects of umbu-cajá (Spondias spp.) fruit processing by-product flour on the human intestinal microbiota. LWT 210: 116764.

Detopoulou, P., Voulgaridou, G., Moschos, P., Levidi, D., Anastasiou, T., Dedes, V. et al. (2023). Artificial intelligence, nutrition, and ethical issues: A mini-review. Clinical Nutrition Open Science 50: 46–56.

Dhingra, D., Michael, M., Rajput, H. and Patil, R. T. (2012). Dietary fibre in foods: a review. Journal of Food Science and Technology 49: 255–266.

Domínguez Díaz, L., Fernández-Ruiz, V. and Cámara, M. (2020). An international regulatory review of food health-related claims in functional food products labeling. Journal of Functional Foods 68: 103896.

Dordevic, D., Capikova, J., Dordevic, S., Tremlová, B., Gajdács, M. and Kushkevych, I. (2023). Sulfur content in foods and beverages and its role in human and animal metabolism: A scoping review of recent studies. Heliyon 9: e15452.

Embling, R., Neilson, L., Mellor, C., Durodola, M., Rouse, N., Haselgrove, A. et al. (2024). Exploring consumer beliefs about novel fortified foods: a focus group study with UK-based older and younger adult consumers. Appetite 193: 107139.

Emon, D. Das, Islam, M. S., Mazumder, M. A. R., Aziz, M. G. and Rahman, M. S. (2025). Recent applications of microencapsulation techniques for delivery of functional ingredient in food products: A comprehensive review. Food Chemistry Advances 6: 100923.

Enríquez, M., Villafuerte, F. and Figueroa, A. (2023). Efectos de los componentes bioactivos de frutas, vegetales, lácteos y plantas medicinales en la nutrición humana. Revista de Ciencias Agropecuarias ALLPA 6(11): 2–24.

Eskinazi, R. H., Aléssio, R. L. dos S. and Santos M. F. de S. (2024). Memória e ancoragem na construção de representações sociais sobre edição genética de embriões humanos. Psicologia & Sociedade 36: e278088.

Essa, M. M., Bishir, M., Bhat, A., Chidambaram, S. B., Al-Balushi, B., Hamdan, H. et al. (2023). Functional foods and their impact on health. Journal of Food Science and Technology 60(3): 820–834.

FAO, IFAD, UNICEF, WFP, & WHO. (2020). The state of food security and nutrition in the world 2020. Food and Agriculture Organization of the United Nations.

Farias, E. R. (2024). Agricultura familiar: produção sustentável na chácara Recanto das Folhas, em Gameleira de Goiás (Monografia de Graduação).

Ferreira, D. S., Figueiredo, A. C., Ferreira, J. M. S., Cabral, S. de O., Gonçalves, D. da C., Dian, V. S. et al. (2023). Melhoramento genético no contexto da produção agroecológica. Epitaya E-Books 1(55): 447–462. https://doi.org/10.47879/ed.ep.2023946p447.

Farhadi, F., Iranshahi, M., Taghizadeh, S. F. and Asili, J. (2020). Volatile sulfur compounds: The possible metabolite pattern to identify the sources and types of asafoetida by headspace GC/MS analysis. Industrial Crops and Products 155: 112827.

Florêncio, R. R. and Santos, A. E. O. dos. (2024). Qualidade pós-colheita de hortaliças biofortificadas: revisão bibliográfica. Revista Ouricuri 14(1): 03–20. jan./jul.

Frumuzachi, O., Flanagan, A., Rohn, S. and Mocan, A. (2025). The dichotomy between functional and functionalized foods – A critical characterization of concepts. Food Research International 208: 116173.

Giampieri, F., Forbes-Hernandez, T. F., Gasparrini, M., Alvarez-Suarez, J. M., Afrin, S. et al. (2015). Strawberry as a health promoter: an evidence-based review. Food & Function 6(5): 1386–1398.

Gibson, G. R., Scott, K. P., Rastall, R. A., Tuohy, K. M., Hotchkiss, A., Dubert-Ferrandon, A. et al. (2010). Dietary prebiotics: current status and new definition Food Science & Technology Bulletin: Functional Foods 7: 1–19.

Gibson, G. R., Hutkins, H., Sanders, M. E., Prescott, S. L., Reimer, R. A., Salminen, S. J. et al. (2017). Expert consensus document: the International Scientific Association for Probiotics and Prebiotics (ISAPP) consensus statement on the definition and scope of prebiotics. Nature Reviews Gastroenterology & Hepatology 14: 491–502.

Giuntini, E. B., Sardá, F. A. H. and de Menezes, E. W. (2022). The effects of soluble dietary fibers on glycemic response: an overview and futures perspectives. Foods 11: 3934.

Gliessman, S. R. (2015). Agroecology: The Ecology of Sustainable Food Systems. 3. ed. Boca Raton: CRC Press.

Gliessman, S. R. (2000). Agroecologia: processos ecológicos em agriculturas sustentáveis. Porto Alegre: Editora da UFRGS.

Gliessman, S. R. (2021). Agroecology: The Ecology of Sustainable Food Systems. 4. ed. Boca Raton: CRC Press.

Guzmán, G. I. (2013). Transición agroecológica: donde confluyen la investigación y la acción. Laboratorio de Historia de los Agroecossistemas 1–12.

Hamad, H. T. (2025). The anti-cancer effectiveness of some heterocyclic compounds containing sulfur atom. Results in Chemistry 15: 102182.

Hashimoto, H., Uragami, C. and Cogdell, R. J. (2016). Carotenoids and photosynthesis. Carotenoids in Nature 79: 111–139.

Helgegren, H., Winkvist, A., Lepp, M. and Larsson, C. (2025). 'There are healthy things inside': Children's thoughts about, experiences with and approaches to vegetable and fruit consumption. Appetite 206: 107819.

Hettihewa, S. K., Hemar, Y. and Rupasinghe, H. A.-O. (2018). Flavonoid-rich extract of Actinidia macrosperma (a wild kiwifruit) inhibits angiotensin-converting enzyme *in vitro*. Foods 7(9): 146, 1–8.

Hou, X., Rivers, J., León, P., McQuinn, R. P. and Pogson, B. J. (2016). Synthesis and function of apocarotenoid signals in plants. Trends in Plant Science 21: 792–803.

Hungria, M. (2024). Segurança alimentar e nutricional: o papel da ciência brasileira no combate à fome. Academia Brasileira de Ciências, Rio de Janeiro.

Islam, F., Muni, M., Mitra, S., Emran, T. B., Chandran, D., Das, R. et al. (2022). Recent advances in respiratory diseases: Dietary carotenoids as choice of therapeutics. Biomedicine & Pharmacotherapy 155: 113786.

Jabbour, Al Maalouf, N. Elia, J., Boutros, F., Balouza, M. and Sawaya, C. (2024). Sustainable practices and their impact on the operations of fresh fruits and vegetables market in the cold chain. Cleaner Logistics and Supply Chain 13: 100196.

Jikah, A. N. and Edo, G. I. (2023). Mechanisms of action by sulphur compounds in *Allium sativum*. A review. Pharmacological Research - Modern Chinese Medicine 9: 100323.

Jurado-Fasoli, L., Mesas-Fernández, A. and Rodríguez-García, C. (2023). Lycopene supplementation for patients under cancer therapy: a systematic review and meta-analysis of randomized controlled trials. Journal of Herbal Medicine 41: 100725.

Kainat, S., Arshad, M. S., Khalid, W., Zubair Khalid, M., Koraqi, H., Afzal, M. F. et al. (2022). Sustainable novel extraction of bioactive compounds from fruits and vegetables waste for functional foods: a review. International Journal of Food Properties 25(1): 2457–2476.

Kazazian, T. (2005). A idade das coisas leves: design e desenvolvimento sustentável. São Paulo: Ed. SENAC.

Kim, H., Banerjee, N., Barnes, R. C., Pfent, C. M., Talcott, S. T., Dashwood, R. H. et al. (2017). Mango polyphenolics reduce inflammation in intestinal colitis-involvement of the miR-126/PI3K/AKT/mTOR axis *in vitro* and *in vivo*. Molecular Carcinogenesis 56: 197–207.

Khangura, R., Ferris, D., Wagg, C. and Bowyer, J. (2023). Regenerative agriculture: a review on practices to improve soil health. Sustainability 15(3): 2338.

Klojdová, I., Milota, T., Smetanová, J. and Stathopoulos, C. (2023). Encapsulation: A strategy to deliver therapeutics and bioactive compounds? Pharmaceuticals 16: 362.

Kopec, R. E. and Failla, M. L. (2018). Recent advances in the bioaccessibility and bioavailability of carotenoids and effects of other dietary lipophiles. Journal of Food Composition and Analysis 68: 16–30.

Köpsel, M., Kostka, T., Niesen, S., Winterhalter, P. and Esatbeyoglu, T. (2025). Influence of fractionation of polyphenols by membrane chromatography on antioxidant, antimicrobial and proliferation-inhibiting effects of red fruit juices. Food Chemistry 463: 141216.

Lie-Piang, A., Möller, A. C., Kollmann, N., Garre, A., Boom, R. and van der Padt, A. (2022). Functionality-driven food product formulation – an illustration on selecting sustainable ingredients that build viscosity. Food Research International 152: 110889.

Limketkai, B. N., Mauldin, K., Manitius, N., Jalilian, L. and Salonen, B. R. (2021). The age of artificial intelligence: use of digital technology in clinical nutrition. Current Surgery Reports 9: 7.

Linseisen, J., Renner, B., Gedrich, K., Wirsam, J., Holzapfel, C., Lorkowski, S. et al. (2025). Data in personalized nutrition: bridging biomedical, psycho-behavioral, and food environment approaches for population-wide impact. Advances in Nutrition 100377.

Liu, M., Li, T., Deng, R., Zhang, Z., Zhou, Y., Duan, X. et al. (2025). Transcription factor AcAGL9 and AcBEH1 synergistically regulate carotenoid biosynthesis during pineapple fruit ripening. Postharvest Biology and Technology 222: 113385.

Liu, R. H. (2013). Health-promoting components of fruits and vegetables in the diet. Advances in Nutrition 4(4): 384S–392S.

Liu, Z. B., Zhang, M., Bhandari, B. and Wang, Y. C. (2017). 3D printing: printing accuracy and application in the food industry. Trends in Food Science & Technology 69: 83–94.

Lu, K., Yu, T., Cao, X., Xia, H., Wang, S., Sun, G. et al. (2023). Effect of viscous soluble dietary fiber on glucose and lipid metabolism in patients with type 2 diabetes mellitus: a systematic review and meta-analysis on randomized clinical trials. Frontiers in Nutrition 10: 1253312.

Luca, S. V., Macovei, I., Bujor, A., Miron, A., Skalicka-Woźniak, K., Aprotosoaie, A. C. et al. (2020). Bioactivity of dietary polyphenols: The role of metabolites. Critical Reviews in Food Science and Nutrition 60: 626–659.

Manoj, P. M., Mohan, J. R., Khasherao, B. Y., Shams, R. and Dash, K. K. (2023). Fruit based probiotic functional beverages: a review. Journal of Agriculture and Food Research 14: 100729.

Mello, R. L. (2007). Agricultura familiar: sustentabilidade social e ambiental. Disponível em http://agro.unitau.br:8080/dspace/handle/2315/137. Acesso em: 10 abr. 2025.

Menezes, M. E. de, Lima, M. G. de and Melo M. S. de. (2016). Alimentos funcionais: uma revisão. Educação, Ciência e Saúde 3(1): 1–15.

Mohammadi, N., Franchin, M., Pressete, C. G., Antunes, L. M. G. and Granato, D. (2024). Green recovery and application of berry anthocyanins in functional gummies: Stability study, plasma and cellular antioxidant and anti-inflammatory activity. Food Research International 196: 115128.

Moreira, R. N., Marinho, L. F. L. and Barbosa, F. L. S. (2015). O Modelo de Produção Sustentável Upcycling: o caso da empresa TerraCycle. XVII ENGEMA-Encontro Internacional Sobre Gestão Empresarial e Meio Ambiente, Desafios da Sustentabilidade na Economia de Baixo Carbono 17: 1–11.

Nowak, D., Gośliński, M. and Kłębukowska, L. (2022). Antioxidant and antimicrobial properties of selected fruit juices. Plant Foods for Human Nutrition 77: 427–435.

Oduro-Obeng, H., Chaudhry, M. M. A., Zogona, D., Paliwal, J., Fu, B. X. and Beta, T. (2024). Carotenoid cellular uptake and antioxidant activity from differently cooked semolina and whole wheat pasta. Journal of Cereal Science 120: 104037.

O'Riordan, K. J., Collins, M. K., Moloney, G. M., Knox, E. G., Aburto, M. R., Fülling, C. et al. (2022). Short chain fatty acids: microbial metabolites for gut-brain axis signalling. Molecular and Cellular Endocrinology 546: 111572.

Palacios-López, L. A. (2025). La agricultura regenerativa como solución para la degradación del suelo a través de investigaciones recientes. Horizon Nexus Journal 3(1): 18–30.

Pathania, S., Narang, R. K. and Rawal, R. K. (2019). Role of Sulphur-heterocycles in medicinal chemistry: an update. European Journal of Medicinal Chemistry 180: 486–508.

Qiu, Y., Wang, R., Zhang, E., Shang, Y., Feng, G., Wang, W. et al. (2024). Carotenoid biosynthesis profiling unveils the variance of flower coloration in Tagetes erecta and enhances fruit pigmentation in tomato. Plant Science 347: 112207.

Ramos, M. G., Samouda, H. and Bohn, T. (2022). Short chain fatty acid metabolism in relation to gut microbiota and genetic variability. Nutrients 14: 5361.

Rezagholizade-shirvan, A., Soltani, M., Shokri, S., Radfar, R., Arab, M. and Shamloo, E. (2024). Bioactive compound encapsulation: Characteristics, applications in food systems, and implications for human health. Food Chemistry 24.

Ribeiro Júnior, J. R. S. (2021). A fome como processo e a reprodução social capitalista. Boletim Paulista de Geografia 1(105): 15–39. Disponível em: <https://publicacoes.agb.org.br/boletim-paulista/article/view/1992>. Acesso em: 8 abr. 2025.

Rodríguez-Cortina, A. and Hernández-Carrión, M. (2025). Amazonian fruits in Colombia: Exploring bioactive compounds and their promising role in functional food innovation. Journal of Food Composition and Analysis 137: 106878.

Rodriguez-Mateos, A., Rendeiro, C., Bergillos-Meca, T., Tabatabaee, S., George, T. W., Heiss, C. et al. (2013). Intake and time dependence of blueberry flavonoid-induced improvements in vascular function: a randomized, controlled, double-blind, crossover intervention study with mechanistic insights into biological activity. The American Journal of Clinical Nutrition 98(5): 1179–1191.

Romero-Martínez, M., Andrade-Pizarro, R. and De Paula, C. (2025). Functional compounds in tropical fruit processing by-products and intrinsic factors affecting their composition: a review. Current Research in Food Science 10: 101028.

Safraid, G. F., Portes, C. Z., Dantas, R. M. and Batista, Â. G. (2024). Perception of functional food consumption by adults: Is there any difference between generations? Brazilian Journal of Food Technology 27.

Sakulrang, S., Razem, M., Mohammadi, N. and Granato, D. (2025). Sustainable, functional food design: Characterizing and utilizing passion fruit by-product extract in anthocyanin-enriched delivery systems. Future Foods 11: 100599.

Samami, E., Starkweather, A., Lyon, D. E. and Kelly, D. L. (2025). Associations between dietary fiber, the gut microbiota, and health outcomes in breast cancer survivors: A scoping review. Clinical Nutrition Open Science 61: 174–189.

Santos, D., Lopes da Silva, J. A. and Pintado, M. (2022). Fruit and vegetable by-products' flours as ingredients: a review on production process, health benefits and technological functionalities. LWT - Food Science and Technology 154: 112707.

Seafood Brasil. (2023). Insegurança alimentar atinge 2,33 bi de pessoas em 2023. Disponível em: https://www.seafoodbrasil.com.br/inseguranca-alimentar-atinge-233-bi-de-pessoas-em-2023. Acesso em: 10 abr. 2025.

Sereti, F., Alexandri, M., Papapostolou, H., Papadaki, A. and Kopsahelis, N. (2025). Recent progress in carotenoid encapsulation: Effects on storage stability, bioaccessibility and bioavailability for advanced innovative food applications. Food Research International 203: 115861.

Sgroi, F., Sciortino, C., Baviera-Puig, A. and Modica, F. (2024). Analyzing consumer trends in functional foods: A cluster analysis approach. Journal of Agriculture and Food Research 15: 101041.

Silva, T. P. de P., Florêncio, R. R. and Santos, A. E. O. dos. (2024). Qualidade pós-colheita de hortaliças biofortificadas: revisão bibliográfica. Revista Ouricuri 14(1): 03–20. jan./jul.

Sosa-Holwerda, A., Park, O. H., Albracht-Schulte, K., Niraula, S., Thompson, L. and Oldewage-Theron, W. (2024). The role of artificial intelligence in nutrition research: a scoping review. In Nutrients 16: 2066.

Souza, M. N. (2021). Tópicos em recuperação de áreas degradadas (Vol. I). Canoas: Mérida Publishers.

Souza, M. N. (2022). Tópicos em recuperação de áreas degradadas (Vol. IV). Canoas, RS: Mérida Publishers.

Souza, M. N. (2023). Tópicos em recuperação de áreas degradadas (Vol. V). Canoas, RS: Mérida Publishers.

Souza, M. N. (2024). Tópicos em recuperação de áreas degradadas (Vol. I). Canoas: Mérida Publishers.

Sui, J., Shen, W., Zhang, Y., Liang, J., Xia, H. and Sun, G. (2025). The protective role of carotenoids in chronic obstructive pulmonary disease: A systematic review and Meta-analysis of observational studies. Journal of Functional Foods 125: 106685.

Sun, T., Yuan, H., Cao, H., Yazdani, M., Tadmor, Y. and Li, L. (2018). Carotenoid metabolism in plants: the role of plastids. Molecular Plant 11: 58–74.

Surya Ulhas, R., Ravindran, R., Malaviya, A., Priyadarshini, A., Tiwari, B. K. and Rajauria, G. (2023). A review of alternative proteins for vegan diets: Sources, physico-chemical properties, nutritional equivalency, and consumer acceptance. Food Research International 173: 113479.

Teibo, J. O., Ayinde, K. S., Olaoba, O. T., Adelusi, T. I., Teibo, T. K. A., Bamikunle, M. V. et al. (2021). Functional foods' bioactive components and their chemoprevention mechanism in cervical, breast, and liver cancers: A systematic review. Functional Foods in Health and Disease 11: 559–585.

Theodore Armand, T. P., Nfor, K. A., Kim, J. I. and Kim, H. C. (2024). Applications of artificial intelligence, machine learning, and deep learning in nutrition: a systematic review. Nutrients 16: 7.

Titonell, P. (2019). Las transiciones agroecológicas: múltiples escalas, niveles y desafíos. Revista de La Facultad de Ciencias Agrarias UNCuyo 51(1): 231–246.

Tomasevic, I., Putnik, P., Valjak, F., Pavlic, B., Sojic, B. and Kovacevic, D. B. (2021). 3D printing as novel tool for fruit-based functional food production. Current Opinion in Food Science 41: 138–145.

Turner, N. D. and Lupton, J. R. (2021). Dietary fiber. Advances in Nutrition 12: 2553–2555.

Vega, F. V. A. (2024). Beneficios de la agricultura regenerativa en la salud del suelo. RECIAMUC 8(2): 665–677.

Vidal, A. M., Dias, D. O., Martins, E. S. de M., Oliveira, R. S., Nascimento, R. M. S. and Correia M. S. da. (2012). A ingestão de alimentos funcionais e sua contribuição para a diminuição da incidência de doenças. Cadernos de Graduação – Ciências Biológicas e da Saúde 1(1): 43–52.

Wan, D., Zhou, Y., Wang, K., Hou, Y., Hou, R. and Ye, X. (2016). Resveratrol provides neuroprotection by inhibiting phosphodiesterases and regulating the cAMP/AMPK/SIRT1 pathway after stroke in rats. Brain Research Bulletin 121: 255–262.

Wang, Y., Wang, C., Shi, J. and Zhang, Y. (2024). Effects of derivatization and probiotic transformation on the antioxidative activity of fruit polyphenols. Food Chemistry: X 23: 101776.

Ward, N. P. and DeNicola, G. M. (2019). Sulfur metabolism and its contribution to malignancy International Review of Cell and Molecular Biology 347: 39–103.

Wenczenovicz, T. J., Bühring, M. A. and Machado, L. D. L. (2024). Direito ambiental, agrário e socioambientalismo II. XXXI Congresso Nacional do CONPEDI, Brasília - DF. Florianópolis.

Wojtunik-Kulesza, K., Oniszczuk, A., Oniszczuk, T., Combrzynski, M., Nowakowska, D. and Matwijczuk, A. (2020). Influence of *in vitro* digestion on composition, bioaccessibility and antioxidant activity of food polyphenols—A non-Systematic review. Nutrients 12: 1401.

Wu, W., Mi, Y., Meng, Q., Li, N., Li, W., Wang, P. et al. (2025). Natural polyphenols as novel interventions for aging and age-related diseases: Exploring efficacy, mechanisms of action and implications for future research. Chinese Herbal Medicines 17: 279–291.

Xu, W., Luo, Y., Yin, J., Huang, M. and Luo. (2023). Targeting AMPK signaling by polyphenols: A novel strategy for tackling aging. Food and Function 14: 56–73.

Xu, X., Li, J., Wang, X., Wang, S., Meng, S., Zhu, Y. et al. (2016). Tomato consumption and prostate cancer risk: a systematic review and meta-analysis. Scientific Reports 6: 37091.

Yahfoufi, N., Alsadi, M., Jambi, M. and Matar, C. (2018). The immunomodulatory and anti-inflammatory role of polyphenols. Nutrients 10: 1618.

Yuan, X., Wang, T., Sun, L., Qiao, Z., Pan, H., Zhong, Y. et al. (2024). Recent advances of fermented fruits: a review on strains, fermentation strategies, and functional activities. Food Chemistry: X 22: 101482.

Yusuf, K., Saha, S. and Umar, S. (2022). Health benefits of dietary fiber for the management of inflammatory bowel disease. Biomedicines 10: 1242.

Zhang, L., Wu, J., Jiang, X., Wu, X. and Wang, X. (2024). Different types of dietary carotenoids improve the color and odor quality of Eriocheir sinensis ovaries. Aquaculture and Fisheries 7: 001.

Zhang, S. and Zhu, H. (2023). Development and prospect of gene-edited fruits and vegetables. College of Food Science and Nutritional Engineering, China Agricultural University, Beijing, China.

Zhang, S. and Zhu, H. (2024). Development and prospect of gene-edited fruits and vegetables. Food Quality and Safety 8: 1–15.

Index

Editors' Biography

Dr. Shandesh Bhattarai is a Senior Scientific Officer at Nepal Academy of Science and Technology, Khumaltar, Lalitpur, Nepal. He holds a PhD in Botany (Ethnopharmacology) from the Central Department of Botany, Tribhuvan University, Nepal. His major fields of research interest are applied ethnobotany/ethnopharmacology, plant systematics, biological invasion, and conservation biology of useful plants. He has research and training experience in over 10 research projects funded by national and international organizations (Flora of Nepal, Nepal, UK, Japan; NUFU, Norway; VW Foundation, Germany; Darwin Initiative, UK; DFID, UK; IUCN Bangkok; EvK2CNR Italy; CESVI Italy; UGC Nepal, etc.). Dr. Bhattarai has nearly 55 research and review papers published in national and international journals. Additionally, he has edited and authored seven books, one conference proceeding, and one report.

Dr. Edivan Carvalho Vieira is Professor at the Department of Chemistry, Federal University of Piaui (UFPI), Teresina, Brazil. He holds a PhD in Analytical Chemistry from UFSCar and completed postdoctoral research at the Universidade de São Paulo. His expertise includes inductively coupled plasma spectrometry and multielemental analysis of agricultural samples. He has worked at Embrapa Pecuária Sudeste, focusing on analytical methods for phosphorus determination. Dr. Vieira has published extensively in leading journals, with notable citations in Web of Science and Scopus, contributing to food science, environmental analysis, and material chemistry through innovative analytical techniques. He was the coordinator of the postgraduate program in Chemistry at UFPI.

Dr. Mahendra Rai is presently a Senior visiting Professor at the Department of Chemistry, Federal University of Piaui (UFPI), Teresina, Brazil. Formerly, he was a Professor and Head of the Department of Biotechnology at SGB Amravati University in Maharashtra, India. He has also held positions as a visiting scientist at several prestigious institutions, including the University of Geneva, Debrecen University in Hungary, the University of Campinas in Brazil, VSB Technical University of Ostrava in the Czech Republic, the National University of Rosario, in Argentina, and the University of

São Paulo. Dr. Rai has an impressive publication record, with over 450 research papers published in national and international journals with 95 h-index. In addition, he has edited/authored more than 70 books and holds 6 patents. Notably, he has been recently featured in Stanford's list of the top 2% of scientists in nanoscience and nanotechnology, and ranks 7th nationally in Biology and Biochemistry, according to Research.com.

For Product Safety Concerns and Information please contact our
EU representative GPSR@taylorandfrancis.com Taylor & Francis
Verlag GmbH, Kaufingerstraße 24, 80331 München, Germany